Commutative Semigroups

Advances in Mathematics

VOLUME 2

The titles published in this series are listed at the end of this volume.

Commutative Semigroups

by

P.A. GRILLET

Tulane University, New Orleans, U.S.A.

SPRINGER-SCIENCE+BUSINESS MEDIA, B.V.

A C.I.P. Catalogue record for this book is available from the Library of Congress.

DOI 10.1007/978-1-4757-3389-1

Printed on acid-free paper

A ma mère

à qui je dois tout

CONTENTS

PREFACE.

The first book on commutative semigroups was Rédei's *The theory of finitely generated commutative semigroups*, published in Budapest in 1956.

Subsequent years have brought much progress. By 1975 the structure of finite commutative semigroups was fairly well understood. Recent results have perfected this understanding and extended it to finitely generated semigroups. Today's coherent and powerful structure theory is the central subject of the present book.

1. Commutative semigroups are more important than is suggested by the standard examples of semigroups, which consist of various kinds of transformations or arise from finite automata, and are usually quite noncommutative. Commutative semigroups provide a natural setting and a useful tool for the study of factorization in rings. Additive subsemigroups of $\mathbb{N}$ and $\mathbb{N}^n$ have close ties to algebraic geometry. Commutative rings are constructed from commutative semigroups as semigroup algebras or power series rings. These areas are all subjects of active research and together account for about half of all current papers on commutative semigroups.

Commutative results also invite generalization to larger classes of semigroups. Archimedean decompositions, a comparatively small part of today's arsenal, have been generalized extensively, as shown for instance in the upcoming books by Nagy [2001] and Ćirić [2002].

In the past few years commutative semigroups have appeared in lattice theory (Wehrung [1998]), global fields (Geroldinger [1997]), functional equations (Brillouet-Belluot [1996], Taylor [1999]), Banach algebras (Esterle [1997]), C*-algebras (Murphy [1996], Ressel & Ricker [1998]), harmonic analysis (Youssfi [1998]), numerical analysis (Markov [1995]), theory of computation (Abdali [1993]), and other unexpected places (see Chapter VIII).

Commutative semigroups thus have a wide range of potential applications. Unfortunately, they are also quite complex and what has been known of their structure has been either insufficient for applications or not readily available from existing books. The present account of current structure theory fills a large gap and will, we hope, ameliorate this situation. Recent results and many older results, now scattered among a number of articles, should also benefit from a systematic exposition.

2. Research on commutative semigroups has a long history. Lawson [1996] makes a good case that the earliest article which would currently receive a semigroup classification is an 1826 paper by Abel which clearly contains cancellative commutative semigroups. It has indeed long been known that these semigroups can be retrieved from abelian groups; some aspects of this difficult process are explored in Chapter II (after the inevitable basic definitions and results in Chapter I).

The "classical" structure theory is more recent and includes notable contributions by Gluskin, McAlister, Petrich, Preston, Rédei, Schein, Š. Schwarz, Shevrin, Tamura, Yamada, and others. It provides a solid foundation for later results and is based upon three main ideas. Semilattice decompositions, discovered by Schwartz [1953] and Tamura & Kimura [1954], and extended to all semigroup by Yamada [1955], provided for many years the only structural information on commutative semigroups. Subdirect decompositions of finite commutative semigroups were found by Ponizovsky [1962]. Group coextensions were developed independently by Grillet [1974] and Leech [1975]. These three tools are explained in Chapters III, IV, and V, and together provide a good first grasp of finite commutative semigroups (more generally, of complete semigroups) as explained in Grillet [1995].

The elaborate theory by Rédei [1956], intended to produce numerical invariants for finitely generated commutative semigroups, brought little structural insight but uncovered the important property, known today as *Rédei's Theorem*, that congruences on these semigroups satisfy the ascending chain condition. The shorter proof of this theorem given by Herzog [1970] and Preston [1975] led to a first understanding of finitely generated commutative semigroups, by means of subdirect decompositions (Grillet [1975P]) and a fourth tool, completions (Grillet [1975C]). This occupies most of Chapter VI.

Very recent developments complete the basic structure theory of finitely generated semigroups (more generally, of subcomplete semigroups) by means of a fifth tool, Ponizovsky families (Grillet [2001S]). This constitutes Chapter VII and completes the first part of current structure theory. The main results in this part build semigroups from cancellative semigroups and nilsemigroups, arranged along semilattices. Nilsemigroups must then be constructed by other methods.

3. The present structure theory has two other parts. The second part, which

constitutes Chapters IX, X, and XI, contains constructions of subcomplete semigroups, including nilsemigroups and finitely generated semigroups, due to the author [1991N], [1993], [1996C], [2001C]. These semigroups are constructed indirectly, as quotients of free semigroups by subcomplete congruences. Unlike Rédei's short and elegant description, this construction of congruences is rather long but is highly detailed and geometric. It includes (and is based on) all the structural information acquired in the first part and has connections to the cohomology in the third part. Such a global construction bypasses all difficulties that arise when a commutative semigroup is reassembled from its archimedean components or Ponizovsky factors. Chapters IX and X highlight two important particular cases, nilsemigroups and group-free semigroups; the general case occupies Chapter XI.

The third part of the current theory, which constitutes Chapters XII, XIII, and XIV, is the cohomology theory for commutative semigroups introduced by the author in [1974] and [1991C] and defined in Chapter XII. A similar cohomology theory for monoids in general was developed independently by Leech [1975]. Both theories arise naturally from group coextensions and are particular instances of triple cohomology. They use abelian group valued functors for coefficients and their second group classifies group coextensions by such functors. The earlier Eilenberg-MacLane cohomology of monoids is a particular case of Leech cohomology in which the abelian group valued functor is constant. The functor Ext for abelian groups is a similar particular case of commutative semigroup cohomology.

Commutative semigroup cohomology is especially interesting because it classifies semigroups and not just coextensions: if S is complete group-free and $\mathbb{G}$ is Schützenberger, then $H^2(S, \mathbb{G})$ classifies complete commutative semigroups T with $T/\mathcal{H} \cong S$ and Schützenberger functor isomorphic to $\mathbb{G}$ (Grillet [1974]). Cohomology is therefore an integral part of the structure theory. Applications had to wait until the overpath method in Chapter XIII provided an efficient way to compute $H^2(S, \mathbb{G})$ (Grillet [1995M]). A number of results can be proved by this method, using the knowledge of congruences acquired in the second part, and will be found in Chapters XIII and XIV.

4. A very large amount of work has been done on commutative semigroups, enough for several volumes. This book covers the structure theory in some detail but surveys other worthwhile topics only briefly. Left out altogether are five areas which are covered elsewhere, or for which short surveys would be inappropriate: factorization theory; semigroup rings; varieties and pseudo-varieties of commutative semigroups (recently determined by Kisielewicz [1991], [1994]); ordered

semigroups; and most algorithmic aspects of the theory. All other topics are surveyed very sketchily, mostly in Chapter VIII, and for these we have tried to provide as many references as possible (about 850 altogether). Such comprehensiveness probably precludes perfection, if only because there is no clear boundary between general and commutative semigroup theory. The author extends his apologies to all authors whose work should have received greater mention or has been inadvertently left out.

5. I wanted this book to be published, like Rédei's, by the Akadémiai Kiadó in Budapest in its "Pure and Applied Mathematics" series, which has recently become the series "Advances in Mathematics", published by Kluwer Academic Publishing. I owe heartfelt thanks to my Hungarian friends, Prof. Szép Jeno, Márki Lászlo, and Szendrei Mária, for their unfailing advice and support, and particularly to Prof. Szép who also sponsored this book. Köszönöm, barátaim! Thanks also go to Prof. Scott Chapman, Franz Halter-Koch, and Boris Schein for helpful communications, and to the folk at Kluwer who have consistently been a pleasure to deal with.

Last, but not least, special thanks are due my beloved wife, for her love, patience, and understanding.

Slidell, Louisiana, March 2001.

Chapter I.

ELEMENTARY PROPERTIES.

This chapter contains basic first properties of commutative semigroups: idempotents, subsemigroups, homomorphisms and congruences, ideals, ideal extensions, $\mathcal{H}$-classes and Schützenberger groups, free commutative semigroups, presentations. It is written for readers who are not familiar with semigroups and can therefore expect a few surprises.

Many results in this chapter are stated for commutative semigroups but extend to all semigroups (with suitable modifications); interested readers should consult Clifford & Preston [1961], Howie [1976], [1995], or Grillet [1995].

1. FIRST RESULTS.

1. A **semigroup** is a set S together with an associative binary operation on S. (This is the only axiom.)

The name "semigroup" suggests a generalization of groups. To disabuse the reader of this erroneous belief we compare the number of distinct (not isomorphic) groups of order n with the number of distinct (not isomorphic or antiisomorphic) semigroups of order n, which has been determined for $n \leqq 8$ (Jürgensen & Wick [1977]; Sato, Yama, & Tokizawa [1991]):

All Semigroups

n	Groups	Semigroups
2	1	4
3	1	18
4	2	126
5	1	1160
6	2	15,973
7	1	836,021
8	5	1,843,120,128

The number of semigroups of order 9 has not been determined; it is known to exceed 52,952,000,000,000, but probably not by much (Jürgensen, Migliorini, & Szép [1991]).

A semigroup is **commutative** when its operation is commutative. (Commutative groups are called *abelian.*) Commutative semigroups are not as numerous as general semigroups but are still much more numerous than abelian groups:

Commutative Semigroups

n	Groups	Semigroups
2	1	3
3	1	12
4	2	58
5	1	325
6	1	2143
7	1	17,291
8	3	221,805
9	2	11,545,843

The number of commutative semigroups of order 9 is from Grillet [1996N].

These large numbers suggest that even commutative semigroups are quite unlike groups and constitute a wholly different kind of algebraic object.

2. There are some natural examples of commutative semigroups.

We denote by $\mathbb{N} = \{0, 1, 2, \ldots, n, \ldots\}$ the additive semigroup of natural numbers. $\mathbb{N}$ is a commutative semigroup under multiplication as well. We also denote by $\mathbb{N}^+ = \{1, 2, \ldots, n, \ldots\}$ the additive semigroup of all positive natural numbers, by $\mathbb{Q}^+$ the additive semigroup of all positive rational numbers, and by $\mathbb{R}^+$ the additive semigroup of all positive real numbers.

Abelian groups are commutative semigroups.

A **semilattice** (short for **lower semilattice**) is a partially ordered set in which any two elements a and b have a greatest lower bound (or meet, or infimum) $a \wedge b$. Every lower semilattice is a commutative semigroup under the binary operation $\wedge$.

Groups and semilattices lie wholly outside semigroup theory. By well-established tradition, we regard as solved any problem which can be stated in terms of groups or semilattices (we dump it onto other unsuspecting mathematicians).

Every commutative ring or algebra is a commutative semigroup under multiplication. In fact, many arithmetic properties of rings are now studied in purely multiplicative terms. (Constructing rings from semigroups is even more fruitful; semigroup algebras are a major employer of commutative semigroups.)

3. The operation $S \times S \longrightarrow S$ on a semigroup S is usually written like a multiplication $(x,y) \longmapsto xy$; sometimes (for instance, in $\mathbb{N}$) like an addition $(x,y) \longmapsto x+y$. In the multiplicative notation, **associativity** is the property

$$x(yz) = (xy)z \text{ for all } x,y,z \in S;$$

commutativity is the property

$$xy = yz \text{ for all } x,y \in S.$$

Associativity and commutativity have a number of consequences which we hope are well known to the reader. Associativity allows products of three or more elements to be written without parentheses. Commutativity implies that a product does not depend on the order of its terms:

$$x_{\sigma 1}\, x_{\sigma 2} \cdots x_{\sigma n} = x_1 x_2 \cdots x_n$$

for every permutation σ of $\{1,2,\dots,n\}$.

Powers are particular cases of products. When a is an element of a semigroup S and n is a positive integer, then a^n is the product of n elements of S all equal to a (with $a^n = a$ if $n = 1$). If S has an identity element 1 one can also define $a^0 = 1$. In a commutative semigroup S, the equalities

$$x^m\, x^n = x^{m+n}, \quad (x^m)^n = x^{mn}, \quad x^n\, y^n = (xy)^n$$

hold for all $x,y \in S$ and $m,n > 0$ ($m,n \geqq 0$ if S has an identity element). In the additive notation, products $x_1 x_2 \cdots x_n$ become sums $x_1 + x_2 + \cdots + x_n$ and powers x^n become positive integer multiples nx.

4. An **identity element** of a semigroup S is an element e of S such that $ea = ae = a$ for all $a \in S$. When an identity element exists, it is unique, and we normally denote it by 1 (by 0 in the additive notation). A semigroup with an identity element is a **monoid**.

In a monoid, **empty** products (= products $a_1 a_2 \cdots a_n$ of elements of S in which $n = 0$) are, by definition, equal to the identity element; for instance, zero powers $a^0 = 1$.

An identity element can be adjoined to any semigroup S: given $1 \notin S$, define an operation on $S \cup \{1\}$ as follows: 1 is an identity element and xy is the same in S and $S \cup \{1\}$ for all $x,y \in S$. Then $S \cup \{1\}$ is a monoid, which is commutative if S is commutative.

The identity element of $S \cup \{1\}$ is not the same as the identity element of S

in case the latter exists. Hence the following construction:

$$S^1 = \begin{cases} S & \text{if } S \text{ has an identity element,} \\ S \cup \{1\} & \text{if } S \text{ does not have an identity element,} \end{cases}$$

which adjoins an identity element only when necessary.

A **zero element** of a semigroup S is an element z of S such that $za = az = z$ for all $a \in S$. When a zero element exists, it is unique, and we normally denote it by 0 (by ∞ in the additive notation).

A zero element can be adjoined to any semigroup S: given $0 \notin S$, define an operation on $S \cup \{0\}$ as follows: 0 is a zero element and xy is the same in S and $S \cup \{0\}$ for all $x, y \in S$. Then $S \cup \{0\}$ is a semigroup with zero, which is commutative if S is commutative.

5. More generally, an **idempotent** of a semigroup S is an element e of S such that $e^2 = e$; then $e^n = e$ for all $n > 0$. We denote by $E(S)$ the set of all idempotents of S.

When S is commutative the **Rees order** on $E(S)$ is defined for all $e, f \in E(S)$ by:

$$e \leqq f \text{ if and only if } ef = e.$$

If an identity element (a zero element) exists, then it is the greatest (the least) element of $E(S)$ under the Rees order.

Proposition 1.1. *When S is a commutative semigroup, then $E(S)$ is a (possibly empty) semilattice under the Rees order, in which $e \wedge f = ef$ for all e and f.*

Proof. First the Rees order is a partial order relation: for all $e, f, g \in E(S)$, $e \leqq e$ since e is idempotent; $e \leqq f \leqq e$ implies $e = ef = fe = f$; and $e \leqq f \leqq g$ implies $e = ef = efg = eg$ and $e \leqq g$. Also $e(ef) = ef$ and $(ef)f = ef$, so that $ef \leqq e, f$. If conversely $g \leqq e$ and $g \leqq f$, then $efg = ef$ and $g \leqq ef$. Thus ef is the infimum of e and f in $E(S)$. □

The Rees order can be defined in any semigroup S and is always a partial order relation on $E(S)$ (Rees [1940]); but then $E(S)$ is not necessarily a semilattice. It was generalized to arbitrary elements by Mitsch [1986], [1994].

Corollary 1.2. *Let S be a commutative semigroup. If every element of S is idempotent, then S is a semilattice under the Rees order, in which $a \wedge b = ab$ for all a and b. If conversely Y is a semilattice, then $(Y, \wedge)$ is a commutative semigroup in which every element is idempotent, and the Rees order on $(Y, \wedge)$ is the given partial order on Y.*

Accordingly, commutative semigroups in which every element is idempotent may be identified with (lower) semilattices, and are, in fact, called semilattices.

6. Subsets $A, B \subseteq S$ of a semigroup S are multiplied by:

$$AB = \{ab \mid a \in A,\, b \in B\}.$$

In particular,

$$Ac = \{ac \mid a \in A\} \text{ and } cB = \{cb \mid b \in B\}$$

for all $A, B \subseteq S$ and $c \in S$. Multiplication of subsets inherits associativity and commutativity from S.

A **subsemigroup** of a semigroup S is a subset T of S which is closed under the operation on S ($xy \in T$ for all $x, y \in T$); equivalently, such that $TT \subseteq T$. For instance, S and the empty set are subsemigroups of S.

Every subsemigroup T of S inherits a semigroup operation $T \times T \longrightarrow T$ from S; this semigroup T is also called a subsemigroup of S. If S is commutative, then so is T.

Every intersection of subsemigroups of S is a subsemigroup of S. Hence there is for every subset X of S a smallest subsemigroup T of S which contains X; T is the intersection of all the subsemigroups of S which contain X, and is the subsemigroup (sometimes denoted by $\langle X \rangle$ or by X^*) **generated by** X.

Proposition 1.3. *The subsemigroup generated by a subset X is the set of all products of one or more elements of X. In a commutative semigroup, the subsemigroup generated by a subset X is the set of all products of positive powers of one or more distinct elements of X.*

Proof. A subsemigroup which contains X must by induction contain all nonempty products of elements of X. Conversely the set T of all nonempty products of elements of X is closed under multiplication and contains all products of one element of X, i.e. contains X.

Every sequence $x_1, x_2, \dots, x_n$ of elements of X can be permuted into a sequence $y_1, \dots, y_1, y_2, \dots, y_2, \dots, y_k, \dots, y_k$, where $y_1, y_2, \dots, y_k$ are the distinct elements of $\{x_1, x_2, \dots, x_n\}$. In a commutative semigroup, the products $x_1 x_2 \cdots x_n$ and $y_1 \cdots y_1 y_2 \cdots y_2 \cdots y_k \cdots y_k$ are equal; the latter is a product of positive powers of distinct elements of X. □

For example, the **cyclic** subsemigroup generated by $X = \{x\}$ consists of all the positive powers of x.

If the subsemigroup generated by X is S itself, then X **generates** S and the elements of X are **generators** of S; this means that every element of S is the product of one or more elements of X. A semigroup is **finitely generated** when it is generated by a finite subset, **cyclic** when it is generated by a single element.

Proposition 1.3 has an analogue for monoids. When S is a monoid, a **submonoid** of S is a subsemigroup T of S which contains the identity element of S; then T is a monoid in its own right, with the same identity element as S.

Proposition 1.4. *In a monoid, the submonoid generated by a subset X is the set of all products of elements of X. In a commutative monoid, the submonoid generated by a subset X is the set of all products of positive powers of distinct elements of X.*

This works since empty products yield the identity element.

2. HOMOMORPHISMS AND CONGRUENCES.

1. Let S and T be semigroups. A **homomorphism** of semigroups of S into T is a mapping $\varphi : S \longrightarrow T$ such that $\varphi(ab) = \varphi(a)\,\varphi(b)$ for all $a, b \in S$. Semigroup homomorphisms preserve all nonempty products:

$$\varphi(a_1 a_2 \cdots a_n) = \varphi(a_1)\,\varphi(a_2) \cdots \varphi(a_n)$$

and preserve positive powers: $\varphi(a^n) = \varphi(a)^n$.

The identity mapping 1_S on a semigroup S is a homomorphism of S onto S. If $\varphi : S \longrightarrow T$ and $\psi : T \longrightarrow U$ are homomorphisms, then so is $\psi \circ \varphi : S \longrightarrow U$. An **isomorphism** of semigroups is a bijective homomorphism; the inverse bijection is also an isomorphism.

When T is commutative, the pointwise product

$$(\varphi.\psi)(a) = \varphi(a)\,\psi(a)$$

of two homomorphisms $\varphi, \psi : S \longrightarrow T$ is a homomorphism $\varphi.\psi : S \longrightarrow T$. With this operation the set $\mathrm{Hom}(S,T)$ of all homomorphisms of S into T becomes a commutative semigroup.

2. Semigroup homomorphisms share a number of basic properties with mappings and with homomorphisms of algebraic systems in general.

When S and T are sets, a mapping $\varphi : S \longrightarrow T$ has a **range** or **image** $\mathrm{im}\,\varphi = \varphi(S) \subseteq T$ and induces an equivalence relation $\ker\varphi$ on S,

$$\ker\varphi = \{(a,b) \in S \times S \mid \varphi(a) = \varphi(b)\}.$$

This provides a quotient set $S/\ker\varphi$ (the set of all equivalence classes) and a **projection** or canonical mapping $S \longrightarrow S/\ker\varphi$, which sends $x \in S$ to its equivalence class. Then φ induces a bijection $S/\ker\varphi \longrightarrow \operatorname{im}\varphi$, which sends the equivalence class of $x \in S$ to $\varphi(x)$, and φ can be reconstructed by composing the projection $S \longrightarrow S/\ker\varphi$, the bijection $S/\ker\varphi \longrightarrow \operatorname{im}\varphi$, and the inclusion mapping $\operatorname{im}\varphi \longrightarrow T$.

$$\begin{array}{ccc} S & \xrightarrow{\varphi} & T \\ \downarrow & & \uparrow \\ S/\ker\varphi & \longrightarrow & \operatorname{im}\varphi \end{array}$$

Homomorphisms of groups have similar properties, with the important difference that quotient groups are constructed from subgroups. As we shall see, semigroups are more like sets than like groups in that, in general, quotient semigroups cannot be constructed from subsets and must be constructed from equivalence relations.

3. First, given a semigroup S and an equivalence relation $\mathcal{E}$ on S, how can we induce an operation on the quotient set $S/\mathcal{E}$? The answer is:

Proposition 2.1. *Let S be a semigroup and $\mathcal{E}$ be an equivalence relation on S. The following conditions are equivalent:*

(1) *there exists an associative operation on $S/\mathcal{E}$ such that the projection $S \longrightarrow S/\mathcal{E}$ is a homomorphism;*

(2) *for all $a,b,c,d \in S$, if $a \;\mathcal{E}\; c$ and $b \;\mathcal{E}\; d$, then $ab \;\mathcal{E}\; cd$.*

When either condition holds, there is only one associative operation on $S/\mathcal{E}$ such that the projection $a \longmapsto E_a$ is a homomorphism; the product of E_a and E_b in $S/\mathcal{E}$ is the equivalence class which contains their product as subsets of S, namely E_{ab}. If S is commutative, then so is $S/\mathcal{E}$.

E_a denotes the $\mathcal{E}$-class of a (= the equivalence class of a modulo $\mathcal{E}$).

Proof. If $a \longmapsto E_a$ is a homomorphism, then $E_a = E_b$, $E_c = E_d$ implies $E_{ac} = E_a.E_c = E_b.E_d = E_{bd}$; thus (1) implies (2).

Conversely let (2) hold. By (2), $c \in E_a$, $d \in E_b$ implies $cd \in E_{ab}$; thus the product $E_a E_b$ of E_a and E_b as subsets of S is contained in the single equivalence class E_{ab}. If the projection $a \longmapsto E_a$ is a homomorphism, then the product $E_a.E_b = E_{ab}$ of E_a and E_b in $S/\mathcal{E}$ is the equivalence class which contains their product $E_a E_b$ as subsets of S; there is only one operation on $S/\mathcal{E}$ with this property, and it is the operation described in the statement. With this operation, $E_a.E_b = E_{ab}$ holds in $S/\mathcal{E}$; hence $S/\mathcal{E}$ is a semigroup:

$$E_a.(E_b.E_c) = E_a.E_{bc} = E_{a(bc)} = E_{(ab)c} = E_{ab}.E_c = (E_a.E_b).E_c$$

and the projection $S \longrightarrow S/\mathcal{E}$ is a homomorphism. Thus (2) implies (1). If S is commutative, then

$$E_b.E_a = E_{ba} = E_{ab} = E_a.E_b$$

and $S/\mathcal{E}$ is commutative. □

A **congruence** on a semigroup S is an equivalence relation $\mathcal{E}$ on S which satisfies condition (2) in Proposition 2.1; then the **quotient semigroup** of S by $\mathcal{E}$ is the semigroup $S/\mathcal{E}$ in Proposition 2.1, such that the projection $S \longrightarrow S/\mathcal{E}$ is a homomorphism. The equivalence relation on S induced by the projection $S \longrightarrow S/\mathcal{E}$ is $\mathcal{E}$ itself.

A congruence on a group is completely determined by the equivalence class of the identity element, so that quotient groups can be constructed from normal subgroups. This nice property does not extend to semigroups; not even to commutative monoids with a zero element. For instance let S be the semilattice (also a monoid) $S = \{0, e, 1\}$ in which $0 < e < 1$. The equivalence relation whose classes are $\{e, 0\}$ and $\{1\}$ is a congruence; so is the equality (whose classes are $\{0\}$, $\{e\}$, and $\{1\}$); thus a congruence on S is not determined by the class of the identity element. Similarly, the equivalence relation whose classes are $\{1, e\}$ and $\{0\}$ is a congruence; hence a congruence on S is not determined by the class of the zero element.

4. Armed with quotient semigroups we can now state the **Homomorphism Theorem** (also known as the **First Isomorphism Theorem**):

Theorem 2.2. *When $\varphi : S \longrightarrow T$ is a homomorphism of semigroups:*

(1) $\operatorname{Im}\varphi = \varphi(S)$ *is a subsemigroup of T;*

(2) $\ker\varphi$ *is a congruence on S;*

(3) *there exists an isomorphism $S/\ker\varphi \longrightarrow \operatorname{Im}\varphi$ such that the diagram*

$$\begin{array}{ccc} S & \xrightarrow{\varphi} & T \\ \downarrow & & \uparrow \\ S/\ker\varphi & \longrightarrow & \operatorname{Im}\varphi \end{array}$$

commutes; in particular $S/\ker\varphi \cong \operatorname{Im}\varphi$. If S and T are commutative, then so are $\operatorname{Im}\varphi$ and $S/\ker\varphi$.

Proof. When $\varphi : S \longrightarrow T$ is a homomorphism, $\operatorname{Im}\varphi = \varphi(S)$ is a subsemigroup of T, since $\varphi(a)\,\varphi(b) = \varphi(ab)$ for all $a, b \in S$; $\mathcal{E} = \ker\varphi$ is a congruence on S, since $\varphi(a) = \varphi(b)$, $\varphi(c) = \varphi(d)$ implies $\varphi(ac) = \varphi(a)\,\varphi(c) =$

$\varphi(b)\,\varphi(d) = \varphi(bd)$; and the bijection $E_a \longmapsto \varphi(a)$ is a homomorphism, since it sends $E_a . E_b = E_{ab}$ to $\varphi(ab) = \varphi(a)\,\varphi(b)$. □

Let S and T be semigroups. By Theorem 2.2, S is isomorphic to a subsemigroup of T if and only if there exists an injective homomorphism (an **embedding**) of S into T; then S can be **embedded** into T. Similarly, T is isomorphic to a quotient semigroup of S if and only if there exists a surjective homomorphism of S onto T; then T is a **homomorphic image** of S.

5. Theorem 2.2 can be deduced from more general results which allow one homomorphism to factor through another and help construct diagrams of semigroups and homomorphisms.

Proposition 2.3. *Let $\varphi : S \longrightarrow T$ and $\psi : U \longrightarrow T$ be homomorphisms of semigroups. If φ is injective, then ψ factors through φ ($\psi = \varphi \circ \xi$ for some homomorphism $\xi : U \longrightarrow S$) if and only if* $\operatorname{Im}\psi \subseteq \operatorname{Im}\varphi$*; and then ψ factors uniquely through φ (ξ is unique). If φ and ψ are injective and* $\operatorname{Im}\psi = \operatorname{Im}\varphi$*, then ξ is an isomorphism.*

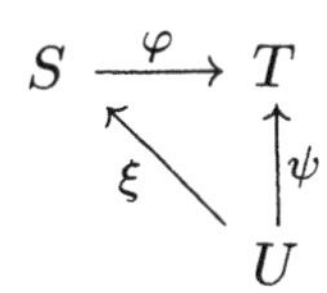

This is clear.

Proposition 2.4. *Let $\varphi : S \longrightarrow T$ and $\psi : S \longrightarrow U$ be homomorphisms of semigroups. If φ is surjective, then ψ factors through φ ($\psi = \xi \circ \varphi$ for some homomorphism $\xi : T \longrightarrow U$) if and only if* $\ker\varphi \subseteq \ker\psi$*; and then ψ factors uniquely through φ (ξ is unique). If φ and ψ are surjective and* $\ker\varphi = \ker\psi$*, then ξ is an isomorphism.*

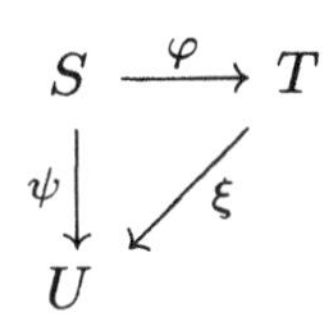

Proof. If $\psi = \xi \circ \varphi$, then $\varphi(a) = \varphi(b)$ implies $\psi(a) = \xi(\varphi(a)) = \xi(\varphi(b)) = \psi(b)$, and $\ker\varphi \subseteq \ker\psi$.

Conversely, assume that φ is surjective and that $\ker\varphi \subseteq \ker\psi$. Let ξ be the set of ordered pairs

$$\xi = \{ (\varphi(a), \psi(a)) \in T \times U \mid a \in S \}.$$

For every $t \in T$, there exists $u \in U$ such that $(t,u) \in \xi$, since φ is surjective; if moreover $(t,u) \in \xi$, $(t',u') \in \xi$, and $t = t'$, then $u = u'$, since $\ker\varphi \subseteq \ker\psi$.

Thus ξ is a mapping of T into U. Also $\xi(\varphi(a)) = \psi(a)$ for all $a \in S$ by definition and

$$\xi(\varphi(a)\,\varphi(b)) = \xi(\varphi(ab)) = \psi(ab) = \psi(a)\,\psi(b) = \xi(\varphi(a))\,\xi(\varphi(b))$$

for all $a,b \in S$, so that ξ is a homomorphism. Thus ψ factors through φ; ψ factors uniquely through φ since any mapping χ such that $\psi = \chi \circ \varphi$ must contain all ordered pairs $(\varphi(a), \psi(a))$ and must coincide with ξ. If moreover ψ is surjective and $\ker\varphi = \ker\psi$, then ξ is injective, since $\xi(\varphi(a)) = \xi(\varphi(b))$ implies $\psi(a) = \psi(b)$ and $\varphi(a) = \varphi(b)$, is surjective, since $\operatorname{Im}\xi = \operatorname{Im}(\xi \circ \varphi) = \operatorname{Im}\psi$, and is an isomorphism. □

Analogues of the (other) two Isomorphism Theorems also hold for semigroups. The most useful employ the following constructions. Let $\varphi : S \longrightarrow T$ be a semigroup homomorphism. The **direct image** under φ of a subsemigroup S' of S is the subset $\varphi(S') = \{\varphi(x) \mid s \in S'\}$ of T. The **inverse image** under φ of a subsemigroup T' of T is

$$\varphi^{-1}(T') = \{x \in S \mid \varphi(x) \in T\}$$

Proposition 2.5. *Let $\varphi : S \longrightarrow T$ be a homomorphism of semigroups and $\mathcal{C} = \ker\varphi$.*

If S' is a subsemigroup of S, then $\varphi(S')$ is a subsemigroup of T.

If T' is a subsemigroup of T, then $\varphi^{-1}(T')$ is a subsemigroup of S and a union of $\mathcal{C}$-classes.

If φ is surjective this defines an order preserving one-to-one correspondence between subsemigroups of T and subsemigroups of S that are unions of $\mathcal{C}$-classes.

Proof. If S' is a subsemigroup of S, then $\varphi(S')$ is a subsemigroup of T, since $\varphi(x)\,\varphi(y) = \varphi(xy) \in \varphi(S')$ for all $x,y \in S'$.

If T' is a subsemigroup of T, then $\varphi^{-1}(T')$ is a union of $\mathcal{C}$-classes and is a subsemigroup of S since $\varphi(x), \varphi(y) \in T'$ implies $\varphi(xy) = \varphi(x)\,\varphi(y) \in T'$.

If φ is surjective, then $\varphi(\varphi^{-1}(T')) = T'$ for every $T' \subseteq T$. Also $S' \subseteq \varphi^{-1}(\varphi(S'))$ for every $S' \subseteq S$; conversely, $x \in \varphi^{-1}(\varphi(S'))$ implies $\varphi(x) = \varphi(s)$ for some $s \in S'$ and $x \in S'$ if $S' \subseteq S$ is a union of $\mathcal{C}$-classes. □

Similarly, the **direct image** under a semigroup homomorphism $\varphi : S \longrightarrow T$ of a congruence $\mathcal{E}$ on S is the binary relation

$$\varphi(\mathcal{E}) = \{(\varphi(a), \varphi(b)) \in T \times T \mid (a,b) \in \mathcal{E}\};$$

equivalently, the direct image of $\mathcal{E} \subseteq S \times S$ under $\varphi \times \varphi : S \times S \longrightarrow T \times T$. The **inverse image** under φ of a congruence $\mathcal{F}$ on T is the binary relation

$$\varphi^{-1}(\mathcal{F}) = \{(a,b) \in S \times S \mid (\varphi(a), \varphi(b)) \in \mathcal{F}\};$$

equivalently, the inverse image of $\mathcal{F}$ under $\varphi \times \varphi$. We also say that $\varphi(\mathcal{E})$, $\varphi^{-1}(\mathcal{F})$ are **induced** by $\mathcal{E}$ and $\mathcal{F}$.

Proposition 2.6. *Let $\varphi : S \longrightarrow T$ be a homomorphism of semigroups and $\mathcal{C} = \ker \varphi$.*

If $\mathcal{F}$ is a congruence on T, then $\varphi^{-1}(\mathcal{F})$ is a congruence on S which contains $\mathcal{C}$; if φ is surjective, then $S/\varphi^{-1}(\mathcal{F}) \cong T/\mathcal{F}$.

If φ is surjective and $\mathcal{E}$ is a congruence on S which contains $\mathcal{C}$, then $\varphi(\mathcal{E})$ is a congruence on T, and $T/\varphi(\mathcal{E}) \cong S/\mathcal{E}$.

If φ is surjective this defines an order preserving one-to-one correspondence between congruences on T and congruences on S that contain $\mathcal{C}$.

Proof. Let $\mathcal{F}$ be a congruence on T and $\rho : T \longrightarrow T/\mathcal{F}$ be the projection, so that $\mathcal{F} = \ker \rho$. We see that $\varphi^{-1}(\mathcal{F}) = \ker(\rho \circ \varphi)$. Therefore $\varphi^{-1}(\mathcal{F})$ is a congruence on S. If φ is surjective, then $S/\varphi^{-1}(\mathcal{F}) \cong \operatorname{Im}(\rho \circ \varphi) = \operatorname{Im} \rho = T/\mathcal{F}$ by Theorem 2.2.

Now let φ be surjective and $\mathcal{E}$ be a congruence on S which contains $\mathcal{C}$. Let $\rho : S \longrightarrow S/\mathcal{E}$ be the projection. By Proposition 2.4, ρ factors through φ: $\rho = \xi \circ \varphi$ for some homomorphism $\xi : T \longrightarrow S/\mathcal{E}$. We have $\varphi(\mathcal{E}) = \ker \xi$: indeed $(a,b) \in \mathcal{E}$ implies $\xi(\varphi(a)) = \rho(a) = \rho(b) = \xi(\varphi(b))$ and $(\varphi(a), \varphi(b)) \in \ker \xi$; if conversely $(t,u) \in \ker \xi$, then $t = \varphi(a)$, $u = \varphi(b)$ for some $a, b \in S$, $(a,b) \in \mathcal{E}$ since $\rho(a) = \xi(t) = \xi(u) = \rho(b)$, and $(t,u) = (\varphi(a), \varphi(b)) \in \varphi(\mathcal{E})$. Therefore $\varphi(\mathcal{E})$ is a congruence; by Theorem 2.2, $T/\varphi(\mathcal{E}) \cong \operatorname{Im} \xi = \operatorname{Im} \rho = S/\mathcal{E}$, since φ is surjective.

If φ is surjective, then $\varphi \times \varphi$ is surjective; therefore $\varphi(\varphi^{-1}(\mathcal{F})) = \mathcal{F}$ for all $\mathcal{F} \subseteq T \times T$. Similarly $\mathcal{E} \subseteq \varphi^{-1}(\varphi(\mathcal{E}))$ for all $\mathcal{E} \subseteq S \times S$. If $\mathcal{E}$ is a congruence on S and $\mathcal{C} \subseteq \mathcal{E}$, and $(a,b) \in \varphi^{-1}(\varphi(\mathcal{E}))$, then $(\varphi(a), \varphi(b)) \in \varphi(\mathcal{E})$, $(\varphi(a), \varphi(b)) = (\varphi(c), \varphi(d))$ for some $(c,d) \in \mathcal{E}$, and, as above, $(a,c) \in \mathcal{E}$ and $(b,d) \in \mathcal{E}$; hence $(a,b) \in \mathcal{E}$, so that $\varphi^{-1}(\varphi(\mathcal{E})) = \mathcal{E}$. □

It follows from Propositions 2.5, 2.6 that the subsemigroups of a quotient semigroup $S/\mathcal{C}$ are precisely the sets of $\mathcal{C}$-classes whose unions are subsemigroups of S; and that the congruences on $S/\mathcal{C}$ are precisely the congruences induced on $S/\mathcal{C}$ by congruences on S that contain $\mathcal{C}$.

6. Similar results hold for monoids. When S and T are monoids, a **homomorphism** of monoids of S into T is a homomorphism of semigroups $\varphi : S \longrightarrow T$ such that $\varphi(1) = 1$. Then φ preserves all products and nonnegative powers.

When S is a monoid and $\mathcal{E}$ is a congruence on S, then

$$E_1.E_a = E_{1a} = E_a = E_{a1} = E_a.E_1$$

for all $a \in S$, so that $S/\mathcal{E}$ is a monoid and the projection $S \longrightarrow S/\mathcal{E}$ is a homomorphism of monoids. If therefore $\varphi : S \longrightarrow T$ is a homomorphism of monoids, then, as in Theorem 2.2, $\operatorname{Im} \varphi$ is a submonoid of T, $\ker \varphi$ is a congruence on S, and there is an isomorphism $S/\ker \varphi \longrightarrow \operatorname{Im} \varphi$ such that the diagram

$$\begin{array}{ccc} S & \xrightarrow{\varphi} & T \\ \downarrow & & \uparrow \\ S/\ker \varphi & \longrightarrow & \operatorname{Im} \varphi \end{array}$$

commutes; in particular $S/\ker \varphi \cong \operatorname{Im} \varphi$.

Results similar to Propositions 2.3, 2.4, 2.5, and 2.6 also hold for monoids; this is left to the reader.

7. We complete this section with some properties of congruences.

Proposition 2.7. *An equivalence relation $\mathcal{C}$ on a commutative semigroup S is a congruence if and only if, for all $a, b, c \in S$, $a \;\mathcal{C}\; b$ implies $ac \;\mathcal{C}\; bc$.*

Proof. If this condition holds, then $a \;\mathcal{C}\; b$, $c \;\mathcal{C}\; d$ implies $ac \;\mathcal{C}\; bc = cb \;\mathcal{C}\; db = bd$, and $\mathcal{C}$ is a congruence. The converse is clear. □

For instance the equality $=$ on a semigroup S is a congruence, and so is the **universal** congruence $\mathcal{U}$, of which S is the only equivalence class; $S/\!= \;\cong S$, whereas $S/\mathcal{U}$ is trivial.

Since congruences on a given semigroup S are subsets of $S \times S$, we can form their unions and intersections in $S \times S$. The following result is straightforward.

Proposition 2.8. *Let S be a semigroup. Every intersection of congruences on S is a congruence on S. The union of a chain of congruences on S is a congruence on S.*

In particular, the **empty intersection** $\bigcap_{i \in \emptyset} \mathcal{C}_i$ of congruences on S can be defined as the universal congruence on S; the **empty union** $\bigcup_{i \in \emptyset} \mathcal{C}_i$ of congruences on S can be defined as the equality on S.

By Proposition 2.8 there is for every binary relation $\mathcal{R} \subseteq S \times S$ a smallest congruence $\mathcal{C}$ on S which contains $\mathcal{R}$; $\mathcal{C}$ is the intersection of all the congruences which contain $\mathcal{R}$ and is the congruence **generated by** $\mathcal{R}$.

Proposition 2.9. *Let S be a commutative semigroup. The congruence $\mathcal{C}$ generated by $\mathcal{R} \subseteq S \times S$ can be constructed as follows. Let*

$$\mathcal{S} = \{(xu, yu) \mid x,y \in S,\ u \in S^1,\ \text{and}\ x \mathrel{\mathcal{R}} y \ \text{ or }\ y \mathrel{\mathcal{R}} x\}.$$

Then $a \mathrel{\mathcal{C}} b$ if and only if there exist $n \geqq 1$ and $s_1, \dots, s_n \in S$ such that $a = s_1$, $s_n = b$, and $s_i \mathrel{\mathcal{S}} s_{i+1}$ for all $1 \leqq i < n$.

Proof. We see that $\mathcal{S}$ contains $\mathcal{R}$ (let $u = 1 \in S^1$ in the definition of $\mathcal{S}$), is symmetric ($a \mathrel{\mathcal{S}} b$ implies $b \mathrel{\mathcal{S}} a$), and admits multiplication ($a \mathrel{\mathcal{S}} b$ implies $ac \mathrel{\mathcal{S}} bc$). Hence $\mathcal{C}$ is symmetric (reverse the sequence $s_1, \dots, s_n$ in the definition of $\mathcal{C}$) and admits multiplication. Moreover $\mathcal{C}$ contains the equality on S (let $n = 1$ in the definition), contains $\mathcal{S}$ (let $n = 2$), and is transitive. Thus $\mathcal{C}$ is a congruence and contains $\mathcal{R}$.

Conversely, a congruence which contains $\mathcal{R}$ must contain $\mathcal{S}$, since a congruence is symmetric and admits multiplication; and a congruence which contains $\mathcal{S}$ must contain $\mathcal{C}$, since a congruence is reflexive and transitive. □

Proposition 2.9 can be stated more simply as follows: every relation $a \mathrel{\mathcal{C}} b$ follows from relations $x \mathrel{\mathcal{R}} y$ by finitely many applications of the following inference rules: $a \mathrel{\mathcal{R}} b$ implies $a \mathrel{\mathcal{C}} b$; $a \mathrel{\mathcal{C}} b$ implies $au \mathrel{\mathcal{C}} bu$, for all $u \in S$; $a \in S$ implies $a \mathrel{\mathcal{C}} a$; $a \mathrel{\mathcal{C}} b$ implies $b \mathrel{\mathcal{C}} a$; $a \mathrel{\mathcal{C}} b$, $b \mathrel{\mathcal{C}} c$ implies $a \mathrel{\mathcal{C}} c$.

3. IDEALS.

1. An **ideal** of a semigroup S is a subset I of S such that $a \in I$ implies $ax \in I$ and $xa \in I$ for all $x \in S$; equivalently, such that $IS \subseteq I$ and $SI \subseteq I$. For instance, S and the empty set are ideals of S. If S is commutative, the condition $SI \subseteq I$ is sufficient.

Proposition 3.1. *Every union of ideals of S is an ideal of S. Every intersection of ideals of S is an ideal of S.*

By Proposition 3.1 there exists, for every subset X of S, an ideal of S which contains X and is contained in every ideal of S which contains X; this is the ideal of S **generated by** X.

Proposition 3.2. *In a commutative semigroup S, the ideal generated by a subset X is the set S^1X of all multiples of elements of X.*

Proof. S^1X is the product in S^1, which is contained in S since either $S^1X = SX$ or $S^1X = SX \cup 1X = SX \cup X \subseteq S$; contains $1X = X$; and is an ideal of S since $SS^1X \subseteq S^1X$. Conversely, an ideal which contains X also contains SX and S^1X. □

In particular (when S is commutative) the ideal generated by one element $a \in S$ is the set S^1a of all multiples of a; such ideals are called **principal**.

Proposition 3.3. *Let S be a commutative semigroup. If K is a minimal nonempty ideal of S, then K is a smallest nonempty ideal of S, and K is a group.*

Proof. Let I be a nonempty ideal. Since I and K are ideals, $IK \subseteq I \cap K$ and $I \cap K$ is a nonempty ideal. Since $I \cap K \subseteq K$ it follows that $K \subseteq I$.

When $a \in K$, then $Ka \subseteq K$ is an ideal of S; hence $Ka = K$. In particular $ea = a$ for some $e \in K$. Since every element of K has the form ax for some $x \in K$ it follows that e is an identity element of K. Then every element a of K has an inverse in K, since $ab = e$ for some $b \in K$, and K is a group. □

The smallest nonempty ideal of S, when it exists, is the **kernel** of S. Every finite commutative semigroup has a kernel; $\mathbb{N}$ does not.

Proposition 3.4. *Let $\varphi : S \longrightarrow T$ be a homomorphism of semigroups and* $\mathcal{C} = \ker \varphi$.

If φ is surjective and I is an ideal of S, then $\varphi(I)$ is an ideal of T.

If J is an ideal of T, then $\varphi^{-1}(J)$ is an ideal of S and a union of $\mathcal{C}$-classes.

If φ is surjective this defines an order preserving one-to-one correspondence between ideals of T and ideals of S that are unions of $\mathcal{C}$-classes.

Proof. If φ is surjective and I is an ideal of S, then $\varphi(I)$ is an ideal of T, since $\varphi(x)\,\varphi(y) = \varphi(xy) \in \varphi(I)$ for all $x \in S$ and $y \in I$.

If J is an ideal of T, then $\varphi^{-1}(J)$ is a union of $\mathcal{C}$-classes and is an ideal of S since $\varphi(y) \in J$ implies $\varphi(xy) = \varphi(x)\,\varphi(y) \in J$ for all $x \in S$.

If φ is surjective, then $\varphi(\varphi^{-1}(J)) = J$ for every $J \subseteq T$, and $\varphi^{-1}(\varphi(I)) = I$ for every $I \subseteq S$ that is a union of $\mathcal{C}$-classes, as in the proof of Proposition 2.5. □

2. Congruences on a group are determined by normal subgroups. In a semigroup, congruences are most easily constructed from ideals. The resulting quotient semigroups, discovered by Rees [1940], are peculiarly different from quotient groups and from quotient rings.

Proposition 3.5. *When I is an ideal of a semigroup S, the relation* $\mathcal{J}$

defined by

$$a \mathfrak{J} b \iff a = b \text{ or } a, b \in I$$

is a congruence on S, *the* **Rees congruence** *of the ideal* I.

Proof. $\mathfrak{J}$ is an equivalence relation, and is a congruence since $a = b$ and $c, d \in I$ implies $ac, bd \in I$; $a, b \in I$ and $c = d$ implies $ac, bd \in I$; and $a, b, c, d \in I$ implies $ac, bd \in I$. □

The quotient semigroup $S/I = S/\mathfrak{J}$ is the **Rees quotient** of S by I. It is standard practice to identify the $\mathfrak{J}$-class $\{x\} \in S/I$ of each $x \notin I$ with $x \in S$. If $I = \emptyset$, then $S/I = S$. If $I \neq \emptyset$, the $\mathfrak{J}$-class $I \in S/I$ is a zero element and is denoted by 0; then $S/I = (S \backslash I) \cup \{0\}$ with the multiplication $\,.\,$ in which 0 is a zero element and

$$x \,.\, y = \begin{cases} xy \in S & \text{if } xy \notin I \\ 0 & \text{if } xy \in I \end{cases}$$

for all $x, y \in S \backslash I$. Thus the Rees quotient is obtained by squeezing I to a zero element (if $I \neq \emptyset$) and leaving $S \backslash I$ untouched.

3. The Rees quotient can be viewed as the completion of a partial semigroup into an authentic semigroup. In general a **partial binary operation** on a set P is a mapping $\mu : D \longrightarrow P$ whose domain D is a subset of $P \times P$: when $x, y \in P$, $\mu(x, y)$ is defined when $(x, y) \in D$ and is undefined otherwise. In the multiplicative notation, $\mu(x, y)$ is denoted by xy. A **partial semigroup** is a set P together with a partial binary operation on P which is **associative** in the sense that $x(yz) = (xy)z$ holds whenever $x, y, z \in P$ and both $x(yz)$ and $(xy)z$ are defined. (Other associativity conditions have been considered; see the book by Lyapin & Evseev [1997].)

When P and Q are partial semigroups, a **partial homomorphism** of P into Q is a mapping $\varphi : P \longrightarrow Q$ which preserves existing products: $\varphi(xy) = \varphi(x)\,\varphi(y)$ whenever xy is defined in P. If Q is a commutative semigroup, then the set $\mathrm{PHom}(P, Q)$ of all partial homomorphisms of P into Q is closed under pointwise addition and is a commutative semigroup; if P is an actual semigroup, then $\mathrm{PHom}(P, Q) = \mathrm{Hom}(P, Q)$.

Every subset A of a semigroup S is a partial semigroup for the partial operation $\,.\,$ induced by S in the obvious way: when $x, y \in A$, then $x \,.\, y$ is defined in A if and only if $xy \in A$, and then $x \,.\, y = xy$. When I is a nonempty ideal of S, the Rees quotient S/I is obtained from the partial semigroup $S \backslash I$ by adjoining a zero element and setting all undefined products to 0.

4. An **ideal extension** of a semigroup S by a semigroup Q with zero is a

semigroup E such that S is an ideal of E and Q is the Rees quotient $Q = E/S$. Ideal extensions were first studied by Clifford [1950].

The **ideal extension problem**, first considered by Clifford [1950], consists in constructing all ideal extensions of a given semigroup S by a given semigroup Q with zero; one may assume $S \cap Q = \emptyset$. This difficult problem is discussed in some detail in Clifford & Preston [1961], Grillet [1995], and especially Petrich [1973]. The particular case of monoids has a very nice solution, due to Clifford [1950]; another case will be seen in Chapter II. More general results are known but have had few applications to commutative semigroups.

When S is a subsemigroup of E, a **retraction** of E onto S is a homomorphism of E into S which is the identity on S.

Proposition 3.6. *Every ideal extension of a monoid S has a retraction $a \longmapsto ea = ae$, where e is the identity element of S.*

Proof. Let e be the identity element of S. In E we have $ea = (ea)e = e(ae) = ae$ for all $a \in E$, since ea and ae are in S. Let

$$\psi(a) = ea = eae = ae \in S$$

for all $a \in E$. Then $\psi(a) = a$ when $a \in S$, and $\psi(ab) = eabe = \psi(a)\,\psi(b)$ for all $a, b \in E$. □

An ideal extension E of S by Q is a **retract** ideal extension when there exists a retraction ψ of E onto S. Then the restriction $\varphi : Q\backslash 0 \longrightarrow S$ of ψ to $Q\backslash 0 = E\backslash S$ is a partial homomorphism, and the operation on E is determined as follows by the operation on S, the partial operation on $Q\backslash 0$, and the partial homomorphism φ. If $a, b \in Q\backslash 0$ and $ab \neq 0$ in Q, then ab is the same in Q and E. If $a, b \in Q\backslash 0$ and $ab = 0$ in $Q = E/S$, then, in E, $ab \in S$ and $ab = \psi(ab) = \varphi(a)\,\varphi(b)$. If $a \in Q\backslash 0$ and $x \in S$, then, in E, $ax = \psi(ax) = \varphi(a)\,x$ and $xa = \psi(xa) = x\,\varphi(a)$. If $x, y \in S$, then xy is the same in S and E.

Lemma 3.7. *Let S be a semigroup and Q be a semigroup with zero such that $S \cap Q = \emptyset$. If φ is a partial homomorphism of $Q\backslash 0$ into S, then the disjoint union $E = S \cup (Q\backslash 0)$, with the multiplication $*$ defined by*

$$\begin{cases} a * b = ab \in Q & \text{if } ab \neq 0 \text{ in } Q, \\ a * b = \varphi(a)\,\varphi(b) & \text{if } ab = 0 \text{ in } Q, \\ a * y = \varphi(a)\,y, \\ x * b = x\,\varphi(b), \\ x * y = xy \in S \end{cases}$$

for all $a,b \in Q\backslash 0$ and $x,y \in S$, is a retract ideal extension of S by Q, and every retract ideal extension of S by Q can be constructed in this fashion. Moreover, E is commutative if and only if S and Q are commutative.

Proof. Associativity in E follows in a long but straightforward manner from associativity in S, associativity in Q, and the hypothesis that φ is a partial homomorphism. Then it is clear that S is an ideal of E and that $E/S = Q$. Moreover the mapping $\psi : E \longrightarrow S$ defined by

$$\psi(a) = \varphi(a) \text{ for all } a \in Q\backslash\{0\}, \ \psi(x) = x \text{ for all } x \in S,$$

is a retraction of E onto S. Conversely we saw that every retract ideal extension can be constructed as in the statement. □

The ideal extension constructed in Lemma 3.7 is **determined by** the partial homomorphism φ. Thus, an ideal extension is a retract ideal extension if and only if it is determined by a partial homomorphism:

Proposition 3.8. *Every ideal extension of a monoid S by a semigroup Q with zero is a retract ideal extension and is therefore determined by a partial homomorphism of $Q\backslash\{0\}$ into S, namely $a \longmapsto ea = ae$, where e is the identity element of S.*

This follows from Lemma 3.7 and Proposition 3.6.

4. DIVISIBILITY.

1. A **preorder** (also called **quasiorder**) is a binary relation which is reflexive and transitive; thus, a preorder which is also antisymmetric is an order relation.

On a commutative semigroup S the **Green's preorder** $\leqq_{\mathcal{H}}$ (also called the **divisibility preorder**) is defined by

$$a \leqq_{\mathcal{H}} b \iff a = tb \text{ for some } t \in S^1 \iff S^1 a \subseteq S^1 b.$$

if for example e and f are idempotents, then $e \leqq_{\mathcal{H}} f$ if and only if $e \leqq f$ in the Rees order: indeed $ef = e$ implies $e \leqq_{\mathcal{H}} f$; conversely, $e = tf$ implies $ef = tff = tf = e$.

Green's relation $\mathcal{H}$ is one of several relations introduced for semigroups in general by Green [1951]:

$$a \,\mathcal{H}\, b \iff a \leqq_{\mathcal{H}} b \text{ and } b \leqq_{\mathcal{H}} a \iff S^1 a = S^1 b.$$

Proposition 4.1. *In a commutative semigroup, $\mathcal{H}$ is a congruence.*

Proof. If $a \mathcal{H} b$, then $a = tb$ and $b = ua$ for some $t, u \in S^1$, $ac = tbc$ and $bc = uac$ for some $t, u \in S^1$, and $ac \mathcal{H} bc$. □

In a sense, $\mathcal{H}$ measures the extent of group-like behavior in semigroups. Thus, multiplication by elements of a semigroup induces permutations of its $\mathcal{H}$-classes:

Lemma 4.2. *Let H be an $\mathcal{H}$-class and $t \in S^1$. If $tH \subseteq H$, then the mapping $g_t : x \longmapsto tx$ of H into H is bijective.*

Proof. Let $h \in H$. Then $th \in H$, $h = uth$ for some $u \in S^1$, and $uH \subseteq H$, since $\mathcal{H}$ is a congruence. If $a \in H$, then $a = hv$ for some $v \in S^1$ and $tua = uta = uthv = hv = a$; thus g_t and g_u are mutually inverse bijections of H onto H. □

2. Actual groups arise from $\mathcal{H}$ in two ways. A **subgroup** of a semigroup S is a subsemigroup G of S which happens to be a group. Necessarily the identity element of G is an idempotent of S. Conversely, every idempotent e of S yields a trivial subgroup $\{e\}$ of S. Less trivially:

Proposition 4.3. *For an $\mathcal{H}$-class H of a commutative semigroup S the following conditions are equivalent:*

(1) *$ab \in H$ for some $a, b \in H$;*

(2) *H is a subsemigroup of S;*

(3) *H contains an idempotent;*

(4) *H is a subgroup of S.*

Proof. (1) implies (2) since $\mathcal{H}$ is a congruence: if a, b, and $ab \in H$, then $x, y \in H$ implies $x \mathcal{H} a$, $y \mathcal{H} b$, $xy \mathcal{H} ab$, and $xy \in H$.

Assume that H is a subsemigroup and let $a \in H$. Then $aH \subseteq H$; by Lemma 4.2, $g_a : x \longmapsto ax$ is a permutation of H. In particular $ae = a$ for some $e \in H$. Then $a = ae = aee$ and $e^2 = e$, since g_a is injective. Thus (2) implies (3).

Now assume that H contains an idempotent e. Then $a, b \in H$ implies $a \mathcal{H} e$, $b \mathcal{H} e$, $ab \mathcal{H} ee = e$ since $\mathcal{H}$ is a congruence, and $ab \in H$; thus H is a subsemigroup of S. For every $a \in H$ we have $a = te$ for some $t \in S^1$ and $ae = tee = te = a$. Moreover $aH \subseteq H$, $g_a : x \longmapsto ax$ is a permutation of H by Lemma 4.2, and $ab = e$ for some $b \in H$. Hence H is a group. Thus (3) implies (4); and (4) implies (1). □

If for instance S is a monoid, then the elements of H_1 are the **units** of S and H_1 is the **group of units** of S.

Proposition 4.4. *In a commutative monoid S, $S \backslash H_1$ is an ideal.*

Proof. If $y \in S$ is not a unit, then there cannot exist $u \in S$ such that $uxy = 1$, and xy is not a unit. □

In general:

Corollary 4.5. *The maximal subgroups of a commutative semigroup S coincide with the $\mathcal{H}$-classes of S which contain idempotents. They are pairwise disjoint. Every subgroup of S is contained in exactly one maximal subgroup.*

Proof. If G is a subgroup of S and e is the identity element of G, then every $x \in G$ satisfies $ex = x$ and $xy = e$ for some $y = x^{-1} \in G \subseteq S$; hence $G \subseteq H_e$. □

The history of Corollary 4.5 goes back to Schwarz [1943] for torsion semigroups and to Wallace [1953] and Kimura [1954] for semigroups in general.

3. In fact Lemma 4.2 yields a group for every $\mathcal{H}$-class H. Let

$$\mathrm{St}\,(H) \;=\; \{t \in S^1 \mid tH \subseteq H\}$$

denote the (left) stabilizer of H. For every $t \in \mathrm{St}\,(H)$, Lemma 4.2 provides a bijection $g_t : H \longrightarrow H$, $x \longmapsto tx$.

Proposition 4.6. *For every $\mathcal{H}$-class H, $\Gamma(H) = \{g_t \mid t \in \mathrm{St}\,(H)\}$ is a simply transitive group of permutations of H, and $t \longmapsto g_t$ is a homomorphism of $\mathrm{St}\,(H)$ onto $\Gamma(H)$. If $H = H_e$ is a maximal subgroup of S, then $\Gamma(H) \cong H$.*

Proof. First $g_t\,(g_u(x)) = tux = g_{tu}(x)$ for all $t, u \in \mathrm{St}\,(H)$ and $x \in H$; thus $t \longmapsto g_t$ is a homomorphism and $\Gamma(H)$ is a semigroup (under composition). Also $1 \in \mathrm{St}\,(H) \subseteq S^1$ and $g_1 = 1_H$ is the identity mapping on H.

Let $g_t \in \Gamma(H)$ and $a \in H$. As in the proof of Lemma 4.2, $a \,\mathcal{H}\, ta$ and $a = uta$ for some $u \in S^1$; in fact $u \in \mathrm{St}\,(H)$, since $\mathcal{H}$ is a congruence. For every $x \in H$ we now have $x = av$ and $utx = utav = av = x$ for some $v \in S^1$; thus g_t has an inverse in $\Gamma(H)$, namely g_u.

If $a, b \in H$, then $b = ta$ for some $t \in S^1$, $t \in \mathrm{St}\,(H)$ since $\mathcal{H}$ is a congruence, and $g_t(a) = b$; thus $\Gamma(H)$ is transitive. In fact $\Gamma(H)$ is simply transitive: if $g_t(a) = g_u(a)$, then $ta = ua$, $tx = tav = uav = ux$ for every $x = av \in H$, and $g_t = g_u$.

If finally $H = H_e$ is a maximal subgroup of S, with identity element e, then $H \subseteq \mathrm{St}\,(H)$ and the homomorphism $h \longmapsto g_h$ of H into $\Gamma(H)$ is bijective, since $g_h(e) = h$ and $\Gamma(H)$ is simply transitive. □

$\Gamma(H)$ is the (left) **Schützenberger group** of H; it was discovered by Schützenberger [1957].

5. FREE COMMUTATIVE SEMIGROUPS.

1. When a commutative semigroup S is generated by a subset X, every element of S is a product of positive powers of one or more distinct elements of X (Proposition 1.3) but can in general be written in this form in several ways. For example $X = S$ generates S, and then every equality $ab = c$ in S equates two distinct products of positive powers of one or more distinct elements of S.

A commutative semigroup S is **free** on a subset X when every element of S can be written uniquely (up to the order of the terms) as a product of positive powers of one or more distinct elements of X. For example, the multiplicative semigroup $\{2,3,\dots,n,\dots\} \subseteq \mathbb{N}$ is free (as a commutative semigroup) on the set of all prime numbers. In the additive notation, powers become positive integer multiples; the additive semigroup $\mathbb{N}^+$ is free on $\{1\}$.

2. For every set X we now construct a commutative semigroup F_X which is free on X.

F_X is one of the few commutative semigroups that we prefer to denote additively. Then products of positive powers of distinct elements of X become sums of positive integer multiples of distinct elements of X, that is, (finite) linear combinations of elements of X with coefficients in $\mathbb{N}^+$. This suggests that we retrieve F_X from the free abelian group G_X on X, which consists of all linear combinations $a = \sum_{x \in X} a_x x$ with integer coefficients $a_x \in \mathbb{Z}$ that are almost all zero (that is, $\{x \in X \mid a_x \neq 0\}$ is finite). (Linear combinations $\sum_{x \in X} a_x x$ can be defined more formally as suitable families $(a_x)_{x \in X}$ of integers.) Addition on G_X is coordinatewise:

$$\textstyle\sum_{x \in X} a_x x \;+\; \sum_{x \in X} b_x x \;=\; \sum_{x \in X} (a_x + b_x)\, x\,.$$

G_X is a partially ordered group, as the coordinatewise partial order

$$\textstyle\sum_{x \in X} a_x x \leqq \sum_{x \in X} b_x x \text{ if and only if } a_x \leqq b_x \text{ for all } x \in X$$

is compatible with the operation (if $a \leqq b$, then $a + c \leqq b + c$).

F_X is the positive cone of G_X, which is a subsemigroup of G_X:

$$F_X \;=\; \{a \in G_X \mid a > 0\};$$

equivalently, F_X is the set of all linear combinations $a = \sum_{x \in X} a_x x$ with integer coefficients a_x such that $a_x = 0$ for almost all x, $a_x \geqq 0$ for all x, and $a_x > 0$ for some x. Note that $a \leqq_{\mathcal{H}} b$ in F_X if and only if $a \geqq b$ in the coordinatewise partial order.

Every $y \in X$ can be written as a linear combination $y = \sum_{x \in X} a_x x \in F_X$ in which $a_y = 1$ and $a_x = 0$ for all $x \neq y$; thus $X \subseteq F_X$. Now every element of F_X can be written uniquely (up to the order of the terms) as a nonempty sum of positive integer multiples of distinct elements of X; hence

Proposition 5.1. *For every set X, F_X is a commutative semigroup which is free on X.*

Sometimes it is better to denote F_X multiplicatively; then every element of F_X is uniquely (up to the order of the terms) a nonempty product $a = \prod_{x \in X} x^{a_x}$ of positive integer powers of distinct elements of X (with $a_x = 0$ for almost all x, $a_x \geqq 0$ for all x, and $a_x > 0$ for some x).

3. The most important property of F_X is its universal property:

Theorem 5.2. *Every mapping f of X into a commutative semigroup S extends uniquely to a homomorphism φ of F_X into S, namely*

$$\varphi(\textstyle\sum_{x \in X} a_x x) \;=\; \prod_{x \in X} f(x)^{a_x} .$$

The image of φ is the subsemigroup of S generated by $f(X)$. If S is generated by $f(X)$, then φ is surjective. If S is free on X, then S is isomorphic to F_X.

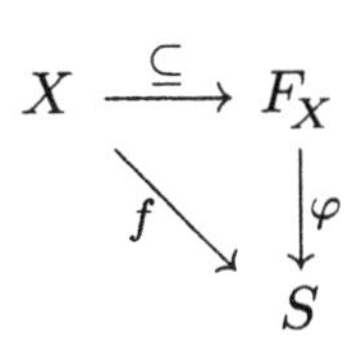

If F_X is denoted multiplicatively, then $\varphi\left(\prod_{x \in X} x^{a_x}\right) = \prod_{x \in X} f(x)^{a_x}$.

Proof. A homomorphism φ transforms sums into products and transforms linear combinations into products of powers:

$$\varphi(a_1 x_1 + a_2 x_2 + \cdots + a_n x_n) \;=\; \varphi(x_1)^{a_1}\, \varphi(x_2)^{a_2} \cdots \varphi(x_n)^{a_n} .$$

If $\varphi : F_X \longrightarrow S$ extends f (if $\varphi(x) = f(x)$ for all i), then, for every $a = \sum_{x \in X} a_x x \in F_X$,

$$\varphi(\textstyle\sum_{x \in X} a_x x) \;=\; \prod_{x \in X} \varphi(x)^{a_x} \;=\; \prod_{x \in X} f(x)^{a_x}$$

($= \prod_{x \in X, a_x \neq 0} f(x)^{a_x}$, which is a finite product). Hence φ is unique.

Conversely define a mapping $\varphi : F_X \longrightarrow S$ by

$$\varphi(\sum_{x\in X} a_x x) = \prod_{x\in X} f(x)^{a_x} .$$

Then φ extends f and is a homomorphism:

$$\begin{aligned} \varphi(\sum_{x\in X} a_x x + \sum_{x\in X} b_x x) &= \prod_{x\in X} f(x)^{a_x+b_x} \\ = \prod_{x\in X} f(x)^{a_x} f(x)^{b_x} &= \Big(\prod_{x\in X} f(x)^{a_x}\Big)\Big(\prod_{x\in X} f(x)^{b_x}\Big). \end{aligned}$$

By Proposition 1.3,

$$\operatorname{Im}\varphi = \{\prod_{x\in X} f(x)^{a_x} \mid a \in F_X\}$$

is the subsemigroup of S generated by $f(X)$. If S is free on X and $f : X \longrightarrow S$ is the inclusion mapping, then φ is an isomorphism. □

It follows from Theorem 5.2 that all commutative semigroups that are generated by a set X are homomorphic images of F_X. Since every semigroup S is generated by some subset $X \subseteq S$ (for instance, by $X = S$), we have:

Corollary 5.3. *Every commutative semigroup is a homomorphic image of a free commutative semigroup. Every finitely generated commutative semigroup is a homomorphic image of a finitely generated free commutative semigroup.*

Commutative semigroups can thus be explored by means of congruences on free commutative semigroups. This approach was pioneered by Rédei [1956] and will be explored in later chapters, and in Proposition 5.8 below.

4. Free commutative semigroups have certain finiteness properties:

Proposition 5.4. *Every free commutative semigroup F satisfies the descending chain condition. If F is finitely generated, then every antichain of F is finite.*

An **antichain** is a subset A which does not contain elements $a < b$.

Proof. By the last part of Theorem 5.2 it suffices to prove these properties for F_X.

When $a = \sum_{x\in X} a_x x \in F$ the positive integer $|a| = \sum_{x\in X} a_x$ is the **length** of a. If $a < b$ in F, then $a_x \leqq b_x$ for all $x \in X$, $a_x < b_x$ for some $x \in X$, and $|a| < |b|$. There cannot exist an infinite descending sequence $a_1 > a_2 > \cdots > a_n > a_{n+1} > \cdots$ of elements of F, for then $|a_1| > |a_2| > \cdots > |a_n| > |a_{n+1}| > \cdots$ would be an infinite descending sequence of positive integers.

Now assume that X is finite. We prove by induction on the number of elements of X that every antichain of F_X is finite. If X is empty, then F_X is empty and so is every antichain of F_X. If X has just one element, then $F_X \cong \mathbb{N}^+$ is a chain and an antichain of F_X has at most one element.

Let X have more than one element and A be an antichain of F_X. For every $y \in X$ and $n \geqq 0$ let $A_{y,n} = \{a \in A \mid a_y = n\}$. Then

$$\{\sum_{x \in X \setminus \{y\}} a_x x \mid a \in A_{y,n}\}$$

is an antichain of $F_{X \setminus \{y\}}$ and $A_{y,n}$ is finite by the induction hypothesis.

For every $x \in X$ let $m(x) = \min\,(a_x \mid a \in A)$ and $M_x = \{a \in A \mid a_x = m(x)\}$. By the above, $M = \bigcup_{x \in X} M_x$ is finite. Let $n(x) = \max\,(a_x \mid a \in M)$. Then $n(x) \geqq m(x)$, since $a_x = m(x)$ for some $a \in M$. If $a \in A$, then $a_x \geqq m(x)$ for all $x \in X$ and $a_x \leqq n(x)$ for some $x \in X$, otherwise $a_x > n(x)$ for all $x \in X$, $a > b$ for all $b \in M \subseteq A$, and A is not an antichain. Hence $A \subseteq \bigcup_{x \in X,\, m(x) \leqq n \leqq n(x)} A_{x,n}$ is finite. □

The second half of Proposition 5.4 is known as **Dickson's Theorem**, after Dickson [1913] who proved it for the free multiplicative subsemigroups of $\mathbb{N}$ generated by finitely many primes. A different proof will be given in Chapter VI along with additional finiteness properties.

5. A commutative monoid S is **free** on a subset X (as a monoid) when every element of S can be written uniquely (up to the order of the terms) as a product of positive powers of distinct elements of X (Proposition 1.4).

The nonnegative cone of G_X is

$$F_X \cup \{0\} = \{a \in G_X \mid a \geqq 0\};$$

equivalently, $F_X \cup \{0\}$ is the set of all linear combinations $a = \sum_{x \in X} a_x x$ with integer coefficients a_x such that $a_x = 0$ for almost all x and $a_x \geqq 0$ for all x. Every element of $F_X \cup \{0\}$ can be written uniquely (up to the order of the terms) as a sum of positive integer multiples of distinct elements of X; hence $F_X \cup \{0\}$ is a free commutative monoid on X. If X is finite, with n elements, then F_X is isomorphic to the direct product $\mathbb{N}^n$. The universal property of $F_X \cup \{0\}$ is:

Proposition 5.5. *Every mapping f of X into a commutative monoid S extends uniquely to a monoid homomorphism φ of $F_X \cup \{0\}$ into S.*

Corollary 5.6. *Every (finitely generated) commutative monoid is a homomorphic image of a (finitely generated) free commutative monoid.*

In later chapters it will be more convenient to denote the free commutative monoid by F_X; then the free commutative semigroup on X is $F_X \backslash \{0\}$.

Similarly we call $F_X \cup \{\infty\}$ the **free** commutative semigroup with zero on the set X, since every element of $F_X \cup \{\infty\}$ is either the zero element ∞ or uniquely a nonempty sum of positive integer multiples of distinct elements of X. Every mapping f of X into a commutative semigroup S with zero extends uniquely to a semigroup homomorphism $\varphi : F_X \cup \{\infty\} \longrightarrow S$ such that $\varphi(\infty) = 0$.

In the multiplicative notation, $F_X \cup \{0\}$ and $F_X \cup \{\infty\}$ become $F_X \cup \{1\}$ and $F_X \cup \{0\}$, respectively.

6. As an application of free commutative semigroups we construct all cyclic semigroups. By Proposition 1.3, a cyclic semigroup S consists of all the powers of its generator x, and is necessarily commutative; hence S is isomorphic to the quotient of $F_{\{x\}}$ by some congruence. Now every element of $F_{\{x\}}$ can be written uniquely in the form nx with $n \in \mathbb{N}^+$; hence $F_{\{x\}} \cong \mathbb{N}^+$. Thus a cyclic semigroup is isomorphic to the quotient of $\mathbb{N}^+$ by some congruence.

Let $\mathcal{C}$ be a congruence on $\mathbb{N}^+$. If $\mathcal{C}$ is not the equality on $\mathbb{N}^+$, the least integer $r > 0$ such that $r \mathrel{\mathcal{C}} t$ for some $t \neq r$ is the **index** of $\mathcal{C}$. Then the least integer $s > 0$ such that $r \mathrel{\mathcal{C}} r+s$ is the **period** of $\mathcal{C}$.

Lemma 5.7. *When $\mathcal{C}$ is a congruence on $\mathbb{N}^+$ of index r and period s, then $a \mathrel{\mathcal{C}} b$ if and only if either $a = b < r$, or $a, b \geqq r$ and $a \equiv b \bmod s$.*

Proof. Since $\mathcal{C}$ is a congruence, $r \mathrel{\mathcal{C}} r+s$ implies $r \mathrel{\mathcal{C}} r+s \mathrel{\mathcal{C}} r+2s \mathrel{\mathcal{C}} \dots \mathrel{\mathcal{C}} r+ks$ for all $k > 0$, $u+r \mathrel{\mathcal{C}} u+r+ks$ for all $k > 0$, $u \geqq 0$, and $a \mathrel{\mathcal{C}} b$ whenever $r \leqq a \leqq b$ and $a \equiv b \bmod s$.

Conversely assume $a \mathrel{\mathcal{C}} b$ with $a < b$. Then $a \geqq r$ by the choice of r. There is an integer $u \geqq 0$ such that $u + a \equiv r \bmod s$, and an integer $k \geqq 0$ such that $a + ks < b \leqq a + ks + s$. Then $t = b - a - ks$ satisfies $0 < t \leqq s$ and $a + ks \mathrel{\mathcal{C}} a \mathrel{\mathcal{C}} b = a + ks + t$; hence

$$r \mathrel{\mathcal{C}} u+a \mathrel{\mathcal{C}} u+a+ks \mathrel{\mathcal{C}} u+a+ks+t \mathrel{\mathcal{C}} r+t.$$

Since $0 < t \leqq s$ it follows from the choice of s that $t = s$. Then $b = a+ks+s \equiv a \bmod s$. □

Proposition 5.8. *Let S be a cyclic semigroup, generated by $x \in S$. Either $S \cong \mathbb{N}^+$, or S is finite and there exist integers $r, s > 0$ (the* **index** *and* **period** *of x) such that $x^i = x^j$ if and only if either $i = j < r$, or $i, j \geqq r$ and $i \equiv j \bmod s$; then every element of S can be written uniquely in the form x^i with $1 \leqq i < r + s$ and*

$$x^i x^j = \begin{cases} x^{i+j} & \textit{if } i+j < r+s, \\ x^k & \textit{if } i+j \geqq r+s, \textit{ where} \\ & \quad r \leqq k < r+s \textit{ and } k \equiv i+j \bmod s; \end{cases}$$

and $\{x^r, x^{r+1}, \dots, x^{r+s-1}\}$ *is a cyclic subgroup of* S.

Proof. We have $S \cong \mathbb{N}^+/\mathcal{C}$ for some congruence $\mathcal{C}$ on $\mathbb{N}^+$. If $\mathcal{C}$ is the equality, then $S \cong \mathbb{N}^+$. Now assume that $\mathcal{C}$ is not the equality. As before, $\mathcal{C}$ has index $r > 0$ and period $s > 0$. By Lemma 5.7, the $\mathcal{C}$-class of $a < r$ is $\{a\}$; the $\mathcal{C}$-classes of $r, r+1, \dots, r+s-1$ are distinct (and infinite); and these are all the $\mathcal{C}$-classes. Since $a \equiv b \bmod s$ implies $a \mathrel{\mathcal{C}} b$ when $a, b \geqq r$ the operation on S is as described in the statement.

Finally, $G = \{x^r, x^{r+1}, \dots, x^{r+s-1}\}$ is a subsemigroup of S and we see from the multiplication on S that $G \cong \mathbb{Z}/s\mathbb{Z}$, the additive group of integers modulo s. □

Proposition 5.8 was first stated (for cyclic semigroups of subsets of a group) by Frobenius [1895], and its Corollary 5.9 below, in its present form, by Moore [1902]. Lemma 5.7 was rediscovered by Chacron [1982]. Tamura [1963] determined all congruences on $\mathbb{Q}^+$.

Corollary 5.9. *Every nonempty finite semigroup contains an idempotent.*

Proof. If S is finite nonempty, then S contains a finite cyclic subsemigroup, which by Proposition 5.8 contains a subgroup and its identity element. □

6. PRESENTATIONS.

Corollary 5.4 suggests that commutative semigroups can be constructed by presentations (= by generators and relations).

1. For this it is more convenient to denote free commutative semigroups multiplicatively. Formally, a commutative semigroup **relation** between elements of a set X is an ordered pair (u, v), normally written as an equality $u = v$, of elements of F_X. (Relations are readily distinguished from actual equalities in F_X, since the latter are all trivial.)

When f is a mapping of X into a commutative semigroup S, we say that the relation $u = v$ **holds in** S **via** f in case the equality $\varphi(u) = \varphi(v)$ holds in S, where $\varphi : F_X \longrightarrow S$ is the homomorphism which extends f.

These somewhat abstract definitions make most sense when X is a subset of S and $f : X \longrightarrow S$ is the inclusion mapping; then φ sends a product $\prod_{x\in X} x^{a_x}$ of elements of X as calculated in F_X to the same product $\prod_{x\in X} x^{a_x}$ calculated in S; hence the relation $\prod_{x\in X} x^{u_x} = \prod_{x\in X} x^{v_x}$ holds in S if and only if the products $\prod_{x\in X} x^{u_x} = \prod_{x\in X} x^{v_x}$ are equal in S.

2. When X is a set and $\mathcal{R} \subseteq F_X \times F_X$ is a set of relations between the elements of X, we denote by $\langle X \,|\, \mathcal{R} \rangle$ the quotient of the free commutative semigroup F_X by the congruence $\mathcal{C}$ generated by $\mathcal{R}$. By Proposition 2.9, $\mathcal{C}$ consists of all the "obvious consequences" of the relations in $\mathcal{R}$.

$\langle X \,|\, \mathcal{R} \rangle$ comes with a canonical mapping $\iota : X \longrightarrow \langle X \,|\, \mathcal{R} \rangle$ which is the composition

$$\iota :\ X \xrightarrow{\subseteq} F_X \longrightarrow F_X/\mathcal{C} = \langle X \,|\, \mathcal{R} \rangle.$$

Proposition 6.1. *$\langle X \,|\, \mathcal{R} \rangle$ is generated by $\iota(X)$ and every relation $(u,v) \in \mathcal{R}$ holds in $\langle X \,|\, \mathcal{R} \rangle$ via ι.*

Proof. $\langle X \,|\, \mathcal{R} \rangle$ is generated by $\iota(X)$, since F_X is generated by X. Moreover the projection $F_X \longrightarrow F_X/\mathcal{C} = \langle X \,|\, \mathcal{R} \rangle$ is the only homomorphism which extends ι; since $\mathcal{R} \subseteq \mathcal{C}$, every relation $(u,v) \in \mathcal{R}$ holds in $\langle X \,|\, \mathcal{R} \rangle$ via ι. □

Accordingly $\langle X \,|\, \mathcal{R} \rangle$ is known as the commutative semigroup **generated by X subject to $\mathcal{R}$**. This is somewhat misleading since X is not a subset of $\langle X \,|\, \mathcal{R} \rangle$ (in fact, ι need not even be injective) and every homomorphic image of $\langle X \,|\, \mathcal{R} \rangle$ has the properties in Proposition 6.1. However, $\langle X \,|\, \mathcal{R} \rangle$ is the "largest" semigroup with these properties:

Proposition 6.2. *Let X be a set and $\mathcal{R}$ be a set of relations between the elements of X. Let S be a commutative semigroup and $f : X \longrightarrow S$ be a mapping such that every relation $u = v$ in $\mathcal{R}$ holds in S via f. There is a unique homomorphism $\varphi : \langle X \,|\, \mathcal{R} \rangle \longrightarrow S$ such that $f = \varphi \circ \iota$. If S is generated by $f(X)$, then φ is surjective.*

$$\begin{array}{ccc} X & \xrightarrow{\ \iota\ } & \langle X \,|\, \mathcal{R} \rangle \\ & {}_{f}\searrow & \downarrow{\scriptstyle \varphi} \\ & & S \end{array}$$

Proof. Let $\mathcal{C}$ be the congruence on F_X generated by $\mathcal{R}$ and $\pi : F_X \longrightarrow F_X/\mathcal{C} = \langle X \,|\, \mathcal{R} \rangle$ be the projection. Let $\psi : F_X \longrightarrow S$ be the homomorphism which extends f. Then $\psi(u) = \psi(v)$ for every relation $u = v$ in $\mathcal{R}$, since $u = v$ holds in S via f; hence $\mathcal{R} \subseteq \ker \psi$ and $\mathcal{C} \subseteq \ker \psi$. By Proposition 2.4, ψ factors through π: $\psi = \varphi \circ \pi$ for some homomorphism $\varphi : \langle X \,|\, \mathcal{R} \rangle \longrightarrow S$.

Then $\varphi \circ \iota = f$.

$$\begin{array}{ccccc} X & \xrightarrow{\subseteq} & F_X & \xrightarrow{\pi} & \langle X \,|\, \mathcal{R} \rangle \\ & {\scriptstyle f} \searrow & \downarrow {\scriptstyle \psi} & \swarrow {\scriptstyle \varphi} & \\ & & S & & \end{array}$$

Let $\varphi' : \langle X \,|\, \mathcal{R} \rangle \longrightarrow S$ be another homomorphism such that $\varphi' \circ \iota = f$. Then $T = \{ a \in \langle X \,|\, \mathcal{R} \rangle \mid \varphi(a) = \varphi'(a) \}$ is a subsemigroup of $\langle X \,|\, \mathcal{R} \rangle$ which contains $\iota(X)$; since $\iota(X)$ generates $\langle X \,|\, \mathcal{R} \rangle$ it follows that $\varphi = \varphi'$. □

3. A **presentation** of a commutative semigroup S consists of a set X, a set $\mathcal{R}$ of commutative semigroup relations between the elements of X, and an isomorphism $S \cong \langle X \,|\, \mathcal{R} \rangle$. By Corollary 5.4, every commutative semigroup S has a presentation, in which X can be any subset of S which generates S, and $\mathcal{R}$ can be any binary relation which generates the congruence induced by $F_X \longrightarrow S$.

For example let $\mathcal{C}$ be the congruence on $\mathbb{N}^+$ of index r and period s. By Lemma 5.7, $\mathcal{C}$ is generated by $(r, r+s)$. Therefore a finite cyclic semigroup S of index r and period s has the presentation $S \cong \langle\, x \mid x^r = x^{r+s} \,\rangle$.

Presentations are associated with a number of logical and computational problems: the **word problem** (deciding when two products of generators are equal); the **isomorphism problem** (deciding when two presentations yield isomorphic semigroups); and recognition problems (recognizing additional properties from a presentation). Algorithms in Rosales & García-Sánchez [1999] solve a number of these problems.

4. Similar definitions apply to commutative monoids and to commutative semigroups with zero. A commutative monoid relation between the elements of a set X is an ordered pair (u, v) (normally written as an equality $u = v$) of elements of the free commutative monoid $F_X \cup \{1\}$ (written multiplicatively); the identity element of $F_X \cup \{1\}$ may appear as u or v. When S is a commutative monoid, a commutative monoid presentation of S consists of a set X, a set $\mathcal{R}$ of commutative monoid relations between the elements of X, and an isomorphism $S \cong \langle X \,|\, \mathcal{R} \rangle$, where $\langle X \,|\, \mathcal{R} \rangle$ now denotes the quotient of $F_X \cup \{1\}$ by the congruence generated by $\mathcal{R}$.

A commutative relation with zero between the elements of a set X is an ordered pair (u, v) (normally written as an equality $u = v$) of elements of the free commutative semigroup with zero $F_X \cup \{0\}$ (written multiplicatively); the zero element of $F_X \cup \{0\}$ may appear as u or v. When S is a commutative semigroup with zero, a presentation of S as a commutative semigroup with zero consists of a set X, a set $\mathcal{R}$ of commutative relations with zero between the elements of X, and an isomorphism $S \cong \langle X \mid \mathcal{R} \rangle$, where $\langle X \mid \mathcal{R} \rangle$ now denotes the quotient of $F_X \cup \{0\}$ by the congruence generated by $\mathcal{R}$.

Chapter II.

CANCELLATIVE SEMIGROUPS.

One of the oldest results in semigroup theory embeds cancellative commutative semigroups into abelian groups, by a construction which also completes $\mathbb{N}$ into $\mathbb{Z}$ and embeds integral domains into fields.

This chapter begins with two generalizations of this construction, semigroups of fractions and universal groups. Then we turn to cancellative commutative semigroups, which later chapters will show are a basic building block for more general commutative semigroups.

For the sake of brevity we often abbreviate "commutative semigroup" as **c.s.** and "commutative monoid" as **c.m.** in what follows.

1. SEMIGROUPS OF FRACTIONS.

The semigroups of fractions in this section are reminiscent of localization in commutative rings and were already known to Vandiver [1940]. They will be used in later chapters to construct completions. More general constructions are noted at the end of this section.

1. Let S be a c.s., written multiplicatively, and C be a nonempty subsemigroup of S. Under the componentwise multiplication $(x,a)(y,b) = (xy, ab)$, $S \times C$ is a c.s. (S^1 and C^1 may be used here instead of S and C, and yield the same fractions.) Let $\equiv$ be the binary relation on $S \subseteq C$ defined by:

$$(x,a) \equiv (y,b) \text{ if and only if } cay = cxy \text{ for some } c \in C.$$

Since S is commutative, it is immediate that $\equiv$ is a congruence on $S \times C$. The **fraction** x/a is the $\equiv$-class of (x,a). The **semigroup of fractions** $C^{-1}S$ is the quotient semigroup $(S \times C)/{\equiv}$. By definition,

$$x/a = y/b \text{ if and only if } cay = cxy \text{ for some } c \in C$$

and $(x/a)(y/b) = xy/ab$, for all $x,y \in S$ and $a,b \in C$.

Proposition 1.1. *Let S be a commutative semigroup and C be a nonempty subsemigroup of S. The fractions x/a with $x \in S$, $a \in C$ constitute a commutative monoid $C^{-1}S$, in which $a/a = 1$ for every $a \in C$. A canonical homomorphism $\alpha : S \longrightarrow C^{-1}S$ is well defined by $\alpha(x) = ax/a$ for any $a \in C$. For every $x \in S$ and $a \in C$, $\alpha(a)$ is a unit of $C^{-1}S$ and $x/a = \alpha(x)\,\alpha(a)^{-1}$.*

Proof. First $a/a = b/b$ for all $a, b \in C$, and $(x/a)(b/b) = xb/ab = x/b$ for all $x \in S$. Hence $C^{-1}S$ has an identity element 1 and $1 = a/a$ for all $a \in C$.

Similarly $ax/a = bx/b$ for all $a, b \in C$, so that $\alpha(x) \in C^{-1}S$ is well defined by $\alpha(x) = ax/a$, for all $x \in S$. For every $a \in C$, $\alpha(a) = a^2/a$ is a unit since $(a^2/a)(a/a^2) = a^3/a^3 = 1$, and $\alpha(a)^{-1} = a/a^2$; hence $\alpha(x)\,\alpha(a)^{-1} = (ax/a)(a/a^2) = a^2x/a^3 = x/a$. □

An element c of a c.s. S is **cancellative** in S when $cx = cy$ implies $x = y$, for all $x, y \in S$. A c.s. S is **cancellative** when every element of S is cancellative in S (when $cx = cy$ implies $x = y$, for all $x, y, c \in S$).

Proposition 1.2. *Let S be a commutative semigroup and C be a nonempty subsemigroup of S. The canonical homomorphism $\alpha : S \longrightarrow C^{-1}S$ is injective if and only if every element of C is cancellative in S.*

Proof. If $\alpha(x) = \alpha(y)$, then $ax/a = by/b$ and $caby = cabx$ for some $a, b, c \in C$; if every element of C is cancellative in S, then $x = y$. Conversely assume that α is injective. If $c \in C$ and $cx = cy$, then $c^3x = c^3y$, $\alpha(x) = cx/c = cy/c = \alpha(y)$, and $x = y$. □

If for instance S is cancellative, then the canonical homomorphism $\alpha : S \longrightarrow S^{-1}S$ is injective; this is the familiar embedding of a cancellative c.s. into an abelian group, which is studied in more detail in the next section.

2. The semigroup of fractions $C^{-1}S$ and its canonical homomorphism enjoy a universal property:

Proposition 1.3. *Let T be a commutative semigroup and $\varphi : S \longrightarrow T$ be a homomorphism. If T is a monoid and $\varphi(c)$ is a unit of T for every $c \in C$, then φ factors uniquely through α: there exists a unique monoid homomorphism $\xi : C^{-1}S \longrightarrow T$ such that $\xi \circ \alpha = \varphi$, namely, $\xi(x/a) = \varphi(x)\,\varphi(a)^{-1}$ for all $x \in S$ and $a \in C$.*

$$\begin{array}{ccc} S & \xrightarrow{\ \alpha\ } & C^{-1}S \\ & {\scriptstyle\varphi}\searrow & \downarrow{\scriptstyle\xi} \\ & & T \end{array}$$

Proof. A monoid homomorphism ξ sends units to units and preserves inverses of units: if u is a unit with inverse $v = u^{-1}$, then $uv = 1$, $\xi(u)\xi(v) = \xi(1) = 1$, and $\xi(u)$ is a unit with inverse $\xi(u^{-1})$. If therefore $\xi : C^{-1}S \longrightarrow T$ is a monoid homomorphism such that $\xi \circ \alpha = \varphi$, then necessarily

$$\xi(x/a) = \xi\big(\alpha(x)\,\alpha(a)^{-1}\big) = \xi\big(\alpha(x)\big)\,\xi\big(\alpha(a)\big)^{-1} = \varphi(x)\,\varphi(a)^{-1}$$

whenever $a \in C$, by Proposition 1.1.

If conversely $x/a = y/b$ in $C^{-1}S$, then $cbx = cay$ for some $c \in C$ and $\varphi(c)\,\varphi(b)\,\varphi(x) = \varphi(c)\,\varphi(a)\,\varphi(y)$; multiplying by $\varphi(a)^{-1}\,\varphi(b)^{-1}\,\varphi(c)^{-1}$ yields $\varphi(a)^{-1}\,\varphi(x) = \varphi(b)^{-1}\,\varphi(y)$. Therefore a mapping $\xi : C^{-1}S \longrightarrow T$ is well defined by $\xi(x/a) = \varphi(x)\,\varphi(a)^{-1}$. Since $\varphi(ab)^{-1} = \varphi(a)^{-1}\,\varphi(b)^{-1}$, it is immediate that ξ is a monoid homomorphism. The first part of the proof then shows that ξ is the only monoid homomorphism $\xi : C^{-1}S \longrightarrow T$ such that $\xi \circ \alpha = \varphi$. □

As a first application of Proposition 1.3 we note:

Proposition 1.4. *If S is a subsemigroup of T, then $C^{-1}S$ is isomorphic to the subsemigroup $\{x/a \in C^{-1}T \mid x \in S\}$ of $C^{-1}T$.*

Proof. Let $\varphi : S \longrightarrow C^{-1}T$ be the restriction to S of the canonical homomorphism $\beta : T \longrightarrow C^{-1}T$. Proposition 1.3 yields a homomorphism $\psi : C^{-1}S \longrightarrow C^{-1}T$ which sends $x/a \in C^{-1}S$ to $\varphi(x)\,\varphi(a)^{-1} = \beta(x)\,\beta(a)^{-1} = x/a \in C^{-1}T$. If $x, y \in S$ and $x/a = y/b$ in $C^{-1}T$, then $cay = cbx$ for some $c \in C$ and $x/a = y/b$ in $C^{-1}S$; thus ψ is injective. We see that $\operatorname{Im}\psi = \{x/a \in C^{-1}T \mid x \in S,\ a \in C\}$. □

If more generally S and $C \neq \emptyset$ are subsemigroups of T, we denote the subsemigroup $\{x/a \in C^{-1}T \mid x \in S,\ a \in C\}$ by $C^{-1}S$. By Proposition 1.4, this subsemigroup may be identified with the semigroup of fractions $C^{-1}S$ in case $C \subseteq S$.

Proposition 1.5. *If I is an ideal of S which contains C, then $C^{-1}I = C^{-1}S$.*

Proof. Let $a \in C$. For every $x/a \in C^{-1}S$ we have $x/a = xc/ac \in C^{-1}I$, since $xc \in I$ and $ac \in C$. □

3. Semigroups of fractions have been studied, mostly following the analogy with localization, by Maury [1958], Bouvier [1969], [1970], and Bouvier & Faisant [1970].

Much more general semigroups of quotients have been defined for all semigroups, most notably by Berthiaume [1964], [1971], McMorris [1971], and Luedemann [1976], following similar constructions in ring theory (Johnson [1951]; Utumi [1956]; Findlay & Lambek [1958]). The author likes the very general construction in Almkvist [1968] and Stenström [1970]. The ingredients are a monoid S and a set $\mathfrak{F}$ of nonempty right ideals of S, such that: $S \in \mathfrak{F}$; $I \in \mathfrak{F}$ implies $I : s = \{x \in S \mid sx \in I\} \in \mathfrak{F}$ for all $s \in S$; and $\mathfrak{F}$ contains every right ideal I such that $I : s \in \mathfrak{F}$ for all s in some $J \in \mathfrak{F}$. Then $\mathfrak{F}$ is a filter ($J \supseteq I \in \mathfrak{F}$ implies $J \in \mathfrak{F}$, and $I, J \in \mathfrak{F}$ implies $I \cap J \in \mathfrak{F}$). A congruence $\mathfrak{T}$ on S is defined by

$$x \;\mathfrak{T}\; y \text{ if and only if there exists } I \in \mathfrak{F} \text{ such that } xs = ys \text{ for all } s \in I.$$

The elements of $S_{\mathfrak{F}}$ are equivalence classes of mappings $f : I \longrightarrow S/\mathfrak{T}$ such that $I \in \mathfrak{F}$ and $f(xs) = f(x)\tau(s)$ for all $x \in I$ and $s \in S$, where $\tau : S \longrightarrow S/\mathfrak{T}$ is the projection; $f : I \longrightarrow S$ and $g : J \longrightarrow S$ are equivalent if and only if f and g agree on some $K \in \mathfrak{F}$, $K \subseteq I \cap J$.

The semigroups $C^{-1}S$ are a particular case. When S is a commutative semigroup and C is a subsemigroup of S, one may let $\mathfrak{F}$ be the set of all nonempty ideals I of S^1 such that $I : x$ intersects C for every $x \in S^1$. Then

$$D = \{s \in S^1 \mid S^1 s \in \mathfrak{F}\} = \{s \in S^1 \mid S^1 s \cap C \neq \emptyset\}$$

is a subsemigroup of S^1 (in fact D is the face of S^1 generated by C). Proposition 6.8 of Stenström [1970] shows that $S^1/\mathfrak{F} \cong D^{-1}S^1$; it is readily verified that $D^{-1}S^1 \cong C^{-1}S$.

Other particular cases are more relevant for semigroups in general and have been studied in commutative cases by McMorris [1972], Hinkle [1972], [1974], Berthiaume [1973], Rompke [1975], and Johnson & McMorris [1977]. Semigroups of quotients were also constructed by Toca [1977], Fountain & Petrich [1986], and Easdown & Gould [1996].

2. UNIVERSAL GROUPS.

Universal groups now dangle from the previous section, ripe for picking.

1. A **universal group** of a semigroup S is a group $G(S)$, together with a homomorphism $\gamma : S \longrightarrow G(S)$, such that every homomorphism $\varphi : S \longrightarrow G$ of S into a group G factors uniquely through γ (= there exists a unique group

homomorphism $\xi : G(S) \longrightarrow G$ such that $\varphi = \xi \circ \gamma$).

$$\begin{array}{ccc} S & \xrightarrow{\ \gamma\ } & G(S) \\ & {\scriptstyle\varphi}\searrow & \downarrow{\scriptstyle\psi} \\ & & G \end{array}$$

This universal property implies that $G(S)$ is unique up to isomorphism if it exists.

That every semigroup S has a universal group follows from the Adjoint Functor Theorem. When S is commutative, Proposition 1.3 provides a simple construction of $G(S)$:

Proposition 2.1. *When S is a nonempty commutative semigroup, then the abelian group $S^{-1}S$, together with the canonical homomorphism $\alpha : S \longrightarrow S^{-1}S$, is a universal group of S: for every homomorphism $\varphi : S \longrightarrow G$ of S into a group G there exists a unique group homomorphism $\xi : S^{-1}S \longrightarrow G$ such that $\varphi = \xi \circ \alpha$, which sends $x/a \in S^{-1}S$ to $\varphi(x)\,\varphi(a)^{-1} \in G$.*

Proof. The c.m. $S^{-1}S$ is a group since every $x/a = \alpha(x)\,\alpha(a)^{-1}$ is a product of units. If $\varphi : S \longrightarrow G$ is a homomorphism of S into a group G, then $\varphi(S)$ generates a commutative subgroup H of G, in which every $\varphi(x)$ is a unit, and Proposition 1.3 yields a unique monoid homomorphism (also a group homomorphism) $\xi : S^{-1}S \longrightarrow H \subseteq G$ such that $\varphi = \xi \circ \alpha$, as in the statement. □

Because of Proposition 2.1 we use the **group of fractions** $S^{-1}S$ of S as its universal group $G(S)$ in what follows. In Section III.2 we shall see that $G(S)$ has the following presentation: when written additively, $G(S)$ is the abelian group generated by the elements of S, subject to all relations $a + b = ab$ with $a, b \in S$.

Proposition 2.2. *Every homomorphism $\varphi : S \longrightarrow T$ of commutative semigroups induces a unique homomorphism $G(\varphi) : G(S) \longrightarrow G(T)$ such that the square*

$$\begin{array}{ccc} S & \xrightarrow{\ \varphi\ } & T \\ {\scriptstyle\alpha}\downarrow & & \downarrow{\scriptstyle\beta} \\ G(S) & \xrightarrow[G(\varphi)]{} & G(T) \end{array}$$

commutes, where $\alpha : S \longrightarrow G(S)$ and $\beta : T \longrightarrow G(T)$ are the canonical homomorphisms; $G(\varphi)$ sends $x/a \in G(S)$ to $\varphi(x)/\varphi(a) \in G(T)$.

Proof. By the universal property of $G(S)$ there is a unique homomorphism $G(\varphi)$ such that $G(\varphi) \circ \alpha = \beta \circ \varphi$; from Proposition 2.1, $G(\varphi)$ sends $x/a \in G(S)$

to $\beta(\varphi(x))\,\beta(\varphi(a))^{-1} = \varphi(x)/\varphi(a)$. □

2. Underlying the construction of universal groups is the following result. A congruence $\mathcal{C}$ on a c.s. S is **cancellative** when the quotient semigroup $S/\mathcal{C}$ is cancellative; equivalently, when $ac \mathrel{\mathcal{C}} bc$ implies $a \mathrel{\mathcal{C}} b$.

Proposition 2.3. *On every commutative semigroup S there is a smallest cancellative congruence $\mathcal{C}$, namely*

$$x \mathrel{\mathcal{C}} y \text{ if and only if } cx = cy \text{ for some } c \in S;$$

then $S/\mathcal{C}$, together with the projection $S \longrightarrow S/\mathcal{C}$, is a universal cancellative semigroup of S, and $G(S/\mathcal{C})$ is a universal group of S.

Proof. $\mathcal{C}$ is the congruence induced by the canonical homomorphism $\gamma : S \longrightarrow G(S)$; it is a cancellative congruence since $S/\mathcal{C} \cong \operatorname{Im}\gamma \subseteq G(S)$ is cancellative. If conversely $\mathcal{E}$ is a cancellative congruence on S, then the homomorphism $S \longrightarrow S/\mathcal{E} \longrightarrow G(S/\mathcal{E})$ factors through γ; hence $\mathcal{C} \subseteq \mathcal{E}$. Thus $\mathcal{C}$ is the smallest cancellative congruence on S. (This can also be proved directly.)

Let $\pi : S \longrightarrow S/\mathcal{C}$ be the projection. If $\varphi : S \longrightarrow T$ is a homomorphism of S into a cancellative semigroup T, then $\operatorname{Im}\varphi$ is cancellative, $\ker\varphi$ is a cancellative congruence, $\mathcal{C} \subseteq \ker\varphi$, and φ factors uniquely through π (Proposition I.2.4). Thus $S/\mathcal{C}$, together with the projection $S \longrightarrow S/\mathcal{C}$, is a universal cancellative semigroup of S.

Let G be a group and $\varphi : S \longrightarrow G$ be a homomorphism. Since G is cancellative, there is a unique homomorphism $\psi : S/\mathcal{C} \longrightarrow G$ such that $\varphi = \psi \circ \pi$, where $\pi : S \longrightarrow S/\mathcal{C}$ is the projection. Then there is a unique homomorphism $\chi : G(S/\mathcal{C}) \longrightarrow G$ such that $\psi = \chi \circ \beta$, where $\beta : S/\mathcal{C} \longrightarrow G(S/\mathcal{C})$ is the canonical homomorphisms. Therefore φ factors uniquely through $\beta \circ \pi$. Thus $G(S/\mathcal{C})$, together with the homomorphism $\beta \circ \pi$, is a universal group of S. □

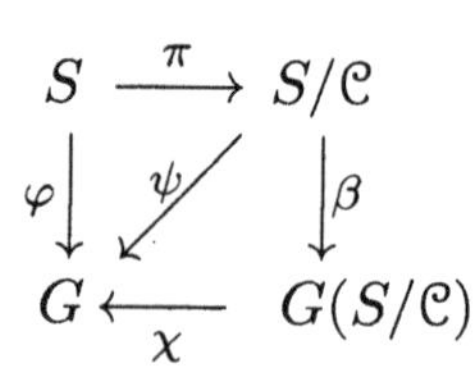

3. By Proposition 1.2, $\gamma : S \longrightarrow G(S)$ is injective if and only if S is cancellative. Semigroups for which γ is surjective were studied by McAlister & O'Carroll [1971] and Tamura & Hamilton [1971]. The latter proved:

Proposition 2.4. *$\gamma : S \longrightarrow G(S)$ is surjective if and only if for every $x \in S$ there exist $y, t \in S$ such that $xyt = t$.*

Proof. Assume that this condition holds and let $x/a \in G(S)$. Then $ayt = t$

for some $y, t \in S$. Hence $axyt = xt$ and $x/a = xyt/t = \gamma(xy)$. Thus γ is surjective.

If conversely γ is surjective and $x \in S$, then $\gamma(x)^{-1} = \gamma(y)$ for some $y \in S$, $x/x^2 = xy/x$, $x^2u = x^3yu$ for some $y, u \in S$, and $xyt = t$ for some y and $t = x^2u \in S$. □

If for instance S is finite, then $\gamma : S \longrightarrow G(S)$ is surjective: for every $x \in S$, some $x^n = e$ is idempotent, and $xye = e$, where $y = x^{n-1}$ if $n \geqq 2$, $y = e$ if $n = 1$; then Proposition 2.5 below implies that $G(S)$ is isomorphic to the kernel of S. If on the other hand $S = \mathbb{N}$, or more generally if S is cancellative but not a group, then γ is not surjective, since it is injective but not bijective.

4. For future reference we note the following properties.

Proposition 2.5. *When I is a nonempty ideal of a commutative semigroup S, then the inclusion homomorphism $I \longrightarrow S$ induces an isomorphism $G(I) \cong G(S)$.*

Proof. By Proposition 2.2 the homomorphism $\xi : G(I) \longrightarrow G(S)$ induced by the inclusion $I \longrightarrow S$ sends $x/a \in G(I)$ to $x/a \in G(S)$. Let $a, b, x, y \in I$. If $x/a = y/b$ in $G(S)$, then $tay = tbx$ for some $t \in S$, $atay = atbx$ with $at \in I$, and $x/a = y/b$ in $G(I)$; thus ξ is injective. Also let $a, x \in S$ and $c \in I$. Then $x/a = xc/ac$ with $ac, xc \in I$; hence ξ is surjective. □

Proposition 2.5 is essentially due to Brameret [1962].

Proposition 2.6. *$G(S \times T) \cong G(S) \times G(T)$ for all nonempty commutative semigroups S and T.*

Proof. The canonical homomorphisms $\alpha : S \longrightarrow G(S)$ and $\beta : T \longrightarrow G(T)$ induce a homomorphism

$$\varphi = \alpha \times \beta : \; S \times T \longrightarrow G(S) \times G(T), \; (x, y) \longmapsto (\alpha(x), \, \beta(y)),$$

which by Proposition 1.3 induces a homomorphism

$$\xi : \; G(S \times T) = (S \times T)^{-1}(S \times T) \longrightarrow G(S) \times G(T)$$

that sends $(x, y)/(a, b) \in (S \times T)^{-1}(S \times T)$ to

$$\begin{aligned} \varphi(x, y)\, \varphi(a, b)^{-1} &= \big(\alpha(x), \beta(y)\big)\big(\alpha(a), \beta(b)\big)^{-1} \\ &= \big(\alpha(x)\, \alpha(a)^{-1}, \; \beta(y)\, \beta(b)^{-1}\big) \; = \; (x/a, \; y/b). \end{aligned}$$

It is clear that ξ is surjective. If $\xi\big((x, y)/(a, b)\big) = \xi\big((z, t)/(c, d)\big)$, then $x/a = z/c$ and $y/b = t/d$, $uaz = ucx$ and $vbt = vdy$ for some $u \in S$ and $v \in T$,

$(u,v)(a,b)(z,t) = (u,v)(c,d)(x,y)$ in $S \times T$, and $(x,y)/(a,b) = (z,t)/(c,d)$; thus ξ is injective. □

3. CANCELLATIVE SEMIGROUPS.

Cancellative commutative semigroups provide natural structures for abstract arithmetic and for general ideal theory. Some also have long been investigated extensively in the guise of lattice ordered abelian groups. Examples include abelian groups and free commutative semigroups. This section contains some general properties and examples. Later sections give a few structural properties.

1. First we show:

Proposition 3.1. *A cancellative c.s. which is finite and nonempty is a group.*

Proof. In any c.s. S, multiplication by $a \in S$ is a mapping $\lambda_a : x \longmapsto ax$ of S into S. If S is cancellative, then every λ_a is injective; if S is also finite, then every λ_a is surjective. This implies that S is a single $\mathcal{H}$-class; then S is a group by Proposition I.4.3. □

Proposition 3.2. *A commutative semigroup S can be embedded into a group if and only if it is cancellative; then the canonical homomorphism $\gamma : S \longrightarrow G(S)$ is injective.*

Proof. Subsemigroups of groups are cancellative. If conversely S is cancellative, then the canonical homomorphism $\gamma : S \longrightarrow G(S)$ is injective by Proposition 1.2, and S can be embedded into the abelian group $G(S)$. □

This provides the familiar embedding of S into an abelian group, namely, the group of fractions or universal group $G(S)$ of S. When S is cancellative it is standard practice to identify $x \in S$ and $\alpha(x) \in G(S)$; then x/a becomes xa^{-1} and S becomes a subsemigroup of $G(S)$.

Universal groups of cancellative c.s. enjoy additional properties.

Proposition 3.3. *Let S be a nonempty cancellative c.s. An abelian group G is isomorphic to the universal group of S if and only if S is isomorphic to a subsemigroup T of G such that every $g \in G$ can be written in the form $g = ab^{-1}$ for some $a, b \in T$.*

Proof. We just saw that every element of $G(S)$ can be written in the form xa^{-1} for some $x, a \in S$. Conversely let G be an abelian group and T be a subsemigroup of G such that every $g \in G$ can be written in the form $g = ab^{-1}$

for some $a, b \in T$ (in particular, $T \neq \emptyset$). Proposition 2.1, applied to the inclusion homomorphism $T \longrightarrow G$, provides a homomorphism $\xi : G(T) \longrightarrow G$, which sends x/a to xa^{-1} for all $a, x \in T$. Then ξ is surjective by the hypothesis, and is injective, since $\xi(x/a) = \xi(y/b)$ implies $xa^{-1} = yb^{-1}$, $bx = ay$, and $x/a = y/b$. □

By Proposition 3.3 we may view cancellative c.s. (up to isomorphism) as generating subsemigroups of abelian groups.

Proposition 3.4. *Let S and T be cancellative c.s. If S is a subsemigroup of T, then $G(S)$ is isomorphic to a subgroup of $G(T)$.*

Proof. By Proposition 2.2 the inclusion homomorphism $S \longrightarrow T$ induces a homomorphism $\xi : S^{-1}S \longrightarrow T^{-1}T$ which sends $x/a \in S^{-1}S$ to $x/a \in T^{-1}T$. Let $a, b, x, y \in S$. If $x/a = y/b$ in $T^{-1}T$, then $ay = bx$ since T is cancellative, and $x/a = y/b$ in $S^{-1}S$; thus ξ is injective. □

Alternately we may regard T as a subsemigroup of $G(T)$, and then the subgroup $\{ab^{-1} \mid a, b \in S\}$ of $G(T)$ generated by S is isomorphic to $G(S)$ by Proposition 3.3, and may be identified with $G(S)$.

Proposition 3.4 implies that $G(S)$ is, up to isomorphism, the smallest abelian group G of which S is a subsemigroup.

These results may lull readers into thinking that cancellative semigroups are easily retrieved from abelian groups and are therefore not terribly complicated. A cure for this particular delusion may be found in the next section, which shows just how easy it is to retrieve all subsemigroups of $\mathbb{Z}$.

2. Let S be a cancellative c.s. If E is a commutative ideal extension of S by Q, then $G(E) \cong G(S)$ by Proposition 2.3, and the canonical homomorphism $E \longrightarrow G(S)$ induces a partial homomorphism of $Q \backslash 0$ into $G(S)$. The ideal extension E can be reconstructed from this partial homomorphism much as in Lemma I.3.7; the details are given below. More general results similarly construct ideal extensions of a weakly reductive semigroup S from partial homomorphisms into the translational hull of S (Clifford [1950]), which when S is a cancellative c.s. is isomorphic to the subsemigroup $\{g \in G(S) \mid gS \subseteq S\}$ of $G(S)$ (Petrich [1973]).

Lemma 3.5. *Let E be an ideal extension of a nonempty cancellative c.s. S. A homomorphism $\tau : E \longrightarrow G(S)$ which is the identity on S is well defined by $\tau(a) = as/s = sa/s$ for any $s \in S$.*

Proof. Let $s, t \in S$ and $a \in E$. In $G(S)$ we have as, $ta \in S$ and $as/s = ta/t$, since $s(ta) = (ta)s = t(as)$. Hence $\tau(a) = as/s = ta/t$ depends neither on s nor on t. For all $a,b \in E$ and $s \in S$ we have $sbs = bs^2$, since s, $bs \in S$, and

$$\tau(a)\,\tau(b) = (as/s)\,(bs/s) = asbs/s^2 = abs^2/s^2 = \tau(ab);$$

hence τ is a homomorphism. Also $\tau(a) = as/s = a$ if $a \in S$. $\square$

If E is commutative, then, as noted above, τ can also be constructed by composing the canonical homomorphism $E \longrightarrow G(E)$ and the isomorphism $G(E) \cong G(S)$ in Proposition 2.3.

In general, let S be a nonempty cancellative c.s. and E be any ideal extension of S by a semigroup Q with zero. We may assume that $Q \cap G(S) = \varnothing$. The homomorphism $\tau : E \longrightarrow G(S)$ in Lemma 3.5 is the **canonical** homomorphism of the ideal extension E of S. (A more general canonical homomorphism sends an ideal extension of any semigroup S into the translational hull of S.)

Let φ be the restriction to $E\backslash S$ of the canonical homomorphism. Then φ is a partial homomorphism of $Q\backslash 0 = E\backslash S$ into $G(S)$. Moreover, $x\,\varphi(a) = \varphi(a)\,x = \tau(a)\,\tau(x) = \tau\,(ax) = ax \in S$ for all $a \in Q\backslash 0$ and $x \in S$, and $\varphi(a)\,\varphi(b) = \tau(a)\,\tau(b) = \tau\,(ab) = ab \in S$ whenever $ab = 0$ in Q. Hence the operation on E is determined by the operation on S, the partial operation on $Q\backslash 0$, and the partial homomorphism φ; in fact E is a subsemigroup of the ideal extension of $G(S)$ determined by φ as in Lemma I.3.7.

The converse is clear:

Lemma 3.6. *Let S be a nonempty cancellative c.s. and Q be a semigroup with zero such that $Q \cap G(S) = \varnothing$. Let φ be a partial homomorphism of $Q\backslash 0$ into $G(S)$, such that $\varphi(a)\,x \in S$ for all $a \in Q\backslash 0$ and $x \in S$ and $\varphi(a)\,\varphi(b) \in S$ whenever $ab = 0$ in Q. Let $E = S \cup (Q\backslash 0)$, with multiplication $*$ defined by*

$$\begin{cases} a * b = ab \in Q & \text{if } ab \neq 0 \text{ in } Q, \\ a * b = \varphi(a)\,\varphi(b) & \text{if } ab = 0 \text{ in } Q, \\ a * y = \varphi(a)\,y, & \\ x * b = x\,\varphi(b), & \\ x * y = xy \in S & \end{cases}$$

for all $a,b \in Q\backslash 0$ and $x,y \in S$. Then E is an ideal extension of S by Q, and a subsemigroup of the ideal extension of $G(S)$ determined by the partial homomorphism φ. Moreover, E is commutative if and only if Q is commutative.

If φ sends $Q\backslash 0$ into $S \subseteq G(S)$, then the ideal extension in Lemma 3.6 coincides with the ideal extension in Lemma I.3.7 which is determined by the partial homomorphism φ. In general we say that the ideal extension E in Lemma 3.6 is **determined** (as an ideal extension of S) **by** the partial homomorphism φ. The remarks before Lemma 3.6 show that all ideal extensions of S can be constructed in this fashion. Thus:

Proposition 3.7. *Every ideal extension of a nonempty cancellative c.s. S by a semigroup Q with zero is determined by a partial homomorphism of $Q\backslash\{0\}$ into $G(S)$, namely the restriction to $Q\backslash 0$ of the canonical homomorphism $a \longmapsto ax/x = xa/x$, where x is any element of S.*

Stronger results were obtained by Heuer & Miller [1966] and Heuer [1971] in case Q is a group with a zero element adjoined.

3. Cancellative c.s. have been studied in various ways. Sankaran [1961] showed that every cancellative c.s. S has a (topological) uniform structure. Schein [1975] showed that the divisibility preorder $\leqq_{\mathcal{H}}$ on S induces a compatible preorder $\leq$ on $G(S)$ and showed that every subsemigroup of S is densely embedded in its idealizer if and only if $G(S)$ is archimedean under this preorder. Hall [1971] constructed all cancellative c.s. with two generators.

Cancellative c.s. with stronger properties are considered in the next sections. Properties weaker than cancellativity have also been studied. A c.s. S is **separative** when $a^2 = ab = b^2$ implies $a = b$; these semigroups are studied in Section III.2. A c.s. S is **reductive** when $ax = bx$ for all $x \in S$ implies $a = b$; equivalently, when it is weakly reductive. Clifford [1950] showed that this is a helpful property when studying ideal extensions of S. Gluskin [1983] studied the weaker property that $ax = bx$ for all $x \in S$ implies $a = b$ when $a \in S^2$.

4. NUMERICAL SEMIGROUPS.

In this section we consider subsemigroups of $\mathbb{Z}$.

1. First we look at positive subsemigroups.

Proposition 4.1. *When S is a nonempty subsemigroup of $\mathbb{N}^+$, there exists an integer $d > 0$ such that S consists of multiples of d and contains all sufficiently large multiples of d. Hence S is finitely generated.*

Proof. By Proposition 3.4, $G(S) \subseteq G(\mathbb{N}^+) = \mathbb{Z}$ and $G(S) = \mathbb{Z}d$ for some $d > 0$. Then every element of $S \subseteq \mathbb{Z}d$ is a multiple of d. Also $d = n - m$ for some $m, n \in S$ by Proposition 3.3, and $m + d \in S$ for some $m \in S$. (Thus d is the greatest common divisor of the elements of S.)

Let $m = kd$. If $l \geqq k^2$, then $l = kq + r$ for some $0 \leqq r < k$ and $q \geqq k$, $l = (q-r)k + r(k+1)$, and $ld = (q-r)m + r(m+d) \in S$. In particular S is generated by m, $m+d$, and every other element $s < k^2 d$ of S. □

Proposition 4.1 has been known for some time. A proof was published by Jensen & Miller [1968].

Corollary 4.2. *A nonempty subsemigroup of $\mathbb{Z}$ either contains only nonnegative integers, or contains only nonpositive integers, or is a subgroup of $\mathbb{Z}$.*

Proof. If S is not trivial or empty, then as above $G(S) = \mathbb{Z}d$ for some $d > 0$. Assume that S contains both positive and negative integers. Then S contain arbitrarily large positive integers and arbitrarily large negative integers. Since d is the difference of two elements of S, d is then the difference of two positive elements of S and the difference of two negative elements of S; hence both parts of S generate the same subgroup $\mathbb{Z}d$ of $\mathbb{Z}$. By Proposition 4.1, S contains all sufficiently large positive multiples of d and all sufficiently large negative multiples of d. Hence S contains both d and $-d$ and $S = \mathbb{Z}d \cong \mathbb{Z}$. □

2. A **numerical** semigroup is a subsemigroup of $\mathbb{N}^+$ which contains all sufficiently large natural numbers; numerical monoids also contain 0 (and are usually called numerical semigroups in the literature.) By Proposition 4.1 every nonempty subsemigroup of $\mathbb{N}^+$ is isomorphic to a numerical semigroup, and numerical semigroups are finitely generated. The subsemigroup of $\mathbb{N}^+$ generated by $a_1, a_2, \dots, a_n$ is numerical if and only if $a_1, a_2, \dots, a_n$ are relatively prime. Presentations of numerical semigroups have been studied by Gastinger, Kunz, & Waldi [1991]; the minimal number of defining relations has been studied by Rosales [1996A], [1996N] and Rosales & García-Sánchez [1998N].

Numerical semigroups have been studied extensively, partly because of a number of open questions, partly because of their connections to algebraic geometry. Higgins [1972] and Kunz [1987] are good surveys of early results.

Numerical semigroups first appear in problems posed by Sylvester [1884] and Frobenius. Sylvester's problem consists in determining the largest integer not in $\langle a,b \rangle$, where a and b are relatively prime. He showed that the answer is $ab - a - b$. According to Brauer [1942], Frobenius "mentioned occasionally in his lectures" the more general problem of determining the largest integer not in $\langle a_1, a_2, \dots, a_n \rangle$. This is also called the **money-changing problem**: given an unlimited supply of coins with relatively prime denominations $a_1, a_2, \dots, a_n$, what is the largest amount that cannot be formed by means of these coins. (We prefer to think of this as the fried chicken problem: given that fried chicken can be ordered in boxes of a_1, a_2, …, a_n pieces, where $a_1, a_2, \dots, a_n$ are relatively prime, what is the largest number of pieces that cannot be ordered.)

Generally the **Frobenius number** of a numerical semigroup S, usually denoted by $g(S)$, is the largest integer $n \notin S$. (The **conductor** of S is $g(S)+1$.) No general formula is known that computes $g(S)$ from the generators of S. Curtis [1990] showed that there is no finite set of polynomial formulas for $g(S)$. Upper bounds and a few exact formulas for $g(S)$ have been found in particular cases (Sathaye [1977], Selmer [1977], Rödseth [1978] for 3-generator semigroups, Fröberg [1994] for 4-generator symmetric semigroups). See also Deddens [1979].

A numerical semigroup is **symmetric** when $g(S)$ is odd and $n > 0$, $n \notin S$ implies $g(S) - n \in S$; equivalently, when $g(S)$ is odd and S is maximal for its Frobenius number. Fröberg, Gottlieb, & Häggkvist [1987] showed that there are at least $2^{\lfloor n/8 \rfloor}$ symmetric semigroups with $g(S) = n$. Backelin [1990] showed that there are $C.2^{n/2}$ numerical semigroups and $C'.2^{n/6}$ symmetric semigroups with $g(S) = n$, where C and C' vary within finite bounds.

When S is generated by relatively prime numbers $a_1 < a_2 < \ldots < a_n$, g.c.d.'s $d_1, d_2, \ldots, d_n$ are defined by induction by $d_1 = a_1$ and $d_i = \text{g.c.d.}(d_{i-1}, a_i)$ if $i > 1$ (then $d_n = 1$). Bertin & Carbonne [1975], [1977] called a numerical semigroup **free** when every element of S can be written in the form $\sum_i n_i a_i$ with $n_i < d_i/d_{i+1}$. They showed that free numerical semigroups are symmetric.

Numerical semigroups generated by finite arithmetic progressions were studied by Juan [1993]. For numerical semigroups generated by intervals, explicit formulas give the elements, Frobenius number, and minimum number of defining relations, and decide whether the semigroup is symmetric or a complete intersection (García-Sánchez & Rosales [1999]). Patil & Singh [1990] studied numerical semigroups generated by almost arithmetic progressions. Other interesting classes are considered by Barucci, Dobbs, & Fontana [1997].

The **multiplicity** $m(S)$ of a numerical semigroup S is the smallest (positive) element m of S. A semigroup of multiplicity m is generated by at most m elements, and has **maximal embedding dimension** when it requires m generators (see below). Rosales [1996N] constructed a one-to-one correspondence between numerical semigroups with Frobenius number g and multiplicity m, and numerical semigroups with Frobenius number $g+m$, multiplicity m, next positive element $> 2m$, and maximal embedding dimension. A similar one-to-one correspondence is constructed in Rosales [1996S]. See also Rosales & García-Sánchez [1999].

3. Numerical semigroups also arise in algebraic geometry. At every point

P of a projective curve C, the rational functions on C which are regular on $C\backslash\{P\}$ but not at P have pole orders at P which constitute a subsemigroup of $\mathbb{N}^+$, the **Weierstrass semigroup** of C at P. (This concept was generalized by Delgado [1990]; see also Kim [1994] and Homma [1996].) The Weierstrass Gap Theorem states that S contains all but g elements of $\mathbb{N}^+$, where g is the genus of C; in particular, S is a numerical semigroup. At an ordinary point, $S = \{g+1, g+2, \dots\}$, but other configurations are possible.

In general, the **genus** of a numerical semigroup $S \subseteq \mathbb{N}^+$, generally denoted by g or γ, is the number of elements of $\mathbb{N}^+\backslash S$ (the number of **gaps**). Selmer [1977] gave a general formula for the genus.

It was long conjectured that every numerical semigroup occurs as a Weierstrass semigroup of some curve (Hurwitz [1893]). The first counterexample was given by Buchweitz [1980]. More were found by Torres [1994], [1995], Kim [1996], and Komeda [1997]; Komeda [1998] constructed counterexamples of arbitrarily high genus. All these counterexamples have genus $\geqq 16$. It is still not known precisely which numerical semigroups are Weierstrass. Komeda showed that Weierstrass semigroups include every numerical semigroup with 2 or 3 generators, every symmetric or almost symmetric ($g(S) = 2g-2$) semigroup with 4 generators [1982], every numerical semigroup of multiplicity 4 [1983] or 5 [1992], every numerical semigroup of genus at most 7, and every numerical semigroup such that $g(S) < 2m(S)$ [1994]. Weierstrass semigroups were also constructed by Maclachlan [1971], Rim & Vitulli [1977], Waldi [1980], Knebl [1984], Tutalar [1987], Coppens [1988], and Komeda [1991].

4. Ring theory provides a more exact relationship between numerical semigroups and certain valuation rings. When R is a discrete valuation ring, and $v : Q \longrightarrow \mathbb{Z}$ is the corresponding valuation on the quotient field Q of R, then the set of values $v(R)$ of v on R is a submonoid of $\mathbb{N}$, which one may assume is a numerical monoid. Conversely let k be a field and $k[X]$ be the polynomial ring with one indeterminate. If $S = \langle a_1, a_2, \dots, a_n \rangle$ is a numerical monoid, then the semigroup ring of S is (up to isomorphism) the subring $k[S] = k[X^{a_1}, X^{a_2}, \dots, X^{a_n}]$ of $k[X]$; moreover there is a valuation v on the field of quotients $k(X)$ such that $v(k[S]) = S$. (The multiplicity of S should thus be interpreted in terms of polynomials in $k[S]$ and not in terms of points on algebraic curves.)

The finitely generated domain $k[S] = k[X^{a_1}, X^{a_2}, \dots, X^{a_n}]$ is isomorphic to a quotient ring $k[X_1, X_2, \dots, X_n]/\mathfrak{p}$ by a prime ideal $\mathfrak{p}$, which in turn corresponds to an n-dimensional affine algebraic variety V whose coordinate ring is $k[S]$. In fact V is a curve, since $k[S]$ has Krull dimension 1. The ideal $\mathfrak{p}$ was studied for generators by Bresinsky [1975], [1988], Patil [1993], Campillo & Pisón [1993], and Herzinger [1999]. The **value semigroup** S of V has been studied extensively since Waldi [1972] showed that two curves are equisingular if and only if they have the same value semigroup.

The **embedding dimension** $e(S)$ of S is the minimal number of generators of S (the smallest n such that $S = \langle a_1, a_2, \dots, a_n \rangle$ for some relatively prime $a_1, a_2, \dots, a_n$); equivalently, $e(S)$ is the dimension of the smallest affine space into which an algebraic variety can be embedded when its coordinate ring is $k[S]$. Since S contains all multiples of $m(S)$, $e(S) \leqq m(S)$ always.

Numerical semigroups have been studied for properties of the corresponding variety, such as being a complete intersection, and for various ring properties of $k[S]$, such as Gorenstein, Buchsbaum, Cohen-Macaulay, that are of interest for the corresponding variety. Kunz [1973] showed that $k[S]$ is Gorenstein if and only if S is symmetric; for a different proof see Huang [1995]. Cavaliere & Niesi [1983] and Bresinsky [1984] gave general characterisations of Buchsbaum and Macaulay semigroups. Herzog [1970], Delorme [1976], and Cavaliere & Niesi [1984] characterized complete intersection semigroups (see also Fischer & Shapiro [1996]). Herzog [1970] also showed that a numerical semigroup with three generators is complete intersection if and only if it is symmetric; this also holds for semigroups with four generators (Bresinsky [1979]). See also Kato [1979]; Ruiz [1985]; Brown & Curtis [1991]; Pfister & Steenbrink [1992]; Campillo & Marijuan [1991]; Barucci, Dobbs, & Fontana [1994]; Torres [1994], [1997]; Anderson & Scherpenisse [1995]; Barucci & Fröberg [1997].

The **type** $t(S)$ of S is the number of positive integers $n \notin S$ such that $n + s \in S$ for all $0 < s \in S$. Fröberg [1994] showed that the type of S is also the Cohen-Macaulay type of $k[S]$. Cavaliere & Niesi [1983] showed that there are 4-generator semigroups of arbitrary type. Fröberg, Gottlieb, & Häggkvist [1987] showed that $(t(S) + 1)(g(S) - g) \geqq g(S) + 1$. "Type" sequences that begin with $t(S)$ were studied by D'Anna & Delfino [1997] and D'Anna [1998].

5. Numerical semigroups arise from algebraic curves in other ways. The **Lüroth semigroup** of a curve, and that of its rational function field $K = k(t)$, was defined by Moh & Heinzer [1982] as the set of all finite degrees $[K : k(x)]$ (where $x \in K$ is transcendental over k). It has been studied by Greco [1990], Greco & Raciti [1991], Paxia, Raciti, & Ragusa [1992], Coppens [1995], and Tokunaga & Yoshihara [1995].

A value semigroup can more generally be assigned to every branch of an algebraic curve (Apéry [1946]). Two branches are equisingular if and only if they have the same value semigroup (see e.g. Zariski [1973]). The resulting numerical semigroups are symmetric (Apéry [1946]) and were characterized by Bresinsky [1972] and Teissier [1973]. Bertin & Carbonne [1975], [1977] gave an alternate construction and showed that these value semigroups are free. See also Angermüller [1991].

5. GENERAL STRUCTURE.

The results in this section analyze cancellative c.s. by means of cancellative congruences that eliminate units, then eliminate torsion; geometric properties of convex sets can be brought to bear on the resulting semigroups. These techniques are used in the next sections.

1. A congruence $\mathcal{C}$ on a c.s. S is **cancellative** in case the quotient semigroup $S/\mathcal{C}$ is cancellative; equivalently, when $ac\ \mathcal{C}\ bc$ implies $a\ \mathcal{C}\ b$, for all $a,b,c \in S$.

Proposition 5.1. *Let S be a nonempty cancellative c.s. Every subgroup H of $G(S)$ induces a cancellative congruence $\mathcal{C}$ on S, for which $x\ \mathcal{C}\ y$ in S if and only if $xy^{-1} \in H$ in $G(S)$; then $G(S/\mathcal{C}) \cong G(S)/H$. Moreover, every cancellative congruence $\mathcal{C}$ on S is induced in this fashion by a subgroup of $G(S)$, namely $R(\mathcal{C}) = \{xy^{-1} \in G(S) \mid x,y \in S,\ x\ \mathcal{C}\ y\}$.*

$R(\mathcal{C})$ is the **Rédei group** of $\mathcal{C}$, after Rédei [1956].

Proof. When H is a subgroup of $G(S)$, the partition of $G(S)$ into cosets of H is a congruence $\mathcal{E}$ on $G(S)$; $G(S)/\mathcal{E}$ is the quotient group $G(S)/H$, and $g\ \mathcal{E}\ h$ if and only if $gh^{-1} \in H$, for all $g,h \in G(S)$. In particular $\mathcal{E}$ is a cancellative congruence on $G(S)$. Hence the restriction $\mathcal{C}$ of $\mathcal{E}$ to S is a cancellative congruence on S. The $\mathcal{C}$-class of $x \in S$ is $Hx \cap S$, and $S/\mathcal{C}$ is isomorphic to the subsemigroup $\{Hx \mid x \in S\}$ of $G(S)/H$. Moreover every coset of H can be written in the form $Hxy^{-1} = (Hx)(Hy^{-1})$ for some $x,y \in S$; hence $G(S/\mathcal{C}) \cong G(S)/H$, by Proposition 3.3.

Conversely let $\mathcal{C}$ be a cancellative congruence on S. Then $R = R(\mathcal{C})$ is a subgroup of $G(S)$. Let $a,b \in S$. If $ab^{-1} \in R$, then $ab^{-1} = xy^{-1}$, where $x,y \in S$ and $x\ \mathcal{C}\ y$, $ax\ \mathcal{C}\ ay = bx$, and $a\ \mathcal{C}\ b$, since $\mathcal{C}$ is cancellative. If conversely $a\ \mathcal{C}\ b$, then $ab^{-1} \in R$. Thus $\mathcal{C}$ is the congruence induced by R. □

When $\mathcal{C}$ is the cancellative congruence on S induced by a subgroup H of $G(S)$, we denote the quotient semigroup $S/\mathcal{C}$ by S/H and regard it as a subsemigroup of $G(S)/H$; then $G(S/H) = G(S)/H$.

2. A cancellative c.s. S has at most one idempotent, which is its identity element: indeed $e^2 = e$ in S implies $e = 1$ in $G(S)$. We call a cancellative c.s. S **reduced** when either S has no identity element, or S has an identity element and its group of units is trivial. These semigroups are also called **group-free** and a variety of other names. Hamilton, Nordahl, & Tamura [1975] extended to these semigroups Tamura's construction of N-semigroups in Section III.4.

The next result shows that reduced cancellative c.s. are really just another name for partially ordered abelian groups. There is a well developed theory of lattice ordered groups; see the recent books by Anderson & Feil [1988], Kopytov & Medvedev [1994], Darnel [1995]. Partially ordered abelian groups in general have not been studied as extensively; the book by Goodearl [1986] concerns partially ordered abelian groups with the Riesz interpolation property.

Proposition 5.2. *A c.s. is cancellative without identity element if and only if it is the positive cone of a partially ordered abelian group. A c.m. is cancellative and reduced if and only if it is the nonnegative cone of a partially ordered abelian group.*

Proof. When S is a reduced cancellative c.m., then the relation

$$g \leqq h \text{ if and only if } g^{-1}h \in S$$

is antisymmetric since S is reduced and is a partial order relation on $G(S)$, which is compatible (if $g \leqq h$, then $gk \leqq hk$ for all $k \in G(S)$). Then $(G(S), \leqq)$ is a partially ordered abelian group, and $S = \{g \in G(S) \mid g \geqq 1\}$ is the nonnegative cone of $G(S)$. Conversely, the nonnegative cone of a partially ordered abelian group $(G, \leqq)$ is a cancellative c.m. S, which is reduced since $\leqq$ is antisymmetric. (However, it may happen that $G(S) \subsetneqq G$.)

If now S is a cancellative c.s. without identity element, then the monoid S^1 is cancellative, commutative, and reduced; hence S^1 is the nonnegative cone of a partially ordered abelian group, whose positive cone is S. Conversely, the positive cone of a partially ordered abelian group is a cancellative c.s. □

A cancellative c.s. S with a group of units U is an ideal extension of $S \backslash U$ by $U \cup \{0\}$, as studied by Heuer & Miller [1966] and Heuer [1971]. The next result provides another approach which anticipates Chapter V.

Proposition 5.3. *Let S be a cancellative c.m. and U be the group of units of S. In $G(S)$, S is a union of cosets of U, which coincide with the $\mathcal{H}$-classes of S; moreover S/U is reduced.*

Proof. S is a union of cosets of U since $Us \subseteq S$ for all $s \in S$; $S/U \subseteq G(S)/U$ is reduced since $s, t \in S$ and $(Us)(Ut) = U$ in $G(S)/U$ implies

$st \in U$, $stu = 1$ for some $u \in U$, and $s,t \in U$.

Also $s \mathcal{H} t$ implies $s = ut$, $t = vs$ for some $u,v \in S$, $s = uvs$, $uv = 1$, $u,v \in U$, and $s \in Ut$. Conversely, $s = ut$ for some $u \in U$ implies $t = u^{-1}s$ and $s \mathcal{H} t$. Thus $H_t = Ut$. □

By Proposition 5.3, a cancellative c.m. S is determined by a subgroup U of $G(S)$ and a reduced subsemigroup of $G(S)/U$. If S is uniquely divisible (and cancellative) then $S \cong U \times S/U$ (Brown & LaTorre [1966]). In general, S/U is a universal cancellative reduced c.m. of S, since every homomorphism of S into a reduced semigroup must send U to the identity element. The **rank** of the cancellative c.m. S is that of $G(S)$; the **reduced rank** of S is the rank of its reduced semigroup S/U, equivalently, the rank of $G(S)/U$.

3. A semigroup S is **power cancellative** in case $x^n = y^n$ implies $x = y$, for all $n > 0$ and $x,y \in S$; c.s. with this property are also called **torsion free**. Free c.s. and torsion free abelian groups are power cancellative. The author's earliest reference to cancellative and power cancellative c.s. is Abellanas [1965].

Proposition 5.4. *When S is a cancellative c.s., then $G(S)$ is torsion free if and only if S is power cancellative.*

Proof. $G(S)$ is torsion free if and only if $x^n y^{-n} = (xy^{-1})^n = 1$ implies $xy^{-1} = 1$, for all $n > 0$ and $x,y \in S$. □

A congruence $\mathcal{C}$ is **power cancellative** when the corresponding quotient semigroup is power cancellative; equivalently, when $x^n \mathcal{C} y^n$ implies $x \mathcal{C} y$, for all $n > 0$.

Proposition 5.5. *Let S be a c.s. There is a smallest power cancellative congruence $\mathcal{T}$ on S, namely*

$$x \mathcal{T} y \text{ if and only if } x^n = y^n \text{ for some } n > 0.$$

Moreover, $S/\mathcal{T}$ has no idempotent if and only if S has no idempotent.

Proof. $\mathcal{T}$ is reflexive, symmetric, and transitive (if $x^m = y^m$ and $y^n = z^n$, then $x^{mn} = y^{mn} = z^{mn}$), and is a congruence (if $x^n = y^n$, then $(xz)^n = (yz)^n$). $\mathcal{T}$ is a power cancellative congruence since $x^m \mathcal{T} y^m$ implies $x^{mn} = y^{mn}$ for some $n > 0$ and $x \mathcal{T} y$. If finally $\mathcal{C}$ is a power cancellative congruence on S, then $x \mathcal{T} y$ implies $x^n = y^n$ for some $n > 0$, $x^n \mathcal{C} y^n$ for some $n > 0$, and $x \mathcal{C} y$; thus $\mathcal{T}$ is the smallest power cancellative congruence on S.

If S has an idempotent, then so does $S/\mathcal{T}$. If conversely the $\mathcal{T}$-class T_x of $x \in S$ is idempotent in $S/\mathcal{T}$, then $x^2 \mathcal{T} x$, $x^{2n} = x^n$ for some $n > 0$, and $x^n \in S$ is idempotent. □

The quotient semigroup $S/\mathfrak{T}$ in Proposition 5.5 is the universal power cancellative semigroup of S: if $\varphi : S \longrightarrow T$ is a homomorphism of S into a power cancellative semigroup T, then $\operatorname{Im}\varphi$ is power cancellative, $\ker\varphi$ is a power cancellative congruence, and φ factors uniquely through the projection $S \longrightarrow S/\mathfrak{T}$ by Proposition I.2.4.

Proposition 5.6. *Let S be a cancellative c.s. The smallest power cancellative congruence $\mathfrak{T}$ on S is induced by the torsion part of $G(S)$. Moreover $S/\mathfrak{T}$ is cancellative; $S/\mathfrak{T}$ has no idempotent if and only if S has no idempotent; if S is reduced, then $S/\mathfrak{T}$ is reduced.*

Proof. Let $T = \{t \in G(S) \mid t^n = 1 \text{ for some } n > 0\}$ be the torsion subgroup of $G(S)$. $\mathfrak{T}$ is the congruence induced by T, since $xy^{-1} \in T$ if and only if $x^n = y^n$ for some $n > 0$. Hence $S/\mathfrak{T} = S/T$ is cancellative. If S is reduced without an identity element, then $S/\mathfrak{T}$ has no idempotent, by Proposition 5.5, and is reduced. If S is reduced with an identity element, then $S/\mathfrak{T}$ has an identity element and is reduced: if the $\mathfrak{T}$-class T_x of $x \in S$ is a unit of $S/\mathfrak{T}$, then $xy \;\mathfrak{T}\; 1$ for some $y \in S$, $x^n y^n = 1$ for some $n > 0$, x is a unit of S, and $x = 1$. □

By Proposition 5.6, every commutative semigroup S has a universal cancellative, power cancellative, and reduced c.s., which is obtained from the universal cancellative semigroup $C(S)$ by dividing $C(S)/U$ by its smallest power cancellative congruence $\mathfrak{T}$. These universal semigroups were constructed by Head [1979] using monoids of quotients. In what follows we abbreviate the mouthful "cancellative, power cancellative, and reduced" as **c.pc.r.**; c.pc.r. semigroups are the positive or nonegative cones of partially ordered torsion free abelian groups, and have been called by a variety of other names. Laplaza [1964] showed that every c.pc.r. c.s. can be embedded into the nonnegative cone of a totally ordered abelian group.

4. The following constructions let basic geometric properties of convex sets be applied to c.pc.r. semigroups.

When S is c.pc.r. we regard the torsion free abelian group $G = G(S)$ as a subgroup of its divisible hull $\overline{G} = \mathbb{Q} \otimes_{\mathbb{Z}} G$, which is a vector space over $\mathbb{Q}$, and which we in turn regard as a divisible subgroup of its real hull $\overline{\overline{G}} = \mathbb{R} \otimes_{\mathbb{Z}} G$, which is a vector space over $\mathbb{R}$. The rank of S is also the dimension of $\overline{G}$ over $\mathbb{Q}$ and the dimension of $\overline{\overline{G}}$ over $\mathbb{R}$. If S is finitely generated, then G is a free abelian group and every basis of G is also a basis of $\overline{G}$ over $\mathbb{Q}$ and a basis of $\overline{\overline{G}}$ over $\mathbb{R}$.

In a real vector space a convex cone is an additive submonoid C such that $c \in C$ implies $\lambda c \in C$ for all $\lambda \in \mathbb{R}$, $\lambda \geqq 0$. Similarly a convex cone in a rational vector space is an additive submonoid C such that $c \in C$ implies $qc \in C$ for all $q \in \mathbb{R}$, $q \geqq 0$. A convex cone is generally called **reduced** when it contains no nonzero subspace; equivalently, when it is reduced as an additive semigroup.

Every subset X of $\overline{G}$ generates a convex cone $\overline{X}$ of $\overline{G}$, namely

$$\overline{X} = \{q_1 x_1 + \cdots + q_k x_k \mid x_1, \ldots, x_k \in X,\ q_1, \ldots, q_k \in \mathbb{Q},\ q_1, \ldots, q_k \geqq 0\}.$$

Similarly, every subset X of $\overline{\overline{G}}$ generates a convex cone $\overline{\overline{X}}$ of $\overline{\overline{G}}$, namely

$$\overline{\overline{X}} = \{\lambda_1 x_1 + \cdots + \lambda_k x_k \mid x_1, \ldots, x_k \in X,\ \lambda_1, \ldots, \lambda_k \in \mathbb{R},\ \lambda_1, \ldots, \lambda_k \geqq 0\}.$$

The convex cone $\overline{S}$ of $\overline{G}$ generated by S is the **divisible hull** of S; we call $\overline{\overline{S}}$ the **real hull** of S. Embeddings into divisible semigroups have been set up by enough authors to gain the status of folk theorems. Minimal divisible embeddings were studied by Tamura [1963].

Proposition 5.7. *When T is a subsemigroup of $\overline{G}$, then*

$$\overline{T} = \{t/n \in \overline{G} \mid t \in T,\ n > 0\} \cup \{0\}$$

and $\overline{T}$ is a uniquely divisible monoid (for every $a \in \overline{T}$ and $m > 0$ there exists a unique $x \in \overline{T}$ such that $mx = a$). If T is reduced, then so is $\overline{T}$.

Proof. Every $\overline{X}$ is a submonoid of $\overline{G}$. Let $a = q_1 t_1 + \cdots + q_k t_k \in \overline{T}$, where $t_1, \ldots, t_k \in T$, $q_1, \ldots, q_k \in \mathbb{Q}$, $q_1, \ldots, q_k \geqq 0$. In $\mathbb{Q}$, $q_1, \ldots, q_k$ have a common denominator $n > 0$; then $t = nq_1 t_1 + \cdots + nq_k t_k \in T$, since T is a semigroup and $nq_1, \ldots, nq_k \in \mathbb{N}$, and $a = t/n$. Conversely $t/n \in \overline{T}$ for all $t \in T$ and $n > 0$.

Let $a = t/n \in \overline{T}$, where $s \in T$ and $n > 0$. Then $na \in T$. If $m > 0$ and $x \in \overline{G}$, then $mx = a$ if and only if $x = t/mn$, and then $x \in \overline{T}$. Thus $\overline{T}$ is uniquely divisible.

Let $a = s/m$ and $b = t/n \in \overline{T}$, with $s,t \in T$ and $m,n > 0$. Then $a + b = (ns + mt)/mn$. If T is reduced, then $a + b = 0$ implies $ns + mt = 0$, $s = t = 0$, and $a = b = 0$; hence $\overline{T}$ is reduced. □

Proposition 5.8. *When T is a reduced subsemigroup of $\overline{G}$, then $\overline{\overline{T}}$ is a reduced cone of $\overline{\overline{G}}$. If T is a cone, then $\overline{\overline{T}} \cap \overline{G} = T$.*

Proof. Assume that $\overline{\overline{T}}$ is not reduced. Then $a + b = 0$ for some $a,b \in \overline{\overline{T}}$, $a,b \neq 0$. We have $a = \lambda_1 t_1 + \cdots + \lambda_k t_k$, $b = \lambda_{k+1} t_{k+1} + \cdots + \lambda_l t_l$ for some

$t_1, \ldots, t_l \in T$ and $\lambda_1, \ldots, \lambda_l \in \mathbb{R}$, $\lambda_1, \ldots, \lambda_l \geqq 0$. We may assume that $t_1, \ldots, t_l \neq 0$ and $\lambda_1, \ldots, \lambda_l > 0$ and still have $l \geqq 2$. In the subspace of $\overline{\overline{G}}$ spanned by $t_1, \ldots, t_l$, the equality

$$\lambda_1 t_1 + \cdots + \lambda_l t_l = 0$$

can then be viewed as a (finite) homogeneous system of linear equations with unknowns $\lambda_1, \ldots, \lambda_l$, which has rational coefficients since $t_1, \ldots, t_l \in \overline{G}$ and has a solution in which all unknowns are positive. By Lemma 5.9 below this system has a solution in which all unknowns are positive rational numbers $q_1, \ldots, q_l$. The equality

$$q_1 t_1 + \cdots + q_l t_l = 0$$

then shows that $\overline{T}$ is not reduced. Hence T is not reduced.

Now let T be a cone in $\overline{G}$. Let $a \in \overline{\overline{T}} \cap \overline{G}$, $a \neq 0$. Then a has rational coordinates in $\overline{\overline{G}}$, and a is a linear combination $a = \lambda_1 t_1 + \cdots + \lambda_l t_l$ of $t_1, \ldots, t_l \in T$ with real coefficients $\lambda_1, \ldots, \lambda_l$ that may be assumed positive. As above the equality

$$\lambda_0 a - \lambda_1 t_1 - \cdots - \lambda_l t_l = 0$$

can be viewed as a (finite) homogeneous system of linear equations with unknowns $\lambda_0, \lambda_1, \ldots, \lambda_l$, which has rational coefficients since $a, t_1, \ldots, t_l \in \overline{G}$ and has a solution in which all unknowns are positive (including $\lambda_0 = 1$). By Lemma 5.9 this system has a solution in which all unknowns are positive rational numbers $q_0, q_1, \ldots, q_l$. The equality

$$q_0 a - q_1 t_1 - \cdots - q_l t_l = 0$$

then shows that $q_0 a \in T$. Hence $a \in T$. Hence $\overline{\overline{T}} \cap \overline{G} = T$. □

Lemma 5.9. *A homogeneous system of linear equations which has rational coefficients and has a solution in which all unknowns are positive real numbers also has a solution in which all unknowns are positive rational numbers.*

Proof. Applying Gauss-Jordan reduction to the given system (S) yields a simpler equivalent system (R). The unknowns can be numbered so that (R) reads

$$\begin{cases} x_1 - q_{1,k+1} x_{k+1} - \cdots - q_{1,n} x_n = 0, \\ x_2 - q_{2,k+1} x_{k+1} - \cdots - q_{2,n} x_n = 0, \\ \vdots \\ x_k - q_{k,k+1} x_{k+1} - \cdots - q_{k,n} x_n = 0; \end{cases}$$

equivalently, $x_1 = \varphi_1(x_{k+1}, \dots, x_n), \dots, x_k = \varphi_k(x_{k+1}, \dots, x_n)$, where

$$\varphi_i(x_{k+1}, \dots, x_n) = q_{i,k+1}\, x_{k+1} + \dots + q_{i,n}\, x_n\,.$$

Here $q_{1,k+1}, \dots, q_{k,n}$ are rational since they are obtained from the rational coefficients of (S) by Gauss-Jordan reduction. Now (S) and (R) have a solution $a_1, \dots, a_n$ in which $a_1, \dots, a_n > 0$. Since $\varphi_1, \dots, \varphi_k$ are continuous there exist neighborhoods $U_{k+1}, \dots, U_n$ of $a_{k+1}, \dots, a_n$ such that $\varphi_i(x_{k+1}, \dots, x_n) > 0$ for all $i \leqq k$ and all $x_{k+1} \in U_{k+1}, \dots, x_n \in U_n$. There exist positive rational numbers $b_{k+1} \in U_{k+1}, \dots, b_n \in U_n$; then $b_1 = \varphi_1(b_{k+1}, \dots, b_n)$, $\dots, b_k = \varphi_k(b_{k+1}, \dots, b_n)$ are positive rational numbers and $b_1, \dots, b_n$ is a solution of (R) and of (S) in which all unknowns are positive rational numbers. □

6. FACES.

In this section we use the techniques of Section 5 to study the faces of cancellative semigroups.

1. A **face** of a commutative semigroup S is a nonempty subsemigroup A of S such that $ab \in A$ implies $a, b \in A$; equivalently, a nonempty subsemigroup of S whose complement is an ideal of S. The name is due to Petrich [1963F]. Faces of semigroups were first considered by Thierrin [1953] but faces of cones go back to antiquity.

When A is a face of S, the ideal $P = S \backslash A \subsetneqq S$ is a (completely) **prime** ideal of S ($ab \in P$ implies $a \in P$ or $b \in P$). If conversely $P \subsetneqq S$ is a proper prime ideal of S, then $S \backslash P$ is a face of S. The following results can therefore be stated in terms of proper prime ideals rather than faces.

Proposition 6.1. *A finitely generated c.s. has finitely many faces.*

Proof. Let X be a generating subset of S and A be a face. Every $a \in A$ is a finite product $a = x_1 \cdots x_n$ of elements of S; since A is a face, $x_i \in A$ for all i. Thus A is generated by $X \cap A$. If X is finite, then there are finitely many subsets $X \cap A$ and finitely many faces of S. □

Proposition 6.2. *Let S be a cancellative c.s.*

(1) *If S is a monoid with group of units U, then a subset of S is a face of S if and only if it is a union of cosets of U that constitute a face of S/U.*

(2) *If S is reduced, then a subset of S is a face of S if and only if it is a union of $\mathfrak{T}$-classes that constitute a face of $S/\mathfrak{T}$.*

Proof. (1). Let A be a face of S. If $a \in A$ and $u \in U$, then $u^{-1}ua = a \in A$ and $ua \in A$. Hence A is a union of cosets of U. Now A/U is a nonempty subsemigroup of S/U. If $Ua, Ub \in S/U$ and $(Ua)(Ub) \in A/U$, then $Uab = (Ua)(Ub) = Uc$ for some $c \in A$, $ab \in Uc \subseteq A$, $a,b \in A$, and Ua, $Ub \in A/U$; thus A/U is a face of S/U. If conversely B is a face of S/U, then $A = \{a \in S \mid Ua \in B\}$ is a nonempty subsemigroup of S (Proposition I.2.5) and $ab \in A$ implies $(Ua)(Ub) = Uab \in B$, Ua, $Ub \in B$, and $a,b \in A$. Thus A is a face of S. We see that $B = A/U$.

(2). Let $T = S/\mathfrak{T}$ and $\tau : S \longrightarrow S/\mathfrak{T}$ be the projection. If A is a face of S and $x \;\mathfrak{T}\; y \in A$, then $x^n = y^n \in A$ for some $n > 0$ and $x \in A$; thus A is a union of $\mathfrak{T}$-classes. Also $\tau(A)$ is a nonempty subsemigroup of T. If $\tau(a)\,\tau(b) \in \tau(A)$, then $ab \;\mathfrak{T}\; c$ for some $c \in A$, $ab \in A$ since A is a union of $\mathfrak{T}$-classes, $a,b \in A$, and $\tau(a)$, $\tau(b) \in \tau(A)$; thus $\tau(A)$ is a face of T. If conversely B is a face of T, then $A = \tau^{-1}(B)$ is a face of S, and $B = \tau(A)$. □

Proposition 6.2 constructs an order-preserving one-to-one correspondence between the faces of a cancellative semigroup and the faces of its universal c.pc.r. semigroup.

2. When S is a c.pc.r. semigroup, we write S additively and let $G = G(S)$ and $\overline{S} \subseteq \overline{G}$ be the divisible hull of S, as in Section 5.

Proposition 6.3. *Let S be a cancellative, power cancellative, and reduced c.s. If T is a face of S, then $\overline{T}$ is a face of $\overline{S}$ and $T = \overline{T} \cap S$. If conversely A is a face of $\overline{S}$, then $G(A)$ is a subspace of $\overline{G}$, $T = A \cap S = G(A) \cap S$ is a face of S, and $\overline{T} = A$.*

By Lemma 5.9, a similar result holds for the real hull $\overline{\overline{S}}$.

Proof. Let T be a face of S. Let $a = s/m \in \overline{S}$, with $s \in S$ and $m > 0$. If $a \in \overline{T}$, then $a = t/n$ for some $t \in T$ and $n > 0$, $ns = mt \in T$, and $s \in T$; hence $s/m \in \overline{T}$ if and only if $s \in T$. With $m = 1$ this yields $\overline{T} \cap S = T$.

Let $a = s/m$ and $b = t/n \in \overline{S}$, with $s,t \in S$ and $m,n > 0$. If $a,b \in \overline{T}$, then $s,t \in T$ and $a + b = (ns + mt)/mn \in \overline{T}$. If conversely $a + b \in T$, then $ns + mt \in T$, $s,t \in T$ since $m,n > 0$, and $a,b \in \overline{T}$. Thus $\overline{T}$ is a face of $\overline{S}$.

Conversely let A be a face of $\overline{S}$. Then $T = A \cap S$ is a face of S. If $a = s/n \in A$, with $s \in S$ and $n > 0$, then $s = na \in A \cap S = T$ and $a \in \overline{T}$. If conversely $a = t/m \in \overline{T}$, with $t \in T$ and $m > 0$, then $ma = t \in A$ and $a \in A$. Thus $A = \overline{T}$.

Let $q = r/s \in \mathbb{Q}^+$, with $r, s > 0$. Then $a \in A$ implies $sqa = ra \in A$ and $qa \in A$. Hence $a - b \in G(A)$ implies $q(a-b) \in G(A)$. Since $G(A)$ is a group this holds for all $q \in \mathbb{Q}$ and $G(A)$ is a subspace of $\overline{G}$. Moreover $G(A) \cap \overline{S} = A$: if $c = a - b \in G(A) \cap \overline{S}$, where $a, b \in A$, then $b + c = a \in A$ with $b, c \in \overline{S}$, and $c \in A$. Hence $G(A) \cap S = A \cap S = T$. □

Proposition 6.4. *In a cancellative c.s. of finite reduced rank* n, *every chain of proper prime ideals has length at most* n.

Proof. First let S be c.pc.r. of rank n. When T is a face of S, then $G(\overline{T})$ is a subspace of $\overline{G}$ and $T = G(\overline{T}) \cap S$, by Proposition 6.4. Hence $T_1 \subsetneqq T_2$ implies $G(\overline{T}_1) \subsetneqq G(\overline{T}_2)$. A chain of faces of S generates a chain of subspaces of $\overline{G}$ of equal length and therefore has length at most n.

Now let S be any cancellative c.s. of finite reduced rank n, so that the universal c.pc.r. c.s. T of S has rank n. By Proposition 6.2, a chain of faces of S induces a chain of faces of T of equal length and has length at most n. Hence a chain of proper prime ideals of S also has length at most n. □

If S is a finitely generated c.m. in Proposition 6.4, then S has a chain of proper prime ideals of length n, since, in $\mathbb{R}^n$, a polyhedral set (= the intersection of finitely many closed half spaces) has faces of every dimension (see e.g. Brøndsted [1983], Corollary 8.6). This property does not extend to infinitely generated semigroups, as the next example shows. Matsuda [1998A] proved additional height properties of prime ideals in cancellative c.s.

Example 6.5. Let $X = \{x, y\}$ and

$$S = \{a_x x + a_y y \in F_X \mid a_x, a_y > 0\}.$$

The only face of S is S itself: if indeed A is a face of S and $a = a_x x + a_y y \in A$, then for any $b = b_x x + b_y y \in S$ we have $b \leqq na$ if $n > 0$ is large enough (namely, if $na_x \geqq b_x$ and $na_y \geqq b_y$), and then $b + c = (n+1)a \in A$ for some $c \in S$ and $b \in A$, for every $b \in S$. □

3. Next we show:

Proposition 6.6. *Let* S *be a finitely generated, cancellative c.s. If* S *has reduced rank* n, *then* S *has at least* n *faces of reduced rank* 1.

Proof. Let S be c.pc.r. of rank n with a finite generating subset X. Then $\overline{X} = \overline{S}$. Let $Y \subseteq X$ be minimal such that $\overline{Y} = \overline{S}$. Then Y spans $\overline{G}$ and has at least n elements. Also $x \notin \overline{Y \backslash \{x\}}$ when $x \in Y$, otherwise $\overline{Y \backslash \{x\}} = \overline{S}$. For every $x \in Y$ we show that the subsemigroup

$$C = \{qx \mid q \in \mathbb{Q},\ q \geqq 0\}$$

of $\overline{S}$ is a face of $\overline{S}$; hence $\overline{S}$ has at least n faces of rank 1, and so does S by Proposition 6.3; then so does every finitely generated cancellative c.s. of reduced rank n, by Proposition 6.2.

Assume that $c = a + b \in C$, where $a, b \in \overline{S}$. If $c = 0$, then $a = b = 0 \in C$, since $\overline{S}$ is reduced. Assume $c \neq 0$. We have $a = \sum_{y \in Y} a_y\, y$, $b = \sum_{y \in Y} b_y\, y$, $c = \sum_{y \in Y} c_y\, y$, with a_y, $b_y \in \mathbb{Q}$, a_y, $b_y \geqq 0$, $c_y = a_y + b_y \geqq 0$, for every $y \in Y$; also $c = qx$, where $q > 0$. If $c_x < q$, then $(q - c_x)\, x = c - c_x\, x \in \overline{Y \backslash \{x\}}$ and $x \in \overline{Y \backslash \{x\}}$, a contradiction by the choice of Y. If $c_x > q$, then $0 = c - qx = (c_x - q)\, x + \sum_{y \neq x} c_y\, y$, a contradiction since $\overline{S}$ is reduced. Therefore $c_x = q$. Then $0 = c - qx = \sum_{y \neq x} c_y\, y$ and $c_y = 0$ for all $y \neq x$, since $\overline{S}$ is reduced. Hence $a = a_x\, x \in C$ and $b = b_x\, x \in C$. □

$\overline{S}$ can be replaced by $\overline{\overline{S}}$ in this argument, and then the result follows from Minkowski's Theorem (see e.g. Brøndsted [1983], Theorem 5.10). Example 6.5 shows that Proposition 6.6 does not extend to infinitely generated semigroups.

If S is finitely generated of reduced rank $n \leqq 2$, then it is immediate that S has exactly n faces of reduced rank 1. The next example shows that a finitely generated semigroup of reduced rank $n \geqq 3$ may have more than n faces of reduced rank 1; similar examples show that there can be any (finite) number of faces of reduced rank 1.

Example 6.7. Let $X = \{x, y, z\}$ and $S \subseteq F_X$ be generated by z, $x + z$, $y + z$, and $x + y + z$. Then S has rank 3, and $a = a_x\, x + a_y\, y + a_z\, z \in \overline{S}$ if and only if $0 \leqq a_x \leqq a_z$ and $0 \leqq a_y \leqq a_z$. It is straightforward that z, $x + z$, $y + z$, and $x + y + z$ all generate faces of S of rank 1. (This also follows from the previous proof.) For instance assume that $a + b \in \langle x + y + z \rangle$, where $a, b \in S$. Then $a + b = n(x + y + z)$ for some $n \geqq 0$, and $a_x + b_x = a_y + b_y = a_z + b_z = n$. Since $0 \leqq a_x \leqq a_z$, $0 \leqq b_x \leqq b_z$ and $0 \leqq a_y \leqq a_z$, $0 \leqq b_y \leqq b_z$, this implies $a_x = a_z$, $b_x = b_z$ and $a_y = a_z$, $b_y = b_z$; hence $a = a_z\,(x + y + z)$ and $b = b_z\,(x + y + z) \in \langle x + y + z \rangle$. □

7. FREE EMBEDDING.

In this section we show that a finitely generated commutative semigroup can be embedded into a free commutative monoid if and only if it is cancellative, reduced, and power cancellative. This was proved by the author [1970E] and Hochster [1972]; a third proof is given in Rosales & García-Sánchez [1999]. We give yet another proof, which uses geometry as in Section 6, and shows that there is an embedding into a free commutative monoid with the same rank. This stronger Free Embedding Theorem appears to be a folk theorem.

1. First we note:

Lemma 7.1. *Let C be a reduced convex cone in a finite dimensional real vector space V. For any inner product on V, there exists $p \in V$ such that $p\,.\,s > 0$ for all $s \in C\backslash 0$.*

Proof. $C\backslash 0$ is convex, since C is reduced. Hence there is a hyperplane H such that $0 \in H$ and $H \cap (C\backslash 0) = \emptyset$ (see e.g. Brøndsted [1983], Lemma 4.4). Then $H = \{s \in V \mid p\,.\,s = 0\}$ for some $p \neq 0$ and $p\,.\,s \neq 0$ for all $s \in C\backslash 0$. Since $C\backslash 0$ is convex, either $p\,.\,s < 0$ for all $s \in C\backslash 0$, or $p\,.\,s > 0$ for all $s \in C\backslash 0$. We can replace p by $-p$ if necessary and arrange that $p\,.\,s > 0$ for all $s \in C\backslash 0$. □

Lemma 7.1 implies that finitely generated reduced cones can be truncated to polytopes and inherit thereby a number of geometric properties, for which the reader is generally referred to Brøndsted [1983].

Corollary 7.2. *Let C be a finitely generated reduced convex cone in a finite dimensional real vector space V. There exists a closed halfspace K, of which 0 is an interior point, such that $C \cap K$ is a polytope. If C spans V, then C is the intersection of closed halfspaces bounded by the hyperplanes that are spanned by the facets of C.*

Proof. Let $H = \{v \in V \mid p\,.\,v = 1\}$ and $K = \{v \in V \mid p\,.\,v \leqq 1\}$, where $p\,.\,s > 0$ for all $s \in C\backslash 0$ as in Lemma 7.1. Then 0 lies in the interior $\{v \in V \mid p\,.\,v < 1\}$ of K. Every nonzero generator c of C has a projection in H, which is the intersection of H and the ray from 0 through c; these projections, together with 0, generate $C \cap K$ as a convex set. Hence $C \cap K$ is a polytope.

If C spans V, then so does $C \cap K$. Hence $C \cap K$ is the intersection of closed halfspaces bounded by the hyperplanes that contain the facets of $C \cap K$ (see e.g. Brøndsted [1983], Corollary 9.6). One of these facets is $C \cap H$ and the corresponding halfspace in K; the remaining halfspaces are bounded by the hyperplanes that are spanned by the facets of C, and their intersection is C. □

2. The main result in this section is:

Theorem 7.3. *For a finitely generated commutative monoid S the following conditions are equivalent:*

(1) *S is cancellative, power cancellative, and reduced;*

(2) *S can be embedded into a finitely generated free commutative monoid;*

(3) *S can be embedded into a finitely generated free commutative monoid with the same rank as S.*

Proof. (3) implies (2), and (2) implies (1). Now let S be a c.m. which is finitely generated by $s_1, \dots, s_m \in S$ and c.pc.r. Let $G = G(S)$, X be a basis of G, and $\overline{G}$, $\overline{\overline{G}}$, and $\overline{\overline{S}}$ be as in Section 5. $\overline{\overline{G}}$ has an inner product

$$\left(\textstyle\sum_{x\in X} a_x x\right) \cdot \left(\textstyle\sum_{x\in X} b_x x\right) \;=\; \textstyle\sum_{x\in X} a_x b_x \,.$$

$\overline{\overline{S}}$ is a reduced cone by Proposition 5.8; by Lemma 7.1 there exists $p \in \overline{\overline{G}}$ such that $p \cdot s > 0$ for all $s \in \overline{\overline{S}}\backslash 0$; equivalently, $p \cdot s > 0$ for all $s \in S\backslash 0$.

We improve on Lemma 7.1 as follows. We may assume that $s_1, \dots, s_m \neq 0$. In $\overline{\overline{G}}$, the nonempty set

$$\{p \in \overline{\overline{G}} \mid p \cdot s > 0 \text{ for all } s \in \overline{\overline{S}}\backslash 0\} = \{p \in \overline{\overline{G}} \mid p \cdot s_i > 0 \text{ for all } i\}$$

is open and contains a point with rational coordinates. Therefore we may assume that $p \in \overline{G}$. Let

$$H \;=\; \{h \in \overline{G} \mid p \cdot h = 0\}.$$

We show that there exists $k \in H$ such that every coordinate of $q = p + k$ is positive. Indeed $p = \sum_{x\in X} p_x x$ has at least one positive coordinate p_t, since $p \cdot s_1 > 0$. If $p_y \leqq 0$, then $|p_y|t + p_t y \in H$; hence

$$h \;=\; \textstyle\sum_{y\in X,\, p_y \leqq 0} \left(|p_y|t + p_t y\right) \;\in\; H$$

has coordinates $h_x \geqq 0$ for all $x \in X$ and $h_y = p_t > 0$ whenever $p_y \leqq 0$. If $\lambda \in \mathbb{Q}$, $\lambda > 0$ is large enough, then $\lambda h \in H$ and every coordinate of $q = p + \lambda h$ is positive.

The semigroup S can now be embedded into the positive cone of $\overline{G}$, as follows. Since p has rational coordinates, $\overline{G}$ is the direct sum $\overline{G} = H \oplus \mathbb{Q}p$ and every $g \in \overline{G}$ can be written uniquely in the form

$$g \;=\; \pi(g) + \varphi(g)\, p$$

with $\pi(g) \in H$ and $\varphi(g) \in \mathbb{Q}$. Then $p \cdot s_i = \varphi(s_i)\,(p \cdot p)$ and $\varphi(s_i) > 0$ for all i. Let

$$T(g) = \pi(g) + \varphi(g)\,\lambda q,$$

where $\lambda \in \mathbb{Q}$, $\lambda > 0$. Since $q \notin H$, T is an injective linear transformation. If λ is large enough, then every coordinate of $T(s_i)$ is positive, since every coordinate of $\varphi(s_i)\,q$ is positive; hence we can choose λ so that every coordinate of $T(s_1), \ldots, T(s_m)$ is positive. Then $T(S) \cong S$ is contained in the positive cone of $\overline{G}$. The coup de grâce is that the positive coordinates of $T(s_1), \ldots, T(s_m)$ all have a common denominator $n > 0$; then $nT(s_1), \ldots, nT(s_m)$ all have positive integer coordinates, and generate a subsemigroup or submonoid $nT(S) \cong S$ of the free commutative monoid on X, which has the same rank as S. □

The hypothesis that S is finitely generated is very necessary in Theorem 7.3. The additive semigroup $\mathbb{Q}^+$ of all positive rationals is cancellative, power cancellative, and reduced, but not finitely generated, and cannot be embedded into a free commutative monoid, let alone one of rank 1.

Corollary 7.4. *Let S be a finitely generated c.pc.r. c.m. The elements of S are separated by finitely many homomorphisms of S into $\mathbb{N}$.*

Proof. When S is a submonoid of F_X, there is for each $x \in X$ an x-th coordinate homomorphism $\sum_{x \in X} a_x x \longmapsto a_x$ of S into $\mathbb{N}$; these homomorphisms separate the elements of S. □

3. A finitely generated c.pc.r. c.m. S can also be embedded into a free c.m. F with the **quasi-universal** property that every homomorphism of S into a free c.m. factors (though perhaps not uniquely) through the embedding $S \longrightarrow F$; we say that F (together with $S \longrightarrow F$) is a **quasi-universal** free c.m. of S. This was proved by the author [1970E] using properties of orthogonality in free abelian groups. Examples show that a true universal property cannot be required in general, and that the quasi-universal free c.m. of S may all have greater rank than S. Consequently, the embedding constructed in Theorem 7.3 does not usually have a universal or quasi-universal property.

4. Finitely generated submonoids of finitely generated free c.m. are known as **affine semigroups**. By the Free Embedding Theorem, these semigroups can be defined abstractly as finitely generated c.pc.r. c.s.

Affine semigroups have been studied from various viewpoints. The papers by Rosales [1997] and Fischer, Morris, & Shapiro [1997] contain (among other things) interesting constructions; Bruns & Gubeladze [1996] and Gubeladze [1998] are especially penetrating. See also Ewald & Wessels [1991], Altman [1996], and Thoma [1996]. Rosales & García-Sánchez developed a "gluing" construction [1995] and gave algorithms that recognize normal affine semigroups, or recognize normality in affine semigroups [1999N]. See also Ljubenova & Cvetkova [1967] and Hoa [1991].

Like numerical semigroups, affine semigroups appear in algebraic geometry. They are related to toric varieties (see e.g. Hochster [1972], Neeb [1992], and the book by Oda [1988]). The semigroup ring $k[S]$ of an affine semigroup S is also the coordinate ring of an algebraic variety and has been studied for various ring properties that are relevant to that variety. Affine semigroups whose semigroup ring is Gorenstein or Cohen-Macaulay were characterized by Trung & Hoa [1986] and studied by Rosales, García-Sánchez, & Urbano-Blanco [1998] and Rosales & García-Sánchez [1998C], [1999F]. Rosales & García-Sánchez [1995] studied complete intersection semigroups. Affine semigroups also appear as Weierstrass semigroups of sets of points (Delgado [1990]) and as semigroups of values for several valuations (Delgado [1988]). See also Bouvier & Gonzales-Springberg [1992], [1995]. This list does not include a number of papers which focus on algebraic geometry rather than semigroups.

8. KRULL MONOIDS.

In this section we show that several constructions yield (up to isomorphism) the same class of semigroups. These semigroups, called Krull monoids, are the central concept in factorization theory and its applications to rings and have been studied by Chapman, Geroldinger, Halter-Koch, Krause, Lettl, and others.

1. A **Krull monoid** is a commutative and cancellative monoid S such that there exists a free c.m. F and a homomorphism $\partial : S \longrightarrow F$ (called a **divisor theory** for S) such that

(D1) $\partial a \leqq_{\mathcal{H}} \partial b$ in F implies $a \leqq_{\mathcal{H}} b$ in S;

(D2) for every $c \in F$ there exist $a_1, \dots, a_n \in S$ such that $\partial a_1 \wedge \dots \wedge \partial a_n = c$.

This concept has roots in the ideal theory of commutative rings. Its first abstract formulation is due to Clifford [1938].

Proposition 8.1. *A cancellative c.m. S with group of units U is a Krull monoid if and only if S/U is a Krull monoid. In fact, there is a one-to-one correspondence between the divisor theories for S and the divisor theories for S/U; the latter are injections.*

Proof. Let S be a Krull monoid and $\partial : S \longrightarrow F$ be a divisor theory for S. Since $a \leqq_{\mathcal{H}} b$ in S implies $\partial a \leqq_{\mathcal{H}} \partial b$ in F, and $\mathcal{H}$ is the equality on F, (D1) implies that $\partial a = \partial b$ if and only if $a \,\mathcal{H}\, b$. By Proposition 5.3, ∂ is injective if S is reduced, and in general factors uniquely through the projection $S \longrightarrow S/U \cong S/\mathcal{H}$. Since $a \leqq_{\mathcal{H}} b$ in S if and only if $Ua \leqq_{\mathcal{H}} Ub$ in S/U, the induced homomorphism $S/U \longrightarrow F$ is a divisor theory for S/U. If conversely $\partial : S/U \longrightarrow F$ is a divisor theory for S/U, then $S \longrightarrow S/U \longrightarrow F$ is a divisor theory for S. □

Thus, divisor theories are basically a matter of c.pc.r. semigroups.

When S has a divisor theory $\partial : S \longrightarrow F$, ∂ and F are unique up to isomorphism (Clifford [1938]). With regard to condition (D2) we note

Lemma 8.2. *Let S be a submonoid of a free c.m. F. The set*

$$\{c \in F \mid \text{ there exist } a_1, \dots, a_n \in S \text{ such that } a_1 \wedge \dots \wedge a_n = c\}$$

is a submonoid of F.

Proof. If $c = a_1 \wedge \dots \wedge a_m$ and $d = b_1 \wedge \dots \wedge b_m$, then $c + d = (a_1 + d) \wedge \dots \wedge (a_m + d)$ and every $a_i + d = (a_i + b_1) \wedge \dots \wedge (a_i + b_m)$; hence $c + d = \bigwedge_{i,j} (a_i + b_j)$. □

2. When F is a free c.m., the intersection of F and a subgroup of $G(F)$ is a submonoid of F. The resulting submonoids can be characterized as follows.

A subsemigroup T of a commutative semigroup S is **unitary** when $a \in T$, $b \in S$, and $ab \in T$ implies $b \in T$. This terminology goes back to Dubreil [1941]. A submonoid S of a free c.m. F is unitary in F if and only if $a, b \in S$ and $a \geqq b$ in F implies $a - b \in S$.

Proposition 8.3. *For a submonoid S of a free c.m. F the following conditions are equivalent:*

(1) *S is unitary (if $a, b \in S$ and $a \geqq b$ in F, then $a - b \in S$);*

(2) *if $a, b \in S$ and $a \leqq_{\mathcal{H}} b$ in F, then $a \leqq_{\mathcal{H}} b$ in S;*

(3) *$S = G(S) \cap F$;*

(4) *$S = H \cap F$ for some subgroup H of $G(F)$.*

If F is finitely generated, then a unitary submonoid of F has only finitely many irreducible elements, and is generated by its irreducible elements.

Proof. First, S is unitary if and only if (2) holds. Then $t \in G(S) \cap F$ implies $t = a - b$ for some $a, b \in S$ and $t \in S$; hence $S = G(S) \cap F$. If conversely

$S = H \cap F$ for some subgroup H of $G(F)$, then $a, b \in S$ and $a \geqq b$ in F implies $a - b \in H \cap F = S$; hence S is unitary.

Let S be unitary. A minimal (nonzero) element m of S (in the usual partial order on F) is irreducible, since $m \neq 0$ and $m = a + b$ with $a, b \in S$ implies $0 \leqq a, b \leqq m$ in F and $0 < a, b < m$ is not possible. Conversely an irreducible element c of S is minimal, since $0 < a < c$ with $a \in S$ would imply $c = a + (c - a)$ with $c - a,\ a \in S$ and $c - a,\ a \neq 0$.

We show that every nonzero element of S is a nonempty sum of minimal (nonzero) elements of S. If this is false, there is, by the descending chain condition (Proposition I.5.4), a nonzero element a of S which is minimal such that a is not a sum of one or more minimal elements of S. In particular a is not a minimal element of S and there is a minimal element $m < a$ of S. Then $a - m \in S$, $a - m < a$, $a - m$ is a sum of minimal elements of S by the choice of a, and so is $a = (a - m) + m$; this is the required contradiction.

The minimal elements of S constitute an antichain of F. If F is finitely generated, then, by Dickson's Theorem (Proposition I.5.4), this antichain is finite and S has only finitely many minimal elements. □

3. Up to isomorphism, a finitely generated free c.m. of rank n is the direct product $F = \mathbb{N}^n$ and consists of all families $a = (a_1, \dots, a_n)$ of nonnegative integers, under pointwise addition. Then $G(F) = \mathbb{Z}^n$ consists of all families of n integers, under pointwise addition and $\overline{G(F)} = \mathbb{Q}^n$. In $\mathbb{Z}^n$ the solutions of a finite homogeneous system of linear equations with integer coefficients

$$c_{i1} x_1 + \dots + c_{in} x_n = 0 \ (i = 1, 2, \dots, m)$$

constitute a subgroup H. The nonnegative solutions of the same system constitute a submonoid $H \cap F$ of F. These submonoids are sometimes called **Diophantine**, as they are solution sets of systems of homogeneous linear Diophantine equations.

Proposition 8.4. *Let $F = \mathbb{N}^n$. For a submonoid S of F the following conditions are equivalent:*

(1) *S is unitary and $ma \in S$, $m > 0$, and $a \in F$ implies $a \in S$;*

(2) *$S = \overline{G(S)} \cap F$;*

(3) *$S = H \cap F$ for some subspace H of $\overline{G(F)}$;*

(4) *$S = K \cap F$ for some pure subgroup K of $G(F)$;*

(5) *S is the nonnegative solution set of a finite homogeneous system of linear equations with integer coefficients.*

As before, $\overline{G(S)}$ is the subspace of $\overline{G(F)}$ generated by $G(S)$. A subgroup K of $G(S)$ is **pure** when $ma \in K$, $m > 0$, and $a \in G(F)$ implies $a \in K$; equivalently, if $G(F)/K$ is torsion-free.

Proof. (1) $\Longrightarrow$ (2). By Proposition 5.7, every element of $\overline{G(S)}$ can be written in the form g/m for some $g \in G(S)$ and $m > 0$, and every element of $\overline{G(F)}$ can be written in the form g/m for some $g \in G(F)$ and $m > 0$. If S is unitary and $ma \in S$, $m > 0$, and $a \in F$ implies $a \in S$, then $t \in \overline{G(S)} \cap F$ implies $mt \in G(S) \cap F$ for some $m > 0$, $mt \in S$, and $t \in S$; hence $S = \overline{G(S)} \cap F$.

(2) $\Longrightarrow$ (3) since $\overline{G(S)}$ is a subspace of $\overline{G(F)}$.

(3) $\Longrightarrow$ (4) since $K = H \cap G(F)$ is a pure subgroup of $G(F)$ when H is a subspace of $\overline{G(F)}$.

(4) $\Longrightarrow$ (3). When K is a pure subgroup of $G(F)$, then $\overline{K}$ is a subspace of $\overline{G(F)}$, and $K = \overline{K} \cap G(F)$, since $t \in \overline{K} \cap G(F)$ implies $mt \in K$ for some $m > 0$ and $t \in K$. Hence $K \cap F = \overline{K} \cap F$.

(3) $\Longrightarrow$ (5). In the rational vector space $\overline{G(F)}$ every subspace H is the solution set of a finite homogeneous system of linear equations with rational coefficients. Multiplying every equation by a common denominator of its coefficients yields an equivalent system with integer coefficients, whose nonnegative solution set is $H \cap F$.

(5) $\Longrightarrow$ (3). The nonnegative solution set S of a homogeneous system of linear equations with integer coefficients is the intersection of F and its solution set in $\overline{G(F)}$, which is a subspace H of $\overline{G(F)}$.

(3) $\Longrightarrow$ (1). If H is a subspace of $\overline{G(H)}$, then S is unitary, as $a \geqq b$, $a, b \in S$ implies $a - b \in H \cap F = S$; and $mt \in S$, $m > 0$, and $t \in F$ implies $mt \in H$ and $t \in H \cap F = S$. □

4. More generally, the integer solutions of a homogeneous system of linear inequalities with integer coefficients

$$c_{i1} x_1 + \ldots + c_{in} x_n \geqq 0 \ (i = 1, 2, \ldots, m)$$

constitute a submonoid of $F = \mathbb{N}^n$.

Proposition 8.5. *Let S be a reduced subsemigroup of $\mathbb{Z}^n$ such that $G(S) = \mathbb{Z}^n$. The following conditions are equivalent:*

(1) *S is finitely generated and $g \in G(S)$, $m > 0$, and $mg \in S$ implies $g \in S$;*

(2) *S is finitely generated and $S = G(S) \cap \overline{S}$;*

(3) *S is finitely generated and $S = G(S) \cap \overline{\overline{S}}$;*

(4) *$S = G(S) \cap C$ for some finitely generated reduced convex cone C of $\mathbb{Q}^n$;*

(5) *$S = G(S) \cap C$ for some reduced convex cone $C \subseteq \mathbb{R}^n$ that is generated by finitely many elements of $\mathbb{Q}^n$;*

(6) *S is the set of integer solutions of a finite homogeneous system of linear inequalities with integer coefficients.*

Then S has only finitely many irreducible elements, and is generated by its irreducible elements.

As in Section 5, $\overline{S} \subseteq \mathbb{Q}^n$ and $\overline{\overline{S}} \subseteq \mathbb{R}^n$ are the convex cones generated by S.

Proof. (1) $\Longrightarrow$ (2). If (1) holds, then $g \in G(S) \cap \overline{S}$ implies $mg \in S$ for some $m > 0$ and $g \in S$; hence $G(S) \cap \overline{S} = S$.

(2) $\Longleftrightarrow$ (3). Let $g \in G(S) \cap \overline{\overline{S}}$. Then there exist $a, b, a_1, \dots, a_k \in S$ such that $g = a - b = \lambda_1 a_1 + \dots + \lambda_k a_k$ for some $\lambda_1, \dots, \lambda_k \in \mathbb{R}$, $\lambda_1, \dots, \lambda_k > 0$. Since $a_1, \dots, a_k \in \mathbb{Z}^n$ it follows from Lemma 5.9 that $q_0(-g) + q_1 a_1 + \dots + q_k a_k = 0$ for some $q_0, q_1, \dots, q_k \in \mathbb{Q}$, $q_0, q_1, \dots, q_k > 0$. Hence $g \in \overline{S}$ and (2) implies (3); the converse holds since $S \subseteq G(S) \cap \overline{S} \subseteq G(S) \cap \overline{\overline{S}}$.

(2) $\Longrightarrow$ (4) since $\overline{S}$ is a reduced cone by Proposition 5.7 and is generated as a convex cone of $\mathbb{Q}^n$ by the generators of S.

(4) $\Longrightarrow$ (5). Let C be a finitely generated reduced convex cone of $\mathbb{Q}^n$. Then $\overline{\overline{C}}$ is a convex cone of $\mathbb{R}^n$ that is generated by finitely many elements of $\mathbb{Q}^n$. By Proposition 5.8, $\overline{\overline{C}}$ is reduced and $\overline{\overline{C}} \cap \mathbb{Q}^n = C$. Hence $G(S) \cap \overline{\overline{C}} = G(S) \cap C$.

(5) $\Longrightarrow$ (6). Let $S = G(S) \cap C$ for some reduced convex cone $C \subseteq \mathbb{R}^n$ that is generated by finitely many elements of $\mathbb{Q}^n$. Then C spans $\mathbb{R}^n$, since $G(S) = \mathbb{Z}^n$. By Corollary 7.2, C is the intersection of finitely many closed halfspaces, bounded by the hyperplanes spanned by the facets of C. Thus C is the solution set of a finite homogeneous system of linear inequalities

$$c_{i1} x_1 + \dots + c_{in} x_n \geqq 0 \ (i = 1, 2, \dots, m).$$

Hence $S = G(S) \in C$ is the set of integer solutions of a finite homogeneous system of linear inequalities

$$c_{i1} x_1 + \dots + c_{in} x_n \geqq 0 \ (i = 1, 2, \dots, m).$$

The hyperplanes spanned by the facets of C have equations with rational coefficients, since they are spanned by generators of C which have rational coordinates in $\mathbb{R}^n$. Hence we can arrange in the above that all coefficients c_{ij} are rational. Multiplying each inequality by a common denominator then yields a finite homogeneous system of linear inequalities with integer coefficients, of which S is the set of integer solutions.

(6) $\Longrightarrow$ (1). Let S be the set of integer solutions of a finite homogeneous system of linear inequalities

$$\gamma_i(x) = c_{i1}x_1 + \ldots + c_{in}x_n \geqq 0 \ (i = 1,2,\ldots,m).$$

with integer coefficients. If $g \in G(S)$ and $mg \in S$ for some $m > 0$, then $\gamma_i(mg) \geqq 0$ for all i, $\gamma_i(g) \geqq 0$ for all i, and $g \in S$. Also $\{x \in \mathbb{Z}^n \mid \gamma_i(x) = 0$ for all $i\}$ is a subgroup of S; since S is reduced, $\gamma_i(x) = 0$ for all i implies $x = 0$. Therefore the homomorphism

$$\gamma : x \longmapsto (\gamma_1(x), \ldots, \gamma_m(x)),\ \mathbb{Z}^n \longrightarrow \mathbb{Z}^m$$

is injective. We see that $\gamma(S) \subseteq F = \mathbb{N}^m$. As in the proof of Theorem 7.3, $\gamma(S)$ is a unitary submonoid of F: if $\gamma(a) \leqq \gamma(b)$ in F, then $\gamma_i(a) \geqq \gamma_i(b)$ for all i, $\gamma_i(a-b) \geqq 0$ for all i, $a - b \in S$, and $\gamma(a) - \gamma(b) = \gamma(a-b) \in \gamma(S)$. By Proposition 8.3, $S \cong \gamma(S)$ has only finitely many irreducible elements, and is generated by its irreducible elements; in particular, S is finitely generated. □

A c.pc.r. semigroup S is **normal** when $ng \in S$ implies $g \in S$ when $n > 0$ and $g \in G(S)$; equivalently, when $a + nb = nc$ with $a,b,c \in S$ and $n > 0$ implies $a = nd$ for some $d \in S$. Unitary submonoids are normal, but normal submonoids need not be unitary. For instance the (free) submonoid S of $F_{\{x,y\}}$ generated by $x + 2y$ and $2x + y$ is normal, but is not unitary, since $2x + 4y$, $2x + y \in S$ but $3y \notin S$; also, $3x + 3y \in S$ but $x + y \notin S$.

Normal c.pc.r. semigroups are also called **integrally closed** and **root closed**; the **integral closure** or **root closure** of S is

$$\{g \in G(S) \mid ng \in S \text{ for some } n > 0\}.$$

(A larger **complete integral closure** $\{g \in G(S) \mid ng + a \in S$ for some $n > 0$ and $a \in S\}$ has been used e.g. by Clifford [1938] and Geroldinger [1996].) Hochster [1972] showed that a c.pc.r. c.m. is finitely generated and normal if and only if its semigroup algebra (over any field K) is a normal domain (is noetherian and integrally closed), and then it is a Cohen-Macaulay ring. A cancellative and power cancellative c.m. is normal if and only if its semigroup algebra (over any field K) is integrally closed (Gilmer [1984]).

5. Preliminary to our main result we show:

Lemma 8.6. *Let S be a reduced subsemigroup of $\mathbb{Z}^n$ such that $G(S) = \mathbb{Z}^n$ and S is the set of integer solutions of a finite homogeneous system of linear inequalities with integer coefficients. There exist homomorphisms $\partial_1, \dots, \partial_r : \mathbb{Z}^n \longrightarrow \mathbb{Z}$ with the following properties:*

(1) $\operatorname{Im} \partial_i = \mathbb{Z}$ *and* $S = \{x = (x_1, \dots, x_n) \in \mathbb{Z}^n \mid \partial_i x \geqq 0$ *for all* $i = 1, 2, \dots, r\}$;

(2) *there exist* $a_{jk} \in S$ $(j, k = 1, 2, \dots, r)$ *such that* $\partial_k a_{kk} = 1$ *and* $\partial_k a_{jk} > 0$, $\partial_j a_{jk} = 0$ *whenever* $j \neq k$;

(3) $\partial_1, \dots, \partial_r$ *span* $\operatorname{Hom}(\mathbb{Z}^n, \mathbb{Q})$;

(4) *the convex cone* $D \subseteq \operatorname{Hom}(\mathbb{Z}^n, \mathbb{Q})$ *generated by* $\partial_1, \dots, \partial_r$ *is reduced;*

(5) $C = D \cap \operatorname{Hom}(\mathbb{Z}^n, \mathbb{Z})$ *has finitely many irreducible elements, and is generated by its irreducible elements;*

(6) $\partial_1, \dots, \partial_r$ *are irreducible in* C;

(7) *the group* $\operatorname{Hom}(\mathbb{Z}^n, \mathbb{Z})$ *is generated by the irreducible elements of* C.

Proof. Let S be the set of integer solutions of a finite homogeneous system of linear inequalities

$$c_{i1} x_1 + \dots + c_{in} x_n \geqq 0 \ (i = 1, 2, \dots, m)$$

with integer coefficients. Then $S = \mathbb{Z}^n \cap P$, where P is the convex cone

$$P = \{(x_1, \dots, x_n) \in \mathbb{R}^n \mid c_{i1} x_1 + \dots + c_{in} x_n \geqq 0 \text{ for all } i = 1, 2, \dots, m\}.$$

P is reduced since S is reduced, and spans $\mathbb{R}^n$ since $G(S) = \mathbb{Z}^n$. By Corollary 7.2, P is the intersection of finitely many distinct closed halfspaces

$$K_i = \{(x_1, \dots, x_n) \in \mathbb{R}^n \mid d_{i1} x_1 + \dots + d_{in} x_n \geqq 0\} \ (i = 1, 2, \dots, r)$$

such that every hyperplane

$$H_i = \{(x_1, \dots, x_n) \in \mathbb{R}^n \mid d_{i1} x_1 + \dots + d_{in} x_n = 0\}$$

contains a facet of B. Since H_i contains a facet of $\overline{\overline{S}}$ and is spanned by elements of S which have rational coordinates in $\mathbb{R}^n$, H_i has an equation with rational coefficients, and we can arrange that all d_{ij} are rational. Multiplying by a common denominator we can then arrange that $d_{i1}, \dots, d_{in}$ are integers; dividing by $\gcd(d_{i1}, \dots, d_{in})$ we can further arrange that $d_{i1}, \dots, d_{in}$ are relatively prime. Then

$$\partial_i : (x_1, \dots, x_n) \longmapsto d_{i1} x_1 + \dots + d_{in} x_n$$

is a homomorphism of $\mathbb{Z}^n$ into $\mathbb{Z}$ and $\operatorname{Im} \partial_i = \mathbb{Z}$. Also

$$S = \{x = (x_1, \dots, x_n) \in \mathbb{Z}^n \mid \partial_i x \geqq 0 \text{ for all } i = 1,2,\dots,r\}$$

since $S = \mathbb{Z}^n \cap P$ and $P = K_1 \cap \dots \cap K_r$, and (1) holds.

(2). If $n = 1$, then $P \subseteq \mathbb{R}$ is a closed halfspace, $r = 1$, ∂_1 is an isomorphism, and (2) holds. Now assume $n \geqq 2$.

If $j, k \leqq r$ and $j \neq k$, then $\partial_k a_{jk} > 0$, $\partial_j a_{jk} = 0$ for some $a_{jk} \in S$: indeed $S \cap H_j \nsubseteq H_k$, since H_j is spanned by $S \cap H_j$ and $H_j \nsubseteq H_k$.

Now let $k \leqq r$. We have $\partial_k g = 1$ for some $g \in G(S)$. Since $S \nsubseteq H_k$ we have $\partial_k a_1 > 0$ for some $a_1 \in S$. Since $H_k \cap P$ is a facet of P and is spanned by elements of S, H_k contains linearly independent elements $a_2, \dots, a_n$ of S. Then $a_1, a_2, \dots, a_n$ are linearly independent and $g = q_1 a_1 + q_2 a_2 + \dots + q_n a_n$ for some $q_1, \dots, q_n \in \mathbb{Q}$. Since $a_2, \dots, a_n \in H_k$ we have $\partial_k(q_1 a_1) = \partial_k g = 1$ and $q_1 > 0$. Let p_j be the greatest integer $p_j \leqq q_j$ and

$$a = q_1 a_1 + (q_2 - p_2) a_2 + \dots + (q_n - p_n) a_n = g - p_2 a_2 - \dots - p_n a_n \in G(S).$$

Then $\partial_i a \geqq 0$ for all i, since $q_1 > 0$ and $q_j - p_j \geqq 0$ for all $j > 1$, and $a \in S$; also $\partial_k a = \partial_k g = 1$, since $a_2, \dots, a_n \in H_i$. Thus $\partial a_{kk} = 1$ for some $a_{kk} \in S$.

(3). Every homomorphism $\gamma : \mathbb{Z}^n \longrightarrow \mathbb{Q}$ extends to a unique linear transformation $\overline{\gamma} : \mathbb{Q}^n \longrightarrow \mathbb{Q}$. Let $V = \mathbb{Q}^n$ and W be the subspace of the dual space V^* spanned by $\overline{\partial}_1, \dots, \overline{\partial}_r$. If $W \subsetneqq V^*$, then $\{u \in \mathbb{Q}^n \mid w(u) = 0$ for all $w \in W\}$ has dimension at least 1 and there exists $u = (u_1, \dots, u_n) \in \mathbb{Q}^n$, $u \neq 0$ such that $\overline{\partial}_i u = 0$ for all i. Multiplying u by a common denominator of $u_1, \dots, u_n$ yields $a \in \mathbb{Z}^n$, $a \neq 0$ such that $\partial_i a = \overline{\partial}_i a = 0$ for all i. Then $a \in S$ and $-a \in S$, a contradiction since S is reduced. Therefore $\overline{\partial}_1, \dots, \overline{\partial}_r$ span V^*; every linear transformation $\overline{\gamma} : \mathbb{Q}^n \longrightarrow \mathbb{Q}$ is a linear combination of $\overline{\partial}_1, \dots, \overline{\partial}_r$ with rational coefficients; and every homomorphism $\gamma : \mathbb{Z}^n \longrightarrow \mathbb{Q}$ is a linear combination of $\partial_1, \dots, \partial_r$ with rational coefficients.

(4). D consists of all linear combinations of $\partial_1, \dots, \partial_r$ with nonnegative rational coefficients. If $q_1 \partial_1 + \dots + q_r \partial_r = 0$ with $q_1, \dots, q_r \in \mathbb{Q}$, $q_1, \dots, q_r \geqq 0$, then $q_1 \partial_1 a + \dots + q_r \partial_r a = 0$ for all $a \in S$, $q_i \partial_i a = 0$ for all $a \in S$ and all i since $q_i \partial_i a \geqq 0$ for all i, and $q_i = 0$ for all i since $\partial_i a > 0$ for some $a \in S$. If now $\gamma \in D$ and $-\gamma \in D$, then $\gamma = q_1 \partial_1 + \dots + q_r \partial_r$ for some $q_1, \dots, q_r \geqq 0$, $-\gamma = p_1 \partial_1 + \dots + p_r \partial_r$ for some $p_1, \dots, p_r \geqq 0$, $(p_1 + q_1) \partial_1 + \dots + (p_r + q_r) \partial_r = 0$, $p_i + q_i = 0$ for all i, $q_i = 0$ for all i, and $\gamma = 0$.

(5). C is the set of all homomorphisms $\gamma : \mathbb{Z}^n \longrightarrow \mathbb{Z}$ that can be written as linear combinations $\gamma = q_1 \partial_1 + \cdots + q_r \partial_r$ with $q_1, \ldots, q_r \in \mathbb{Q}$ and $q_1, \ldots, q_r \geqq 0$. Then $G(C) \subseteq \operatorname{Hom}(\mathbb{Z}^n, \mathbb{Z})$ is, like $\operatorname{Hom}(\mathbb{Z}^n, \mathbb{Z})$, a finitely generated free abelian group. Also C is the intersection of $G(C)$ and the convex cone of $\overline{G(C)}$ generated by $\partial_1, \ldots, \partial_r$. By Proposition 8.5, C has finitely many irreducible elements, and is generated by its irreducible elements.

(6). We show that $\partial_j = q_1 \partial_1 + \cdots + q_r \partial_r$, with $q_1, \ldots, q_r \in \mathbb{Q}$ and $q_1, \ldots, q_r \geqq 0$, implies $q_j = 1$ and $q_i = 0$ for all $i \neq j$; hence ∂_j cannot be written as a sum $\partial_j = \gamma_1 + \gamma_2$ with $\gamma_1, \gamma_2 \in C$ and $\gamma_1, \gamma_2 \neq 0, \partial_j$, and is irreducible in C.

Let $\partial_j = q_1 \partial_1 + \cdots + q_r \partial_r$ as above. By (2) there is for every $k \neq j$ some $a_{jk} \in S$ such that $\partial_k a_{jk} > 0 = \partial_j a_{jk}$. Then $q_1 \partial_1 a_{jk} + \cdots + q_r \partial_r a_{jk} = \partial_j a_{jk} = 0$, $q_i \partial_i a_{jk} = 0$ for all i, $q_k \partial_k a_{jk} = 0$, and $q_k = 0$. Hence $q_i = 0$ for all $i \neq j$; then $q_j = 1$, since $\partial_j a > 0$ for some $a \in S$.

(7). Let $\partial_1, \ldots, \partial_r, \partial_{r+1}, \ldots, \partial_s$ be all the irreducible elements of C. By (3) every homomorphism $\gamma : \mathbb{Z}^n \longrightarrow \mathbb{Z}$ is a linear combination of $\partial_1, \ldots, \partial_r$ with rational coefficients. Therefore there exist rational numbers $q_1, \ldots, q_r \geqq 0$ such that $\gamma + q_1 \partial_1 + \cdots + q_r \partial_r \in D$. For any integers $p_1 \geqq q_1, \ldots, p_r \geqq q_r$ we now have $\gamma + p_1 \partial_1 + \cdots + p_r \partial_r \in D \cap \operatorname{Hom}(\mathbb{Z}^n, \mathbb{Z}) = C$. Hence $\gamma + p_1 \partial_1 + \cdots + p_r \partial_r$ is a linear combination of $\partial_1, \ldots, \partial_s$ with nonnegative integer coefficients, and γ is a linear combination of $\partial_1, \ldots, \partial_s$ with integer coefficients. Thus $\partial_1, \ldots, \partial_s$ generate $\operatorname{Hom}(\mathbb{Z}^n, \mathbb{Z})$. □

6. The main result in this section is:

Theorem 8.7. *Let S be a reduced subsemigroup of $\mathbb{Z}^n$ such that $G(S) = \mathbb{Z}^n$. The following conditions are equivalent:*

(1) *S is a Krull monoid (= S has a divisor theory);*

(2) *S is finitely generated and normal (if $g \in G(S)$, $m > 0$, and $mg \in S$, then $g \in S$);*

(3) *S is the set of integer solutions of a finite homogeneous system of linear inequalities with integer coefficients;*

(4) *S is isomorphic to the nonnegative solution set of a finite homogeneous system of linear equations with integer coefficients.*

(5) *S is isomorphic to a unitary submonoid of a finitely generated free com-*

mutative monoid.

Proof. (1) $\Longrightarrow$ (2). Let $\partial : S \longrightarrow F$ be a divisor theory for S. Let $g \in G(S)$, so that $g = a - b$ for some $a, b \in S$. If $mg \in S$ for some $m > 0$, then $ma = mb + mg \leqq_{\mathcal{H}} mb$ in S, $\partial(ma) \geqq \partial(mb)$ in F, $\partial a \geqq \partial b$, $a = b + c$ for some $c \in S$ by (D1), and $g = c \in S$. Thus S is normal.

Since $G(S) = \mathbb{Z}^n$ there exist $a_1, \dots, a_n \in S$ that are linearly independent in $\mathbb{Q}^n$. Let $a = a_1 + \cdots + a_n$. Every $b \in S$ is a linear combination of $a_1, \dots, a_n$ with rational coefficients. If $m \in \mathbb{N}^+$ is large enough, then $ma - b \in \overline{S}$, $ma - b \in G(S) \cap \overline{S} = S$, and $\partial(b) \leqq \partial(ma)$ in F. By (D2), every generator x of F can be written in the form $x = \partial b_1 \wedge \dots \wedge \partial b_k$ for some $b_1, \dots, b_k \in S$; hence $x \leqq \partial(ma)$ for some $m > 0$ and $x \leqq \partial a$. Therefore F is finitely generated. Then ∂S is finitely generated by Proposition 8.3 (since $a, b \in \partial S$ and $a \leqq_{\mathcal{H}} b$ in F implies $a \leqq_{\mathcal{H}} b$ in ∂S), and so is $S \cong \partial S$.

(2) $\Longrightarrow$ (3) by Proposition 8.5.

(3) $\Longrightarrow$ (1). Let S be the set of integer solutions of a finite homogeneous system of linear inequalities with integer coefficients. Lemma 8.6 provides homomorphisms $\partial_1, \dots, \partial_r : \mathbb{Z}^n \longrightarrow \mathbb{Z}$ such that $\operatorname{Im} \partial_i = \mathbb{Z}$,

$$S = \{x = (x_1, \dots, x_n) \in \mathbb{Z}^n \mid \partial_i x \geqq 0 \text{ for all } i = 1, 2, \dots, r\},$$

and other wonderful properties. We show that the homomorphism

$$\partial : x \longmapsto (\partial_1 x, \dots, \partial_r x)$$

of S into the free c.m. $F = \mathbb{N}^r$ is a divisor theory for S.

Let $a, b \in S$. If $\partial a \geqq \partial b$ in F, then $\partial_i(a - b) \geqq 0$ for all i and $a - b \in S$; hence (D1) holds.

Part (2) of Lemma 8.6 provides $a_{jk} \in S$ such that $\partial a_{kk} = 1$ and $\partial_k a_{jk} > 0$, $\partial_j a_{jk} = 0$ whenever $j \neq k$. Then

$$e_k = \partial a_{1k} \wedge \dots \wedge \partial a_{rk} \in F$$

has the following properties. The k-th coordinate of e_k is

$$\min\left(\partial_k a_{1k} \wedge \dots \wedge \partial_k a_{rk}\right) = 1,$$

since $\partial_k a_{kk} = 1$ and $\partial_k a_{jk} > 0$ for all $j \neq k$. But, for all $j \neq k$, the j-th coordinate of e_k is

$$\min\left(\partial_j a_{1k} \wedge \dots \wedge \partial_j a_{rk}\right) = 0,$$

since $\partial_j a_{jk} = 0$. Thus e_k is the k-th standard generator of F, and every generator of F is an infimum of elements of ∂S. Then (D2) follows from Lemma 8.2.

(3) $\Longrightarrow$ (4). Let $\partial_1, \dots, \partial_r$ be as above, D be the convex cone generated by $\partial_1, \dots, \partial_r$ in $\mathrm{Hom}(\mathbb{Z}^n, \mathbb{Q})$, and $C = D \cap \mathrm{Hom}(\mathbb{Z}^n, \mathbb{Z})$. By Lemma 8.6, C has finitely many irreducible elements $\partial_1, \dots, \partial_r, \partial_{r+1}, \dots, \partial_s$, including $\partial_1, \dots, \partial_r$. The homomorphism

$$\delta : a \longmapsto (\partial_1 a, \dots, \partial_s a),\ \mathbb{Z}^n \longrightarrow \mathbb{Z}^s$$

is injective on S: if $a, b \in S$ and $\delta a = \delta b$, then $\delta(a-b) = 0$, $\partial_i(a-b) = 0$ for all $i \leqq r$, and $a - b \in S$ by part (1) of Lemma 8.6; similarly $b - a \in S$, and $a = b$ since S is reduced.

We show that $\mathrm{Im}\,\delta$ is a pure subgroup of $\mathbb{Z}^s$. Indeed assume that $mc = \delta a$, where $a = (a_1, \dots, a_n) \in \mathbb{Z}^n$, $c = (c_1, \dots, c_s) \in \mathbb{Z}^s$, and $m > 0$. Then m divides $\partial_i a$ for all i. Then m divides γa for every homomorphism $\gamma : \mathbb{Z}^n \longrightarrow \mathbb{Z}$, since $\partial_1, \dots, \partial_s$ generate $\mathrm{Hom}(\mathbb{Z}^n, \mathbb{Z})$ by Lemma 8.6. In particular m divides every a_j. Hence $a = mb$ for some $b \in \mathbb{Z}^n$, $mc = \delta(mb)$, and $c = \delta b$.

Finally we show that $\delta S = \mathrm{Im}\,\delta \cap \mathbb{N}^s$: if $c = (c_1, \dots, c_s) \in \mathbb{Z}^s$ and $c_i \geqq 0$ for all i, and $c = \delta a$ for some $a \in \mathbb{Z}^n$, then $\partial_i a \geqq 0$ for all i and $a \in S$. Hence it follows from Proposition 8.4 that $\delta S \cong S$ is the nonnegative solution set of a finite homogeneous system of linear equations with integer coefficients.

(4) $\Longrightarrow$ (5) by Proposition 8.4.

(5) $\Longrightarrow$ (2). Let T be a unitary submonoid of a finitely generated c.m. F. Then T is finitely generated by Proposition 8.3. If $g \in G(T)$, $m > 0$, and $mg \in T$, then $g = a - b$ for some $a, b \in T$, $ma \geqq mb$ in F, $a \geqq b$, and $g = a - b \in T$. Thus T is normal. □

Theorem 8.7 reached its present shape in Chapman, Krause, & Oeljeklaus [2001]. Its proof is due to several authors. The equivalence of (2) and (5) was proved by Hochster [1972]. The equivalence of (1), (2), and (3) is due to Halter-Koch [1993] and Lettl [1988]. The proof that (3) implies (4) is from Kainrath & Lettl [2000]. The semigroup C in this proof is isomorphic to $\mathrm{Hom}(S, \mathbb{N})$, so that $\delta_{|S} : S \longrightarrow F$ is the quasi-universal free c.m. of S in Grillet [1970E]. The purity of $\mathrm{Im}\,\delta$ was shown by John [1974].

Chapter III.

SEMILATTICE DECOMPOSITIONS.

Semilattice decompositions were introduced in the 1950's and provide the earliest structural insight into commutative semigroups in general, showing, for instance, that finite commutative semigroups are assembled from groups and nilsemigroups arranged along a semilattice. They have been the mainstay of commutative semigroup theory for many years (see e.g. Petrich [1977]).

It turns out that every commutative semigroup has a semilattice decomposition into semigroups called archimedean that lack proper decompositions. There are two cases, Clifford semigroups and separative semigroups, where a semigroup can be effectively reconstructed from its archimedean components. Unfortunately this reassembly problem is in general very difficult. Hence much of this chapter is about some classes of archimedean semigroups that are comparatively simple in structure. Finitely generated archimedean semigroups will be considered in Chapter VI.

Some of these results extend to more general semigroups; interested readers are referred to Clifford & Preston [1961], Petrich [1973], Grillet [1995], Nagy [2001], and Ćirić [2002].

As before we abbreviate "commutative semigroup" into "c.s." and "commutative monoid" into "c.m.".

1. GENERAL RESULTS.

1. This chapter studies c.s. (= commutative semigroups) through their homomorphisms into semilattices. When φ is a homomorphism of a semigroup S into a semilattice Y, then $S_a = \{x \in S \mid \varphi(x) = a\}$ is a subsemigroup of S for every $a \in Y$, and $S_a\, S_b = S_{ab}$ $(= S_{a \wedge b})$ for all $a, b \in Y$; this gives an overall picture of the multiplication on S.

In general, S is a **semilattice of semigroups** when there exists a semilattice Y and a partition $S = \bigcup_{a \in Y} S_a$ of S into subsemigroups S_a (one for every

$a \in Y$) such that $S_a S_b \subseteq S_{ab}$ for all $a, b \in Y$; equivalently, when there exists a congruence $\mathcal{C}$ on S (a **semilattice** congruence, or **semilattice decomposition**) such that $S/\mathcal{C}$ is a semilattice (then $Y = S/\mathcal{C}$ and the subsemigroups S_a are the $\mathcal{C}$-classes). This concept is due to Clifford [1954].

Other decompositions into direct unions have been studied, most notably by Croisot [1953], Poyatos [1966], [1967], Szép [1969], [1970], [1973], and Migliorini & Szép [1979], [1981] (see also Jürgensen, Migliorini, & Szép [1991]). These decompositions are typically more detailed, at some cost in multiplicative properties.

2. A semilattice congruence provides the most detailed picture of the multiplication on S when it has the most classes; this also tends to make the classes smaller and simpler in structure. It turns out that there is always a smallest semilattice congruence and that its classes have no further semilattice decompositions. This result was first obtained for finite (more generally torsion) c.s. by Schwarz [1953] and Thierrin [1954] and extended to all c.s. by Tamura and Kimura [1954] and to all semigroups by Yamada [1955]. For other proofs, see Clifford and Preston [1961]; Petrich [1963F], [1973]; Grillet [1995].

When S is commutative, a clue to the smallest semilattice congruence is provided by the following result, which constructs a preorder $\leqq_{\mathcal{N}}$ that contains the divisibility preorder $\leqq_{\mathcal{H}}$. Recall that $a \leqq_{\mathcal{H}} b$ if and only if $a \in S^1 b$, if and only if every ideal which contains b also contains a. An ideal P of S is (completely) **prime** when $xy \in P$ implies $x \in P$ or $y \in P$.

Lemma 1.1. *For elements a and b of a c.s. S the following conditions are equivalent:*

(1) *every prime ideal which contains b also contains a;*

(2) *$a^n = bt$ for some $n > 0$ and some $t \in S^1$;*

(3) *$a^n = bt$ for some $n > 0$ and some $t \in S$;*

(4) *$\varphi(a) \leqq \varphi(b)$ for every homomorphism φ of S into a semilattice.*

We write $a \leqq_{\mathcal{N}} b$ when a and b satisfy these conditions; $\leqq_{\mathcal{N}}$ is a preorder and contains $\leqq_{\mathcal{H}}$ (if $a \leqq_{\mathcal{H}} b$, then $a \leqq_{\mathcal{N}} b$).

Proof. (1) $\Longrightarrow$ (2). If (2) does not hold, then the principal ideal $S^1 b$ contains no positive power a^n of a. Let P be the union of all ideals of S which contain no positive power of a. Then P is an ideal of S, by Proposition I.3.1, and P contains $b \in S^1 b$ but contains no positive power a^n of a. If $x \notin P$ and $y \notin P$, then $S^1 x$ must contain some positive power a^m of a, otherwise we would have

$S^1x \subseteq P$; S^1y must contain some positive power a^n of a, otherwise we would have $S^1y \subseteq P$; hence $S^1xy = S^1x\,S^1y$ contains a positive power a^{m+n} of a, $S^1xy \nsubseteq P$, and $xy \notin P$. Thus P is a prime ideal. Since P contains b but not a, (1) does not hold. (The similar argument in commutative rings uses Zorn's Lemma; for semigroups, Proposition I.3.1 makes this unnecessary.)

(2) $\Longrightarrow$ (3). If $a^n = bt$ for some $t \in S^1$, then $a^{n+1} = bat$ with $at \in S$.

(3) $\Longrightarrow$ (4). If $a^n = bt$ for some $n > 0$ and $t \in S$, and φ is a homomorphism of S into a semilattice Y, then $\varphi(a) = \varphi(a)^n = \varphi(b) \wedge \varphi(t) \leqq \varphi(b)$ in Y.

(4) $\Longrightarrow$ (1). Let P be a prime ideal of S. Then $Q = S \backslash P$ is a subsemigroup of S and $QQ \subseteq Q$, $PQ \subseteq P$, $PP \subseteq P$. Hence there is a homomorphism φ of S into the semilattice $\{0,1\}$ which sends every $x \in P$ to 0 and every $y \in Q$ to 1. Then $b \in P$ implies $\varphi(a) \leqq \varphi(b) = 0$ by (4) and $a \in P$. □

3. A c.s. S is **archimedean** when $a \leqq_{\mathcal{N}} b$ holds for every $a, b \in S$; equivalently, when there exist for every $a, b \in S$ some $n > 0$ and $t \in S$ such that $a^n = tb$. (This terminology originated with ordered abelian groups: the positive cone of a totally ordered abelian group G is archimedean as a semigroup if and only if G is archimedean as a totally ordered abelian group.)

Theorem 1.2. *On every commutative semigroup S there is a smallest semilattice congruence $\mathcal{N}$, under which*

$$\begin{aligned} a \,\mathcal{N}\, b &\iff a \leqq_{\mathcal{N}} b \text{ and } b \leqq_{\mathcal{N}} a \\ &\iff a^m = tb \text{ and } b^n = ua \text{ for some } m, n > 0 \text{ and } t, u \in S. \end{aligned}$$

Every homomorphism of S into a semilattice factors uniquely through the projection $S \longrightarrow S/\mathcal{N}$. Moreover, every $\mathcal{N}$-class is archimedean.

Proof. $a \leqq_{\mathcal{N}} b$ implies $ac \leqq_{\mathcal{N}} bc$, since $\varphi(a) \leqq \varphi(b)$ implies $\varphi(a)\,\varphi(c) \leqq \varphi(b)\,\varphi(c)$ for every homomorphism φ of S into a semilattice. Hence $\mathcal{N}$ is a congruence. Every element of $S/\mathcal{N}$ is idempotent, since $a^2 \leqq_{\mathcal{N}} a$ and $a \leqq_{\mathcal{N}} a^2$ for every $a \in S$; hence $\mathcal{N}$ is a semilattice congruence.

$\mathcal{N}$ is contained in every semilattice congruence, by (4) in Lemma 1.1. When φ is a homomorphism of S into a semilattice Y, then the subsemigroup $\operatorname{im} \varphi$ of Y is a semilattice and $\ker \varphi$ is a semilattice congruence. Hence $\mathcal{N} \subseteq \ker \varphi$ and φ factors uniquely through the projection $S \longrightarrow S/\mathcal{N}$ by Proposition I.2.4.

Finally, let A be an $\mathcal{N}$-class of S. Let $a, b \in A$. Then $a^n = tb$ for some $n > 0$ and $t \in S$, and $a^{n+1} = (at)\,b$, with $at \in A$ since $a \leqq_{\mathcal{N}} at$ and $at \leqq_{\mathcal{N}} a$. Thus A is archimedean. □

By Theorem 1.2, $\mathcal{N}$ is a semilattice decomposition of S into semigroups with no further semilattice decompositions. The $\mathcal{N}$-class of x is the **archimedean component** of x, which we denote by $A(x)$ (rather than N_x). Since $\mathcal{H} \subseteq \mathcal{N}$, every archimedean component of S is a union of $\mathcal{H}$-classes.

Proposition 1.3. *An archimedean semigroup contains at most one idempotent.*

Proof. When e and f are idempotent, $e \leqq_{\mathcal{N}} f$ implies $e = e^n \leqq_{\mathcal{H}} f$ and $e \leqq f$. Hence $e \,\mathcal{N}\, f$ implies $e = f$. □

Archimedean semigroups are considered in more detail in the next sections.

4. The semilattice $Y(S) = S/\mathcal{N}$ of archimedean components of S is also called the **universal semilattice** of S, owing to its universal property in Theorem 1.2.

Corollary 1.4. *For every c.s. S there is an injective homomorphism $E(S) \longrightarrow Y(S)$, which is an isomorphism when S is finite.*

Proof. Restricting the projection $S \longrightarrow Y(S) = S/\mathcal{N}$ to $E(S) \subseteq S$ yields a homomorphism $E(S) \longrightarrow Y(S)$ which is injective by Proposition 1.3. If S is finite, then every archimedean component of S contains an idempotent, by Corollary I.5.9, and the homomorphism $E(S) \longrightarrow Y(S)$ is surjective as well. □

Petrich [1962] showed that $Y(S \times T) \cong Y(S) \times Y(T)$ for all c.s. S and T.

When $\varphi : S \longrightarrow T$ is a homomorphism, the homomorphism $S \longrightarrow T \longrightarrow Y(T)$ factors uniquely through $Y(S)$:

$$\begin{array}{ccc} S & \xrightarrow{\ \varphi\ } & T \\ \downarrow & & \downarrow \\ \mathrm{Y(S)} & \xrightarrow[Y(\varphi)]{} & \mathrm{Y(T)} \end{array}$$

Thus φ induces a homomorphism $Y(\varphi) : Y(S) \longrightarrow Y(T)$ such that the diagram above commutes; when A is an archimedean component of S, then $\varphi(A)$ is archimedean and $Y(\varphi)$ sends A to the archimedean component of T which contains $\varphi(A)$.

5. The **semilattice composition problem** is the problem of reconstructing a semigroup S from a semilattice decomposition $S = \bigcup_{a \in Y} S_a$ of S. This amounts to constructing the multiplication maps $S_a \times S_b \longrightarrow S_{ab}$ which determine the operation on S.

The semilattice composition problem is discussed in some detail in Petrich [1973] and Grillet [1995]. It is even more difficult that the ideal extension problem: indeed it requires the simultaneous solution of many ideal extension problems, since, in a semilattice of semigroups $(S_a)_{a \in Y}$, every S_{ab} is an ideal of $S_a \cup S_b \cup S_{ab}$. Even the reassembly of a commutative semigroup from its archimedean components seems quite difficult, and has been solved only in highly particular cases: when the archimedean components are groups (Clifford [1941]), cancellative (Hewitt & Zuckerman [1956]), or cyclic (Arendt and Stuth [1970]). The first two cases are considered in the next section. See also Yamada [1965] and Tamura [1971].

Archimedean semigroups appear in every commutative semigroup but can be expected to be simpler to some extent. The extent however is not so large as to have borne many general results. Maximal cancellative subsemigroups of archimedean semigroups were studied by Putcha [1975], [1980], Tamura [1976], [1977S] and Gale & Tamura [1976]. See also Putcha [1973], [1976], Lord [1978], and the survey by Tamura [1978]. A general construction in terms of trees and integer-valued factor sets was given by Tamura [1966], [1968] (see also Grillet [1995]). But the study of archimedean semigroups has generally been limited to smaller classes.

In this chapter we construct three such classes. Other classes were constructed by Brown & LaTorre [1965] (uniquely divisible archimedean semigroups) and Tully [1966], [1974] (archimedean semigroups on which $\leqq_{\mathcal{H}}$ is a tree order); nilsemigroups will be considered in Chapter IX.

2. CLIFFORD SEMIGROUPS.

Propositions I.3.8 and II.3.7 suggest that the semilattice composition problem can be solved when the components are monoids or are cancellative.

1. A **semilattice of groups**, also called a **Clifford semigroup**, is a semigroup with a semilattice decomposition $S = \bigcup_{a \in Y} G_a$ in which every semigroup G_a is a group. Commutative Clifford semigroups are also called **commutative regular** semigroups and **commutative inverse** semigroups.

Clifford [1941] showed that the multiplication on a semilattice of groups is determined by a functor. A semilattice Y can be regarded as a small category, whose objects are the elements of Y and whose morphisms are the ordered pairs $(a,b) : a \longrightarrow b$ such that $a \geqq b$ in Y; composition is given by $(b,c) \circ (a,b) = (a,c)$ and the identity morphism on a is (a,a). A **group valued functor** $\mathbb{G} = (G, \gamma)$ on a semilattice Y is a functor on this category: it assigns a group G_a to every element a of Y, and assigns a homomorphism $\gamma^a_b : G_a \longrightarrow G_b$ to every pair

(a,b) of elements of Y such that $a \geqq b$, so that γ_a^a is the identity on G_a and $\gamma_c^b \circ \gamma_b^a = \gamma_c^a$ whenever $a \geqq b \geqq c$. The following result is due to Clifford [1941]. Petrich [1973] proved a similar result for semilattices of monoids in general.

Theorem 2.1 (Clifford). *Let $\mathbb{G} = (G,\gamma)$ be a group valued functor on a semilattice Y. Let S be the disjoint union $S = \bigcup_{a \in Y} (G_a \times \{a\})$ with multiplication*

$$(g,a)(h,b) = (\gamma_{ab}^a(g)\, \gamma_{ab}^b(h),\, ab)$$

for all $a,b \in Y$, $g \in G_a$, $h \in G_b$. Then S is a semilattice of groups, and, up to isomorphism, every semilattice of groups can be constructed in this fashion. Moreover, S is commutative if and only if G_a is abelian for every $a \in Y$.

Proof. The multiplication in the statement is associative:

$$\begin{aligned}
(g,a)\big((h,b)(k,c)\big) &= (g,a)\big(\gamma_{bc}^b(h)\, \gamma_{bc}^c(k),\, bc\big) \\
&= \big(\gamma_{abc}^a(g)\, \gamma_{abc}^{bc}(\gamma_{bc}^b(h)\, \gamma_{bc}^c(k)),\, abc\big) \\
&= \big(\gamma_{abc}^a(g)\, \gamma_{abc}^b(h)\, \gamma_{abc}^c(k),\, abc\big) \\
&= \big((g,a)(h,b)\big)(k,c).
\end{aligned}$$

Also, the subsemigroup $S_a = \{(g,a) \in S \mid g \in G_a\}$ of S is isomorphic to G_a, and $S_a S_b \subseteq S_{ab}$. Thus S is a semilattice of groups.

Conversely let S be a semilattice $S = \bigcup_{a \in Y} G_a$ of groups G_a. Let e_a be the identity element of G_a. When $a > b$ in Y, $G_a \cup G_b$ is an ideal extension of G_b and is by Proposition 1.3 determined by a (partial) homomorphism $\gamma_b^a : G_a \longrightarrow G_b$, namely $\gamma_b^a(g) = g e_b = e_b g$ for all $g \in G_a$. In particular $e_a e_b = e_b e_a = \gamma_b^a(e_a) = e_b$. Also γ_a^a is the identity on G_a and $\gamma_a^a(g) = g e_a = e_a g$ for all $g \in G_a$. If $a \geqq b \geqq c$, then $\gamma_c^b(\gamma_b^a(g)) = g e_b e_c = g e_c = \gamma_c^a(g)$. Now (G,γ) is a group valued functor on Y.

When $g \in G_a$ and $h \in G_b$, then the products gh and $h e_{ab}$ in S are in G_{ab}, so that $gh = g h e_{ab} = g e_{ab} h e_{ab} = \gamma_{ab}^a(g)\, \gamma_{ab}^b(h)$ as calculated in G_{ab}. Therefore S is isomorphic to the semigroup $\bigcup_{a \in Y} (G_a \times \{a\})$ in the statement. $\square$

A semilattice is directed downward; a group valued functor on a semilattice can be regarded as a direct system. Tamura & Hamilton [1971] showed that the direct limit $\varinjlim G_a$ is the universal group of the corresponding Clifford semigroup.

Congruences on Clifford semigroups were considered by Jarek [1964] and others. Chapter VIII lists a number of additional results: due to Theorem 2.1, Clifford semigroups have been studied extensively and have provided a testing ground for many ideas and problems.

2. A **semilattice of cancellative semigroups** is a semigroup with a semilattice decomposition $S = \bigcup_{a \in Y} C_a$ in which every semigroup C_a is cancellative. These semigroups can be characterized as follows:

Proposition 2.2. *For a c.s. S the following are equivalent:*

(1) *S is a semilattice of cancellative semigroups;*

(2) *every archimedean component of S is cancellative;*

(3) *in S, $x^2 = xy = y^2$ implies $x = y$;*

(4) *for every $n > 0$, $xy^n = y^{n+1}$ and $x^n y = x^{n+1}$ implies $x = y$.*

Proof. (2) $\Longrightarrow$ (1) is clear.

(1) $\Longrightarrow$ (3). Let $S = \bigcup_{a \in Y} C_a$ be a semilattice of cancellative semigroups C_a. Let $x^2 = xy = y^2$ in S. If $x \in C_a$ and $y \in C_b$, then $x^2 \in C_a$, $y^2 \in C_b$, $a = b$, and $x = y$ since C_a is cancellative.

(3) $\Longrightarrow$ (4). This is clear if $n = 1$. For $n > 1$ we proceed by induction on n. Let $ab^n = b^{n+1}$ and $ba^n = a^{n+1}$, where $n > 1$. Then (with $b^0 = 1 \in S^1$ if $n = 2$)

$$\begin{aligned} (ab^{n-1})^2 &= ab^{n-2}\, ab^n = ab^{n-2}\, b^{n+1} = (ab^{n-1})\, b^n, \\ (ab^{n-1})\, b^n &= ab^n\, b^{n-1} = b^{n+1}\, b^{n-1} = (b^n)^2, \end{aligned}$$

and $ab^{n-1} = b^n$ by (3). Exchanging a and b yields $a^{n-1}b = a^n$. Then $a = b$ by the induction hypothesis.

(4) $\Longrightarrow$ (2). Let A be an archimedean component of S. Assume that $a, b, c \in A$ satisfy $ac = bc$. Since A is archimedean we have $a^m = cx$ and $b^n = cy$ for some $m, n > 0$ and $x, y \in A$. We may assume that $m = n$: if, say, $m < n$, then $a^n = cx\,a^{n-m}$, with $xa^{n-m} \in A$. Then

$$ab^n = acy = bcy = b^{n+1} \text{ and } ba^n = bcx = acx = a^{n+1}$$

and $a = b$ by (4). $\square$

Proposition 2.2 implies a result of Kimura & Tsai [1972] that power cancellative archimedean semigroups are cancellative, which was first showed by Brown & LaTorre [1966] for uniquely divisible semigroups.

A c.s. S is **separative** when it satisfies the equivalent conditions in Proposition 2.2. (We will show in the next chapter that the characters of S then separate the elements of S.) Separative semigroups were introduced by Hewitt & Zuckerman [1956] and Schwarz [1956]. Gluskin [1971] determined their dense extensions (such that every homomorphism which is injective on the semigroup is injective on the extension), which are also separative. Rompke [1975] studied the semigroups of fractions of reductive separative c.s.

3. The following result is due to Hewitt & Zuckerman [1956].

Proposition 2.3. *Every commutative semilattice $S = \bigcup_{a \in Y} C_a$ of cancellative semigroups C_a can be embedded into a semilattice $T = \bigcup_{a \in Y} G(C_a)$ of abelian groups. If every C_a is archimedean, then every homomorphism of S into a semilattice of groups extends uniquely to T.*

Proof. We may assume that the groups $G(C_a)$ are pairwise disjoint. If $x/y = x'/y'$ in $G(C_a)$ and $z/t = z'/t'$ in $G(C_b)$ (with $a, b \in Y$ and $x, y, x', y' \in C_a$, $z, t, z', t' \in C_b$), then $xz, yt, x'z', y't' \in C_{ab}$, $xy' = x'y$, $zt' = z't$, $xzy't' = x'z'yt$ in C_{ab}, and $xz/yt = x'z'/y't'$ in $G(C_{ab})$. Therefore a multiplication on $T = \bigcup_{a \in Y} G(C_a)$ is well defined by $(x/y)(z/t) = xz/yt$. This makes T into a c.s., in which $G(C_a)\, G(C_b) \subseteq G(C_{ab})$; thus T is a semilattice of abelian groups. S is a subsemigroup of T if we identify $x \in C_a$ and $x^2/x \in G(C_a)$, so that $(x^2/x)(y^2/y) = (xy)^2/xy$.

Now assume that every C_a is archimedean and let $\varphi : S \longrightarrow U$ be a homomorphism of S into a semilattice of groups $U = \bigcup_{c \in Z} G_c$. Every $\varphi(C_a)$ is archimedean and is contained in a single group G_c. The restriction of φ to C_a then extends uniquely to a homomorphism $G(C_a) \longrightarrow G_c$, which sends $x/y \in G(C_a)$ to $\varphi(x)\, \varphi(y)^{-1}$, where $\varphi(y)^{-1}$ is the inverse in G_c (Proposition II.2.1). This provides a mapping $\psi : x/y \longmapsto \varphi(x)\, \varphi(y)^{-1}$ of T into U.

We show that ψ is a homomorphism; then ψ is the only homomorphism which extends φ. Let $x, y \in C_a$ and $z, t \in C_b$. Let $\varphi(C_a) \subseteq G_c$, $\varphi(C_b) \subseteq G_d$ and e_c denote the identity element of G_c. Then $\varphi(y)$ and $\varphi(t)$ have inverses in G_c and G_d and commute with each other in G_{cd}, since S is commutative. Also $\varphi(y)$ commutes with e_d in the Clifford semigroup U. Hence

$$\begin{aligned} \varphi(y)\, \varphi(t)^{-1}\, \varphi(t)\, e_{cd} &= \varphi(y)\, \varphi(t)^{-1}\, \varphi(t) = \varphi(y)\, e_d = e_d\, \varphi(y) \\ &= \varphi(t)^{-1}\, \varphi(t)\, \varphi(y) = \varphi(t)^{-1}\, \varphi(y)\, \varphi(t) = \varphi(t)^{-1}\, \varphi(y)\, \varphi(t)\, e_{cd}, \end{aligned}$$

which in G_{cd} implies $\varphi(y)\, \varphi(t)^{-1} = \varphi(t)^{-1}\, \varphi(y)$. The same calculation with $\varphi(y)$ replaced by $\varphi(y)^{-1}$ (which commutes with $\varphi(t)$) shows that $\varphi(y)^{-1}$ commutes with $\varphi(t)^{-1}$. Hence

$$\varphi(y)\,\varphi(t)\,\varphi(y)^{-1}\,\varphi(t)^{-1} \;=\; \varphi(y)\,\varphi(y)^{-1}\,\varphi(t)\,\varphi(t)^{-1} \;=\; e_c e_d \;=\; e_{cd}$$

and $\varphi\,(yt)^{-1} = \varphi(y)^{-1}\,\varphi(t)^{-1}$ in G_{cd}. Then

$$\begin{aligned}\psi\,(xz/yt) &= \varphi\,(xz)\,\varphi(yt)^{-1} \;=\; \varphi(x)\,\varphi(z)\,\varphi(y)^{-1}\,\varphi(t)^{-1}\\ &= \varphi(x)\,\varphi(y)^{-1}\,\varphi(z)\,\varphi(t)^{-1} \;=\; \psi(x/y)\,\psi(z/t)\end{aligned}$$

and ψ is a homomorphism. □

By Clifford's Theorem the operation on T is determined by homomorphisms $\gamma^a_b : G(C_a) \longrightarrow G(C_b)$ $(a \geqq b)$, namely, $\gamma^a_b(x/y) = (x/y)(z/z) = xz/yz$, for all $x, y \in C_a$ and $z \in C_b$. We see that γ^a_b is induced by the canonical homomorphism $\tau : C_a \longrightarrow G(C_b)$ of the ideal extension $C_a \cup C_b$ of C_b in Lemma II.3.5.

4. When S is separative, Proposition 2.4 provides a **universal semilattice of groups** $T = \bigcup_{A \in Y(S)} G(A)$ of S; that is, T is a semilattice of groups, S is a subsemigroup of T, and every homomorphism of S into a semilattice of groups extends uniquely to T. In fact, every c.s. has a universal semilattice of groups.

A congruence $\mathcal{C}$ on a c.s. S is **separative** when $S/\mathcal{C}$ is separative; equivalently, when $x^2 \;\mathcal{C}\; xy \;\mathcal{C}\; y^2$ implies $x \;\mathcal{C}\; y$.

Proposition 2.4. *On every c.s. there is a smallest separative congruence* $\mathcal{S}$, *for which*

$$x \;\mathcal{S}\; y \textit{ if and only if } x^n y = x^{n+1} \textit{ and } xy^n = y^{n+1} \textit{ for some } n > 0.$$

Every homomorphism of S *into a separative semigroup factors uniquely through the projection* $S \longrightarrow S/\mathcal{S}$.

Proof. If S is separative, then $\mathcal{S}$ is the equality on S by Proposition 2.2. In general, $\mathcal{S}$ is reflexive and symmetric. Let $x \;\mathcal{S}\; y$ and $y \;\mathcal{S}\; z$. Then $x^m y = x^{m+1}$, $xy^m = y^{m+1}$, $y^n z = y^{n+1}$, $yz^n = z^{n+1}$ for some $m, n > 0$, and

$$\begin{aligned} xz^{mn+m+n} &= xz^{mn+m}\, z^n \;=\; x\,(yz^n)^m\, z^n \;=\; xy^m\, z^{mn+n}\\ &= y^{m+1}\, z^{mn+n} \;=\; (yz^n)^{m+1} \;=\; z^{mn+m+n+1}. \end{aligned}$$

Exchanging x and z yields $zx^{mn+n+m} = x^{mn+n+m+1}$. Hence $x \;\mathcal{S}\; z$. Thus $\mathcal{S}$ is transitive.

Let $x \;\mathcal{S}\; y$ and $z \in S$. Then $x^n y = x^{n+1}$, $xy^n = y^{n+1}$ for some $n > 0$ and

$$(xz)(yz)^n \;=\; xy^n\, z^{n+1} \;=\; y^{n+1}\, z^{n+1} \;=\; (yz)^{n+1}.$$

Exchanging x and y yields $(yz)(xz)^n = (xz)^{n+1}$. Hence $xz \mathcal{S} yz$. Thus $\mathcal{S}$ is a congruence.

Let $x^2 \mathcal{S} xy$ and $xy \mathcal{S} y^2$. Then $x^{2m+1}y = (x^2)^m(xy) = (x^2)^{m+1} = x^{2m+2}$ and $xy^{2n+1} = (xy)(y^2)^n = (y^2)^{n+1} = y^{2n+2}$ for some $m, n > 0$. We may assume that $m = n$: if, say, $m < n$, then multiplying the first equality by x^{2n-2m} yields $x^{2n+1}y = x^{2n+2}$. Hence $x \mathcal{S} y$. Thus $\mathcal{S}$ is separative.

Finally let $\varphi : S \longrightarrow T$ be a homomorphism, where T is separative. If $x \mathcal{S} y$ in S, then $x^n y = x^{n+1}$ and $xy^n = y^{n+1}$ for some $n > 0$, $\varphi(x)^n \varphi(y) = \varphi(x)^{n+1}$ and $\varphi(x)\varphi(y)^n = \varphi(y)^{n+1}$, and $\varphi(x) = \varphi(y)$. Hence $\mathcal{S} \subseteq \ker \varphi$ and φ factors uniquely through the projection $S \longrightarrow S/\mathcal{S}$ by Proposition I.2.4. If in particular $\mathcal{C}$ is a separative congruence on S and $\varphi : S \longrightarrow S/\mathcal{C}$ is the projection, then $\mathcal{S} \subseteq \ker \varphi = \mathcal{C}$. □

In Proposition 2.4, $S/\mathcal{S}$ is the **universal separative semigroup** of S. It should be denoted by $S(S)$, but this would not improve the notation.

Corollary 2.5. *Every c.s. has a universal semilattice of groups, which is also the universal semilattice of groups of its universal separative semigroup.*

Proof. Semilattice of groups are separative. Hence every homomorphism φ of a c.s. S into a semilattice of groups factors uniquely through the projection $S \longrightarrow S/\mathcal{S}$, and thence extends uniquely to the universal semilattice of groups T of $S/\mathcal{S}$. Hence φ factors uniquely through $S \longrightarrow S/\mathcal{S} \longrightarrow T$. □

The universal separative and Clifford semigroups of a c.s. have been known implicitly since Hewitt & Zuckerman [1956] and Schwarz [1956]. C.s. that map onto their universal Clifford semigroup were investigated by Head [1971H] and McAlister & O'Carroll [1971].

3. COMPLETE ARCHIMEDEAN SEMIGROUPS.

By Proposition 1.3, an archimedean semigroup has at most one idempotent. We call an archimedean semigroup **complete** when it has an idempotent. For instance, finite archimedean semigroups are complete; an interesting smaller class was studied by Trueman [1979], [1980].

Complete archimedean semigroups are comparatively simple in structure.

1. A **nilsemigroup** is a semigroup S with a zero element in which every element x is **nilpotent**: $x^n = 0$ for some $n > 0$. Commutative nilsemigroups are archimedean and contain an idempotent; so do abelian groups. Conversely:

Proposition 3.1. *A c.s. S which contains an idempotent e (for instance, a finite c.s.) is archimedean if and only if it is an ideal extension of a group by a nilsemigroup; then S has a kernel $K = H_e = eS$ and S/K is a nilsemigroup.*

Proof. Let S be archimedean. Let $a \in H_e$ and $x \in S$. Since S is archimedean we have $e = e^n = axy$ for some $n > 0$ and $y \in S$. Also $ax = eax$. Hence $ax \in H_e$. Thus the maximal subgroup H_e is an ideal of S; then $H_e = eS$. For every $x \in S$ we now have $x^n = ey \in H_e$; hence the Rees quotient S/H_e is a nilsemigroup. Every nonempty ideal of S contains all powers of some element, contains an element of H_e, and contains H_e; thus H_e is the kernel of S.

Conversely let the c.s. S be an ideal extension of a group G by a nilsemigroup N. If $a \in S \backslash G$, then $a^n = 0$ for some $n > 0$ in $S/G = N$, and $a^n \in G$. If x is the inverse of a^n in G, then the equalities $a^n = ea^n$, $e = a^n x$ show that $a \leqq_{\mathcal{N}} e$ and $e \leqq_{\mathcal{N}} a$. Hence S is archimedean. □

If in the above G is trivial, then $S \cong N$ is a nilsemigroup; if $N = 0$ is trivial, then $S = G$ is a group.

As a first application of Proposition 3.1, we prove:

Corollary 3.2. *For a c.s. S the following conditions are equivalent:*

(1) *every archimedean component of S contains an idempotent;*

(2) *every element of S has a power in some subgroup of S;*

then the homomorphism $E(S) \longrightarrow Y(S)$ is an isomorphism.

Proof. If (1) holds, then the archimedean component A of $x \in S$ contains an idempotent f, and $x^m = ft$, $f^n = f = x^m u$ for some $m, n > 0$ and $t, u \in S$; hence x^m belongs to the subgroup H_f of G, which proves (2). Conversely, (2) implies that every archimedean component A contains an element of a subgroup of S; since a subgroup is contained in a single $\mathcal{H}$-class, which is contained in a single $\mathcal{N}$-class, A contains the entire subgroup and contains its identity element. Condition (1) implies that the injective homomorphism $E(S) \longrightarrow Y(S)$ in Proposition 1.4 is surjective. □

Semigroups with property (2) have been called π**-regular**, **pseudo-invertible**, and **epigroups**; groups, nilsemigroups, and finite semigroups have this property.

2. The ideal extensions in Proposition 3.1 are determined by partial homomorphisms, as in Proposition I.3.8. Our next result constructs the partial homomorphisms themselves from homomorphisms of abelian groups; this constructs all archimedean semigroups with idempotent from nilsemigroups and groups.

Proposition 3.3. *Let P be a commutative partial semigroup. There exist an* *abelian group $G(P)$ and a partial homomorphism $\gamma : P \longrightarrow G(P)$ such that every partial homomorphism φ of P into an abelian group G factors uniquely through γ ($\varphi = \xi \circ \gamma$ for some unique group homomorphism $\xi : G(P) \longrightarrow G$);*

$$\begin{array}{ccc} P & \xrightarrow{\gamma} & G(P) \\ & \searrow^{\varphi} & \downarrow \xi \\ & & G \end{array}$$

in the additive notation, $G(P)$ is the abelian group generated by the elements of P, subject to one relation $a + b = ab$ for every $a, b \in P$ such that ab is defined in P. Then the group $\mathrm{PHom}(P, G)$ of all partial homomorphisms of P into G is isomorphic to the group $\mathrm{Hom}(G(P), G)$ of all group homomorphisms of $G(P)$ into G.

Because of its universal property we call $G(P)$ the **universal** (abelian) **group** of P. If P is an actual semigroup, then $G(P)$ is the universal group of P defined in Section II.2, and Proposition 3.3 provides a presentation of $G(P)$.

Proof. Let $G(P)$ be the abelian group generated by P, subject to $a + b = ab$ whenever ab is defined in P. Let $\gamma : P \longrightarrow G(P)$ be the canonical mapping. Every defining relation $a + b = ab$ holds in $G(P)$ via γ; that is, $\gamma(a) + \gamma(b) = \gamma(ab)$ whenever ab is defined in P. Thus γ is a partial homomorphism.

Let G be any abelian group, written additively, and $\varphi : P \longrightarrow G$ be a partial homomorphism. Then $\varphi(ab) = \varphi(a) + \varphi(b)$ whenever ab is defined in P and every defining relation holds in G via φ. Therefore φ factors uniquely through γ ($\varphi = \xi \circ \gamma$ for some unique group homomorphism $\xi : G(P) \longrightarrow G$.)

Since G is abelian, $\mathrm{PHom}(P, G)$ and $\mathrm{Hom}(G(P), G)$ are abelian groups under pointwise addition. If $\xi : G(P) \longrightarrow G$ is a group homomorphism, then $\xi \circ \gamma : P \longrightarrow G$ is a partial homomorphism. This yields a homomorphism $\xi \longmapsto \xi \circ \gamma$ of $\mathrm{Hom}(G(P), G)$ into $\mathrm{PHom}(P, G)$, which is bijective by the above. □

3. When S is a semigroup with zero, presentations of S readily yield presentations of $G(S \backslash 0)$:

Proposition 3.4. *If S is generated, as a commutative semigroup or monoid with zero, by a set X subject to relations $u_i = v_i$ ($i \in I$) and $w_j = 0$ ($j \in J$), where $u_i = v_i \neq 0$ and $x \neq 0$ in S for all $x \in X$ and $i \in I$, then $G(S \backslash 0)$ is isomorphic to the abelian group generated by X subject to all relations $u_i = v_i$ ($i \in I$).*

Proof. Let $F = F_X$ be the free c.s. (or c.m.) on the set X, written multiplicatively. Up to isomorphism, $S = (F \cup \{0\})/\mathcal{C}$, where $\mathcal{C}$ is the congruence on $F \cup \{0\}$ generated by all pairs (u_i, v_i) and $(w_j, 0)$. The $\mathcal{C}$-class of 0 is an ideal of $F \cup \{0\}$ and is the union of $\{0\}$ and an ideal Z of F. By the hypothesis, Z does not contain u_i, v_i, or any $x \in X$; in particular the canonical mapping $\iota : X \longrightarrow S$ sends X into $S \backslash 0$.

There is a smaller congruence $\mathcal{B}$ on F which is generated by the pairs (u_i, v_i); the quotient semigroup $T = F/\mathcal{B}$ is the c.s. (or c.m.) generated by X subject only to the relations $u_i = v_i$ $(i \in I)$. When $a, b \in F \backslash Z$ we show that $a \,\mathcal{C}\, b$ if and only if $a \,\mathcal{B}\, b$; in particular $F \backslash Z$ is a union of $\mathcal{B}$-classes. If indeed $a \,\mathcal{C}\, b$, then by Proposition I.2.9 there is a sequence $x_1, \dots, x_n \in F \cup \{0\}$ such that $a = x_1$, $x_n = b$, and every pair (x_k, x_{k+1}) has the form (tu_i, tv_i), (tv_i, tu_i), $(tw_j, 0)$, or $(0, tw_j)$, where $t \in F^1 \cup \{0\}$. Since $x_1, \dots, x_n$ are in the equivalence class of a and b it i not possible that $x_k = 0$ or $x_k \in Z$. Therefore every pair (x_k, x_{k+1}) has the form (tu_i, tv_i) or (tv_i, tu_i), and $a \,\mathcal{B}\, b$.

Now $S \backslash 0$, which is the set of all $\mathcal{C}$-classes contained in $F \backslash Z$, is a partial subsemigroup of T, and the canonical mappings $\iota : X \longrightarrow S \backslash 0$, $\kappa : X \longrightarrow T$ agree on X.

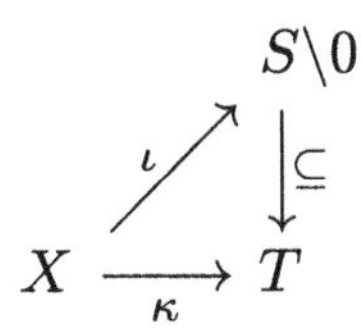

Let G be the abelian group (written multiplicatively) generated by X subject to all relations $u_i = v_i$ $(i \in I)$, and $\eta : X \longrightarrow G$ be the canonical mapping. Since the defining relations of T hold in G via η, there is a homomorphism $\tau : T \longrightarrow G$ such that the following diagram commutes:

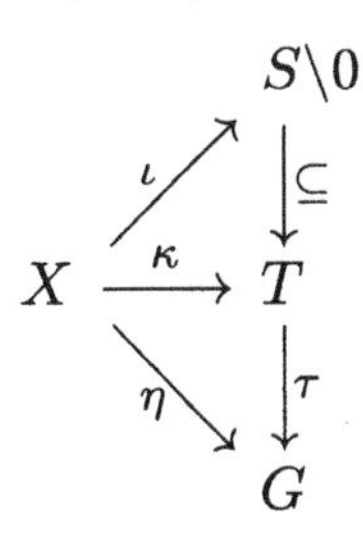

The restriction of τ to $S \backslash 0$ is a partial homomorphism γ of $S \backslash 0$ into G.

When $\varphi : S \backslash 0 \longrightarrow A$ is a partial homomorphism of $S \backslash 0$ into an abelian group A, then every relation $u_i = v_i$ holds in A via $\varphi \circ \iota$; hence there is a unique homomorphism $\xi : G \longrightarrow A$ such that $\varphi \circ \iota = \xi \circ \eta$.

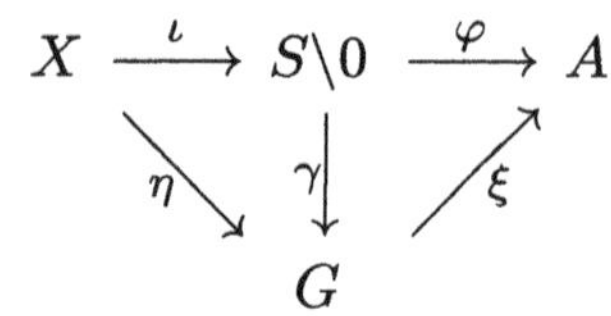

Since $\iota(X)$ generates S we have $\xi \circ \gamma = \varphi$. Thus G (together with γ) is isomorphic to the universal group of $S\backslash 0$. □

Example 3.5. Let N be the commutative nilsemigroup

$$N = \langle a, b \mid a^8 = a^5 b^4 = b^6 = 0,\ a^2 b^4 = a^5 b^2 \rangle.$$

By Proposition 3.4, $G(N\backslash 0)$ is the abelian group

$$\langle a, b \mid a^2 b^4 = a^5 b^2 \rangle = \langle a, b \mid b^2 = a^3 \rangle$$

or (in the additive notation) $(\mathbb{Z} \oplus \mathbb{Z})/K$, where K is the subgroup generated by $(3, -2)$. Now $(3, -2)$, $(-1, 1)$ is a basis of $\mathbb{Z} \oplus \mathbb{Z}$ (as shown by its determinant); hence $G(N\backslash 0) \cong \mathbb{Z}$.

The partial homomorphism $\gamma : N\backslash 0 \longrightarrow \mathbb{Z}$ is found as follows. The images of a and b in $\mathbb{Z} \oplus \mathbb{Z}$ are $(1, 0) = (3, -2) + 2(-1, 1)$ and $(0, 1) = (3, -2) + 3(-1, 1)$; in $\mathbb{Z} \cong (\mathbb{Z} \oplus \mathbb{Z})/K$, the images of a and b are 2 and 3. Thus $\gamma(a) = 2$ and $\gamma(b) = 3$. For any abelian group G, $\mathrm{PHom}(N\backslash 0, G) \cong \mathrm{Hom}(\mathbb{Z}, G) \cong G$; a partial homomorphism $\varphi : N\backslash 0 \longrightarrow G$ sends a and b to g^2 and g^3 for some unique $g \in G$. □

Example 3.6. Let N be the commutative nilsemigroup

$$N = \langle a, b \mid a^3 = a^2 b = a b^2 = b^4 = 0,\ a^2 = ab = b^3 \rangle.$$

This example was communicated to the author by Prof. Volkov. By Proposition 3.4, $G(N\backslash 0)$ is the abelian group

$$\langle a, b \mid a^2 = ab = b^3 \rangle.$$

We see that $a = b$ holds in G, and then $a^2 = b^3$ yields $a = b = 1$. Thus $G(N\backslash 0)$ is trivial (even though $N\backslash 0$ itself is not trivial). For any abelian group G, the only partial homomorphism $\varphi : N\backslash 0 \longrightarrow G$ is the trivial partial homomorphism $\varphi(a) = \varphi(b) = 1$. □

4. N-SEMIGROUPS.

An **N-semigroup** is an archimedean semigroup without idempotent which is cancellative. N-semigroups have also been called $\mathcal{N}$-semigroups, η-semigroups, and $\mathfrak{N}$-semigroups. This section contains a structure theorem, due to Tamura [1957], which constructs N-semigroups in terms of abelian groups and factor sets and started numerous papers on N-semigroups. A similar construction applies to all reduced cancellative c.s. (Hamilton, Nordahl, & Tamura [1975]).

1. First we prove two general properties. By Proposition II.5.5, there is on any c.s. S a smallest power cancellative congruence $\mathcal{T}$, given by

$$x \mathcal{T} y \text{ if and only if } x^n = y^n \text{ for some } n > 0.$$

The following result is due to Kimura & Tsai [1972].

Proposition 4.1. *When S is archimedean, then $S/\mathcal{T}$ is archimedean and cancellative; moreover S has no idempotent if and only if $S/\mathcal{T}$ has no idempotent.*

Proof. When S is archimedean, then $T = S/\mathcal{T}$ is archimedean, power cancellative, separative (since $x^2 = y^2$ implies $x = y$), and cancellative by Proposition 2.2. By Proposition II.5.5, S has no idempotent if and only if T has no idempotent. □

N-semigroups thus provide a basic framework for all archimedean semigroups without idempotent.

Corollary 4.2. *When S is archimedean without idempotent, then $x \neq xy$ for all $x, y \in S$.*

Proof. If $x = xy$, then $xy = xy^2$, $xy \mathcal{T} xy^2$, $y \mathcal{T} y^2$ since $S/\mathcal{T}$ is cancellative, $S/\mathcal{T}$ contains an idempotent, and S contains an idempotent by Proposition 4.1. □

2. Now let S be an N-semigroup. We regard S as a subsemigroup of its group of fractions $G(S) = S^{-1}S$, whose elements can then be written in the form xy^{-1} with $x, y \in S$.

Lemma 4.3. *Let S be an N-semigroup and $a \in S$. Every element of $G(S)$ can be written in the form $a^n p$ for some unique $n \in \mathbb{Z}$ and $p \in S \backslash aS$. Every element of $G(S)$ can be written in the form $a^n p$ for some unique $n \in \mathbb{Z}$ and $p \in S \backslash aS$.*

Proof. Let $s \in S$. Since S is archimedean, $a^k = st$ for some $k > 0$ and $t \in S$. Then $s \notin a^k S$, otherwise $a^k = a^k tu$ for some $u \in S$, contradicting Corollary 4.2. Thus s is not "infinitely divisible" by a: there is a greatest $n \geqq 0$ such that $s = a^n p$ for some $p \in S$, and then $p \notin aS$ by the choice of n.

Now let $g = xy^{-1} \in G(S)$, where $x, y \in S$. Since S is archimedean we have $a^n = yt$ for some $n > 0$ and $t \in S$ and $xy^{-1} = xta^{-n}$ in $G(S)$. Therefore $g = a^m p$ for some $m \in \mathbb{Z}$ and $p \in S \backslash aS$.

Assume that $a^m p = a^n q$ holds in $G(S)$, where $m, n \in \mathbb{Z}$ and $p, q \in S \backslash aS$. If $m < n$, then $p = a^{n-m} q \in aS$, a contradiction. If similarly $m > n$, then $q = a^{m-n} p \in aS$, a contradiction. Therefore $m = n$, and then $p = q$. □

Let $a \in S$ and A be the cyclic subgroup of $G(S)$ generated by a. Corollary 4.2 implies that a has infinite order in $G(S)$, so that $A \cong \mathbb{Z}$. The abelian group $G = G(S)/A$ is the **structure group** of S relative to a. By Schreier's Theorem on group extensions, $G(S)$ can be reconstructed from $G = G(S)/A$ by means of an integer valued factor set on G. Lemma 4.3 implies that this construction also yields S. (The role of group extensions in his 1957 construction was known to Tamura [1970F].) The details are as follows.

By Lemma 4.3 the projection $\pi : G(S) \longrightarrow G(S)/A = G$ induces a bijection $S \backslash aS \longrightarrow G$. Let $u \longmapsto p_u$ be the inverse bijection. For instance $p_1 = a$, since $a \notin aS$ by Corollary 4.2 and π sends $a \in A$ to the identity element of G. By Lemma 4.3, every element of S can be written in the form $a^n p_u$ for some unique $n \geqq 0$ and $u \in G$; then $\pi(a^n p_u) = \pi(p_u) = u$.

Let $u, v \in G$. Since $\pi(p_u) = u$, $\pi(p_v) = v$ we have $\pi(p_u p_v) = uv$ and

$$p_u\, p_v \;=\; a^{\sigma_{u,v}}\, p_{uv}$$

for some unique nonnegative integer $\sigma_{u,v}$. (Tamura denotes $\sigma_{u,v}$ by $I(\alpha, \beta)$.) Then

$$(a^m p_u)(a^n p_v) \;=\; a^{m+n} p_u p_v \;=\; a^{m+n+\sigma_{u,v}} p_{uv}\,.$$

Thus S is isomorphic to $\mathbb{N} \times G$ when multiplication on the latter is

$$(m, u)(n, v) \;=\; (m + n + \sigma_{u,v}\,,\; uv).$$

(More generally, Tamura [1970F] showed that every factor set on G with values in $\mathbb{Z}$ is equivalent to a nonnegative factor set; Nordahl [1977] showed that all additive subgroups of $\mathbb{R}$ have this property.)

Lemma 4.4. *In an N-semigroup the following holds for all* $u, v, w \in G$*:*

(1) $\sigma_{1,u} = 1$*;*

(2) $\sigma_{v,u} = \sigma_{u,v}$*;*

(3) $\sigma_{u,v} + \sigma_{uv,w} = \sigma_{u,vw} + \sigma_{v,w}$*;*

(4) *for every* $u \in G$ *there exists* $k > 0$ *such that* $\sigma_{u,u^k} > 0$.

Proof. (1) holds since $p_1 p_u = a p_u$; (2) holds since $p_v p_u = p_u p_v$; (3) holds since

$$\begin{aligned} (p_u p_v) p_w &= a^{\sigma_{u,v}} p_{uv} p_w &= a^{\sigma_{u,v}+\sigma_{uv,w}} p_{(uv)w}, \\ p_u (p_v p_w) &= a^{\sigma_{v,w}} p_u p_{vw} &= a^{\sigma_{u,vw}+\sigma_{v,w}} p_{u(vw)}. \end{aligned}$$

In fact, properties (2) and (3) hold in every abelian group extension.

Also $p_u^2 = a^{\sigma_{u,u}} p_{u^2}$, $p_u^3 = p_u p_{u^2} = a^{\sigma_{u,u}+\sigma_{u,u^2}} p_{u^3}$, and, by induction,

$$p_u^n = a^{s_n} p_{u^n}, \text{ where } s_n = \textstyle\sum_{0<k<n} \sigma_{u,u^k}.$$

Since S is archimedean we have $p_u^n \in aS$ for some $n > 0$. Then $s_n > 0$ and $\sigma_{u,u^k} > 0$ for some $0 < k < n$. □

Theorem 4.5 (Tamura). *Let G be an abelian group and $\sigma : G \times G \longrightarrow \mathbb{N}$ satisfy the conditions in Lemma 4.4. Let $S = \mathbb{N} \times G$ with multiplication given by*

$$(m,u)(n,v) = (m+n+\sigma_{u,v},\ uv).$$

Then S is an N-semigroup and, up to isomorphism, every N-semigroup can be constructed in this fashion.

Proof. S is a semigroup by (3) since

$$\begin{aligned} \big((m,u)(n,v)\big)\,(p,w) &= (m+n+\sigma_{u,v},\ uv)\,(p,w) \\ &= (m+n+p+\sigma_{u,v}+\sigma_{uv,w},\ uvw), \\ (m,u)\,\big((n,v)(p,w)\big) &= (m,u)\,(n+p+\sigma_{v,w},\ vw) \\ &= (m+n+p+\sigma_{u,vw}+\sigma_{v,w},\ uvw); \end{aligned}$$

S is commutative by (2); and S is cancellative: if $(m,u)(n,v) = (m,u)(p,w)$, then $(m+n+\sigma_{u,v},\ uv) = (m+p+\sigma_{u,w},\ uw)$, $uv = uw$, and $v = w$; then $m+n+\sigma_{u,v} = m+p+\sigma_{u,w}$ yields $n = p$. If $(n,u) \in S$ is idempotent, then $(n+n+\sigma_{u,u},\ u^2) = (n,u)^2 = (n,u)$, $u^2 = u$, $u = 1$, and $n+n+\sigma_{u,u} = n$ implies $n = 0$ and $\sigma_{1,1} = 0$, contradicting (1); therefore S contains no idempotent. We show that S is archimedean.

Let $a = (0,1) \in S$. We have $a\,(m,u) = (m+1,\ u)$ by (1) and, by induction, $a^n\,(m,u) = (m+n,\ u)$. In particular $a^{n+1} = (n,1)$ and $(m,u)(1,u^{-1}) = (n,1) = a^{n+1}$ for some $n > 0$, so that $a \leqq_{\mathcal{N}} (m,u)$. On the other hand, $(n,u) = a\,(n-1,\ u) \leqq_{\mathcal{N}} a$ whenever $n > 0$. If $n = 0$, then $p = (0,u)$ satisfies

$p^2 = (\sigma_{u,u}, u^2)$, $p^3 = (\sigma_{u,u} + \sigma_{u,u^2}, u^3)$, and, by induction,

$$p^n = (s_n, u^n), \text{ where } s_n = \textstyle\sum_{0<k<n} \sigma_{u,u^k};$$

by (4), $s_n > 0$ for some $n > 0$, and then $p^n \leqq_{\mathcal{N}} a$ and $p \leqq_{\mathcal{N}} a$. Thus $(m,u) \,\mathcal{N}\, a$ for all $(m,u) \in S$ and S is archimedean. Hence S is an N-semigroup. Conversely we saw that, up to isomorphism, every N-semigroup can be constructed as in the statement. □

Completing Theorem 4.5, Biggs, Sasaki, & Tamura [1965] determined all factor sets that satisfy the conditions in Lemma 4.3 when G is finitely generated. Sasaki [1965], [1966], Higgins [1966], [1969I], and Lord [1979] investigated when two N-semigroups are isomorphic. Tamura [1974B], [1975] also proved a second structure theorem for N-semigroups, in terms of one abelian group G and a positive mapping from G to $\mathbb{R}$. He also obtained descriptions as pullbacks (Tamura [1975], [1977T]). More precise constructions exist when the structure group is $\mathbb{Z}$ (Tamura [1977N]) or $\mathbb{Z}^n$ (Gale & Tamura [1980]).

N-semigroups have been studied extensively. Congruences on N-semigroups (especially, N-semigroup congruences) were investigated by Dickinson [1972], Tamura [1973C], and Hamilton [1978N]. Hall [1972] and Kobayashi [1982] studied ideal extensions of N-semigroups. Kobayashi [1974] also studied their homomorphisms into $\mathbb{R}$. Homomorphisms of N-semigroups were studied by Tamura [1974B], [1977T]. Simons [1963] proved a Mackey-like homomorphism theorem.

Some smaller classes have also been studied. Power joined N-semigroups are considered in Section 6 and finitely generated N-semigroups in Section VI.6. Tamura [1973C] called an N-semigroup **irreducible** when it has no proper nontrivial N-semigroup congruence; equivalently, when it is isomorphic to a subsemigroup of $\mathbb{R}^+$; a construction is given in Tamura [1974I]. Tamura [1973S] also studied N-semigroups that cannot be embedded as proper ideals in other N-semigroups. Arendt [1975] studied N-semigroups in which the intersection of all maximal congruences is the equality.

N-semigroups have been generalized by Dickinson [1976] and Sasaki [1978], [1982], [1983], [1984].

5. SUBCOMPLETE ARCHIMEDEAN SEMIGROUPS.

We call an archimedean semigroup S **subcomplete** when it is isomorphic to a nonempty subsemigroup of a complete archimedean semigroup. These semigroups are somewhat simpler in structure than archimedean semigroups in general. The results in this section are due to Grillet [1975C].

1. First we show:

Proposition 5.1. *For a nonempty archimedean semigroup S the following conditions are equivalent:*

(1) *S is subcomplete (S can be embedded into a complete archimedean semigroup);*

(2) *S contains a nonempty cancellative ideal;*

(3) *S is an ideal extension of a nonempty cancellative semigroup by a nilsemigroup;*

(4) *S can be embedded into the direct product of an abelian group and a nilsemigroup.*

Proof. (1) $\Longrightarrow$ (2). Let S be a subsemigroup of an archimedean semigroup T that contains an idempotent e. By Proposition 3.1, eT is a group. If $a \in S$, then $a^n = et$ for some $n > 0$ and $t \in T$, and $S \cap eT \neq \varnothing$ is cancellative semigroup and an ideal of S.

(2) $\Longrightarrow$ (3). Let C be a nonempty cancellative ideal of S. Let $c \in C$. If $a \in S$, then $a^n = ct \in C$ for some $n > 0$ and $t \in S$; hence S/C is a nilsemigroup.

(3) $\Longrightarrow$ (4). When S is an ideal extension of a nonempty cancellative semigroup C, Lemma II.3.5 provides a homomorphism $\tau : S \longrightarrow G(C)$ which is the identity on C. Let $\pi : S \longrightarrow S/C = N$ be the projection. The homomorphism $s \longmapsto (\tau(s), \pi(s))$ of S into $G(C) \times N$ is injective, since τ is injective on C and π is injective on $S \backslash C$.

(4) $\Longrightarrow$ (1). The direct product of an abelian group G and a nilsemigroup N is archimedean, since both G and N are archimedean, and contains an idempotent $(1,0)$. $\square$

2. Not all archimedean semigroups are subcomplete.

Example 5.2 Let $T = \mathbb{N}^+ \times \mathbb{N}$ with the associative multiplication

$$(m,p)\,(n,q) \;=\; (m+n,\, p \vee q)$$

where $p \vee q = \max\,(p,q)$ is the max operation. In particular, $(m,p)^k = (km,p)$; thus T consists of all powers of all $t_p = (1,p) \in T$, with $t_p^m\, t_q^n = t_{p \vee q}^{m+n}$.

Define $(m,p) \equiv (n,q)$ if and only if $(m,p) = (n,q)$ or $m = n > p \vee q$; $\equiv$ is an equivalence relation on T since $m = n > p \vee q$, $n = k > q \vee r$ implies

$m = k > \max\,(p \vee q,\, q \vee r) \geqq p \vee r$. In fact, $\equiv$ is a congruence on T: if $m = n > p \vee q$, then, for any $(k,r) \in T$, either $r \geqq p \vee q$, in which case $p \vee r = r = q \vee r$ and $(m,p)(k,r) = (n,q)(k,r)$, or $r < p \vee q$, in which case $m + k = n + k > (p \vee r) \vee (q \vee r)$.

Let $S = T/{\equiv}$; S consists of all positive powers s_p^n, where $p \geqq 0$ and s_p is the $\equiv$-class of $(1,p)$; then $s_p^m\, s_q^n = s_{p\vee q}^{m+n}$ and $s_p^n = s_q^n$ when $n > p,q$. Hence S is **power-joined**: any two elements x and y of S have a common power $x^m = y^n$. In particular S is archimedean.

Let I be a nonempty ideal of S. Some $s_p^k \in I$; hence $s_0^{n-1} \in I$ if n is large enough. Then $s_0^n \in I$ and $s_n^n = s_0^{n-1}\, s_n \in I$. But $s_n^n\, s_0^n = s_n^n\, s_n^n$, whereas $s_0^n \neq s_n^n$ in S. Thus I is not cancellative, and S has no nonempty cancellative ideal. □

3. On the other hand, a subcomplete archimedean semigroup has what may be called a completion, in spite of certain shortcomings, for instance that the completion depends on the choice of a cancellative ideal.

Proposition 5.3 *Let S be an archimedean semigroup with a nonempty cancellative ideal C. The restriction to $S \backslash C$ of the canonical homomorphism $\tau : S \longrightarrow G(C)$ is a partial homomorphism and determines an ideal extension T of $G(C)$ by S/C which is a complete archimedean semigroup and contains S as a subsemigroup.*

Let $\varphi : S \longrightarrow U$ be a homomorphism. If U is archimedean with an idempotent e and $\varphi(C) \subseteq H_e$, then φ extends uniquely to a homomorphism of T into U.

Proof. By definition, the multiplication $.$ on T is as follows: for all $s,t \in S \backslash C$ and $g \in G(C)$, $g.s = s.g = \tau(s)\, g$ and $s.t = \tau(s)\, \tau(t)$, in case $st \notin S \backslash C$ in S. In the latter case, $\tau(s)\, \tau(t) = \tau(st) = st$ since τ is a homomorphism and is the identity on C; if $c \in C$, then $s.c = \tau(s)\, c = (sc)\, c^{-1} c = sc$. Hence S is a subsemigroup of T (cf. Lemma II.3.6). The latter is a complete archimedean semigroup by Proposition 3.1, since S/C is a nilsemigroup.

Let U be an archimedean semigroup with an idempotent e and $\varphi : S \longrightarrow U$ be a homomorphism such that $\varphi(C) \subseteq H_e$. The restriction of φ to C extends uniquely to a homomorphism $\xi : G(C) \longrightarrow H_e$, and $\xi\,(ab^{-1}) = \varphi(a)\, \varphi(b)^{-1} \in H_e$ for all $a,b \in C$ (Proposition II.2.1). Let $\psi : T \longrightarrow U$ be defined by: $\psi(g) = \xi(g)$ for all $g \in G(C)$, $\psi(s) = \varphi(s)$ for all $s \in S \backslash C$. Then ψ extends φ. That ψ is a homomorphism is tedious but straightforward: for all $g,h \in G(C)$ and $s,t \in S \backslash C$,

$\psi(g.h) = \psi(g)\,\psi(h)$ and $\psi(s.t) = \psi(s)\,\psi(t)$ are clear when $st \in S\backslash C$;

if $s.t \in C$, then $s.t = st$ and $\psi(s)\,\psi(t) = \varphi(st) = \varphi(s.t) = \psi(s.t)$;

if $g = ab^{-1}$, with $a, b \in C$, then $s.g = \tau(s)\,g = (sa)\,a^{-1}g = sab^{-1}$ and $\psi(s)\,\psi(g) = \varphi(s)\,\xi(g) = \varphi(s)\,\varphi(a)\,\varphi(b)^{-1} = \xi(sab^{-1}) = \psi(s.g)$.

Any homomorphism $\chi : T \longrightarrow U$ which extends φ must send $G(C)$ into the only subgroup H_e of U; therefore the restriction of χ to $G(C)$ must be ξ, whereas the restriction of χ to $T\backslash G(C) = S\backslash C$ is that of φ; hence $\chi = \psi$. □

4. One could wish for a better completion than Proposition 5.3 provides. Alas, examples show that this may be impossible, that is, a universal complete archimedean semigroup may not exist: when S is subcomplete, there may be no complete archimedean semigroup $T \supseteq S$ such that *every* homomorphism of S into a complete archimedean semigroup extends uniquely to T.

Example 5.4. Let $S \cong \mathbb{N}^+$ be an infinite cyclic semigroup $S = \{a, a^2, \dots, a^n, \dots\}$. This mischievous semigroup is archimedean and has a nonempty cancellative ideal $C_m = \{a^m, a^{m+1}, \dots\}$ for every $m > 0$.

For every $m > 0$, Proposition 5.3 provides a complete archimedean semigroup $T_m = (S\backslash C_m) \cup G(C_m) \supseteq S$. Let $G = \{b^n \mid n \in \mathbb{Z}\} \cong \mathbb{Z}$ be an infinite cyclic group. If we identify a^n and b^n for every $n \geqq m$, then C_m becomes a subsemigroup $C_m = \{b^m, b^{m+1}, \dots\}$ of G and Proposition II.2.5 yields $G(C_m) = G$. In Lemma II.3.5, $\tau(a^k) = (a^k a^m)(a^m)^{-1} = b^{k+m}\,(b^m)^{-1} = b^k$. Hence $T_m = \{a, a^2, \dots, a^{m-1}\} \cup \{b^n \mid n \in \mathbb{Z}\}$ with multiplication $a^i a^j = a^{i+j}$ if $i + j < m$, $a^i a^j = a^{i+j} = b^{i+j}$ if $i + j \geqq m$, $a^i b^j = b^{i+j}$.

Now assume that there is an embedding, or, more generally, a homomorphism $\varphi : S \longrightarrow T$ of S into a complete archimedean semigroup T, such that every homomorphism of S into a complete archimedean semigroup factors uniquely through φ. Let e be the idempotent of T. Since T is archimedean we have $\varphi(a^k) = et$ for some $k > 0$ and $t \in T$.

Let $m > k$. If the inclusion homomorphism $S \longrightarrow T_m$ factors through φ, that is, if there is a homomorphism $\psi : T \longrightarrow T_m$ such that $\psi \circ \varphi$ is the identity on S:

$$\begin{array}{ccc} S & \xrightarrow{\ \varphi\ } & T \\ & \underset{\subseteq}{\searrow} & \downarrow{\scriptstyle \psi} \\ & & T_m \end{array}$$

then $\psi(eT)$ is a subgroup of T_m, since eT is a subgroup of T, $\psi(eT) \subseteq G$, and $a^k = \psi(\varphi(a^k)) = \psi(et) \in G$. This is not possible when $m > k$. □

6. POWER JOINED SEMIGROUPS.

A commutative semigroup S is **power joined** when, for every $a, b \in S$ there exist positive integers m, n such that $a^m = b^n$; equivalently, when every two cyclic subsemigroups of S intersect. These semigroups were first considered by Abellanas [1965] (for cancellative semigroups only) and McAlister [1968], who called then **rational**, due to Corollary 6.4 below.

1. Power joined semigroups are archimedean; the abelian group $\mathbb{Z}$ is archimedean but not power joined. In fact, an abelian group is power joined if and only if it is torsion. More generally,

Proposition 6.1. *An archimedean semigroup with an idempotent is power joined if and only if its kernel is torsion.*

Proof. Let S be archimedean with an idempotent e. If S is power joined, then H_e is power joined and is torsion. If conversely H_e is torsion, then, for every $x, y \in S$, we have x^m, $y^n \in H_e$ for some $m, n > 0$, and then $(x^m)^k = e = (y^n)^l$ for some $k, l > 0$; thus S is power joined. □

For semigroups without idempotent we first look at N-semigroups. The following result is due to Levin [1968].

Proposition 6.2. *An N-semigroup is power joined if and only if its structure group (relative to any element) is torsion.*

Proof. The structure group of an N-semigroup S, relative to $a \in S$, is the quotient $G = G(S)/A$ of $G(S)$ by the cyclic subgroup $A \cong \mathbb{Z}$ generated by a.

Assume that S is power joined. For any $x, y \in S$ we have $x^m = a^p$ and $y^n = a^q$ for some $m, n, p, q > 0$; then $x^{mn} = a^{np}$, $y^{mn} = a^{mq}$, and $(xy^{-1})^{mn} = a^{np-mq}$ in $G(S)$; hence $G = G(S)/A$ is torsion.

If conversely G is torsion, then for any $x, y \in S$ we have $(xa^{-1})^m \in A$, $x^m = a^p$ for some integers $m > 0$ and p, and $y^n = a^q$ for some integers $n > 0$ and q. Then $p, q > 0$, otherwise S contains an idempotent $1 = x^m a^{-p}$ or $1 = y^n a^{-q}$, and $x^{mq} = a^{pq} = y^{np}$. Thus S is power joined. □

For the general case we use Proposition II.5.5, which provides a smallest power cancellative congruence $\mathcal{T}$, for which

$$x \,\mathcal{T}\, y \text{ if and only if } x^n = y^n \text{ for some } n > 0.$$

When S is archimedean without idempotent, then $S/\mathcal{T}$ an N-semigroup, by Proposition 4.1. Also, the universal group of $S/\mathcal{T}$ is torsion-free.

Proposition 6.3. *An archimedean semigroup S without idempotent is power joined if and only if the universal group of $S/\mathcal{T}$ has rank* 1.

Proof. $T = S/\mathcal{T}$ is an N-semigroup and $G(T)$ has a subgroup $A \cong \mathbb{Z}$ such that $G = G(T)/A$ is the structure group of T. If S is power joined, then $T = S/\mathcal{T}$ is power joined, G is torsion by Proposition 6.2, and $G(T)$ has rank 1. If conversely $G(T)$ has rank 1, then G is torsion and T is power joined, by Proposition 6.2; for all $x, y \in S$, we have $x^m \,\mathcal{T}\, y^n$ for some $m, n > 0$ and $x^{km} = y^{kn}$ for some $k > 0$; hence S is power joined. □

Corollary 6.4. *A c.s. S without idempotent is power joined if and only if $S/\mathcal{T}$ is isomorphic to a subsemigroup of $\mathbb{Q}^+$.*

Proof. If S is power joined without idempotent, then, by Proposition 6.3, $G(S/\mathcal{T})$ is a torsion free abelian group of rank 1 and can be embedded into the additive group $\mathbb{Q}$; hence $S/\mathcal{T}$ is isomorphic to an additive subsemigroup of $\mathbb{Q}$. This subsemigroup cannot contain positive numbers and negative numbers, since $S/\mathcal{T}$ contains no idempotent. Hence $S/\mathcal{T}$ is isomorphic to an additive subsemigroup of $\mathbb{Q}^+$.

If conversely $S/\mathcal{T}$ is isomorphic to a subsemigroup of $\mathbb{Q}^+$, then $S/\mathcal{T}$ is power joined without idempotent. Hence S has no idempotent, by Proposition II.5.5. Also S is power joined: if $x, y \in S$, then $x^k \,\mathcal{T}\, y^l$ for some $k, l > 0$, since $S/\mathcal{T}$ is power joined, and $x^{kn} = y^{ln}$ for some $n > 0$. □

Corollary 6.4 was obtained by McAlister [1968] and Levin [1968]. Other characterizations of power joined semigroups were given by Levin & Tamura [1970].

2. Another characterization of power joined N-semigroups was given by McAlister & O'Carroll [1970] and Sasaki & Tamura [1971].

Lemma 6.5. *Let S be a commutative semigroup. For every $a \in S$ a congruence $\mathcal{A}$ on S is defined by*

$$x \,\mathcal{A}\, y \text{ if and only if } a^m x = a^n y \text{ for some } m, n > 0.$$

If S is archimedean, then $S/\mathcal{A}$ is a group. If S is an N-semigroup, then $S/\mathcal{A}$ is the structure group of S.

Proof. It is immediate that $\mathcal{A}$ is a congruence; in fact, $\mathcal{A}$ is one of the reversible congruences of Dubreil [1941]. We have $x \mathcal{A} ax$ for all $x \in S$, so the $\mathcal{A}$-class of a is the identity element of $S/\mathcal{A}$. If S is archimedean, then for every $x \in S$ there exist $t \in S$ and $m > 0$ such that $xt = a^m \mathcal{A} a$; therefore $S/\mathcal{A}$ is a group.

If S is an N-semigroup, then $\mathcal{A}$ is also the congruence induced on S by the partition of $G(S)$ into cosets of $A = \langle a \rangle$. Hence $S/\mathcal{A}$ is a subsemigroup of the structure group $G = G(S)/A$. In fact $S/\mathcal{A}$ is all of G: for all $x, y \in S$ there exist $t \in S$ and $m > 0$ such that $yt = a^m$, and then $xy^{-1}A = xta^{-m}A = xtA$; thus every coset of A contains an element of S. □

The following result anticipates Chapter IV.

Proposition 6.6. *An N-semigroup S is a subdirect product of its structure group and $S/\mathcal{T}$; hence an N-semigroup is power joined if and only if it is a subdirect product of a torsion group and a subsemigroup of $\mathbb{Q}^+$.*

Proof. Let $a \in S$ and $\mathcal{A}$ be the congruence in Lemma 6.5, so that $S/\mathcal{A} = G$, the structure group of S. If $x \mathcal{A} y$ and $x \mathcal{T} y$, then $x^n = y^n$ and $a^r x = a^s y$ for some $n, r, s > 0$, $a^{nr} x^n = a^{ns} y^n$, $nr = ns$ by Corollary 4.2, $r = s$, and $x = y$ since S is cancellative. Thus $\mathcal{A} \cap \mathcal{T}$ is the equality on S. Hence the projections $S \longrightarrow S/\mathcal{A}$ and $S \longrightarrow S/\mathcal{T}$ separate the elements of S and S is a subdirect product of $G = S/\mathcal{A}$ and $S/\mathcal{T}$, by Proposition IV.1.1.

If S is power joined, then G is torsion by Proposition 6.2 and $S/\mathcal{T}$ is isomorphic to a subsemigroup of $\mathbb{Q}^+$ by Corollary 6.4. If conversely S is contained in the direct product $G \times Q$ of a torsion group G and a subsemigroup Q of $\mathbb{Q}^+$, then every element of S has a power in $\{1\} \times Q$; since Q is power joined, any two elements of S have a common power and S is power joined. □

Power joined N-semigroups were first considered by Levin [1968] and Tamura [1970F]. Tamura [1970C] showed that every N-semigroup is a subdirect product of an abelian group and a positive additive subsemigroup of $\mathbb{R}$. Tamura [1973C] also showed that every N-semigroup is a subdirect product of an abelian group and an irreducible N-semigroup; this leads to the second structure theorem (Tamura [1974B]).

3. Every c.s. has power joined components, that are smaller than but similar to its archimedean components.

Proposition 6.7. *Every c.s. is a disjoint union of power joined subsemigroups.*

Proof. Let S be a c.s. When $x, y \in S$, let

$$x \mathcal{P} y \text{ if and only if } x^m = y^n \text{ for some } m,n > 0.$$

Then $\mathcal{P}$ is reflexive, symmetric, and transitive (if $x^m = y^n$ and $y^p = z^q$, then $x^{mp} = y^{np} = z^{nq}$). Moreover every $\mathcal{P}$-class is a subsemigroup of S: if indeed $x^m = y^n$, then $(xy)^n = x^{m+n}$. □

The $\mathcal{P}$-classes are the **power joined components** of S. They were discovered by Tamura [1970] (and rediscovered by Radelecki [1986]). They are contained in but generally smaller than the archimedean components of S. Moreover $\mathcal{P}$ is not in general a congruence.

Example 6.8. Let $S = \{a^n \mid n \in \mathbb{Z}\} \cong \mathbb{Z}$ be an infinite cyclic group. S is archimedean but has three power joined components: $U = \{1\}$, $P = \{a^n \mid n > 0\}$, and $N = \{a^n \mid n < 0\}$. $\mathcal{P}$ is not a congruence since $PN = S = U \cup N \cup P$. □

Example 6.9. Let $X = \{x,y\}$ and $S = \{a_x x + a_y y \mid a_x, a_y > 0\} \subseteq F_X$; in fact S is an archimedean component of F_X. In S, which is written additively, $a_x x + a_y y \;\mathcal{P}\; b_x x + b_y y$ if and only if $m(a_x x + a_y y) = n(b_x x + b_y y)$ for some $m,n > 0$, if and only if $a_x/a_y = b_x/b_y$. Therefore S has infinitely many power joined components, one component $\{kpx + kqy \mid k > 0\}$ for every pair (p,q) of relatively prime positive integers. □

Tamura [1970P] proved that an archimedean semigroup has either 1 or 3 or infinitely many power joined components. McAlister [1968] proved that a commutative semigroup S has the character extension property (all characters of subsemigroups of S can be extended to characters of S) if and only if its archimedean components are power joined.

Chapter IV.

SUBDIRECT DECOMPOSITIONS.

Subdirect decomposition is a general construction method, which goes back to Birkhoff [1944] and was applied to semigroups by Thierrin [1956], Malcev [1958], Ponizovsky [1962], and Lesohin ([1963A] and later articles). Ponizovsky decompositions provide another way to assemble finite commutative semigroups from groups and nilsemigroups arranged along a semilattice, and are particularly fruitful for later chapters.

Finding all subdirect products of a given family of semigroups is generally not an easy task. Hence subdirect decompositions, like semilattice decompositions, yield general descriptions rather than precise structure theorems. They also uncover two other important classes of commutative semigroups: complete semigroups and elementary semigroups.

This chapter contains basic properties of subdirect products (Section 1), separative semigroups (Section 2), nilsemigroups (Section 3), complete semigroups (Section 4) and elementary semigroups (Section 5). Ponizovsky decompositions are in Section 4. The study of subdirectly irreducible semigroups, begun in Sections 1, 3, and 4, requires additional techniques and is continued in the next chapters.

1. SUBDIRECT PRODUCTS.

1. The **direct product** of a set $(S_i)_{i\in I}$ of semigroups is their cartesian product $\prod_{i\in I} S_i$, with componentwise multiplication:

$$(x_i)_{i\in I}\,(y_i)_{i\in I} \;=\; (x_i\,y_i)_{i\in I}\,.$$

This is the only operation on $\prod_{i\in I} S_i$ such that the **projection** $\pi_i : \prod_{i\in I} S_i \longrightarrow S_i$, $(x_i)_{i\in I} \longmapsto x_i$ is a homomorphism for every $i \in I$. If $I = \varnothing$, the **empty** product $\prod_{i\in I} S_i$ contains only the empty family and is trivial ($\prod_{i\in I} S_i = \{\varnothing\}$).

If in general $S_i \neq \emptyset$ for all i, then $\prod_{i \in I} S_i \neq \emptyset$ and the projections π_i are all surjective.

A **subdirect product** of a set $(S_i)_{i \in I}$ of semigroups is a subsemigroup P of the cartesian product $\prod_{i \in I} S_i$ such that $\pi_i(P) = S_i$ for all $i \in I$. Any semigroup which is isomorphic to such a subsemigroup is also called a subdirect product of the semigroups $(S_i)_{i \in I}$. For instance, a direct product of nonempty semigroups is a subdirect product of these semigroups; there usually are many others.

A **subdirect decomposition** of a semigroup S describes S as a subdirect product of a set $(S_i)_{i \in I}$ of semigroups, the **components** of the decomposition. For example the empty semigroup and the trivial semigroup $\{\emptyset\}$ have trivial subdirect decompositions as subdirect products of the empty family. We are mostly interested in other subdirect decompositions.

When S is isomorphic to a subdirect product $P \subseteq \prod_{i \in I} S_i$, the **projections**

$$p_i : S \xrightarrow{\cong} P \xrightarrow{\subseteq} \textstyle\prod_{i \in I} S_i \xrightarrow{\pi_i} S_i$$

are surjective homomorphisms and **separate** the elements of S in the sense that $x \neq y$ in S implies $p_i(x) \neq p_i(y)$ for some $i \in I$. The converse holds:

Proposition 1.1. *A semigroup S is a subdirect product of semigroups $(S_i)_{i \in I}$ if and only if there exist surjective homomorphisms $\varphi_i : S \longrightarrow S_i$ which separate the elements of S.*

Proof. If the homomorphisms $\varphi_i : S \longrightarrow S_i$ separate the elements of S, then the homomorphism $\varphi(x) = (\varphi_i(x))_{i \in I}$ of S into $\prod_{i \in I} S_i$ is injective, and $S \cong \operatorname{Im} \varphi$. If every φ_i is surjective, then $\pi_i(\operatorname{Im} \varphi) = \operatorname{Im} \varphi_i = S_i$ for every i and $S \cong \operatorname{Im} \varphi$ is a subdirect product of the semigroups $(S_i)_{i \in I}$. □

2. A semigroup S is **subdirectly irreducible** in case S has at least two elements and, whenever S is (isomorphic to) a subdirect product of semigroups S_i ($i \in I$), at least one of the projections $S \longrightarrow S_i$ is an isomorphism. (Thus S is not a subdirect product of "smaller" semigroups.) Every semigroup with two elements is subdirectly irreducible.

Proposition 1.2. *A semilattice is subdirectly irreducible if and only if it has two elements.*

Proof. Let Y be a semilattice. For every $a \in Y$, the mapping $\varphi_a : Y \longrightarrow \{0,1\}$ defined by

$$\varphi_a(x) = 1 \text{ if } x \geqq a, \ \varphi_a(x) = 0 \text{ if } x \ngeqq a$$

is a homomorphism. The homomorphisms φ_a separate the elements of Y: if $a \neq b$ in Y, then, say, $b \not\geqq a$, since $a \geqq b$ and $b \geqq a$ are not both possible, and then $\varphi_a(a) = 1$, $\varphi_a(b) = 0$. Hence Y is a subdirect product of two-element semilattices. If Y is subdirectly irreducible, then Y has just two elements. □

Subdirectly irreducible semigroups in general can be characterized by their congruences. Call a congruence **proper** when it is not the equality.

Proposition 1.3. *A semigroup S with at least two elements is subdirectly irreducible if and only if the equality on S is not the intersection of proper congruences, if and only if S has a smallest proper congruence.*

Proof. This follows from Proposition 1.1. When S is a subdirect product of semigroups S_i $(i \in I)$, then the projections $\pi_i : S \longrightarrow S_i$, which separate the elements of S, induce congruences on S whose intersection is the equality on S. If S is not subdirectly irreducible, then one can arrange that none of the π_i is an isomorphism; then the equality on S is the intersection of proper congruences. If conversely the equality on S is the intersection of proper congruences $\mathcal{C}_i$ $(i \in I)$, then S is a subdirect product of the quotient semigroups $S/\mathcal{C}_i$, but none of the projections $S \longrightarrow S/\mathcal{C}_i$ is an isomorphism, so S is not subdirectly irreducible.

If the equality on S is not the intersection of proper congruences, then the intersection of all the proper congruences on S is not the equality, and is the smallest proper congruence on S. If conversely S has a smallest proper congruence $\mathcal{C}$, then the equality on S is not the intersection of proper congruences, since the latter contains $\mathcal{C}$. □

3. Subdirectly irreducible semigroups are of interest because of a general property of algebraic systems, due to Birkhoff [1944].

Theorem 1.4 (Birkhoff). *Every semigroup is a subdirect product of subdirectly irreducible semigroups; every commutative semigroup is a subdirect product of subdirectly irreducible commutative semigroups.*

Proof. The union of a chain of congruences on a semigroup S is a congruence on S (Proposition I.2.8). For every $a, b \in S$ with $a \neq b$, there exists, by Zorn's Lemma, a congruence $\mathcal{C}_{a,b}$ on S which is maximal such that $a \mathrel{\mathcal{C}_{a,b}} b$ does not hold. By definition, $a \mathrel{\mathcal{C}} b$ for every congruence $\mathcal{C} \supsetneqq \mathcal{C}_{a,b}$; hence $\mathcal{C}_{a,b}$ is not the intersection of congruences $\mathcal{C} \supsetneqq \mathcal{C}_{a,b}$. Then the equality on the quotient $S/\mathcal{C}_{a,b}$ is not the intersection of proper congruences, by Proposition I.2.6, and $S/\mathcal{C}_{a,b}$ is subdirectly irreducible. The intersection of all $\mathcal{C}_{a,b}$ is the equality on S; hence the projections $S \longrightarrow S/\mathcal{C}_{a,b}$ separate the elements of S and S is a subdirect product of the semigroups $S/\mathcal{C}_{a,b}$. If S is commutative, then every $S/\mathcal{C}_{a,b}$ is commutative. □

If for instance S is a semilattice, then Birkhoff's Theorem states that S is a subdirect product of two element semilattices, and can be embedded into the lower semilattice of all subsets of a set (under intersection). The proof of Proposition 1.3 provides a ready embedding $x \longmapsto (\varphi_a(x))_{a\in S}$ of S into $\prod_{a\in S}\{0,1\}$.

4. Birkhoff's Theorem provides subdirect decompositions whose components are as simple as possible. Decompositions with fewer but more general components can also be valuable. Thus we call a semigroup S **finitely subdirectly irreducible** in case S has at least two elements and, whenever S is (isomorphic to) a subdirect product of finitely many semigroups S_i $(i \in I)$, at least one of the projections $S \longrightarrow S_i$ is an isomorphism.

Every subdirectly irreducible semigroup is finitely subdirectly irreducible. The converse is false; Propositions 1.7 and 1.8 below provides a simple counterexample, $\mathbb{N}^+$.

Finitely subdirectly irreducible semigroups can be characterized by their congruences:

Proposition 1.5. *A semigroup S with at least two elements is finitely subdirectly irreducible if and only if the equality on S is not the intersection of finitely many proper congruences.*

This is proved like Proposition 1.3.

Corollary 1.6. *A finite semigroup which is finitely subdirectly irreducible is in fact subdirectly irreducible.*

5. A look at abelian groups and cancellative c.s. will illustrate the various concepts in this section.

Proposition 1.7. *An abelian group is subdirectly irreducible if and only if it is a nontrivial cyclic or quasicyclic p-group for some prime p. A cancellative c.s. is subdirectly irreducible if and only if it is a nontrivial cyclic or quasicyclic p-group for some prime p.*

An abelian group is *quasicyclic* when it is isomorphic to $\mathbb{Z}(p^\infty)$ for some prime p. The second part of the statement is due to Schein [1965].

Proof. If the abelian group G is subdirectly irreducible, then G has a smallest nontrivial subgroup K, since G has a smallest proper congruence. If $c \in K$, $c \neq 1$, then every group homomorphism $\varphi : G \longrightarrow H$ such that $c \notin \operatorname{Ker}\varphi$ is injective. Thus G is cocyclic; by Theorem 3.1 of Fuchs [1970], G is a nontrivial cyclic or quasicyclic p-group. Conversely, these groups have a smallest nontrivial subgroup and are subdirectly irreducible.

By the above, a subdirectly irreducible abelian group G is torsion; hence every nonempty subsemigroup of G is a subgroup of G and is itself a cyclic or quasicyclic p-group. Now let S be a cancellative c.s. By Birkhoff's Theorem, $G(S)$ is a subdirect product of subdirectly irreducible abelian groups. Hence S is a subdirect product of subsemigroups of subdirectly irreducible abelian groups, and is a subdirect product of subdirectly irreducible abelian groups. If therefore S is subdirectly irreducible, then S is a nontrivial cyclic or quasicyclic p-group; conversely we saw that nontrivial cyclic or quasicyclic p-groups are subdirectly irreducible. □

Proposition 1.8. *A finitely generated abelian group is finitely subdirectly irreducible if and only if it is either a nontrivial cyclic p-group for some prime p, or an infinite cyclic group. A finitely generated cancellative c.s. is finitely subdirectly irreducible if and only if it is either a nontrivial cyclic p-group for some prime p, or isomorphic to $\mathbb{Z}$ or to a subsemigroup of $\mathbb{N}$.*

Proof. Let G be a finitely generated abelian group which is finitely subdirectly irreducible. Then G is a direct product of finitely many cyclic groups. Therefore G is cyclic. If G is finite, then G is subdirectly irreducible by Corollary 1.6, and is a nontrivial cyclic p-group for some prime p, by Proposition 1.7. Conversely, nontrivial cyclic p-groups are subdirectly irreducible by Proposition 1.7, and $\mathbb{Z}$ is finitely subdirectly irreducible by Proposition 1.5, since the intersection of finitely many nontrivial subgroups of $\mathbb{Z}$ is nontrivial.

Now let S be a finitely generated cancellative c.s. which is finitely subdirectly irreducible. Then $G(S)$ is finitely generated and is a direct product of finitely many cyclic p-groups (for various primes p) and infinite cyclic groups, and S is a subdirect product of finitely many cyclic p-groups and subsemigroups of $\mathbb{Z}$. Therefore S is either a nontrivial cyclic p-group or isomorphic to a subsemigroup of $\mathbb{Z}$; by Corollary II.4.2, a subsemigroup of $\mathbb{Z}$ is isomorphic to $\mathbb{Z}$ or to a subsemigroup of $\mathbb{N}$.

Conversely we saw that nontrivial cyclic p-groups and infinite cyclic groups are finitely subdirectly irreducible. Now let S be a subsemigroup of $\mathbb{N}$. If $\mathcal{C}$ is a congruence on S, then

$$K(\mathcal{C}) \;=\; \{a-b \in \mathbb{Z} \mid a,b \in S \text{ and } a \;\mathcal{C}\; b\}$$

is a subgroup of $\mathbb{Z}$. Moreover, $K(\mathcal{C}) = \{0\}$ if and only if $\mathcal{C}$ is the equality on S, and $K(\bigcap_{i\in I} \mathcal{C}_i) = \bigcap_{i\in I} K(\mathcal{C}_i)$ if I is finite. If now $\mathcal{C}$ is the intersection of finitely many proper congruences on S, then $K(\mathcal{C})$ is the intersection of finitely many nonzero subgroups of $\mathbb{Z}$, $K(\mathcal{C}) \neq \{0\}$, and $\mathcal{C}$ is not the equality on S; thus S is finitely subdirectly irreducible. □

By Corollary II.7.4, a finitely generated c.s. which is cancellative, power cancellative, and reduced is a subdirect product of finitely many subsemigroups of $\mathbb{N}$; that is, affine semigroups are subdirect products of finitely many numerical semigroups or monoids. The proof of Proposition 1.8 also shows that every finitely generated cancellative c.s. is a subdirect product of finitely many, finitely subdirectly irreducible semigroups; we shall see in Chapter VI that this property extends to all finitely generated c.s.

6. The next sections contain well-known results on subdirect decompositions of separative semigroups, nilsemigroups, and finite c.s.; decompositions of finitely generated and subcomplete c.s., including Malcev's Theorem [1958] that finitely generated c.s. which are subdirectly irreducible are finite, will be found in Chapters VI and VII.

Results on subdirectly irreducible c.s. in general have been published by Schein [1962], [1966] and McNeil [1971], [1972], but the proofs have some gaps.

Lesohin [1963A] called a semigroup S **approximable** by a class $\mathcal{C}$ of semigroups in case homomorphisms of S into members of $\mathcal{C}$ separate the elements of S; equivalently, S is a subdirect product of subsemigroups of members of $\mathcal{C}$. Semigroups that are approximable by finite semigroups are **finitely approximable**; these semigroups are more usually called **residually finite** (or **profinite**, in the case of groups). Lesohin [1963A] also called a semigroup S **separable** by a class $\mathcal{C}$ of semigroups in case homomorphisms of S into members of $\mathcal{C}$ separate the elements of S from its subsemigroups: if T is a subsemigroup of S and $a \in S\backslash T$, then there is a homomorphism φ of S into a member of $\mathcal{C}$ such that $\varphi(a) \notin \varphi(T)$. A number of papers by Lesohin and others investigate these concepts and variations thereof: Lesohin [1963A], [1966A], [1967A], [1968A], [1969], [1971A], [1972], [1974A], [1974I], [1976], [1983], Lesohin & Golubov [1966], Golubov [1969], [1970A], [1970S], Kublanovsky & Lesohin [1976], Popyrin [1986]. Residually finite semigroups are studied in Lesohin [1963A], Lesohin & Golubov [1966], Lesohin [1966A], [1968A], [1974I], [1976], and Golubov [1969], [1970A]. Approximability by various other classes, including character semigroups, is considered in Lesohin [1963A], [1966A], [1967A], [1969], [1974A], Lomadze [1992A], [1992B], and Ignat'eva [1996].

2. SEPARATIVE SEMIGROUPS.

The main results in this section were discovered independently by Schwarz [1954A], [1954B] and Hewitt & Zuckerman [1955], [1956].

1. In what follows, $\mathbb{C}$ is the multiplicative semigroup of all complex numbers and $\mathbb{S}$ is the subgroup of $\mathbb{C}$ of all complex numbers of modulus 1 (the **circle** group or 1-sphere).

In group theory, a (complex irreducible) **character** of an abelian group G is a homomorphism of G into $\mathbb{S}$; these mappings arise as traces of irreducible representations of abelian groups by complex matrices.

A **character** of a c.s. S is variously defined as a nonzero homomorphism χ of S into either $\mathbb{S} \cup \{0\}$ or $\mathbb{C}$. We use the former definition and call a nonzero homomorphism of S into $\mathbb{C}$ a **semicharacter** of S. If S is a group, then the ideal $\chi^{-1}(0)$ of S is empty, otherwise $\chi = 0$; hence a character of an abelian group is a character in the group theory sense.

As it turns out, semigroups don't much care whether we use characters or semicharacters. There is a "radial" homomorphism $\rho : 0 \longmapsto 0$, $z \longmapsto z/|z|$ if $z \neq 0$, of $\mathbb{C}$ onto $\mathbb{S} \cup \{0\}$. Hence every semicharacter χ yields a character $\chi \circ \rho$; conversely, every character is a semicharacter.

2. Characters are basically a matter of Clifford semigroups. We saw (Corollary IV.2.5) that every c.s. S has a universal semilattice of groups T, which comes with a homomorphism $\tau : S \longrightarrow T$ through which factors uniquely every homomorphism of S into a semilattice of groups. Since $\mathbb{S} \cup \{0\}$ (and $\mathbb{C}$) are semilattices of groups, every character (or semicharacter) of S factors uniquely through τ and has the form $\chi \circ \tau$ for some unique character (or semicharacter) χ of T. Thus the characters and semicharacters of S are determined by those of its universal Clifford semigroup.

Proposition 2.1. *The characters of a Clifford semigroup separate its elements.*

Proof. Let G be an abelian group. By Proposition 1.7, the homomorphisms of G into nontrivial cyclic and quasicyclic p-groups separate the elements of G. Now $\mathbb{S}$ notoriously contains a copy of every $\mathbb{Z}(p^\infty)$, which consists of all complex p^k-th roots of unity with $k \geqq 0$, and contains a copy of every cyclic p-group. Therefore the characters of G separate its elements.

Now let $S = \bigcup_{a \in Y} (G_a \times \{a\})$ be a semilattice of groups, where Y is a semilattice, (G, γ) is an abelian group valued functor on Y, and the multiplication on S is given, as in Clifford's Theorem III.2.1, by

$$(x,a)(y,b) = (\gamma^a_{ab}x\,\gamma^b_{ab}y,\ ab).$$

For any two elements $(t,c) \neq (u,d)$ of S we construct a character χ of S such that $\chi(t,c) \neq \chi(u,d)$.

If $c \neq d$, then, as in the proof of Proposition 1.2, there is a homomorphism φ of Y onto $\{0,1\}$ such that $\varphi(c) \neq \varphi(d)$. Then $(x,a) \longmapsto \varphi(a)$ is a homomorphism of S into $\{0,1\} \subseteq \mathbb{S} \cup \{0\}$ and $\chi(t,c) \neq \chi(u,d)$.

If $c = d$, then there is a homomorphism $\varphi : G_c \longrightarrow \mathbb{S}$ such that $\varphi(t) \neq \varphi(u)$. Define $\chi : S \longrightarrow \mathbb{S} \cup \{0\}$ by:

$$\chi(x,a) = 0 \text{ if } a \ngeqq c,\ \chi(x,a) = \varphi(\gamma^a_c(x)) \text{ if } a \geqq c.$$

Then χ is a homomorphism: if $a,b \geqq c$, then

$$\begin{aligned}\chi\big((x,a)(y,b)\big) &= \varphi\big(\gamma^{ab}_c(\gamma^a_{ab}(x)\gamma^b_{ab}(y))\big)\\ &= \varphi\big(\gamma^a_c(x)\big)\,\varphi\big(\gamma^b_c(y)\big) = \chi(x,a)\,\chi(y,b);\end{aligned}$$

otherwise $\chi\big((x,a)(y,b)\big) = \chi(x,a)\,\chi(y,c) = 0$. Also $\chi(t,c) = \varphi(t) \neq \varphi(u) = \chi(u,d)$. □

3. A **group with zero** is a group with a zero element adjoined. For instance, $\mathbb{C}$ and $\mathbb{S} \cup \{0\}$ are groups with zero. A **cancellative semigroup with zero** is a cancellative semigroup with a zero element adjoined. The following result is due to Hewitt & Zuckerman [1956] and Schein [1965]. Its proof catches subdirect decompositions smuggling structural information into semigroups.

Proposition 2.2. *For a c.s. S the following properties are equivalent:*

(1) *the characters of S separate the elements of S;*

(2) *the semicharacters of S separate the elements of S;*

(3) *S is a subdirect product of groups and/or groups with zero;*

(4) *S is a subdirect product of cancellative semigroups and/or cancellative semigroups with zero;*

(5) *S is separative.*

Proof. (1) $\Longrightarrow$ (2) since every character is a semicharacter.

(2) $\Longrightarrow$ (3). (2) implies that S is a subdirect product of the images $\operatorname{Im}\chi$ of its semicharacters. Every $\operatorname{Im}\chi \subseteq \mathbb{C}$ is either cancellative or cancellative with zero; by Proposition 1.7 there are homomorphisms of $\operatorname{Im}\chi$ onto groups or onto groups with zero that separate the elements of $\operatorname{Im}\chi$. This provides homomorphisms of S onto groups or groups with zero that separate the elements of S.

(3) $\Longrightarrow$ (4) since groups are cancellative.

(4) $\Longrightarrow$ (5). When S is a subdirect product of cancellative semigroups C_i and/or cancellative semigroups with zero $C_i \cup \{0\}$, then the projection of S onto C_i or $C_i \cup \{0\}$ sends an archimedean component A of S into an archimedean component of C_i or $C_i \cup \{0\}$, that is, into a cancellative semigroup; therefore every archimedean component of S is cancellative.

(5) $\Longrightarrow$ (1). By Proposition IV.2.3, a separative semigroup S can be embedded into its universal semilattice of groups T. If $a \neq b$ in S, then by Proposition 2.1 there is a character χ of T which separates a and b ($\chi(a) \neq \chi(b)$); then the restriction of χ to S is a character of S which separates a and b. □

4. Additional properties of semigroup characters and semicharacters can be found in Clifford & Preston [1961] and in the literature. A **bounded** semicharacter satisfies $|\chi(s)| \leqq M$ for all $s \in S$; then $|\chi(s)| \leqq 1$ for all $s \in S$, since $|\chi(s)| > 1$ implies that $|\chi(s^n)| = |\chi(s)|^n$ is unbounded. Thus a c.s. S has several character and semicharacter semigroups, all usually denoted by S^*.

Schwarz [1954C] and Iseki [1957] related **semiprime** ideals of S (such that $a^n \in I$ implies $a \in I$) to ideals of its character semigroup S^*. Comfort [1960] investigated semicharacters that are isolated in the pointwise convergence topology. Hewitt & Zuckerman [1960] and Parízek & Schwarz [1961] determined all characters of the multiplicative semigroup $\mathbb{Z}_n$. Petrich [1962], [1963C] studied bounded semicharacters, showing in particular that, for bounded semicharacters, $(S \times T)^* \cong S^* \times T^*$. Lesohin [1968A], [1969], [1974S] (and other papers) investigated approximability and separability by characters and by character semigroups. All semigroups S^* are semilattices of groups; the actual groups and semilattices were described by Warne & Williams [1961], Fulp [1966], McAlister [1968], and Anderson [1979A]. Lesohin [1970] gave an abstract characterization of character semigroups. Lesohin [1969], [1971C] also investigated when two semigroups have isomorphic character semigroups, and when S^* is finite or torsion. Dress & Grabmeier [1991] studied sums of characters.

The problem of extending characters or semicharacters of a subsemigroup to the entire semigroup was studied by Ross [1959], Comfort & Hill [1966], Hill [1966] for bounded semicharacters; Kobayashi & Tamura [1977] for bounded semicharacters that preserve suitable preorders; Ross [1961] and Fulp [1967] for characters; Hill [1967] and McAlister [1968], [1970] for semicharacters in general.

There is a canonical homomorphism $S \longrightarrow S^{**}$. Necessary and sufficient conditions that it be an isomorphism were found by Lesohin [1965] for characters, by Fulp & Hill [1965] and Pondeliček [1966] for semicharacters. More general forms of duality were investigated by Lesohin [1966D], [1968D].

3. NILSEMIGROUPS.

This section contains general properties of commutative nilsemigroups, and a nice result of Schein [1966] which determines when the finite ones are subdirectly irreducible.

1. **Proposition 3.1.** *In a commutative nilsemigroup,* $xy \neq x$ *whenever* $x \neq 0$*; hence Green's preorder* $\leqq_{\mathcal{H}}$ *is a partial order relation, and Green's relation* $\mathcal{H}$ *is the equality.*

Proof. If $xy = x$, then $x = xy = xy^2 = \cdots = xy^n = 0$; hence $xy \neq x$ whenever $x \neq 0$. If now $a \,\mathcal{H}\, b$ and $a \neq b$, then $a = ub$, $b = va$ for some $u, v \in N$ and then $a = uva$ implies $a = 0$ and $b = va = 0 = a$, a contradiction. Hence $\mathcal{H}$ is the equality on N and $\leqq_{\mathcal{H}}$ is antisymmetric and is a partial order relation. □

In a commutative nilsemigroup we denote $\leqq_{\mathcal{H}}$ by just $\leqq$. By Proposition 3.1, $xy < x$ whenever $x, y \in N$ and $x \neq 0$.

A congruence $\mathcal{C}$ on a nilsemigroup is **pure** when $\{0\}$ is a $\mathcal{C}$-class. The next result goes back to Teissier [1951] and Croisot [1953].

Proposition 3.2. *On any commutative nilsemigroup* N *there is a greatest pure congruence* $\mathcal{P}$*, namely*

$$x \,\mathcal{P}\, y \iff (\forall t \in N^1)(xt = 0 \iff yt = 0).$$

Equivalently, $x \,\mathcal{P}\, y \iff 0 : x = 0 : y$*, where* $0 : x = \{t \in N^1 \mid xt = 0\}$.

Proof. $\mathcal{P}$ is an equivalence relation. Since $(0 : x) : z = 0 : xz$, $\mathcal{P}$ is in fact a congruence: if $0 : x = 0 : y$, then $0 : xz = (0 : x) : z = (0 : y) : z = 0 : yz$. Also $\{0\}$ is a $\mathcal{P}$-class, as $x = 0$ if and only if $0 : x = N^1$.

Conversely let $\mathcal{C}$ be a congruence on N such that $\{0\}$ is a $\mathcal{C}$-class. If $x \,\mathcal{C}\, y$, then $xt \,\mathcal{C}\, yt$ and $xt = 0$ if and only if $yt = 0$; thus $\mathcal{C} \subseteq \mathcal{P}$. □

Example 3.3. Let N be the commutative nilsemigroup

$$N = \langle a, b \mid a^3 = a^2 b = a b^2 = b^3 = 0 \rangle.$$

We have $N = \{a, b, a^2, ab, b^2, 0\}$, and

$$\begin{aligned} 0:a &= \{a^2,\, ab,\, b^2,\, 0\}; \\ 0:b &= \{a^2,\, ab,\, b^2,\, 0\}; \\ 0:a^2 &= 0:ab = 0:b^2 = N; \\ 0:0 &= N^1. \end{aligned}$$

Therefore the $\mathcal{P}$-classes of N are: $\{a,b\}$, $\{a^2,\, ab,\, b^2\}$, and $\{0\}$. We see that $N/\mathcal{P}$ is cyclic. □

Example 3.4. Let N be the commutative nilsemigroup

$$N = \langle\, a,b \mid a^3 = a^2b = ab^2 = b^4 = 0 \,\rangle.$$

We have $N = \{a,\, b,\, a^2,\, ab,\, b^2,\, b^3,\, 0\}$ and

$$\begin{aligned} 0:a &= \{a^2,\, ab,\, b^2,\, b^3,\, 0\}; \\ 0:b &= \{a^2,\, ab,\, b^3,\, 0\}; \\ 0:b^2 &= \{a,\, a^2,\, ab,\, b^2,\, b^3,\, 0\}; \\ 0:a^2 &= 0:ab = 0:b^3 = N; \\ 0:0 &= N^1. \end{aligned}$$

Therefore the $\mathcal{P}$-classes are: $\{a\}$, $\{b\}$, $\{b^2\}$, $\{a^2,\, ab,\, b^3\}$, and $\{0\}$. We see that $N/\mathcal{P}$ is the Volkov nilsemigroup Example IV.3.6. □

2. We now look at some finiteness conditions for nilsemigroups. A semigroup N with zero is **nilpotent** if $N^k = 0$ for some $k > 0$ (equivalently, if every product of k or more elements is zero, for some $k > 0$).

An element m of a nilsemigroup N is **minimal** if $m > 0$ and there is no $m > x > 0$. Say that N **has enough minimal elements** if every nonzero element of N lies above some minimal element (if for every $a > 0$ in N there exists a minimal element $m \leqq a$). Finite nilsemigroups have enough minimal elements.

Proposition 3.5. *Let N be a commutative nilsemigroup.*

(1) *If N is finitely generated, then N is finite.*

(2) *If N is finite, then N is nilpotent.*

(3) *If N is nilpotent, then N has enough minimal elements.*

Proof. (1). If N is generated by $x_1, x_2, \dots, x_n$, then every element of N is a product of positive powers of $x_1, x_2, \dots, x_n$. Since N is a nilsemigroup, $x_1, x_2, \dots, x_n$ have only finitely many distinct positive powers; therefore N is finite.

(2). Let $N = \{x_1, x_2, \dots, x_n\}$. Let $k = k_1 + k_2 + \cdots + k_n$, where $x_i^{k_i} = 0$. In any product $x_1^{j_1} x_2^{j_2} \cdots x_n^{j_n}$ of k or more elements of N we have $\sum_i j_i \geqq k$, $j_i \geqq k_i$ for some i, and $x_1^{j_1} x_2^{j_2} \cdots x_n^{j_n} = 0$. Thus $N^k = 0$.

(3). Let N be nilpotent with $N^k = 0$. In N every strictly descending chain $x_1 > x_2 > \cdots > x_n$ has length at most k: indeed $x_{i+1} = x_i u_{i+1}$ for some $u_{i+1} \in N$, so that $x_k = x_1 u_2 \cdots u_k = 0$ and $x_k > x_{k+1}$ is not possible. In particular the descending chain condition holds in N and N has enough minimal elements. □

3. Schein's result applies to nilsemigroups with enough minimal elements.

Proposition 3.6 (Schein). *For a commutative nilsemigroup $N \neq 0$ with enough minimal elements the following conditions are equivalent:*

(1) *N is subdirectly irreducible;*

(2) *N is finitely subdirectly irreducible;*

(3) *the greatest pure congruence $\mathcal{P}$ on N is the equality.*

Proof. Already (1) $\Longrightarrow$ (2).

(2) $\Longrightarrow$ (3). Since $N \neq 0$, the set M of all minimal elements of N is not empty. Assume that N is finitely subdirectly irreducible. By Proposition 1.5, the equality on S is not the intersection of finitely many proper congruences. For every $m \in M$ let $\mathcal{R}_m$ be the Rees congruence of the ideal $\{0, m\}$ of N. If M had at least two elements m and n, then $\mathcal{R}_m \cap \mathcal{R}_n$ would be the equality on N; therefore M has just one element m. Next, 0 and m are not equivalent modulo $\mathcal{P}$; hence $\mathcal{P} \cap \mathcal{R}_m$ is the equality on N and $\mathcal{P}$ is the equality on N.

(3) $\Longrightarrow$ (1). Assume that $\mathcal{P}$ is the equality on N. Since M is a $\mathcal{P}$-class we again have $M = \{m\}$. Let $\mathcal{C}$ be a proper congruence on N, so that $a \mathcal{C} b$ for some $a, b \in N$, $a \neq b$. Then $a \mathcal{P} b$ does not hold and there exists $u \in N^1$ such that, say, $ua \neq 0$ and $ub = 0$. Since N has enough minimal elements we have $ua \geqq m$. Then $vua = m$ for some $v \in N^1$ and $vub = 0$. Therefore $m \mathcal{C} 0$. Thus every proper congruence on N contains the Rees congruence $\mathcal{R}_m$ and N is subdirectly irreducible by Proposition 1.3. □

Corollary 3.7. *When $N \neq 0$ is a commutative nilsemigroup with enough minimal elements, then $N/\mathcal{P}$ is subdirectly irreducible with enough minimal elements.*

Proof. $N/\mathcal{P}$ is a commutative nilsemigroup with one minimal element, the $\mathcal{P}$-class M of all minimal elements of N; since N has enough minimal elements,

$P \geqq M$ for every nonzero $\mathcal{P}$-class P, and $N/\mathcal{P}$ has enough minimal elements. If P_a, P_b are nonzero $\mathcal{P}$-classes, then $P_a P_x = 0$ if and only if $ax = 0$, and $P_b P_x = 0$ if and only if $bx = 0$; hence $P_a \mathcal{P} P_b$ in $N/\mathcal{P}$ implies $a \mathcal{P} b$ in N and $P_a = P_b$. Thus $\mathcal{P}$ is the equality on $N/\mathcal{P}$. □

Example 3.4 shows that the Volkov nilsemigroup Example IV.3.6 is subdirectly irreducible.

4. PONIZOVSKY DECOMPOSITIONS.

The Ponizovsky decomposition of a finite commutative semigroup S is an explicit subdirect decomposition of S with remarkable properties, due to Ponizovsky [1962].

1. The Ponizovsky decomposition results from two properties of finite c.s. First, every archimedean component contains an idempotent. Second:

Lemma 4.1. *Let S be a finite c.s. and $x \in S$. When $e \in E(S^1)$ is idempotent, then $x \leqq_{\mathcal{H}} e$ in S^1 if and only if $ex = x$, and there exists a least such idempotent $\epsilon(x)$.*

Proof. $ex = x$ implies $x \leqq_{\mathcal{H}} e$; conversely, $x \leqq_{\mathcal{H}} e$ implies $x = et$ for some $t \in S^1$ and $ex = eet = et = x$.

Let $e_1, e_2, \dots, e_n$ be the idempotents e_i of S^1 such that $e_i x = x$. The product $e_1 e_2 \cdots e_n$ is the least such idempotent. □

The Ponizovsky decomposition extends readily to any semigroup with these two properties. Formally we call a c.s. S **complete** when

(1) every archimedean component of S contains an idempotent; equivalently, by Corollary IV.3.2, every element of S has a power in a subgroup of S;

(2) for every $x \in S$ there exists a least idempotent e of S^1 such that $x \leqq_{\mathcal{H}} e$; equivalently, a least idempotent e of S^1 such that $ex = x$; we denote it by $\epsilon(x)$.

Condition (2) is trivial when S has only finitely many idempotents: then, as in the proof of Lemma 4.1, S^1 has only finitely many idempotents $e_1, e_2, \dots, e_n$ such that $e_i x = x$, and then the product $e = e_1 e_2 \cdots e_n$ is the least such idempotent.

Finite c.s. are complete, by Lemma 4.1.

2. Complete semigroups are incorrectly defined in Grillet [1995], where condition (2) above was replaced by

(3) the semilattice $E(S)$ is **complete** (every nonempty subset A of $E(S)$ has an infimum $\bigwedge_{e \in A} e$)

in the mistaken belief that (1) and (3) imply (2). A counterexample is provided by Example 4.2 below. The author is highly embarrassed by this error and hopes that no such oversight has found unwanted refuge in the present text. The correct definition and counterexample are from Grillet [2001S].

Example 4.2. Let

$$S = \{e_1, e_2, \ldots, e_n, \ldots, a, e\}$$

with commutative multiplication

	e_j	a	e
e_i	$e_{i \vee j}$	a	e
a	a	e	e
e	e	e	e

where $i \vee j = \max\,(i,j)$. Associativity is straightforward: $(xy)\,z = x\,(yz)$ holds in S whenever x, y, or z equals e, and when $x, y, z = e_i, e_j, e_k$; also $e_i\,(e_j\,a) = a = (e_i\,e_j)\,a$, $e_i\,(a\,e_j) = a = (e_i\,a)\,e_j$, $e_i\,(aa) = e = (e_i\,a)\,a$, $a\,(e_j\,a) = e = (a\,e_j)\,a$, and commutativity yields the remaining cases.

We see that $S = S^1$ and that $E(S)$ is the chain $e_1 \geqq e_2 \geqq \ldots \geqq e_n \geqq \ldots e$; it is a complete semilattice in which $e = \bigwedge_{n>0} e_n$. The archimedean components of S are $\{a, e\}$ and all $\{e_n\}$. Also $e_n\,a = a$ for all n, but $ea \neq a$; thus (2) does not hold for a. $\square$

3. When S is a complete semigroup, the **Ponizovsky factors** of S are the semigroups

$$P_e = Se/\big(\textstyle\bigcup_{f \in E(S),\, f<e} Sf\big),$$

one for every idempotent $e \in E(S^1)$. In particular, if $1 \notin S$ (if S is not a monoid), then $P_1 = S/S\,E(S)$; if there is a least idempotent e_0 (if for instance S is finite), then $P_{e_0} = Se_0$.

The reader will observe that the kernel H_e of the archimedean component of e is the group of units of its Ponizovsky factor.

Composing multiplication by e, which is a homomorphism $S \longrightarrow Se$, with the projection $Se \longrightarrow P_e$ to the Rees quotient yields a canonical surjective homorphism, the **projection** $\pi_e : S \longrightarrow P_e$. By definition,

$$\pi_e(x) \;=\; ex \text{ if } \epsilon(ex) = e, \quad \pi_e(x) = 0 \text{ otherwise,}$$

since $x \in Se\backslash\left(\bigcup_{f\in E(S),\, f<e} Sf\right)$ if and only if $e = \epsilon(x)$ is the least idempotent $e \geqq_{\mathcal{H}} x$.

Theorem 4.3. *Every complete (for instance, finite) commutative semigroup is a subdirect product of its Ponizovsky factors.*

Proof. We show that the canonical projections π_e $(e \in E(S^1))$ separate the elements of S. Assume that $\pi_e(x) = \pi_e(y)$ for every $e \in E(S^1)$. Let $E = E(S)$. If $x \notin SE$, then $1 \notin S$, $\pi_1(x) \neq 0$, $\pi_1(y) \neq 0$, $y \notin SE$, and $\pi_1(x) = \pi_1(y)$ yields $x = y$. Now let $x \in SE$. As above, $y \in SE$. Let $e = \epsilon(x)$ and $f = \epsilon(y)$. If $ef < e$, then $\pi_e(y) = 0$ in P_e, since $ey \leqq ef < e$, whereas $\pi_e(x) = x$, since $e = \epsilon(x)$. Therefore $ef = e$. Similarly $ef = f$. Hence $e = f$, and $x = \pi_e(x) = \pi_f(y) = y$. □

The Ponizovsky decomposition is unique among subdirect decompositions in being also a partition. The **partial Ponizovsky factors** of a complete semigroup S are the partial semigroups

$$P_e^* \;=\; Se\backslash\left(\bigcup_{f\in E(S),\, f<e} Sf\right),$$

one for every idempotent $e \in E(S^1)$. In particular, if $1 \notin S$, then $P_1^* = S\backslash S\,E(S)$ may be empty; if there is a least idempotent e_0, then $P_{e_0}^* = Se_0$. In general, the ideals Se and $\bigcup_{f\in E(S),\, f<e} Sf$ are unions of $\mathcal{H}$-classes; hence every partial Ponizovsky factor is a union of $\mathcal{H}$-classes.

Proposition 4.4. *Let S be a complete (for instance, finite) c.s. and $e \in E(S^1)$. Then $x \in P_e^*$ if and only if $e = \epsilon(x)$. Hence S is the union of its partial Ponizovsky factors, which are pairwise disjoint. If S is a monoid, then $(P_e^*)_{e\in E(S)}$ is a partition of S.*

4. A c.s. S is **elementary** in case it is the disjoint union $S = G \cup N$ of a group G and a nilsemigroup N, in which N is an ideal of S, the identity element of G is the identity element of S, and the zero element of N is the zero element of S. (This last condition may be omitted.) In particular S has an identity element, a zero element, and no other idempotent.

Proposition 4.5. *In a complete (for instance, finite) c.s. S: if there is a least idempotent e_0, then $P_{e_0} = Se_0$ is a group and the kernel of S; if $1 \notin S$, then P_1 is a nilsemigroup; all other Ponizovsky factors are elementary.*

Proof. If there is a least idempotent e_0, then every nonempty ideal of S contains some $x \in S$, contains some x^n in a subgroup H_f of S, contains H_f and the idempotent f, contains $e_0 \leqq f$, and contains Se_0; thus $P_{e_0} = Se_0$ is the kernel of S.

Let $x \in S$ and $e = \epsilon(x) \in E(S^1)$, so that $x \in P_e^*$. Since S is complete, x has an power x^n in a subgroup H_f of S. If $x \notin H_e$ (for instance, if $e = 1 \notin S$), then $f < e$, since $f \leqq_{\mathcal{H}} x <_{\mathcal{H}} e$, so that $x^n \in \bigcup_{f \in E(S),\, f<e} Sf$ and $x^n = 0$ in P_e. Therefore:

if $e = 1 \notin S$, then P_e is a nilsemigroup;

if e is the least idempotent of S (if such exists), then $f < e$ is not possible, $x \in H_e$, and $P_e = P_e^* = H_e$;

otherwise $e \in S$, $\bigcup_{f \in E(S),\, f<e} Sf \neq \emptyset$, P_e has a zero element, and an element of P_e is either in H_e or nilpotent. Then P_e is the disjoint union $P_e = G \cup N$ of a group $G = H_e$ and a nilsemigroup N which consists of all nilpotent elements of P_e; N is an ideal of P_e, since $y \leqq_{\mathcal{H}} x$ and $x^n = 0$ implies $y^n = 0$; the identity element e of $G = H_e$ is the identity element of P_e; and the zero element of N is the zero element of P_e. Thus P_e is elementary. □

Corollary 4.6. *A complete (for instance, finite) subdirectly irreducible c.s. is either a group or a nilsemigroup or elementary.*

Elementary semigroups are studied in more detail in the next section. The subdirectly irreducible semigroups in Corollary 4.6 will be determined in the next chapter.

5. The Ponizovsky factors of a complete c.s. are easier to reassemble than its archimedean components; this is another remarkable feature of the Ponizovsky decomposition. The author [1972] showed that a finite c.s. can be reconstructed from its Ponizovsky factors by means of partial homomorphisms. This extends immediately to complete semigroups.

For all $e, f \in E(S^1)$ let $\rho^e_f : P_e^* \longrightarrow P_f$ be the restriction of π_f to P_e^*; ρ^e_f is a partial homomorphism ($\rho^e_f(xy) = \rho^e_f(x)\, \rho^e_f(y)$ whenever $xy \neq 0$ in P_e).

Lemma 4.7. *The following holds:*

(1) $\rho^e_f(\pi_e(x)) = \pi_f(x)$ *whenever* $e \geqq f$ *and* $\pi_e(x) \neq 0$;

(2) *for every* $x \in S$ *there exists a greatest* $e \in E(S^1)$ *such that* $\pi_e(x) \neq 0$ *in* P_e.

Proof. Let $e \geqq f$ and $\pi_e(x) \neq 0$, so that $\pi_e(x) = ex$ and $\epsilon(ex) = e$. Then $\rho^e_f(\pi_e(x)) = \pi_f(ex) = \pi_f(x)$, since $fex = fx$. This proves (1).

Let $e = \epsilon(x)$. Then $\pi_e(x) = x \neq 0$ in P_e. If conversely $\pi_f(x) \neq 0$, then $\epsilon(fx) = f$ and $f \leqq e$, since $fx \leqq_{\mathcal{H}} fe \leqq_{\mathcal{H}} e$. Thus (2) holds (with $e = \epsilon(x)$). □

Proposition 4.8. *Every complete (for instance, finite) c.s. S is determined up to isomorphism by its Ponizovsky factors and the partial homomorphisms $\rho^e_f : P^*_e \longrightarrow P_f$ with $e, f \in E(S^1)$ and $e \geqq f$. Namely, S is isomorphic to the semigroup P of all $(x_e)_{e \in E(S^1)} \in \prod_{e \in E(S^1)} P_e$ such that*

(A) $\rho^e_f(x_e) = x_f$ *whenever* $e \geqq f$ *and* $x_e \neq 0$;

(B) *there exists a greatest* $e \in E(S^1)$ *such that* $x_e \neq 0$.

Proof. When $x \in S$ let $\theta(x) = (\pi_e(x))_{e \in E(S^1)}$. Then $\theta(x) \in P$ by Lemma 4.7, and $\theta : S \longrightarrow P$ is an injective homomorphism by Theorem 4.3.

Conversely let $y = (y_f)_{f \in E(S^1)} \in P$; let $e \in E(S^1)$ be greatest such that $y_e \neq 0$. Let $x = y_e \in P^*_e$. Then $\pi_e(x) = x = y_e$. If $f \leqq e$, then $\pi_f(x) = \rho^e_f(\pi_e(x)) = \rho^e_f(y_e) = y_f$ by Lemma 4.7. If $f \nleqq e$, then $y_f = 0$, $fx \leqq fe$, $\epsilon(fx) \leqq fe < f$, and $\pi_f(x) = 0 = y_f$. Thus $y = \theta(x)$, which shows that θ is surjective. □

The families $(\rho^e_f)_{e,f \in E(S^1),\, e \geqq f}$ of partial homomorphisms that arise from complete semigroups can be characterized by the following properties (Grillet [1972]):

(i) if $f \leqq e \in S$, then $\rho^e_f(e) = f$;

(ii) $\rho^e_e(x) = x$ whenever $x \in P^*_e$, and $\rho^f_g(\rho^e_f(x)) = \rho^e_g(x)$ whenever $e \geqq f \geqq g$ and $\rho^e_f(x) \neq 0$;

(iii) for every $x \in P^*_e$ and $y \in P^*_f$ there exists a greatest $g \leqq e, f$ such that $\rho^e_g(x)\, \rho^f_g(y) \neq 0$ in P_g.

This does not quite solve the subdirect product retrieval problem, since, even with good constructions of the semigroups P_e, it is not immediately clear how to construct all families of partial homomorphisms with properties (i), (ii), and (iii). Much less, however, can be said about the reassembly of archimedean components.

5. ELEMENTARY SEMIGROUPS.

In this section we construct elementary semigroups from groups and nilsemigroups. Another construction is given in the next chapter.

1. An elementary semigroup $S = G \cup N$ is an ideal extension of the nilsemigroup N by a group with zero $G \cup \{0\}$. In this case the ideal extension problem has a reasonable solution.

When $S = G \cup N$ is elementary, the group G acts on the set S by left multiplication ($g \,.\, x = gx$).

Proposition 5.1. *In an elementary semigroup $S = G \cup N$, the orbits under the action of G coincide with the $\mathcal{H}$-classes of S.*

Proof. If $a \,\mathcal{H}\, b$ holds in T, then either $a = b = 0$, or $b = ta \neq 0$, $a = bu \neq 0$ for some $t, u \in S^1 = S$. If $t \in N$ or $u \in N$, then $a \,\mathcal{H}\, b$ in N and $a = b$ by Proposition 3.1. Otherwise $b = ga$ for some $g \in G$; this also holds if $a = b$. If conversely $b = ga$, then $a = g^{-1}b$ and $a \,\mathcal{H}\, b$. □

The semigroup $S/\mathcal{H}$ is the **monoid of orbits** of S; it is a commutative **nilmonoid**, that is, a commutative nilsemigroup with an identity element adjoined.

2. When $S = G \cup N$ is elementary, the multiplication on $S = G \cup N$ is completely determined by the multiplications on G and N and by the action of G on N. The latter satisfies $g(xy) = (gx)y$ for all $g \in G$ and $x, y \in N$. Conversely, any such group action extends to an elementary semigroup multiplication on $G \cup N$: indeed $g0 = g(00) = (g0)0 = 0$ and all other conditions are satisfied.

Lemma 5.2. *In an elementary semigroup $S = G \cup N$, every $g \in G$ permutes every $\mathcal{P}$-class of N.*

Proof. $\mathcal{P}$ is the largest pure congruence on N (Proposition 3.2). When $x \in N$, then $xt = 0$ implies $(gx)t = 0$, and $(gx)t = 0$ implies $xt = g^{-1}gxt = 0$; thus $gx \,\mathcal{P}\, x$, and multiplication by g permutes the $\mathcal{P}$-class of x. □

Proposition 5.3. *Let N be a commutative nilsemigroup. Let $U(N)$ be the group, under pointwise multiplication, of all families $s = (s_P)_{P \in N/\mathcal{P}}$ such that each s_P is a permutation of the $\mathcal{P}$-class P and*

$$(s_P x)y = s_Q(xy) \quad \text{whenever} \quad P, Q \in N/\mathcal{P},\ x \in P,\ xy \in Q. \qquad (1)$$

For every abelian group G disjoint from N, there are one-to-one correspondences between: (i) elementary semigroups $G \cup N$; (ii) actions of G of N such *that $g(xy) = (gx)y$ for all $g \in G$ and $x,y \in N$; (iii) homomorphisms of G into $U(N)$.*

$U(N)$ is the group of units of the semigroup of left translations of N as defined in, say, Clifford and Preston [1961].

Proof. The one-to-one correspondence between (i) and (ii) was noted above. We construct the one-to-one correspondence between (i) and (iii). Let $U(N)$ be defined as in the statement, with the pointwise operation $(st)_P x = s_P(t_P x)$ for all $x \in P$, $P \in N/\mathcal{P}$. Then $U(N)$ is a monoid, with identity element $(1_P)_{P\in N/\mathcal{P}}$. Moreover, $U(N)$ is closed under pointwise inverses. Indeed let $s = (s_P)_{P\in N/\mathcal{P}} \in U(N)$ and $t = (s_P^{-1})_{P\in N/\mathcal{P}}$. Let $x \in P$, $xy \in Q$ and $z = s_P^{-1} x$. Then $z \in P$, $zy \in Q$, $xy = (s_P z)y = s_Q(zy)$ by (1), and $(s_P^{-1} x)y = zy = s_Q^{-1}(xy)$. Thus $t \in U(N)$. Therefore $U(N)$ is a group, in which t is the inverse of s.

When $S = G \cup N$ is elementary, left multiplication by $g \in G$ induces a permutation g_P of every $\mathcal{P}$-class P, namely $g_P x = gx$ for all $x \in P$ (Lemma 5.2). Moreover $(g_P x)y = gxy = g_Q(xy)$ whenever $x,y \in N$ and $x \in P$, $xy \in Q$, and $(gh)_P = g_P h_P$ (that is, $(gh)_P x = g_P(h_P x)$ for all $x \in P$) for all $g,h \in G$. To an elementary semigroup $G \cup N$ thus corresponds a homomorphism $g \longmapsto (g_P)_{P\in N/\mathcal{P}}$ of G into $U(N)$.

If conversely $\varphi : G \longrightarrow U(N)$ is a homomorphism, then for every $g \in G$ and every $\mathcal{P}$-class P of N we have a permutation $\varphi(g)_P$ of P, and an action of $g \in G$ on $x \in N$ is defined by: $gx = \varphi(g)_P x$, whenever $x \in P$. This is a group action, since φ is a homomorphism, and $(gx)y = g(xy)$ holds for all $x,y \in N$ by definition of $U(N)$. With this action we now have an elementary semigroup multiplication on $G \cup N$. It is immediate that $g \longmapsto (g_P)_{P\in N/\mathcal{P}}$ is the given homomorphism φ. □

3. We conclude this section with two examples.

Example 5.4. Let N be the commutative nilsemigroup (Example 3.3)

$$N = \langle a,b \mid a^3 = a^2b = ab^2 = b^3 = 0 \rangle,$$

$N = \{a, b, a^2, ab, b^2, 0\}$. We saw that the $\mathcal{P}$-classes are $A = \{a, b\}$, $B = \{a^2, ab, b^2\}$, and $\{0\}$. If s_A is the identity on A, then the condition

$$(s_P x)y = s_Q(xy), \quad \text{whenever } x \in P,\ xy \in Q \tag{1}$$

implies that s_B is the identity on B. If s_A is the transposition $(a\ b)$, then condition (1) yields $s_B(ab) = (s_A a)\, b = b^2$ and $s_B(ab) = (s_A b)\, a = a^2$; this contradiction shows that s_A is the identity on A. Therefore $U(N)$ is trivial. In any elementary semigroup $G \cup N$, $gx = x$ for all $g \in G$ and $x \in N$. □

Example 5.5. Let N be the commutative nilsemigroup

$$N = \langle\, a,b \mid a^3 = a^2 b = a b^2 = b^3 = 0,\ a^2 = b^2 \,\rangle,$$

$N = \{a,\ b,\ a^2,\ ab,\ 0\}$. The $\mathcal{P}$-classes are $A = \{a, b\}$, $B = \{a^2,\ ab\}$, and $\{0\}$. Hence $U(N)$ has two elements t, u: u_A, u_B are identity permutations, and t_A, t_B are transpositions. If for instance $G = \{1, g\}$ is cyclic of order 2, then there are two elementary semigroups $G \cup N$: one in which $gx = x$ for all $x \in N$, and one in which $ga = b$, $gb = a$, $ga^2 = ab$, and $gab = a^2 = b^2$. □

Chapter V.

GROUP COEXTENSIONS.

Group coextensions were developed independently by Grillet [1974] and Leech [1975] for semigroups in general. They yield precise constructions of complete commutative semigroups in terms of abelian groups and group-free semigroups. This leads to the semigroup cohomology which is studied in the last chapters of this book.

This chapter contains general properties and, as an application, constructs all finite commutative semigroups that are subdirectly irreducible.

1. DIVIDING BY $\mathcal{H}$.

This section reveals an intimate relationship, first explored by Kolibiarová [1958] and others, between a c.s. S and the quotient semigroup $S/\mathcal{H}$.

1. **Proposition 1.1.** *In a commutative semigroup S, $a \leqq_{\mathcal{H}} b$ holds in S if and only if $H_a \leqq_{\mathcal{H}} H_b$ holds in $S/\mathcal{H}$. Hence $\mathcal{H}$ is the equality on $S/\mathcal{H}$.*

Proof. Let $H_1 = 1 \in (S/\mathcal{H})^1$ in case $S \neq S^1$ does not have an identity element. If $a = bt$ for some $t \in S^1$, then $H_a = H_b H_t$ in $(S/\mathcal{H})^1$ and $H_a \leqq_{\mathcal{H}} H_b$ in $S/\mathcal{H}$. If conversely $H_a \leqq_{\mathcal{H}} H_b$ holds in $S/\mathcal{H}$, then $H_a = H_b H_t = H_{bt}$ for some $t \in S^1$, $a \mathcal{H} bt \leqq_{\mathcal{H}} b$ holds in S, and $a \leqq_{\mathcal{H}} b$.

Hence $H_a \mathcal{H} H_b$ in $S/\mathcal{H}$ implies $a \mathcal{H} b$ in S and $H_a = H_b$. □

Proposition 1.2. *Let S be a commutative semigroup. The projection $S \longrightarrow S/\mathcal{H}$ induces an isomorphism of $E(S)$ onto $E(S/\mathcal{H})$.*

Proof. If e is idempotent in S, then H_e is idempotent in $S/\mathcal{H}$, since $\mathcal{H}$ is a congruence; hence the projection $S \longrightarrow S/\mathcal{H}$ induces a semigroup homomorphism $\theta : E(S) \longrightarrow E(S/\mathcal{H})$. If H is idempotent in $S/\mathcal{H}$, then H is a subsemigroup of S, is a subgroup of S by Proposition I.4.3, and contains exactly one idempotent of S; therefore θ is bijective. □

Corollary 1.3. *A commutative semigroup S is a monoid if and only if $S/\mathcal{H}$ is a monoid; hence $S^1/\mathcal{H} = (S/\mathcal{H})^1$ for every commutative semigroup S.*

Proof. If S has an identity element 1, then $S/\mathcal{H}$ has an identity element H_1. If conversely $S/\mathcal{H}$ has an identity element H_e, then, for every $x \in S$, $H_x \leqq_{\mathcal{H}} H_e$, $x \leqq_{\mathcal{H}} e$ by Proposition 1.2, $x = et$ for some $t \in S^1$, and $ex = eet = et = x$; thus e is an identity element of S.

If S is a monoid, then $S^1/\mathcal{H} = S/\mathcal{H} = (S/\mathcal{H})^1$. If S is not a monoid, then $\{1\}$ is an $\mathcal{H}$-class of S^1 and $S^1/\mathcal{H} = (S/\mathcal{H}) \cup \{1\} = (S/\mathcal{H})^1$, since $S/\mathcal{H}$ is not a monoid. □

2. The universal semilattice and archimedean components of S can be lifted from $S/\mathcal{H}$:

Proposition 1.4. *When S is a commutative semigroup: $a \leqq_{\mathcal{N}} b$ in S if and only if $H_a \leqq_{\mathcal{N}} H_b$ in $S/\mathcal{H}$; $a \,\mathcal{N}\, b$ in S if and only if $H_a \,\mathcal{N}\, H_b$ in $S/\mathcal{H}$; and the projection $S \longrightarrow S/\mathcal{H}$ induces an isomorphism of $Y(S)$ onto $Y(S/\mathcal{H})$.*

Proof. If $H_a \leqq_{\mathcal{N}} H_b$ holds in $S/\mathcal{H}$, then $H_{a^n} \leqq_{\mathcal{H}} H_b$ for some $n > 0$, $a^n \leqq_{\mathcal{H}} b$ for some $n > 0$ by Proposition 1.1, and $a \leqq_{\mathcal{N}} b$ holds in S. The converse is clear. Hence $a \,\mathcal{N}\, b$ holds in S if and only if $H_a \,\mathcal{N}\, H_b$ holds in $S/\mathcal{H}$. Thus the surjective homomorphisms $S \longrightarrow Y(S)$ and $S \longrightarrow S/\mathcal{H} \longrightarrow Y(S/\mathcal{H})$ induce the same congruence on S, and Proposition I.2.4 yields an isomorphism $Y(S) \cong Y(S/\mathcal{H})$ which makes the following square commute:

$$\begin{array}{ccc} S & \longrightarrow & S/\mathcal{H} \\ \downarrow & & \downarrow \\ Y(S) & \xrightarrow[\cong]{} & Y(S/\mathcal{H}) \end{array} \quad \square$$

Corollary 1.5. *A commutative semigroup S is a Clifford semigroup if and only if $S/\mathcal{H}$ is a semilattice.*

Corollary 1.6. *A commutative semigroup S is a complete archimedean semigroup if and only if $S/\mathcal{H}$ is a nilsemigroup.*

Proof. By Propositions 1.4, 1.2, S is archimedean if and only if $S/\mathcal{H}$ is archimedean, and S contains an idempotent if and only if $S/\mathcal{H}$ contains an idempotent. Since $\mathcal{H}$ is the equality on $S/\mathcal{H}$, $S/\mathcal{H}$ is complete and archimedean if and only if it is a nilsemigroup, by Proposition III.3.1. □

3. The Ponizovsky decomposition of S can also be lifted from $S/\mathcal{H}$:

Proposition 1.7. *A commutative semigroup S is complete if and only if*

$S/\mathcal{H}$ is complete; and then, for all $x \in S$ and $e \in E(S^1)$, $\epsilon(H_x) = H_{\epsilon(x)}$, so that $x \in P_e^*$ in S if and only if $H_x \in P_{H_e}^*$ in $S/\mathcal{H}$.

In this statement we regard H_1 as $1 \in S^1/\mathcal{H}$ in case $1 \notin S$.

Proof. By Corollary 1.3, $(S/\mathcal{H})^1 = S^1/\mathcal{H}$. The projection $S \longrightarrow S/\mathcal{H}$ induces an isomorphism $E(S^1) \cong E((S/\mathcal{H})^1)$, by Proposition 1.2, and sends the archimedean components of S onto the archimedean components of $S/\mathcal{H}$, by Proposition 1.4. Hence every archimedean component of $S/\mathcal{H}$ contains an idempotent if and only if every archimedean component of S contains an idempotent. By Propositions 1.1, 1.2, S^1 has a least idempotent $e \geqq_{\mathcal{H}} x$ if and only if $(S/\mathcal{H})^1$ has a least idempotent $H_e \geqq_{\mathcal{H}} H_x$; and then $\epsilon(H_x) = H_{\epsilon(x)}$. Thus S is complete if and only if $S/\mathcal{H}$ is complete. The last part of the statement now holds since $x \in P_e^*$ if and only if $e = \epsilon(x)$ (Proposition IV.4.4). □

Corollary 1.8. *A commutative semigroup S is elementary if and only if S has a zero element and $S/\mathcal{H}$ is a nilmonoid.*

Proof. If S is elementary, then $S/\mathcal{H}$ is a nilmonoid by Proposition IV.5.1.

Conversely assume that S has a zero element and that $S/\mathcal{H}$ is a nilmonoid. Then S has two idempotents $e > f$, by Proposition 1.2; e is an identity element of S by Corollary 1.3, so f is the zero element. Now $G = H_e$ is a subgroup of S, the identity element e of G is the identity element of S, and $N = S\backslash G$ is an ideal of S, since its projection $(S/\mathcal{H})\backslash 1$ in $S/\mathcal{H}$ is an ideal of S/SH. Every $x \in N$ is nilpotent: since H_x is nilpotent in $S/\mathcal{H}$, we have $x^n \,\mathcal{H}\, 0$ for some $n > 0$ and $x^n = 0$. Thus S is elementary. □

2. SCHÜTZENBERGER FUNCTORS.

Continuing the previous section we show that the Schützenberger groups of a commutative semigroup can be arranged into a functor. The results in this section are from Grillet [1974] (see also Grillet [1995]).

1. Recall that the Schützenberger group of an $\mathcal{H}$-class H is $\Gamma(H) = \{g_t^H \mid t \in \mathrm{St}(H)\}$, where

$$\mathrm{St}(H) = \{t \in S^1 \mid tH \subseteq H\}$$

and $g_t^H a = ta$ for all $t \in \mathrm{St}(H)$ and $a \in H$; it is a simply transitive group of permutations of H.

Lemma 2.1. *Let A and B be $\mathcal{H}$-classes of a c.s. S. If $A \geqq_{\mathcal{H}} B$ in $S/\mathcal{H}$, then $\mathrm{St}(A) \subseteq \mathrm{St}(B)$ and there is a unique homomorphism $\gamma_B^A : \Gamma(A) \longrightarrow \Gamma(B)$ such that the square*

$$\begin{array}{ccc} \mathrm{St}(A) & \xrightarrow{\subseteq} & \mathrm{St}(B) \\ {\scriptstyle g^A}\downarrow & & \downarrow{\scriptstyle g^B} \\ \Gamma(A) & \xrightarrow[\gamma_B^A]{} & \Gamma(B) \end{array}$$

commutes; then

$$(ga)\,s \;=\; (\gamma_B^A g)(as)$$

whenever $g \in \Gamma(A)$, $a \in A$, $s \in T^1$, and $as \in B$. Moreover γ_A^A is the identity on $\Gamma(A)$, and $\gamma_C^B \circ \gamma_B^A = \gamma_C^A$ whenever $A \geqq_{\mathcal{H}} B \geqq_{\mathcal{H}} C$ in $S/\mathcal{H}$.

Proof. Assume $A \geqq_{\mathcal{H}} B$ in $S/\mathcal{H}$ and let $a \in A$, $b \in B$. Then $a \geqq_{\mathcal{H}} b$ in S and $b = as$ for some $s \in S^1$. Since $\mathcal{H}$ is a congruence, $As \subseteq B$. If $t \in \mathrm{St}(A)$, then $ta \in A$, $tb = tas \in B$, and $t \in \mathrm{St}(B)$ since $\mathcal{H}$ is a congruence. Thus $\mathrm{St}(A) \subseteq \mathrm{St}(B)$.

If $t, u \in \mathrm{St}(A)$ and $g_t^A = g_u^A$, then $ta = ua$, $tas = uas$, and $g_t^B = g_u^B$ (since $\Gamma(B)$ is simply transitive). Therefore there is a unique homomorphism $\gamma_B^A : \Gamma(A) \longrightarrow \Gamma(B)$ such that the square in the statement commutes. The uniqueness of γ_B^A with this property yields the last part of the statement. If finally $g \in \Gamma(A)$, $a \in A$, $s \in S^1$, and $as \in B$, then $g = g_t^A$ for some $t \in \mathrm{St}(A) \subseteq \mathrm{St}(B)$, $\gamma_B^A g = g_t^B$, and $(ga)s = tas = (\gamma_B^A g)(as)$. □

Let $H(S)$ be the category whose objects are the elements of S and where a morphism from a to b is an ordered pair (a,t) with $t \in S^1$ and $at = b$ (necessarily $b \leqq_{\mathcal{H}} a$). With morphisms "written on the left", composition is given by $(at,u) \circ (a,t) = (a, tu)$. The identity morphism on a is $(a,1)$. $H(S)$ is the commutative version of the categories $L(S)$ and $R(S)$ in Leech [1975].

An **abelian group valued functor on** a c.s. S is a functor $\mathbb{G} = (G,\gamma)$ from $H(S)$ to the category of abelian groups. Thus $\mathbb{G}$ assigns an abelian group G_a to every element a of S and a homomorphism $\gamma_{a,t} : G_a \longrightarrow G_{at}$ to each pair $(a,t) \in S \times S^1$, such that $\gamma_{a,1}$ is the identity on G_a and $\gamma_{at,u} \circ \gamma_{a,t} = \gamma_{a,tu}$.

An abelian group valued functor $\mathbb{G} = (G,\gamma)$ on S is **thin** in case $\gamma_{a,t} = \gamma_{a,u}$ whenever $at = au$; then $\mathbb{G}$ provides only one homomorphism $G_a \longrightarrow G_b$ when

$a \geqq_{\mathcal{H}} b$, which we will generally denote by γ_b^a. A thin functor $\mathbb{G}$ can be viewed as a functor from the simpler category arising from the preordered set S, preordered by $\leqq_{\mathcal{H}}$. Thin functors suffice in this chapter, but the more general functors defined above are needed later.

The **Schützenberger functor** of a c.s. S is the abelian group valued functor $\mathbb{\Gamma} = (\Gamma, \gamma)$ on $S/\mathcal{H}$ in which $\Gamma(A)$ is the Schützenberger group of A and $\gamma_{A,T} = \gamma_{AT}^A$ is provided by Lemma 2.1.

2. The Schützenberger functor of a c.s. S has additional properties when S is finite (Grillet [1974]) or, more generally, complete.

Lemma 2.2. *Let S be a complete c.s., $T = S/\mathcal{H}$, and $A \in T$ be an $\mathcal{H}$-class of S.*

(1) *If $A \notin TE(T)$, then $\Gamma(A)$ is trivial.*

(2) *If $A \in TE(T)$ and $E = \epsilon(A)$ holds in T, then γ_A^E is surjective.*

If S is finite, then $\Gamma(A)$ is finite.

Proof. T is complete by Proposition 1.7. Let $a \in A$ and $t \in \mathrm{St}(A)$, $t \in S$. Since S is complete, some power t^n of t belongs to a subgroup H_f of S. Then $t^n A \subseteq A$ and $fA \subseteq A$ (since $f \,\mathcal{H}\, t^n$), and $FA = A$ in T, where $F = H_f \in E(T)$. If $A \notin TE(T)$, then this is impossible; therefore $\mathrm{St}(A) = \{1\}$ and $\Gamma(A)$ is trivial, which proves (1).

Now assume $A \in TE(T)$. Then $a \in SE(S)$, by Propositions 1.1 and 1.2. Let $e = \epsilon(a)$, so that $E = H_e = \epsilon(A)$ (Proposition 1.7). In the above, $fA \subseteq A$ implies $a \leqq_{\mathcal{H}} f$ and $e \leqq_{\mathcal{H}} f$. Now $e \in \mathrm{St}(A)$, since $ea = a$, and $et \in \mathrm{St}(A)$; moreover $g_{et}^A = g_t^A$ and $et \in H_e$, since $e = ef \leqq_{\mathcal{H}} et \leqq_{\mathcal{H}} e$. Hence $g_t^A = g_{et}^A = \gamma_A^E g_{et}^E$; this proves (2). □

Let $\mathbb{G} = (G, \gamma)$ be an abelian group valued functor on a c.s. S. We call $\mathbb{G}$ **finite** when every G_a is finite. When S is complete, we call $\mathbb{G}$ **surjecting** if it has properties (1) and (2) in Lemma 2.2: that is, G_a is trivial if $a \notin SE(S)$ and $\gamma_{e,a}$ is surjective whenever $e = \epsilon(a) \in S$. (Surjecting functors are called **surjective** in the author's papers, which is not very felicitous.) Thus the Schützenberger functor of a complete semigroup is thin and surjecting; the Schützenberger functor of a finite semigroup is thin, surjecting, and finite.

We note some other consequences of Lemma 2.2.

Proposition 2.3. *For a complete (for instance, finite) c.s. S the following conditions are equivalent:*

(1) *all subgroups of* S *are trivial;*

(2) $\mathcal{H}$ *is the equality on* S*;*

(3) $\leqq_{\mathcal{H}}$ *is an order relation on* S.

This follows from Lemma 2.2 and previous results. A c.s. is **group-free** when all its subgroups are trivial; these semigroups have also been called (in the finite case) **combinatorial** and **aperiodic**. Commutative semigroups on which $\mathcal{H}$ is the equality, so that $\leqq_{\mathcal{H}}$ is an order relation, are also called **naturally partially ordered** and **holoids**. Lemma 2.2 and Proposition 1.1 yield:

Proposition 2.4. *When* S *is a complete (for instance, finite) c.s., then* $S/\mathcal{H}$ *is group-free, and the Schützenberger functor of* S *is thin and surjecting.*

3. COEXTENSIONS.

Continuing the previous sections we now define $\mathcal{H}$-coextensions and group coextensions and study the relationships between these two concepts. Group coextensions will be constructed in the next section. The results in this section are from Grillet [1974] (see also Grillet [1995]).

1. Generally, an **extension** of a semigroup S is a semigroup E of which S is a subsemigroup; or, more generally, an ordered pair (E, μ) of a semigroup E and an injective homomorphism $\mu : S \longrightarrow E$.

A (commutative) **coextension** of a (commutative) semigroup S is an ordered pair (E, π) of a (commutative) semigroup E and a surjective homomorphism $\pi : E \longrightarrow S$; or just the semigroup E, if it is clear what π is. Then the sets

$$E_a = \pi^{-1} a = \{x \in E \mid \pi x = a\}$$

with $a \in S$, which are the equivalence classes of the congruence $\ker \pi$ induced by π, constitute a partition of E; moreover, $E_a E_b \subseteq E_{ab}$ for all $a, b \in S$. For example, every c.s. T is a commutative coextension of its universal semilattice $Y(T)$; then the sets T_a are the archimedean components of T.

Commutative coextensions in which $\pi : E \longrightarrow S$ is a retraction were considered by Schmidt [1975], [1977]. Nguyen [1981] studied monoid coextensions of S by $B = \pi^{-1}(1)$, including Baer sums and a commutative monoid $\operatorname{Ext}(S, B)$ when B is central. This chapter deals with two related types of coextensions, defined as follows.

A **commutative** $\mathcal{H}$**-coextension** of a (necessarily commutative) semigroup S

is a commutative coextension (E,π) of S such that $\ker\pi \subseteq \mathcal{H}$, and is **exact** if $\ker\pi = \mathcal{H}$, in which case $S \cong E/\mathcal{H}$. Thus every c.s. T is an exact $\mathcal{H}$-coextension of $T/\mathcal{H}$; then the sets T_a are the $\mathcal{H}$-classes of T. Clifford semigroups are exact $\mathcal{H}$-coextensions of semilattices (Corollary 1.5). Archimedean semigroups with an idempotent are exact $\mathcal{H}$-coextensions of nilsemigroups (Corollary 1.6). Elementary semigroups are exact $\mathcal{H}$-coextensions of nilmonoids (Corollary 1.8).

When $\mathcal{E} = (E,\pi)$ is an exact commutative $\mathcal{H}$-coextension of a c.s. S, then $S \cong E/\mathcal{H}$ and the categories $H(S)$, $H(E/\mathcal{H})$ are isomorphic; composing the Schützenberger functor of E with the isomorphism $H(S) \longrightarrow H(E/\mathcal{H})$ yields an abelian group valued functor on S, the **Schützenberger functor** $\mathbb{\Gamma} = (\Gamma,\gamma)$ of $\mathcal{E}$. Thus $\Gamma(a) = \Gamma(A)$ and $\gamma_{a,u} = \gamma_{A,U}$ for every $a \in S$ and $u \in S^1$, where $A = E_a = \pi^{-1}a$ and $U = E_u$ if $u \in S$, $U = 1 \in (E/\mathcal{H})^1$ if $u = 1 \in S^1$. Then $\Gamma(a)$ acts simply and transitively on E_a, and

$$(g \centerdot x)\, y \;=\; (\gamma_{a,b}g) \centerdot xy$$

whenever $x,y \in E$, $\pi x = a$, $\pi y = b$, and $g \in G_a$, by Lemma 2.1. These properties suggest the next definition.

2. A **commutative group coextension** of a c.s. S by an abelian group valued functor $\mathbb{G} = (G,\gamma)$ on S is an ordered triple $\mathcal{E} = (E,\pi,\centerdot)$ of a c.s. E, a surjective homomorphism $\pi : E \longrightarrow S$ and, for every $a \in S$, a simply transitive group action $\centerdot$ of G_a on the set $E_a = \{x \in E \mid \pi x = a\}$, such that

$$(g \centerdot x)\, y = (\gamma_{a,b}g) \centerdot xy$$

whenever $x,y \in E$, $\pi x = a$, $\pi y = b$, and $g \in G_a$. We also call the semigroup E a commutative group coextension of S by $\mathbb{G}$ when π and $\centerdot$ are clear.

By Lemma 2.1, every c.s. T is a commutative group coextension of $T/\mathcal{H}$ by its Schützenberger functor; more generally:

Proposition 3.1. *An exact commutative $\mathcal{H}$-coextension is a commutative group coextension by the Schützenberger functor of that coextension.*

For another example, let B be an abelian group and A be a subgroup of B, both written multiplicatively. Then the projection $B \longrightarrow B/A$ is a surjective homomorphism, A acts by left multiplication ($a \centerdot x = ax$) on every one of its cosets, this action is simply transitive, and $(a \centerdot x)\, y = a \centerdot xy$ holds for all $a \in A$ and $x,y \in B$. Hence

Proposition 3.2. *Every abelian group extension of an abelian group A by an abelian group C is a commutative group coextension of C by the constant functor A on C.*

3. The converse of Proposition 3.1 does not hold in general: not every commutative group coextension $(E,\pi,.)$ of S by G is an exact $\mathcal{H}$-coextension, even when $\mathbb{G}$ is the Schützenberger functor of some c.s.; that is, the congruence induced by $\pi : E \longrightarrow S$ on E does not necessarily coincide with $\mathcal{H}$.

Example 3.3. Let T be the semigroup

$$T = \{a_i, b_i, 0 \mid i \in \mathbb{Z}\}$$

in which

$$a_i a_j = a_{i+j}, \quad a_i b_j = b_j a_i = b_{i+j}, \quad b_i b_j = 0,$$

and 0 is a zero element. T is a c.s.: indeed $a_i (a_j a_k) = a_{i+j+k} = (a_i a_j) a_k$, $a_i (a_j b_k) = b_{i+j+k} = (a_i a_j) b_k$, and all other products of three elements are 0. The $\mathcal{H}$-classes of T are all $\{a_i\}$, $B = \{b_i \mid i \in \mathbb{Z}\}$, and $\{0\}$; $\Gamma(B) \cong \mathbb{Z}$ and the other Schützenberger groups are trivial. This specifies the Schützenberger functor $\mathbb{\Gamma}$.

Now let E be the semigroup

$$E = \{a_i, b_i, 0 \mid i \in \mathbb{Z}\}$$

in which

$$a_i a_j = a_{i+j}, \quad a_i b_j = b_j a_i = b_j, \quad b_i b_j = 0,$$

and 0 is a zero element. We see that E is a group coextension of $S = T/\mathcal{H}$ by $\mathbb{\Gamma}$: $\Gamma(B) \cong \mathbb{Z}$ acts simply and transitively on B as before, and $(g \,.\, x)\, y = (\gamma_{a,b} g) \,.\, xy$ holds in E since either $g = 1$, or $xy = 0$ in E, or $x, xy \in B$ and $\gamma_{a,b} g = g$. But $\mathcal{H}$ is the equality on E, so E is not an $\mathcal{H}$-coextension of $T/\mathcal{H}$ by $\mathbb{\Gamma}$, let alone an exact $\mathcal{H}$-coextension. □

4. However, the Schützenberger functors of *complete* semigroups have additional properties which force group coextensions to be exact $\mathcal{H}$-coextensions.

Theorem 3.4. *If T is a complete (for instance, finite) commutative semigroup, then $T/\mathcal{H}$ is complete and group-free and the Schützenberger functor of T is thin and surjecting.*

Conversely let S be a complete group-free commutative semigroup, $\mathbb{G}$ be a thin and surjecting abelian group valued functor on S, and $\mathcal{E} = (E,\pi,.)$ be a group coextension of S by $\mathbb{G}$. Then E is complete, $\ker \pi = \mathcal{H}$, *and the Schützenberger functor of $\mathcal{E}$ is naturally isomorphic to $\mathbb{G}$.*

Proof. The direct part follows from Propositions 1.7 and 2.3 and Lemma 2.2. Conversely let S be a complete group-free c.s., $\mathbb{G}$ be a thin and surjecting abelian group valued functor on S, and $\mathcal{E} = (E,\pi,.)$ be a group coextension of S by $\mathbb{G}$.

Anticipating on later results we note that

$$\begin{aligned}(g \cdot x)(h \cdot y) &= (\gamma_{a,b}g) \cdot (x(h \cdot y)) = (\gamma_{a,b}g) \cdot ((h \cdot y)x) \\ &= (\gamma_{a,b}g) \cdot ((\gamma_{b,a}h) \cdot yx) = ((\gamma_{a,b}g)(\gamma_{b,a}h)) \cdot xy\end{aligned}$$

whenever $x \in E_a$, $y \in E_b$, $g \in G_a$, $h \in G_b$.

If $x \mathcal{H} y$ holds in E, then $\pi x \mathcal{H} \pi y$ holds in S and $\pi x = \pi y$. Conversely, assume $\pi x = \pi y = a \in S$, so that $y = g \cdot x$ for some $g \in G_a$. If $a \notin SE(S)$, then $G_a = \{1\}$ since $\mathbb{G}$ is surjecting, $g = 1$, $x = y$, and $x \mathcal{H} y$. Now let $a \in SE(S)$, so that $e = \epsilon(x) \in S$. Take any $p \in E_e$. Since $ea = a$ we have $px \in E_a$ and $px = k \cdot x$ for some $k \in G_a$. Since $\mathbb{G}$ is surjecting, $\gamma_{e,a}$ is surjective, and $g = (\gamma_{e,a}h)\, k$ for some $h \in G_e$. Then

$$(h \cdot p)x = (\gamma_{e,a}h) \cdot px = g \cdot x = y;$$

thus $y \leqq_{\mathcal{H}} x$. Similarly $x \leqq_{\mathcal{H}} y$. Thus $\ker \pi = \mathcal{H}$. Then E is complete, by Proposition 1.7.

Now every set E_a is an $\mathcal{H}$-class. Note that $t \in \mathrm{St}(E_a)$ if and only if either $t = 1 \in E^1$ or $t \in E_c$ with $ac = a$. Since $\mathbb{G}$ is thin, $\gamma_{a,c}g = \gamma_{a,1}g = g$ for all $g \in G_a$, and $(g \cdot x)\,t = g \cdot xt$ for all $x \in E_a$ and $t \in E_c$. Thus $(g \cdot x)t = g \cdot xt$ for all $x \in E_a$ and $t \in \mathrm{St}(E_a)$.

Let $x \in E_a$. If $g_t \in \Gamma(E_a)$, then $g_t x = tx = g \cdot x$ for some unique $g \in G_a$. Moreover g does not depend on the choice of x: if $y \in E_a$, then $y = xu$ for some $u \in \mathrm{St}(E_a)$, and

$$g \cdot y = g \cdot xu = (g \cdot x)u = txu = ty = g_t y$$

by the above. Hence there is a mapping $\theta_a : \Gamma(E_a) \longrightarrow G_a$ such that $g_t x = \theta_a g_t \cdot x$ for all $x \in E_a$ and $t \in \mathrm{St}(E_a)$. If $t, u \in \mathrm{St}(E_a)$, then, for any $x \in E_a$,

$$\theta_a(g_t g_u) \cdot x = g_t g_u x = \theta_a g_t \cdot g_u x = \theta_a g_t \cdot (\theta_a g_u \cdot x),$$

and $\theta_a(g_t g_u) = (\theta_a g_t)(\theta_a g_u)$. Thus θ_a is a homomorphism.

The homomorphism θ_a is injective, since $g_t\, p_a = g_u\, p_a$ implies $g_t = g_u$. It is surjective since both G_a and $\Gamma(E_a)$ act transitively on E_a: if $g \in G_a$, then $g \cdot p_a = g_t p_a = \theta_a g_t \cdot p_a$ for some $g_t \in \Gamma(E_a)$, and $g = \theta_a g_t$.

Finally we show that θ_a is a natural isomorphism: the square

$$\begin{array}{ccc} \Gamma(a) & \xrightarrow{\theta_a} & G_a \\ {\scriptstyle \gamma_b^a}\downarrow & & \downarrow{\scriptstyle \gamma_{a,c}} \\ \Gamma(b) & \xrightarrow[\theta_b]{} & G_b \end{array}$$

commutes whenever $b = ac \leqq_{\mathcal{H}} a$. This is trivial if $c = 1 \in S^1$. If $c \in S$, take any $x \in E_a$ and $z \in E_c$, so that $xz \in E_b$. Then

$$(g_t x)\, z \;=\; (\theta_a g_t \cdot x)\, z \;=\; \gamma_{a,c} \theta_a g_t \cdot xz\,,$$

for all $g_t \in \Gamma(E_a)$, whereas, by Proposition 3.1,

$$(g_t x)\, z \;=\; (\gamma_b^a g_t)\, xz \;=\; \theta_b \gamma_b^a g_t \cdot xz\,. \;\Box$$

Thus, when S is complete (or finite) and group-free, exact $\mathcal{H}$-coextensions of S coincide with commutative group coextensions of S by thin surjecting functors. Group coextensions are therefore especially interesting in this case. We saw that Theorem 3.4 does not extend to arbitrary commutative semigroups; nor are there similar results for finite semigroups in general.

Corollary 3.5. *Let T be a complete group-free semigroup. An abelian group valued functor on T is (naturally isomorphic to) the Schützenberger functor of a c.s. if and only if it is thin and surjecting. If T is finite, then an abelian group valued functor on T is (naturally isomorphic to) the Schützenberger functor of a finite c.s. if and only if it is thin, finite, and surjecting.*

When T is complete group-free, we call an abelian group valued functor on T **Schützenberger** when it is thin and surjecting.

Corollary 3.6. *A c.s. is a complete archimedean semigroup if and only if it is a commutative group coextension of a nilsemigroup by a thin surjecting functor.*

When $\mathbb{G}$ is a thin surjecting functor on a nilsemigroup N, then G_a is trivial for every $a \neq 0$ in N, and only G_0 can be nontrivial.

Proof. If S is a complete archimedean semigroup, then S is an ideal extension of a group by a nilsemigroup N and $S/\mathcal{H}$ is a nilsemigroup. If conversely S is a commutative group coextension of a nilsemigroup N by a thin surjecting functor, then S is complete by Theorem 3.4 and archimedean by Proposition 1.5, since $S/\mathcal{H} \cong N$ is archimedean. $\Box$

Corollary 3.7. *A c.s. S is elementary if and only if S is a commutative group coextension of a nilmonoid by a thin surjecting functor G such that G_0 is trivial.*

Proof. The Schützenberger functor of an elementary semigroup has these properties. Conversely let S is a commutative group coextension of a nilmonoid by a thin surjecting functor $\mathbb{G}$ such that G_0 is trivial. By Theorem 3.4, S is complete and $S/\mathcal{H}$ is a nilmonoid. By Proposition 1.2, S has two idempotents $e > f$. Now $H_f = \{f\}$, since $\Gamma(H_f) \cong G_0$ is trivial, and $\{f\}$ is an ideal of S since it is the inverse image of $\{0\}$ under the projection $S \longrightarrow S/\mathcal{H}$; thus f is a zero element of S and S is elementary by Corollary 1.8. □

When $\mathbb{G}$ is a thin surjecting functor on a nilmonoid N^1, then $\gamma_a^1 : G_1 \longrightarrow G_a$ is surjective for every $a \neq 0$ in N, so that every group $G_a \neq G_0$ is a homomorphic image of G_1.

4. GROUP COEXTENSIONS.

Group coextensions are very similar to group extensions and are constructed in much the same way; this completes the results in Section 3. The results in this section are from Grillet [1974]; similar results for semigroups in general were obtained independently by Leech [1975] and the author [1974] (see also Grillet [1995]). Schreier-like extensions were first considered, for semigroups in general, by Rédei [1952] and Strecker [1969]; for commutative semigroups, by Inasaridze [1965], [1967] and Lugowski [1966].

1. Two commutative group coextensions $\mathcal{E} = (E, \pi, .)$ and $\mathcal{E}' = (E', \pi', .)$ of S by $\mathbb{G}$ are **equivalent** in case there is an isomorphism $\theta : E \longrightarrow E'$ which preserves projection to S and the action of $\mathbb{G}$: that is, $\pi' \theta x = \pi x$ for all $x \in E$, so that $\theta E_a = E'_a$ for all a, and $\theta(g \,.\, x) = g \,.\, \theta x$ whenever $x \in E_a$ and $g \in G_a$. Group coextensions need only be constructed up to equivalence.

When $\mathcal{E} = (E, \pi, .)$ is a commutative group coextension of S by $\mathbb{G} = (G, \gamma)$, it is convenient to denote $\gamma_{a,t}(g)$ by g^t; then

$$(g \,.\, x)\, y = g^b \,.\, xy \tag{1}$$

when $x \in E_a$, $y \in E_b$, and $g \in G_a$. Also

$$1^a = 1, \quad (gh)^a = g^a\, h^a, \quad g^1 = g, \quad \text{and} \quad (g^a)^b = g^{ab},$$

whenever defined, since $\gamma_{a,t}$ is a homomorphism and (G, γ) is a functor.

Group coextensions are constructed, like group extensions, by Schreier's method [1926]. A **cross section** of $\mathcal{E}$ (actually, of the congruence $\ker \pi$ induced by π) is a family $(p_a)_{a \in S}$ of elements of E such that $p_a \in E_a$ ($\pi p_a = a$) for all $a \in S$. Since G_a acts simply and transitively on E_a, every element x of E can be written uniquely in the form $x = g \cdot p_a$ with $a = \pi x \in S$ and $g \in G_a$.

For every $a, b \in S$ we have $p_a\, p_b \in E_{ab}$ and

$$p_a\, p_b = s_{a,b} \cdot p_{ab} \tag{2}$$

for some unique $s_{a,b} \in G_{ab}$. The family $s = (s_{a,b})_{a,b \in S}$ is the **factor set** of the group coextension $\mathcal{E}$ **relative to** the cross section p. For all $g \cdot p_a$ and $h \cdot p_b \in E$, (1) and (2) yield:

$$p_a\,(h \cdot p_b) = (h \cdot p_b)\, p_a = h^a \cdot (p_b\, p_a) = h^a \cdot (p_a\, p_b) = (h^a\, s_{a,b}) \cdot p_{ab}$$

and

$$(g \cdot p_a)(h \cdot p_b) \;=\; g^b \cdot (p_a\,(h \cdot p_b)) \;=\; (g^b\, s_{a,b}\, h^a) \cdot p_{ab} \tag{M}$$

for all $a, b \in S$ and $g \in G_a$, $h \in G_b$. Thus the operation on E is determined by the factor set and we obtain the following construction, which is similar to Schreier's Theorem.

Theorem 4.1. *Let S be a commutative semigroup and $\mathbb{G} = (G, \gamma)$ be an abelian group valued functor on S.*

Let $s = (s_{a,b})_{a,b \in S}$ be a family such that $s_{a,b} \in G_{ab}$ and

$$s_{b,a} \;=\; s_{a,b} \tag{C}$$

$$s_{a,b}^c\, s_{ab,c} \;=\; s_{a,bc}\, s_{b,c}^a \tag{A}$$

for all $a, b, c \in S$. Let $\mathcal{E}(s) = (E(s), \pi, .)$, where $E(s) = \bigcup_{a \in S} G_a \times \{a\}$ is the set of all ordered pairs (g, a) with $a \in S$ and $g \in G_a$, with multiplication defined for all $a, b \in S$ and $g \in G_a$, $h \in G_b$ by:

$$(g, a)(h, b) \;=\; (g^b\, s_{a,b}\, h^a,\ ab), \tag{3}$$

and $\pi\,(g, a) \;=\; a$, $g \cdot (h, a) \;=\; (gh, a)$. Then $\mathcal{E}(s)$ is a commutative group coextension of S by $\mathbb{G}$.

If conversely $\mathcal{E}$ is a commutative group coextension of S by $\mathbb{G}$ and s is the factor set of $\mathcal{E}$ relative to any cross section, then (A) *and* (C) *hold and $\mathcal{E}$ is equivalent to $\mathcal{E}(s)$.*

Proof. In $E(s)$ we have

$$\begin{aligned}
\big((g,a)(h,b)\big)(k,c) &= (g^b\, s_{a,b}\, h^a,\, ab)(k,c) \\
&= (g^{bc}\, s_{a,b}^c\, h^{ac} s_{ab,c}\, k^{ab},\ (ab)\,c), \\
(g,a)\big((h,b)(k,c)\big) &= (g,a)(h^c\, s_{b,c}\, k^b,\, bc) \\
&= (g^{bc}\, s_{a,bc}\, h^{ac}\, s_{b,c}^a\, k^{ab},\ a\,(bc)).
\end{aligned}$$

Hence associativity in $E(s)$ is equivalent to (A). Commutativity follows from (C). We see that G_a acts simply and transitively on $E(s)_a = \{(g,a) \mid g \in G_a\}$, and that

$$\big(g \mathbin{.} (h,a)\big)(k,b) = (gh,a)(k,b) = (g^b\, h^b\, s_{a,b}\, k^a,\, ab) = g^b \mathbin{.} \big((h,a)(k,b)\big);$$

thus (1) holds and $\mathcal{E}(s)$ is a commutative group coextension of S by $\mathbb{G}$.

Conversely let $\mathcal{E} = (E, \pi, .)$ be any commutative group coextension of S by $\mathbb{G}$ and s be its factor set relative to a cross section p. We already proved (2) and (M). By (M),

$$\begin{aligned}
p_a\,(p_b\, p_c) &= (1 \mathbin{.} p_a)(s_{b,c} \mathbin{.} p_{bc}) &= s_{a,bc}\, s_{b,c}^a \mathbin{.} p_{abc}\,, \\
(p_a\, p_b)\, p_c &= (s_{a,b} \mathbin{.} p_{ab})(1 \mathbin{.} p_c) &= s_{a,b}^c\, s_{ab,c} \mathbin{.} p_{abc}\,;
\end{aligned}$$

hence (A) holds. (C) holds since E is commutative. Every element of E can be written in the form $g \mathbin{.} p_a$ for some unique $a \in S$ and $g \in G_a$; hence the mapping $\theta : E(s) \longrightarrow E$, $(g,a) \longmapsto g \mathbin{.} p_a$ is bijective. Then (M) and (3) show that θ is an isomorphism, which we see preserves projection to S and the action of $\mathbb{G}$. Hence the group coextensions $\mathcal{E}$ and $\mathcal{E}(s)$ are equivalent. □

A **commutative factor set** on S with values in $\mathbb{G}$ is a family $s = (s_{a,b})_{a,b \in S}$ such that $s_{a,b} \in G_{ab}$ for all $a, b \in S$ and (A) and (C) hold. Relative to $p_a = (1,a)$, the factor set of $\mathcal{E}(s)$ is s, since $(1,a)(1,b) = (s_{a,b},\, ab) = s_{a,b} \mathbin{.} (1,ab)$. By Theorem 4.1, s is a commutative factor set if and only if it is the factor set of a group coextension relative to some cross section.

Commutative factor sets sometimes turn up in unexpected places. Factor sets with values in $\mathbb{R}$ (more precisely, in the constant abelian group valued functor $\mathbb{G}$ on S with $G_a = \mathbb{R}$ for all $a \in S$) determine nonprobabilistic information measures on S^n (Ebanks [1979]). Factor sets also turn up in the construction of certain semigroup algebras. Normally the product $a.b$ of $a, b \in S$ in $K[S]$ is their product ab in S. One may define a different multiplication on $K[S]$ by $a.b = s_{a,b}\, ab$, where $s_{a,b} \in K$; this yields a commutative ring if and only if $s_{b,a} = s_{a,b}$ and $s_{a,b}\, s_{ab,c} = s_{a,bc}\, s_{b,c}$ for all $a, b, c \in S$. This construction has been used by Kunz [1966] and others.

2. Equivalence of group coextensions can be detected from their factor sets:

Proposition 4.2. *Two group coextensions of S by $\mathbb{G}$ are equivalent if and only if there exists a family $u = (u_a)_{a \in S}$ with $u_a \in G_a$ for all $a \in S$ such that their factor sets s and t (relative to any cross sections) satisfy*

$$t_{a,b} = s_{a,b}\, u_a^b\, u_{ab}^{-1}\, u_b^a \qquad \text{(E)}$$

for all $a,b \in S$; then both s and t are factor sets of either extension relative to different cross sections.

Proof. Let $\theta : E(t) \longrightarrow E(s)$ be an equivalence of group coextensions. Since θ preserves projection to S we have $\theta(1,a) = (u_a, a)$ for every $a \in S$, where $u_a \in G_a$; then

$$\theta(g,a) = (gu_a, a),$$

since θ preserves the action of G_a. In $E(t)$, $(1,a)(1,b) = (t_{a,b}, ab)$; since θ is a homomorphism,

$$\begin{aligned}(t_{a,b}\, u_{ab}, ab) &= \theta(t_{a,b}, ab) = \theta\big((1,a)(1,b)\big) \\ &= \theta(1,a)\,\theta(1,b) = (u_a, a)(u_b, b) = (u_a^b\, s_{a,b}\, u_b^a, ab)\end{aligned}$$

for all $a,b \in S$. Thus (E) holds. By Theorem 4.1, (E) then holds whenever s and t are the factor sets of two equivalent coextensions of S by $\mathbb{G}$, relative to any cross sections.

The calculation above also shows that $(u_a, a)(u_b, b) = t_{a,b} \cdot (u_{ab}, ab)$ in $\mathcal{E}(s)$, so that t is the factor set of $\mathcal{E}(s)$ relative to the cross section $q_a = (u_a, a)$. By the same calculation, if s is the factor set of a group coextension relative to a cross section p, then t is the factor set of the same coextension relative to the cross section $q_a = u_a \cdot p_a$.

If conversely (E) holds, then $\theta : E(t) \longrightarrow E(s)$, defined by $\theta(g,a) = (gu_a, a)$, is a bijection which preserves projection to S and the action of $\mathbb{G}$. Moreover,

$$\begin{aligned}\theta(g,a)\,\theta(h,b) &= (gu_a, a)(hu_b, b) = (g^b\, u_a^b\, s_{a,b}\, u_b^a\, h^a, ab) \\ &= (g^b\, t_{a,b}\, u_{ab}\, h^a, ab) = \theta\big((g,a)(g,b)\big),\end{aligned}$$

and θ is an equivalence of coextensions. By Theorem 4.1, any two coextensions with these factor sets must then be equivalent. □

Two commutative factor sets s and t on S with values in $\mathbb{G}$ are **equivalent** in case (E) holds, i.e. there exists a family $u = (u_a)_{a \in S}$ with $u_a \in G_a$ for all $a \in S$ such that

$$t_{a,b} = s_{a,b}\, u_a^b\, u_{ab}^{-1}\, u_b^a$$

for all $a, b \in S$. By Proposition 4.2, two factor sets are equivalent if and only if they are the factor sets of two equivalent group coextensions (relative to any cross sections), if and only if they arise from two cross sections of the same group coextension.

3. A commutative group coextension **splits** when it satisfies the equivalent conditions in the next result.

Corollary 4.3. *For a commutative group coextension E of S by $\mathbb{G}$ the following conditions are equivalent:*

(1) *the projection $\pi : E \longrightarrow S$ splits (there exists a homomorphism $\mu : S \longrightarrow E$ such that $\pi \circ \mu = 1_S$);*

(2) *there exists a cross section of E which is a subsemigroup;*

(3) *there exists a cross section of E relative to which $s_{a,b} = 1$ for all $a, b \in S$;*

(4) *relative to any cross section of E the factor set satisfies*

$$s_{a,b} = u_a^b\, u_{ab}^{-1}\, u_b^a \tag{B}$$

for some family $u = (u_a)_{a \in S}$ with $u_a \in G_a$ for all a.

Proof. If $\pi \circ \mu = 1_S$, then $\operatorname{Im} \mu = (\mu a)_{a \in S}$ is a cross section and a subsemigroup, so (1) $\Longrightarrow$ (2). If a cross section $p = (p_a)_{a \in S}$ is a subsemigroup, then $p_a\, p_b = p_{ab}$, since $p_a\, p_b = p_c$ for some c and $\pi(p_a\, p_b) = ab$, so that (2) $\Longrightarrow$ (1), and then $s_{a,b} = 1$, so that (2) $\Longrightarrow$ (3); conversely (3) $\Longrightarrow$ (2). Finally, (3) $\Longleftrightarrow$ (4) by Proposition 4.2. □

A commutative factor set **splits** when (B) holds, i.e. there exists a family $u = (u_a)_{a \in S}$ such that $u_a \in G_a$ for all $a \in S$ and

$$s_{a,b} = u_a^b\, u_{ab}^{-1}\, u_b^a$$

for all $a, b \in S$; equivalently, when it is the factor set of a split coextension.

For any c.s. S and abelian group valued functor $\mathbb{G}$ on S, a split group coextension of S by $\mathbb{G}$ always exists: indeed the trivial factor set $s_{a,b} = 1$ satisfies conditions (A) and (C). By Proposition 4.2, for any family $u = (u_a)_{a \in S}$ such that $u_a \in G_a$ for all $a \in S$, the formula $s_{a,b} = u_a^b\, u_{ab}^{-1}\, u_b^a$ then defines a commutative factor set; that is, every family $u_a \in G_a$ gives rise to a split factor set. This can also be verified directly.

Proposition 4.4. *Every commutative group coextension of a semilattice by a thin functor is a semilattice of groups and hence splits.*

Proof. Let $\mathcal{E} = (E, \pi, .)$ be a commutative group coextension of a semilattice S by a thin functor. Every thin functor on a semilattice is surjecting; by Theorem 3.4, E is an exact $\mathcal{H}$-coextension of the semilattice S and is a semilattice of groups by Corollary 1.5. Then Clifford's Theorem III.2.1 shows that E splits; more directly, if p_e is the idempotent of the group E_e, then $p_e\, p_f$ is the idempotent of E_{ef} and $p_e\, p_f = p_{ef}$. □

Free c.s. S also have the property that every commutative group coextension of S splits. Semigroups with this property are studied in more detail in Chapter XIV.

4. A commutative factor set s is **normalized** when $s_{e,a} = s_{a,e} = 1$ whenever $e^2 = e$ and $ea = a$. This property may always be assumed if $\mathbb{G}$ is thin:

Proposition 4.5. *Every commutative factor set with values in a thin functor is equivalent to a normalized factor set.*

Two normalized commutative factor sets s and t on S with values in a thin functor $\mathbb{G}$ are equivalent if and only if there exists a family $u = (u_a)_{a \in S}$ such that $u_a \in G_a$ for all $a \in S$, $u_e = 1$ whenever e is idempotent, and

$$s_{a,b} \;=\; t_{a,b}\, u_a^b\, u_{ab}^{-1}\, u_b^a \tag{E}$$

for all $a, b \in S$.

Proof. Let E be a group coextension of S by some thin abelian group valued functor $\mathbb{G}$. As in Proposition 1.2, every idempotent e of S lifts to an idempotent of E: indeed $g^e = g^1 = g$ for all $g \in G_e$, since $ee = e = e1$, so that, relative to any cross section p, $s_{e,e}^{-1} \cdot p_e \in E_e$ is idempotent by (M). Hence there is a cross section q so that q_e is idempotent whenever e is idempotent. The factor set t relative to q is then equivalent to s and satisfies $t_{e,e} = 1$ whenever e is idempotent. Then t is normalized: if $ea = a$, then $g^e = g$ for all $g \in G_a$, since $\mathbb{G}$ is thin; hence (A), applied to e, e, and a, yields $t_{e,e}^a\, t_{ee,a} = t_{e,ea}\, t_{e,a}^e$ and $t_{e,a} = 1$.

Now let s and t be normalized and satisfy (E). When e is idempotent, then $1 = t_{e,e} = s_{e,e}\, u_e^e\, u_e^{-1}\, u_e^e = u_e$, since $\mathbb{G}$ is thin; hence the last part of the statement follows from Proposition 4.2. □

5. Theorem 4.1 and Proposition 4.2 imply that the equivalence classes of commutative group coextensions of S by $\mathbb{G}$ constitute an abelian group $\mathrm{Ext}\,(S,\mathbb{G})$. Indeed commutative factor sets constitute an abelian group Z under pointwise multiplication $(st)_{a,b} = s_{a,b}\, t_{a,b}$, since this operation preserves (A) and (C). Split factor sets (of the form $u_a^b\, u_{ab}^{-1}\, u_b^a$) constitute a subgroup B of Z. By Proposition 4.2 there is a one-to-one correspondence between the cosets of B in Z and the equivalence classes of commutative group coextensions of S by $\mathbb{G}$. Therefore the latter are the elements of an abelian group, the **extension group** $\mathrm{Ext}\,(S,\mathbb{G}) \cong Z/B$.

$\mathrm{Ext}\,(S,\mathbb{G})$ generalizes the usual Ext for abelian groups:

Proposition 4.6. *If C is an abelian group, then every abelian group valued functor $\mathbb{G}$ on C is naturally isomorphic to a constant functor A, and then* $\mathrm{Ext}\,(C,\mathbb{G})$ *(in the semigroup sense) coincides with* $\mathrm{Ext}\,(C,A)$ *(in the abelian group sense).*

Naturally we identify the constant functor A and its constant value A.

Proof. C has an identity element 1, so that $C = C^1$. By definition, $\gamma_{a,1}$ is the identity on G_a, and then $\gamma_{at,u}$ and $\gamma_{a,t}$ are mutually inverse isomorphisms whenever $u = t^{-1}$ in C. Hence $\gamma_{1,a} : G_1 \longrightarrow G_a$ is a natural isomorphism from $\mathbb{G}$ to the constant functor G_1. The result then follows from Proposition 3.2. □

When S is not an abelian group, $\mathrm{Ext}\,(S,\mathbb{G})$ is most interesting when S is group-free and $\mathbb{G}$ is thin and surjecting, for then $\mathrm{Ext}\,(S,\mathbb{G})$ classifies c.s. rather than just coextensions; that is, $\mathrm{Ext}\,(S,\mathbb{G})$ classifies the commutative semigroups T with (up to isomorphisms) $T/\mathcal{H} = S$ and Schützenberger functor $\mathbb{G}$ (Theorem 3.4).

6. Results in the last chapters will simplify the generally daunting task of finding all factor sets on a given c.s. with values in a given functor, or at least their equivalence classes. For now we give one example where all factor sets are readily found.

By Corollary 3.6, a complete archimedean semigroup is a commutative group coextension of a nilsemigroup N by a thin surjecting functor $\mathbb{G}$; then G_a is trivial for every $a \neq 0$ in N, and only G_0 can be nontrivial.

More generally, an abelian group valued functor $\mathbb{G}$ on a semigroup with zero S is **almost null** when $\mathbb{G}$ is thin and G_a is trivial for every $a \in S\backslash 0$; then $\mathbb{G}$ is surjecting as well.

Proposition 4.7. *Let S be a semigroup with zero and $\mathbb{G}$ be an almost* *null abelian group valued functor on S. Up to isomorphism, a c.s. is a group coextension of S by $\mathbb{G}$ if and only if it is an ideal extension of G_0 by S, and then there is a one-to-one correspondence between normalized factor sets on S with values in $\mathbb{G}$ and partial homomorphisms from $S\backslash 0$ to G_0. Hence* $\mathrm{Ext}(S,\mathbb{G}) \cong \mathrm{PHom}(S\backslash 0,\, G_0) \cong \mathrm{Hom}(G(S\backslash 0), G_0)$.

Proof. Let s be a normalized factor set on S with values in $\mathbb{G}$. In $E(s)$, $H = \{(g,0) \mid g \in G_0\}$ is a subgroup of which is isomorphic to G_0, since $s_{0,0} = 1$, and an ideal. The Rees quotient $E(s)/H$ is isomorphic to S, since $\mathbb{G}$ is almost null. Every commutative group coextension of S by $\mathbb{G}$ is isomorphic to some $E(s)$ and is likewise an ideal extension of G_0 by S.

Conversely, every ideal extension T of G_0 by S comes with a projection $\pi : T \longrightarrow T/G_0 \cong S$. Also G_0 acts on itself by left multiplication, and every trivial G_a acts trivially (how else) on $\pi^{-1}(a) = \{a\}$. This makes T a group coextension of S by $\mathbb{G}$.

Let s be a normalized factor set on S with values in $\mathbb{G}$. Then $s_{a,b} \in G_0$ when $a,b \in S\backslash 0$ and $ab = 0$, and $s_{a,0} \in G_0$; also $s_{0,0} = 1$ and $s_{a,b} = 1$ whenever $ab \neq 0$ in S, for then $G_{ab} = \{1\}$. When $a,b \in S\backslash 0$, (A) reads:

$$s^0_{a,b}\, s_{ab,0} \;=\; s_{a,b0}\, s^a_{b,0} \quad \text{or} \quad s_{ab,0} \;=\; s_{a,0}\, s_{b,0}$$

if $ab \neq 0$, since $\mathbb{G}$ is thin, and

$$s^0_{a,b}\, s_{ab,0} \;=\; s_{a,b0}\, s^a_{b,0} \quad \text{or} \quad s_{a,b} \;=\; s_{a,0}\, s_{b,0}$$

if $ab = 0$. Thus

$$\varphi(a) \;=\; s_{a,0}$$

is a partial homomorphism of $S\backslash 0$ into G_0. Moreover, s is completely determined by φ: indeed $s_{a,0} = \varphi(a)$ for all $a \neq 0$; $s_{a,b} = \varphi(a)\,\varphi(b)$ whenever $a,b \neq 0$ and $ab = 0$; all other values of s are 1.

Conversely let $\varphi : S\backslash 0 \longrightarrow G_0$ be a partial homomorphism. Let $s_{a,0} = s_{0,a} = \varphi(a)$ for all $a \neq 0$, $s_{a,b} = \varphi(a)\,\varphi(b)$ whenever $a,b \neq 0$ and $ab = 0$, and $s_{a,b} = 1$ whenever $ab \neq 0$. The calculations above show that s is a commutative factor set.

We now have a one-to-one correspondence between commutative factor sets on S with values in $\mathbb{G}$, and partial homomorphisms $S\backslash 0 \longrightarrow G_0$, which we see preserves pointwise multiplication. By Proposition 4.4, two normalized factor sets on S with values in $\mathbb{G}$ are equivalent if and only if they are equal. Hence

$\mathrm{Ext}(S, \mathbb{G}) \cong \mathrm{PHom}(S\backslash 0,\ G_0)$, and $\mathrm{PHom}(S\backslash 0,\ G_0) \cong \mathrm{Hom}(G(S\backslash 0), G_0)$ by Proposition III.3.3. □

In the above the coextension $E(s)$ can be constructed as follows. In $E(s)$ we can identify $(1, a)$ and a for every $a \in S\backslash 0$, and identify $(g, 0)$ and g for every $g \in G_0$. Then $E(s) = (S\backslash 0) \cup G_0$, with multiplication $*$ given for all $a, b \in S\backslash 0$ and $g, h \in G_0$ by: $a * b = ab$ if $ab \neq 0$ in S; $a * b = s_{a,b} = \varphi(a)\varphi(b)$ if $ab = 0$ in S; $g * a = a * g = s_{a,0}\, g = \varphi(a) g$; $g * h = gh$. This is precisely the construction in Lemma I.3.7. Thus, group coextensions and ideal extensions give equivalent constructions in this case.

7. **Summary.** The results obtained so far construct every complete (for instance, finite) commutative semigroup S in three ways.

(1). S is a semilattice of archimedean semigroups containing idempotents. Such an archimedean semigroup is an ideal extension of an abelian group G by a commutative nilsemigroup N and is determined by a partial homomorphism of $N\backslash 0$ into G, or by a homomorphism of $G(N\backslash 0)$ into G. Group coextensions give essentially the same construction. However, S is not in general readily reassembled from its archimedean components.

(2). S is a subdirect product of its Ponizovsky factors, which consist of an abelian group, a commutative nilsemigroup, and elementary semigroups. An elementary semigroup $G \cup N$ is an ideal extension of a commutative nilsemigroup N and is determined by a suitable action of the abelian group G on N; equivalently, $G \cup N$ is a group coextension of its monoid of orbits, which is a commutative nilmonoid. Then S can be reassembled from its Ponizovsky factors by means of partial homomorphisms, which may, however, be difficult to construct.

(3). S is a commutative group coextension of the complete group-free semigroup $S/\mathcal{H}$ by a thin surjecting functor, which Theorem 4.1 then constructs from factor sets on $S/\mathcal{H}$. The group-free semigroup $S/\mathcal{H}$ can be assembled, with the usual difficulties, from nilsemigroups: its archimedean components are nilsemigroups; its Ponizovsky factors are the trivial group, a commutative nilsemigroup, and nilmonoids. Finite nilsemigroups or group-free semigroups have also been constructed, by induction or from trees, by Yamada [1964], Tamura [1968], John [1973], and others.

5. SUBDIRECTLY IRREDUCIBLE SEMIGROUPS.

As an application of the results in this chapter we determine all subdirectly irreducible finite commutative semigroups. The results are from Grillet [1977].

1. From Chapter IV we know:

A complete c.s. which is subdirectly irreducible is either a group, or a nilsemigroup, or an elementary semigroup (Corollary IV.4.6).

An abelian group is subdirectly irreducible if and only if it is a nontrivial cyclic or quasicyclic p-group (Proposition IV.1.7).

A commutative nilsemigroup N with enough minimal elements is subdirectly irreducible if and only if $N \neq 0$ and its greatest pure congruence $\mathcal{P}$ is the equality on N (Proposition IV.3.6). This result can be restated as follows. Let M denote the set of all minimal elements of N. Define

$$a \,\mathcal{M}\, b \iff (\forall u \in N^1)(ua \in M \iff ub \in M \implies ua = ub).$$

Then $\mathcal{M}$ is a congruence on N. In fact, $\mathcal{M}$ is a pure congruence, so that $\mathcal{M} \subseteq \mathcal{P}$. Call N **weakly irreducible** in case $\mathcal{M}$ is the equality.

Proposition 5.1. *A commutative nilsemigroup N with enough minimal elements is subdirectly irreducible if and only if it is weakly irreducible and has just one minimal element.*

Proof. Both conditions are necessary, since M is a $\mathcal{P}$-class and $\mathcal{M} \subseteq \mathcal{P}$, and $\mathcal{P}$ is the equality when N is subdirectly irreducible (Proposition IV.3.6). Conversely assume that N is weakly irreducible and has just one minimal element m. Then N has at least two elements. Let $a, b \in N$, $a \neq b$. Since $a \,\mathcal{M}\, b$ does not hold there exists $u \in N^1$ such that, say, $ua = m$, $ub \neq m$. If $ub = 0$, then $a \,\mathcal{P}\, b$ does not hold. If $ub \neq 0$, then $ub > m$, $m = vub$ for some $v \in N$, whereas $vua < ua = m$ and $vua = 0$; hence again $a \,\mathcal{P}\, b$ does not hold. Thus $\mathcal{P}$ is the equality and N is subdirectly irreducible. □

2. We now turn to elementary semigroups $S = G \cup N$. The cases where G or N are trivial are covered by:

Proposition 5.2. *If S has at least two elements, then S is (finitely) subdirectly irreducible if and only if S^0 is (finitely) subdirectly irreducible, if and only if S^1 is (finitely) subdirectly irreducible.*

When S is trivial, then $S \cup \{0\}$ and $S \cup \{1\}$ have two elements and are subdirectly irreducible.

Proof. Assume that S does not have a zero element. Every congruence $\mathcal{C}$ on S extends to a congruence $\mathcal{C}^0$ on S^0, whose classes are $\{0\}$ and all the $\mathcal{C}$-classes. If $\mathcal{C}$ is proper then $\mathcal{C}^0$ is proper.

Conversely let $\mathcal{C}$ be a congruence on S^0 whose restriction to S is the equality on S. If $\mathcal{C}$ is not the equality on S^0, then $0 \mathcal{C} z$ for some $z \in S$, and then $0 \mathcal{C} xz$ and $xz = z$ for all $x \in S$, contradicting the hypothesis. Therefore a proper congruence on S^0 restricts to a proper congruence on S.

If S^0 is not subdirectly irreducible, then the equality on S^0 is the intersection of proper congruences on S^0, and the equality on S is the intersection of their restrictions to S, which are proper congruences; hence S is not subdirectly irreducible. If conversely S is not subdirectly irreducible, then the equality on S is the intersection of proper congruences on S, and the equality on S^0 is the intersection of their extensions to S^0, which are proper congruences; hence S^0 is not subdirectly irreducible. Similarly, S is finitely subdirectly irreducible if and only if S^0 is finitely subdirectly irreducible.

The proof is similar for S^1. Assume that S does not have an identity element. Every proper congruence $\mathcal{C}$ on S extends to a proper congruence $\mathcal{C}^1$ on S^1, whose classes are $\{1\}$ and all the $\mathcal{C}$-classes. Conversely let $\mathcal{C}$ be a congruence on S^1 whose restriction to S is the equality on S. If $\mathcal{C}$ is not the equality on S^1, then $1 \mathcal{C} e$ for some $e \in S$, and then $x \mathcal{C} xe$ and $xe = x$ for all $x \in S$, contradicting the hypothesis. Therefore a proper congruence on S^1 restricts to a proper congruence on S. As above it follows that S is (finitely) subdirectly irreducible if and only if S^1 is (finitely) subdirectly irreducible. □

3. We now let $S = G \cup N$ be an elementary semigroup in which G and N have at least two elements and N has enough minimal elements. We saw in Section IV.5 that the group G acts on S by left multiplication; under this action, the orbits are precisely the $\mathcal{H}$-classes of S (Proposition IV.5.1); they constitute a commutative nilmonoid $\Omega^1 = S/\mathcal{H}$.

Lemma 5.3. *If S is finitely subdirectly irreducible, then N is weakly irreducible and M is an orbit.*

Proof. The definition of $\mathcal{M}$ shows that $\{0\}$ and every $\{m\} \subseteq M$ is an $\mathcal{M}$-class. Extend $\mathcal{M}$ to a congruence $\mathcal{M}'$ on S as follows: $a \mathcal{M}' b$ if and only if either $a = b \in G$ or $a, b \in N$ and $a \mathcal{M} b$. The intersection of $\mathcal{M}'$ and the Rees congruence of the ideal $M \cup \{0\}$ is the equality on S; therefore $\mathcal{M}'$ is the equality on S, and N is weakly irreducible.

Every $\mathcal{P}$-class is a union of orbits (Lemma IV.5.2); hence M is a union of orbits. If M contains two distinct orbits M_1 and M_2, then $M_1 \cup \{0\}$ and $M_2 \cup \{0\}$ are ideals of S, whose Rees congruences have the equality on S as intersection. When S is finitely subdirectly irreducible this cannot happen, and M is a single orbit. □

An elementary semigroup $S = G \cup N$ is **homogeneous** when G acts simply on $N \backslash 0$; equivalently, when $gx = hx \neq 0$ implies $g = h$ when $g, h \in G$; equivalently, when $gx = x \neq 0$ implies $g = 1$ when $g \in G$; equivalently, when, in the Schützenberger functor of S, γ_A^1 is an isomorphism for every nonzero orbit A.

Lemma 5.4. *If S is finitely subdirectly irreducible, then it is homogeneous.*

Proof. Let $g \in G$ and $c \in N$ satisfy $gc = c \neq 0$. Since N has enough minimal elements there is a minimal element $n \leqq c$, $n = uc$ for some $u \in N^1$. For every $m \in M$ we now have $m = hn$ for some $h \in G$ by Lemma 5.3 and $gm = hugc = huc = m$.

Let $\mathcal{C}$ be the congruence on S defined by: $a \,\mathcal{C}\, b$ if and only if $g^i a = b$ for some $a \in \mathbb{Z}$; the $\mathcal{C}$-classes are the orbits under the action of $\langle g \rangle \subseteq G$. We see that $\mathcal{C}$ is the equality on $M \cup \{0\}$, since $gx = x$ for all $x \in M \cup \{0\}$. Hence the intersection of $\mathcal{C}$ and the Rees congruence of the ideal $M \cup \{0\}$ is the equality on S. Since S is finitely subdirectly irreducible, it follows that $\mathcal{C}$ is the equality on S. But $g \,\mathcal{C}\, 1$, so $g = 1$. □

Lemma 5.5. *If S is subdirectly irreducible, or finitely subdirectly irreducible, then so is G.*

Proof. For every (normal) subgroup H of G, let H^* be the congruence on S defined by: $a \,H^*\, b$ if and only if $b \in Ha$. We see that H^* is a pure congruence on S, and that $g \,H^*\, h$ if and only if $g^{-1}h \in H$ (so that H and H^* induce the same congruence on G).

Assume that $\{1\}$ is the intersection of subgroups $(H_i)_{i \in I}$ of G. If $a \,H_i^*\, b$ for all i, then either $a = b = 0$, or for each $i \in I$ there is some $h_i \in H_i$ such that $b = h_i a \neq 0$. By Lemma 5.4 there is only one $g \in G$ such that $b = ga \neq 0$, and then $g = h_i \in H_i$ for all i, $g = 1$, and $a = b$. Thus $\bigcap_{i \in I} H_i^*$ is the equality on S. If S is subdirectly irreducible, then some H_i^* is the equality on S, and then $H_i = \{1\}$. This shows that G is subdirectly irreducible. If similarly S is finitely subdirectly irreducible, then G is finitely subdirectly irreducible. □

Theorem 5.6. *Let $S = G \cup N$ be an elementary semigroup, in which G and N have at least two elements and N has enough minimal elements. Then S is (finitely) subdirectly irreducible if and only if G is (finitely) subdirectly irreducible, N is weakly irreducible, the minimal elements of N form an orbit, and S is homogeneous.*

Proof. These conditions are necessary by Lemmas 5.3, 5.4, 5.5. Conversely assume that S has all four properties.

Let $\mathcal{C}$ be a proper congruence on S, so that $a \mathcal{C} b$ for some $a, b \in S$, $a \neq b$. We show that $\mathcal{C}$ is not the equality on M. If $a, b \in G$, then for any $m \in M$ we have $am, bm \in M$, $am \mathcal{C} bm$, and $am \neq bm$ since S is homogeneous. If $a \in G$ and $b \in N$, then for any $m \in M$ we have $am \in M$, $bm = 0$, and $n \mathcal{C} 0$ for some $n \in M$; since M is an orbit it follows that $n \mathcal{C} 0$ for all $n \in M$; since S is homogeneous and G is nontrivial we have $m \mathcal{C} n$ for some $m \neq n$ in M. Finally assume $a, b \in N$. Since N is weakly irreducible there exists $u \in N^1$ such that, say, $ua \in M$ and either $ub \in M$, $ub \neq ua$, or $ub \notin M$. In the first case, $\mathcal{C}$ is not the equality on M. Now assume $ub \notin M$. If $ub = 0$, then $ua \mathcal{C} 0$ with $ua \in M$, and as above $\mathcal{C}$ is not the equality on M. If $ub \neq 0$, then $ub > m$ for some $m \in M$, $m = vub$ for some $v \in N$, $vua = 0$ since $ua \in M$, $m = vub \mathcal{C} vua = 0$, and again $\mathcal{C}$ is not the equality on M.

Choose any $m \in M$. For every congruence $\mathcal{C}$ on S define a congruence $\mathcal{C}'$ on G by: $g \mathcal{C}' h$ if and only if $gm \mathcal{C} hm$. If $\mathcal{C}$ is a proper congruence, then by the above $\mathcal{C}$ is not the equality on M and $\mathcal{C}'$ is a proper congruence. Also $\mathcal{C} = \bigcap_{i \in I} \mathcal{C}_i$ implies $\mathcal{C}' = \bigcap_{i \in I} \mathcal{C}'_i$: indeed $g \mathcal{C}'_i h$ for all i is equivalent to $gm \mathcal{C}_i hm$ for all i, to $gm \mathcal{C} hm$, and to $g \mathcal{C}' h$. Since G is (finitely) subdirectly irreducible, the equality on S cannot be the intersection of (finitely many) proper congruences on S, for then the equality on G would be the intersection of (finitely many) proper congruences on G. Thus S is (finitely) subdirectly irreducible. $\square$

Corollary 5.7. *A finite c.s. is subdirectly irreducible if and only if it is one of the following:*

a nontrivial cyclic p-group;

a finite nilsemigroup on which $\mathcal{P}$ is the equality;

$\{0, 1\}$*; a nontrivial cyclic p-group with a zero element adjoined; a finite nilmonoid on which $\mathcal{P}$ is the equality; a homogeneous elementary semigroup $S = G \cup N$ in which G and N have at least two elements, G is a cyclic p-group, N is weakly irreducible, and the minimal elements of N form an orbit.*

4. The elementary semigroup $S = G \cup N$ is a group coextension of its monoid of orbits Ω^1, as in Corollary 3.7. We note the following:

N has enough minimal elements if and only if Ω does;

M is an orbit if and only Ω has only one minimal element (namely, M); then Ω has enough minimal elements if and only if M is the least nonzero element of Ω;

S is homogeneous if and only if its Schützenberger functor is naturally isomorphic to the **almost constant** functor (G, γ), in which $G_A = G$ for every nonzero orbit A, $G_0 = \{1\}$, and $\gamma_{A,T}$ is the identity on G if $AT \neq 0$.

Proposition 5.8. *Let $S = G \cup N$ be a homogeneous elementary semigroup, in which G and N have at least two elements, N has enough minimal elements, and the minimal elements of N form an orbit M. Let s be any factor set of S as a group coextension of its monoid of orbits Ω^1. Then N is weakly irreducible if and only if, for every $A, B \in \Omega$ with $A \mathcal{M} B$ and $A \neq B$, the function $f(T) = s_{A,T}\, s_{B,T}^{-1} \in G$ is not constant on $\mathcal{T} = \{T \in \Omega \mid AT = M\} = \{T \in \Omega \mid BT = M\}$.*

Proof. As in Theorem 4.1 we may assume that $S = E(s)$. Then S consists of 0 and all pairs (g, A) with $g \in G$, $A \in \Omega^1$, $A \neq 0$, with multiplication $(g, A)(h, B) = (gh\, s_{A,B}, AB)$ if $AB \neq 0$, $(g, A)(h, B) = 0$ if $AB = 0$. Also $(g, A) \in M$ if and only if $A = M$, $(g, A) \in G$ if and only if $A = 1$ (this identifies $(g, 1)$ and g).

Assume that N is weakly irreducible. Let $A, B \in \Omega$ satisfy $A \mathcal{M} B$ and $A \neq B$; in particular, $A, B \neq 0, 1, M$, $A, B > M$, and $\mathcal{T} = \{T \in \Omega \mid AT = M\} = \{T \in \Omega \mid BT = M\} \neq \emptyset$. Take any $T \in \mathcal{T}$ and let $a = (s_{B,T}, A)$ and $b = (s_{A,T}, B)$. Then $a \neq b$ and $a \mathcal{M} b$ does not hold. However, $au \in M$ if and only if $bu \in M$, for all $u \in N^1$, by the choice of A and B. Therefore there exists $u \in N^1$ such that $au, bu \in M$ and $au \neq bu$. In fact $u \in N$, since $A, B > M$. Hence $u = (g, U)$ for some $U \in \Omega$; then $AU = BU = M$ and

$$(s_{T,B}\, s_{A,U}\, g,\ M) \;=\; au \;\neq\; bu \;=\; (s_{T,A}\, s_{B,U}\, g,\ M);$$

therefore $s_{A,U}\, s_{B,U}^{-1} \neq s_{A,T}\, s_{B,T}^{-1}$ and f is not constant on $\mathcal{T}$.

Conversely, assume that f is not constant on $\mathcal{T}$, whenever $A, B \in \Omega$, $A \mathcal{M} B$, and $A \neq B$. Let $a, b \in N$, $a \neq b$. If either of a or b is zero or minimal, then $a \mathcal{M} b$ cannot hold in N. Let a and b be neither zero nor minimal, so that $a = (g, A)$, $b = (h, B)$, with $A, B > M$. If $A = B$, then $g \neq h$, $at \neq bt$ whenever $t \neq 0$, and $a \mathcal{M} b$ does not hold. Now assume $A \neq B$. If $A \mathcal{M} B$ does not hold in Ω, then $a \mathcal{M} b$ does not hold in N. Let $A \mathcal{M} B$.

Let $T \in \mathcal{T}$ and $t = (1, T)$. Then $at = (g\, s_{A,T}, M)$, $bt = (h\, s_{B,T}, M)$, and $at, bt \in M$. If $at \neq bt$, then $a \mathcal{M} b$ does not hold. If $at = bt$, then we use the hypothesis that f is not constant: $s_{A,U}\, s_{B,U}^{-1} \neq s_{A,T}\, s_{B,T}^{-1}$ for some $U \in \mathcal{T}$. Then $g\, s_{A,U} \neq h\, s_{B,U}$ and $u = (1, U)$ satisfies $au \neq bu$ and $au, bu \in M$;

hence $a \mathcal{M} b$ does not hold. Thus $\mathcal{M}$ is the equality on N. □

By Proposition 5.8, if Ω is weakly irreducible then so is N. In fact, if S splits, then s can be chosen so that $s_{A,T}\, s_{B,T}^{-1} = 1$ for all T, and N is weakly irreducible if and only if Ω is. But Proposition 5.8 does not in general imply that Ω is weakly irreducible.

Example 5.9. Let $\Omega = \{A, B, M, 0\}$ with $A^2 = AB = B^2 = M$ and all other products equal to 0. In particular, all products of three or more elements of Ω are zero. We have $A \mathcal{M} B$, with $\mathcal{T} = \{A, B\}$, so that Ω is not weakly irreducible. On the other hand, a normalized factor set on Ω^1 consists of $s_{A,A}$, $s_{A,B}$, $s_{B,A}$, and $s_{B,B} \in G$ such that $s_{B,A} = s_{A,B}$. If G is not trivial there is a factor set s such that $s_{A,A}\, s_{B,A}^{-1} \neq s_{A,B}\, s_{B,B}^{-1}$. Then N is weakly irreducible in the corresponding group coextension $S = G \cup N$, by Proposition 5.8. □

Chapter VI.

FINITELY GENERATED SEMIGROUPS.

This chapter contains basic properties of finitely generated commutative semigroups: Rédei's Theorem, various finiteness properties, and the Completion Theorem. The results indicate that these semigroups are assembled somehow from cancellative semigroups and nilsemigroups, arranged along a finite semilattice. This similarity to finite semigroups is pursued much further in the next chapter.

This chapter also studies subelementary semigroups, subdirectly irreducible semigroups, and other smaller classes.

For the sake of brevity we often abbreviate "finitely generated" as **f.g.** in what follows.

1. RÉDEI'S THEOREM.

Rédei's Theorem states that the congruences on a finitely generated free c.s. are themselves finitely generated. This section contains Preston's proof and some consequences of the Theorem. Other proofs were devised by Drbohlav [1963] and Budach [1964] and still another proof will be found in Chapter XIII.

1. Let X be a set. To every element $a = \sum_{x \in X} a_x x$ of the free c.s. F_X corresponds a monomial $X^a = \prod_{x \in X} x^{a_x}$ in the polynomial ring $\mathbb{Z}[X]$ with commuting indeterminates in X. We see that $X^a X^b = X^{a+b}$ for all $a, b \in F_X$.

Freyd [1968] and Preston [1975] observed that every ideal I of $\mathbb{Z}[X]$ **induces** a congruence $\mathcal{C}$ on F_X, for which $a \mathcal{C} b$ if and only if $X^a - X^b \in I$; and that, conversely, every congruence on F_X is induced by an ideal of $\mathbb{Z}[X]$:

Lemma 1.1. *Let $\mathcal{C}$ be a congruence on F_X. Let $I(\mathcal{C})$ be the ideal of $\mathbb{Z}[X]$ generated by all $X^a - X^b$ with $a \mathcal{C} b$. Then $a \mathcal{C} b$ if and only if $X^a - X^b \in I(\mathcal{C})$.*

Proof. Since $\mathcal{C}$ is a congruence, $I(\mathcal{C})$ is the set of all linear combinations $\sum_{1 \leqq j \leqq r} m_j \left(X^{a_j} - X^{b_j} \right)$ in which $r \geqq 0$, m_j is a positive integer, $a_j \neq b_j$ in F_X, and $a_j \; \mathcal{C} \; b_j$, for all j. If $X^a - X^b \in I(\mathcal{C})$, where $a \neq b$ in F_X, then

$$X^a - X^b = \sum_{1 \leqq j \leqq r} m_j \left(X^{a_j} - X^{b_j} \right), \qquad (*)$$

with $m_j > 0$, $a_j \neq b_j$, and $a_j \; \mathcal{C} \; b_j$, for all j. Then $r > 0$, since $X^a \neq X^b$. We prove by induction on $m = \sum_{1 \leqq j \leqq r} m_j > 0$ that $(*)$ implies $a \; \mathcal{C} \; b$.

If $m = 1$, then $r = 1$, $m_1 = 1$, $a = a_1$, $b = b_1$, and $a \; \mathcal{C} \; b$. Now let $m > 1$. The coefficient of X^a in the right hand side of $(*)$ is $\sum_{a_j = a} m_j - \sum_{b_j = a} m_j = 1$; therefore $a_k = a$ for some k. If $b_k = b$, then $a \; \mathcal{C} \; b$ and we are done. If $b_k \neq b$, then the coefficient of X^{b_k} in the right hand side of $(*)$ is $\sum_{a_j = b_k} m_j - \sum_{b_j = b_k} m_j = 0$, since $b_k \neq a = a_k$ and $b_k \neq b$. Now

$$\sum_{a_j = b_k} m_j = \sum_{b_j = b_k} m_j \geqq m_k > 0;$$

hence $b_k = a_l$ for some l. Then $a_k \; \mathcal{C} \; b_l$. In the right hand side of $(*)$ we can now replace $\left(X^{a_k} - X^{b_k} \right) + \left(X^{a_l} - X^{b_l} \right)$ by $\left(X^{a_k} - X^{b_l} \right)$ if $a_k \neq b_l$, by 0 if $a_k = b_l$. This decreases m by 1 or by 2. The induction hypothesis then yields $a \; \mathcal{C} \; b$. Conversely, $a \; \mathcal{C} \; b$ implies $X^a - X^b \in I(\mathcal{C})$ by definition. □

Theorem 1.2 (Rédei). *Congruences on a finitely generated free commutative semigroup satisfy the ascending chain condition.*

Proof. If

$$\mathcal{C}_1 \subsetneqq \mathcal{C}_2 \subsetneqq \cdots \subsetneqq \mathcal{C}_n \subsetneqq \cdots$$

is a strictly ascending sequence of congruences on F_X, then

$$I(\mathcal{C}_1) \subsetneqq I(\mathcal{C}_2) \subsetneqq \cdots \subsetneqq I(\mathcal{C}_n) \subsetneqq \cdots$$

is a strictly ascending sequence of ideals of $\mathbb{Z}[X]$, by Lemma 1.1. This contradicts the Hilbert Basis Theorem if X is finite. □

Corollary 1.3. *Every ideal of a f.g. free c.s. is finitely generated. Every antichain of a f.g. free c.s. is finite.*

An **antichain** of a partially ordered set is a subset A such that $a < b$ never happens when $a, b \in A$. The second part of Corollary 1.3 is Dickson's Theorem (Proposition I.5.4).

Proof. When F_X is finitely generated, the Rees congruences on F_X satisfy the ascending chain condition by Rédei's Theorem; therefore the ideals of F_X satisfy the ascending chain condition, and are finitely generated.

Now let A be an antichain of F_X. The ideal of F_X generated by A is

$$I = \{t \in F_X \mid t \geqq a \text{ for some } a \in A\}.$$

Since A is an antichain, every element of A is a minimal element of I, and must belong to any subset which generates I as an ideal; therefore A is finite. □

Corollary 1.4. *When S is a f.g.c.s., then the congruences on S and the ideals of S satisfy the ascending chain condition.*

Proof. By Corollary I.5.6 there is a surjective homomorphism $\pi : F \longrightarrow S$ where F is finitely generated and free. If there was a strictly ascending sequence of congruences (or of ideals) of S, their inverse images under π would constitute a strictly ascending sequence of congruences (or of ideals) of F, contradicting Theorem 1.2 (or Corollary 1.3). □

By Corollary 1.4, every congruence (every ideal) of a f.g.c.s. is finitely generated as a congruence (as an ideal). The converse holds: a c.s. all of whose congruences are finitely generated is itself finitely generated (Budach [1964]); a c.s. all of whose ideals are finitely generated is itself finitely generated (Satyanarayana [1977]).

On the other hand, the subsemigroups of a f.g.c.s. do not in general satisfy the ascending chain condition, and are not in general finitely generated. F.g.c.s. whose every subsemigroup is finitely generated are studied in Section 7.

Example 1.5. Let F be the free c.s. on $X = \{a, b\}$, written multiplicatively. F has three archimedean components: $A = \{a^m \mid m > 0\}$, $B = \{b^n \mid n > 0\}$, and $C = \{a^m b^n \mid m, n > 0\}$. Then $C^2 = \{a^m b^n \mid m, n \geqq 2\}$. Every subset of C which generates C must contain

$$C \backslash C^2 = \{a^m b^n \mid m = 1,\ n \geqq 2, \text{ or } m \geqq 2,\ n = 1\}$$

and is infinite; hence C is not finitely generated.

2. By Rédei's Theorem, every finitely generated c.s. is **finitely presented** (= has a presentation with finitely many generators and finitely many defining relations). Algorithms which recognize from a finite presentation of S whether S has an identity element, is cancellative, or is a group, were given by Emeličev [1962], [1966] and Birjukov [1966], [1967]. Algorithmic solutions of the **word problem** (deciding when two products of generators are equal) were given by Halezov [1966] for cancellative c.s. and Birjukov [1967] for c.s. in general. The

isomorphism problem (deciding when two finite presentations yield isomorphic semigroups) was solved by Taiclin [1974], [1980]. See also Shleifer [1981]. A number of efficient algorithms for these and other properties can be found in Rosales & García-Sánchez[1999], along with additional references.

2. SUBDIRECT DECOMPOSITIONS.

Rédei's Theorem yields subdirect decompositions of finitely generated commutative semigroups which are like Ponizovsky decompositions, with groups and elementary semigroups replaced by cancellative and subelementary semigroups. This result, due to Grillet [1975P], opens the door to the inner structure of f.g.c.s., beginning in this section with their archimedean components.

1. Call a congruence $\mathcal{C}$ intersection **irreducible** if it has more than one class and is not the intersection of finitely many congruences $\mathcal{C}_i \supsetneqq \mathcal{C}$. For simplicity's sake, we call a c.s. S **irreducible** when it is finitely subdirectly irreducible.

Proposition 2.1. *Let S be a f.g.c.s. Every congruence on S is the intersection of finitely many irreducible congruences. Hence S is a subdirect product of finitely many f.g. irreducible semigroups.*

Proof. The first part is proved by noetherian induction. We regard the universal congruence $\mathcal{U}$ on S (of which S is the only equivalence class) as the empty intersection of congrences on S, and the intersection of just one congruence as that congruence itself. If there exists a congruence on S which is not the intersection of finitely many irreducible congruences, then, by the ascending chain condition, there exists a congruence $\mathcal{M}$ on S which is maximal with this property. In particular $\mathcal{M} \neq \mathcal{U}$ and $\mathcal{M}$ is not irreducible. Therefore $\mathcal{M}$ is the intersection $\mathcal{M} = \bigcap_{i \in I} \mathcal{C}_i$ of finitely many congruences $\mathcal{C}_i \supsetneqq \mathcal{M}$. By the maximality of $\mathcal{M}$ every $\mathcal{C}_i$ is the intersection of finitely many irreducible congruences. But then so is $\mathcal{M} = \bigcap_{i \in I} \mathcal{C}_i$; this is the required contradiction.

In particular the equality on S is the intersection of finitely many irreducible congruences $\mathcal{C}_i$. Then S is a subdirect product of the finitely many semigroups $S/\mathcal{C}_i$. Since $\mathcal{C}_i$ is irreducible, the equality on $S/\mathcal{C}_i$ is not the intersection of finitely many proper congruences, by Proposition I.2.5, and every $S/\mathcal{C}_i$ is finitely subdirectly irreducible. Also $S/\mathcal{C}_i$ is finitely generated like S. □

2. A **subelementary** semigroup is a c.s. which is the disjoint union $S = C \cup N$ of two nonempty subsemigroups C and N, such that N is a nilsemigroup and an ideal of S, the zero element of N is the zero element of S, and every

element of C is cancellative in S ($cx = cy$ implies $x = y$, for all $x, y \in S$ and $c \in C$). In particular C is cancellative.

Theorem 2.2. *A finitely generated irreducible semigroup is either cancellative or nil or subelementary. Hence every finitely generated commutative semigroup is a subdirect product of a cancellative semigroup, a nilsemigroup, and finitely many subelementary semigroups.*

Proof. Let S be a f.g. irreducible semigroup. There exists a finite set X and a surjective homomorphism $\pi : F_X \longrightarrow S$. The congruence $\mathcal{C} = \ker \pi$ is irreducible: indeed $\mathcal{C}$ has more than one class and is not the intersection of finitely many congruences $\mathcal{C}_i \supsetneqq \mathcal{C}$, otherwise the equality on S would be the intersection of finitely many proper congruences, by Proposition I.2.5, and S would not be irreducible.

By Lemma 1.1, $\mathcal{C}$ is induced by an ideal I of $\mathbb{Z}[X]$: $\pi(a) = \pi(b)$ if and only if $X^a - X^b \in I$. We show that $\mathcal{C}$ is induced by a primary ideal Q (if $fg \in Q$ and $f \notin Q$, then $g^n \in Q$ for some $n > 0$). Indeed, in the noetherian ring $\mathbb{Z}[X]$, I is the intersection $I = Q_1 \cap Q_2 \cap \dots \cap Q_r$ of finitely many primary ideals $Q_1, \dots, Q_r$. Then $\mathcal{C}$ is the intersection of the congruences $\mathcal{Q}_1, \dots, \mathcal{Q}_r$ induced by $Q_1, \dots, Q_r$. Since $\mathcal{C}$ is irreducible, we have $\mathcal{C} = \mathcal{Q}_j$ for some j and $\mathcal{C}$ is induced by a primary ideal.

Now that $\mathcal{C}$ is induced by a primary ideal Q, let

$$P = \{f \in \mathbb{Z}[X] \mid f^n \in Q \text{ for some } n > 0\}$$

be the radical of Q, which is a prime ideal of $\mathbb{Z}[X]$ (if $fg \in P$ and $f \notin P$, then $g \in P$). Let $a \in F$. If $X^a \in Q$, then $X^a X^b - X^a \in Q$ and $\pi(a)\pi(b) = \pi(a)$ for all $b \in F$, so that $\pi(a)$ is a zero element of S. If $X^a \in P$, then $(X^a)^n \in Q$ for some $n > 0$, $\pi(a)^n$ is the zero element of S, and $\pi(a)$ is nilpotent in S. If $X^a \notin P$, then $\pi(a)\pi(b) = \pi(a)\pi(c)$ implies $X^a(X^b - X^c) \in Q$, $X^b - X^c \in Q$ (since Q is primary), and $\pi(b) = \pi(c)$; thus $\pi(a)$ is cancellative in S. Thus S is the disjoint union $S = C \cup N$, where

$$C = \{\pi(a) \in S \mid X^a \notin P\} \text{ and } N = \{\pi(a) \in S \mid X^a \in P\}.$$

Either set may be empty. If $N = \emptyset$, then $S = C$ is cancellative. If $C = \emptyset$, then $S = N$ is a nilsemigroup. If $C, N \neq \emptyset$, then S has a zero element $0 \in N$, and $S = C \cup N$ is subelementary. This proves the first part of the Theorem.

Now let S be any f.g.c.s. By Proposition 2.1 and the above, S is a subdirect product of finitely many cancellative semigroups $C_1, \dots, C_k$, nilsemigroups $N_1, \dots, N_l$, and subelementary semigroups $S_1, \dots, S_m$. We may assume only one cancellative component (which may be trivial): if indeed $k > 1$, then the projection C of S in $C_1 \times \cdots \times C_k$ is cancellative, and S is a subdirect product of C and the remaining components, since $S \longrightarrow C$ and the remaining projections still separate the elements of S. Similarly, we may assume only one nilsemigroup component (which may also be trivial), since a nonempty subsemigroup of the finite product $N_1 \times \ldots \times N_l$ is a nilsemigroup. □

It follows from this proof that every congruence on F_X is the intersection of finitely many "primary" congruences $\mathcal{Q}$ (such that $F_X/\mathcal{Q}$ is subelementary). Drbohlav [1963], [1964] proved this directly, for congruences on any f.g.c.s.

3. Theorem 2.2 implies finiteness properties of archimedean components. First we show:

Lemma 2.3. *In a f.g. subelementary semigroup $S = C \cup N$, C is f.g. and N is nilpotent ($N^m = 0$ for some $m > 0$).*

Proof. Let $X = \{x_1, \ldots, x_n\}$ generate S. We may assume that $X \cap N = \{x_1, \ldots, x_k\}$, where $k \leqq n$. Then C is f.g., by $X \cap C = \{x_{k+1}, \ldots, x_n\}$. Every element of N can now be written in the form $x_1^{t_1} \cdots x_k^{t_k} c$ where $c \in C$, $t_1, \ldots, t_k \geqq 0$, and $t_i > 0$ for some i.

Since N is a nilsemigroup every $x_i \in N$ has a power $x_i^{m_i} = 0$. If $m \geqq m_1 + \cdots + m_k$, then every product $x_1^{t_1} \cdots x_k^{t_k} c$ of m elements of N satisfies $t_1 + \cdots + t_k \geqq m$, $t_i \geqq m_i$ for some i, and $x_1^{t_1} \cdots x_k^{t_k} c = 0$. Thus $N^m = 0$. The same argument shows that a nilsemigroup N which is finitely generated (equivalently, finite) is nilpotent (Proposition IV.3.5). □

Proposition 2.4. *When S is a f.g.c.s., then $Y(S)$ is finite; every archimedean component A of S has a cancellative power A^m; and $G(A)$ is f.g.*

In Proposition 2.4, A^m is a nonempty cancellative ideal of A, and A is an ideal extension of the cancellative semigroup A^m by the nilpotent semigroup A/A^m.

Proof. First $Y(S)$ is finite since it is a f.g. semilattice: if $\{A_1, \ldots, A_n\}$ generate $Y(S)$, then every element of $Y(S)$ can be written in the form $A_1^{t_1} \cdots A_n^{t_n}$ with $t_i = 0$ or $t_i = 1$ for all i since $A_1, \ldots, A_n$ are idempotent; therefore $Y(S)$ is finite (with at most 2^n elements).

By Theorem 2.2, S is a subdirect product of a cancellative c.s. C, a nilsemigroup N, and finitely many subelementary semigroups $S_i = C_i \cup N_i$, all of which are homomorphic images of S and are finitely generated. By Lemma 2.3 there exists $m > 0$ such that $N^m = N_i^m = 0$ for all i. Let A be an archimedean component of S. The projection of A in S_i is archimedean and is either contained in C_i or contained in N_i. Hence A is contained in the direct product of C, N, some of the C_i's, and some of the N_i's. Then A^m is contained in a direct product of cancellative semigroups and trivial semigroups, and is cancellative. Thus A^m is a cancellative ideal of A.

By Propositions II.2.6 and II.3.4, $G(A^m)$ is (up to isomorphism) contained in the direct product of $G(C)$ and some of the $G(C_i)$'s, all of which are f.g. since C and all C_i are f.g. by Lemma 2.3. Therefore $G(A^m)$ is f.g. By Proposition II.2.5, $G(A) \cong G(A^m)$ is f.g. □

In particular, a f.g. archimedean semigroup A has a cancellative ideal A^m, and $G(A)$ is f.g. These semigroups are studied in Section 6. But Example 1.5 shows that the archimedean components of a f.g.c.s. need not themselves be f.g. So does the next example.

Example 2.5. Let $S = G \cup N$ be the elementary semigroup in which $G = \{a^n \mid n \in \mathbb{Z}\} \cong \mathbb{Z}$, $N = \{b_n \mid n \in \mathbb{Z}\} \cup \{0\}$, and $a^i b_j = b_{i+j}$, $b_i b_j = 0$ for all i, j. We see that S is finitely generated (by a, a^{-1}, and b_0) and that the archimedean component N of S is not finitely generated. □

Let S be a c.s. If there is an integer $m > 0$ such that A^m is cancellative for every archimedean component $A \in Y(S)$ of S, then the archimedean cancellativity **index** of S is the smallest such integer; otherwise the index of S is infinite. Thus, when $m > 0$ ($m \neq \infty$), S has index $\leqq m$ if and only if A^m is cancellative for every $A \in Y(S)$.

Lemma 2.6. *Let A be archimedean with an idempotent e. If A^m is cancellative, then $A^m = eA = H_e$.*

Proof. $eA = H_e$ by Proposition III.3.2, and $H_e \subseteq A^m$ since $x = e^m x$ for every $x \in H_e$. Conversely, $y \in A^m$ implies $ey = e(ey)$ and $y = ey \in eA$, since A^m is cancellative. □

For instance, Example III.5.2, which is archimedean but has no nonempty cancellative ideal, has infinite index. If on the other hand S is finitely generated, then, by Proposition 2.4, $Y(S)$ is finite and every archimedean component $A \in Y(S)$ of S has a cancellative power A^m; therefore

Proposition 2.7. *Every f.g.c.s. has finite archimedean cancellativity index.*

3. SUBELEMENTARY SEMIGROUPS.

This section contains basic properties of subelementary semigroups.

1. Every subelementary semigroup can be completed to an elementary semigroup:

Proposition 3.1. *Let $S = C \cup N$ be a subelementary semigroup. Then $C^{-1}S = G(C) \cup C^{-1}N$ is an elementary semigroup, and the canonical homomorphism $\iota : S \longrightarrow C^{-1}S$ is injective.*

Proof. In $C^{-1}S$, $0/a = 0/b$ for all $a,b \in C$, and this provides the zero element of $C^{-1}S$, since $(x/c)(0/a) = 0/ac$ for all $x/c \in C^{-1}S$.

Let $x/a \in C^{-1}S$, where $a \in C$. If $x \in C$, then $x/a = \iota(x)\,\iota(a)^{-1}$ is a unit of $C^{-1}S$. If $x \in N$, then $(x/a)^n = x^n/a^n = 0$ for some $n > 0$ and x/a is nilpotent. Also $c/a \neq x/b$ with $c \in C$, $x \in N$, since $ax \in N$ and $bc \in C$. Hence $C^{-1}S$ is the disjoint union $C^{-1}S = C^{-1}C \cup C^{-1}N$ of a group $G(C) = C^{-1}C$ and a nilsemigroup $C^{-1}N$. Then $C^{-1}S$ is elementary, since $C^{-1}N$ is an ideal of $C^{-1}S$, the identity element of $C^{-1}C$ is the identity element of $C^{-1}S$, and the zero element of $C^{-1}N$ is the zero element of $C^{-1}S$; ι is injective by Proposition II.1.2. □

Since $\iota : S \longrightarrow C^{-1}S$ is injective we identify $x \in S$ and $\iota(x) \in C^{-1}S$, so that S becomes a subsemigroup of $C^{-1}S$. Subelementary semigroups can thus be retrieved from elementary semigroups, much as cancellative semigroups can be retrieved from abelian groups.

2. When S is subelementary, the orbits of $C^{-1}S$ induce a partition of S.

Proposition 3.2. *Let $S = C \cup N$ be a subelementary semigroup. Every $\mathcal{H}$-class of $C^{-1}S$ intersects S. The restriction to S of the congruence $\mathcal{H}$ on $C^{-1}S$ is the congruence*

$$x \;\mathcal{O}\; y \text{ if and only if } cx = dy \text{ for some } c,d \in C$$

on S, and $S/\mathcal{O} \cong (C^{-1}S)/\mathcal{H}$. Moreover no two elements of an $\mathcal{O}$-class are comparable in N.

Proof. Let $x,y \in S$. If $x \;\mathcal{H}\; y$ in $C^{-1}S$, then $y = gx$ for some $g = c/d \in C^{-1}C$ (by Proposition IV.5.1) and $cx = dy$ holds in S for some $c,d \in C$. Conversely, $cx = dy$ in S implies $y = (c/d)x$ and $x \;\mathcal{H}\; y$ in $C^{-1}S$. Thus $\mathcal{O}$ is the restriction of $\mathcal{H}$ to S and is therefore a congruence on S.

For every $x/a \in C^{-1}S$, $x \mathcal{H} x/a$ holds in $C^{-1}S$, since $x/a = \iota(x)\, \iota(a)^{-1}$ and $\iota(a)$ is a unit of $C^{-1}S$; thus every $\mathcal{H}$-class of $C^{-1}S$ contains an element of S. Therefore $S/\mathcal{O} \cong (C^{-1}S)/\mathcal{H}$; the isomorphism $(C^{-1}S)/\mathcal{H} \longrightarrow S/\mathcal{O}$ sends an $\mathcal{H}$-class of $C^{-1}S$ to its intersection with S, which is an $\mathcal{O}$-class.

Assume that $cx = dy$ for some $c, d \in C$. If $x > y$ in N, then $y = ux$ for some $u \in N$, there is a greatest $n \geqq 0$ such that $u^n x \neq 0$, and $cu^n x = du^n y = du^{n+1} x = 0 = c0$, a contradiction since $c \in C$ is cancellative in S. Hence no two elements of an $\mathcal{O}$-class are comparable in N. □

When $S = C \cup N$ is subelementary, the $\mathcal{O}$-class O_x of $x \in S$ is the **orbit** of x. For instance, C and $\{0\}$ are orbits. The nilmonoid $S/\mathcal{O}$ is the **nilmonoid of orbits** of S. By Proposition 3.2, the nilmonoid of orbits of S is that of the elementary semigroup $C^{-1}S$.

Proposition 3.3. *When $S = C \cup N$ is finitely generated and subelementary, then C is finitely generated; $C^{-1}S$ is finitely generated; $S/\mathcal{O}$ is finite; N is nilpotent; if $N \neq 0$, the set M of all minimal elements of N is the union of all the minimal orbits.*

Proof. When $S = C \cup N$ is f.g., then C is f.g. by Lemma 2.3 and $C^{-1}S$ is f.g., by the generators of S and the inverses of the generators of C. The nilmonoid of orbits is f.g. and is finite by Proposition IV.3.5. (Hence $N/\mathcal{O}$ is nilpotent and so is N, as in Lemma 2.3.) If $x > y$ in N, then $O_x \geqq O_y$ in $S/\mathcal{O}$, in fact $O_x > O_y$, otherwise $O_x = O_y$ contains two comparable elements of N. In particular m is minimal in N if and only if O_m is minimal in $N/\mathcal{O}$. □

4. THE COMPLETION THEOREM.

The Completion Theorem, due to Grillet [1975C], states that every finitely generated commutative semigroup can be embedded into a finitely generated complete semigroup with the same universal semilattice and a certain universal property. This completes all the archimedean components at the same time. A simpler version (without the universal property) is given in Grillet [1995].

1. We begin with a simpler result. First, a f.g.c.s. S is complete if and only if every archimedean component of S contains an idempotent (since S has finitely many archimedean components and therefore finitely many idempotents).

Proposition 4.1. *Every f.g.c.s. can be embedded into a complete f.g.c.s.*

Proof. By Theorem 2.2, a f.g.c.s. S is a subdirect product of a cancellative semigroup C_0, a nilsemigroup N_1, and subelementary semigroups $S_i = C_i \cup N_i$ ($i = 2,3,\dots,r$). Then S can be embedded into the direct product T of the abelian group with zero $T_0 = G(C_0) \cup \{0\}$, the nilmonoid $T_1 = N_1^1$, and the finitely many elementary semigroups $T_i = C_i^{-1} S_i$ ($i = 2,3,\dots,r$). Since S is f.g., N_1, C_0, S_i, C_i, and every T_i are f.g. Since all T_i are monoids, every element t of T can be written as a product $t = t_0 t_1 \cdots t_r$ with $t_i \in T_i$ for all i. Therefore T is finitely generated.

All T_i are elementary. Let $T_i = G_i \cup N_i'$. Then an archimedean component of T is a (finite) direct product of subsemigroups $A_i \subseteq T_i$, where $A_i = G_i$ or $A_i = N_i'$ for every i: indeed these semigroups are archimedean and constitute a partition of T. Hence every archimedean component of T contains an idempotent. □

2. Next we construct semigroups of fractions, using the following subsets.

Lemma 4.2. *Let S be f.g., $A \in Y(S)$, and $m > 0$. Let $K = A^m$ and L be the ideal of S generated by all B^m with $B \in Y(S)$ and $B \leqq A$. Then $L \cap A = K$. If φ is a homomorphism of S into a complete semigroup T of index $\leqq m$, and g is the idempotent of the archimedean component C of T which contains $\varphi(A)$, then $\varphi(K) \subseteq gC = H_g$ and $\varphi(L) \subseteq gT$.*

Proof. First $K = A^m \subseteq L \cap A$. Let $x \in L \cap A$, so that $x \in A$ and $x = tb$ for some $t \in S^1$ and $b \in B^m$, where $B \leqq A$ in $Y(S)$. Since $x = tb \leqq_{\mathcal{H}} b$ we also have $A \leqq B$ and $B = A$. If $t \in S$ and $C = A(t)$ is the archimedean component of t, then $A \leqq C$ since $x = tb \leqq_{\mathcal{H}} t$, $CA \subseteq A$, $CA^m \subseteq A^m$, and $x = tb \in CA^m \subseteq K$. If $t = 1 \in S^1$, then $x = b \in B^m = K$. Thus $L \cap A = K$.

Now let φ, T, g, and C be as in the statement. Then $gC = H_g = C^m$ by Lemma 2.6. Hence $\varphi(K) = \varphi(A)^m \subseteq C^m = H_g$. If similarly $B \in Y(S)$, $B \leqq A$, $\varphi(B) \subseteq D \in Y(T)$, and h is the idempotent of D, then $\varphi(B^m) \subseteq hD$. But $D \leqq C$, since $B \leqq A$, and $h \leqq g$, since $E(T) \cong Y(T)$ (Corollary III.3.2); hence $\varphi(B^m) \subseteq hD \subseteq gT$. Therefore $\varphi(L) \in gT$. □

Lemma 4.3. *Let $S' = (S \backslash L) \cup K^{-1}L$, where K and L are as in Lemma 4.2, with multiplication $\centerdot$ given for all $s,t \in S\backslash L$, $x,y \in L$, $a,b \in K$ by*

$$\begin{cases} s \centerdot t = st \textit{ if } st \notin L,\ s \centerdot t = \alpha(s)\alpha(t) = \alpha(st) = ast/a \textit{ if } st \in L; \\ s \centerdot (x/a) = (x/a) \centerdot s = sx/a; \\ (x/a) \centerdot (y/b) = (x/a)(y/b) = xy/ab. \end{cases}$$

Then S' is a f.g.c.s. The canonical mapping $\beta : S \longrightarrow S'$ defined for all $s \in S \backslash L$, $x \in L$, $a \in K$ by

$$\beta(s) = s \text{ if } s \in S\backslash L,\ \beta(x) = \alpha(x) = ax/a \text{ if } x \in L$$

is a homomorphism. Every homomorphism of S into a complete semigroup T of index $\leqq m$ factors uniquely through β.

$$\begin{array}{ccc} S & \xrightarrow{\beta} & S' \\ & \searrow^{\varphi} & \downarrow \chi \\ & & T \end{array}$$

Proof. In Lemma 4.2, L is an ideal of S which contains K. Hence $K^{-1}L = K^{-1}S$: if $x/a \in K^{-1}S$ (with $x \in S$ and $a \in K$), then $x/a = ax/a^2 \in K^{-1}L$ since $ax \in L$. Then the restriction to $S\backslash L$ of the canonical homomorphism $\alpha : S \longrightarrow K^{-1}S$ is a partial homomorphism $S\backslash L \longrightarrow K^{-1}L$. We see that S' is the retract ideal extension of $K^{-1}L$ by S/L determined by this partial homomorphism. Hence S' is a c.s.

K is an ideal of A since $K = L \cap A$; hence $G(K) \cong G(A)$ by Proposition II.2.5 and is finitely generated as an abelian group by Proposition 2.4. Hence $G(K)$ is finitely generated as a c.s. and S', which is generated by S and by $G(K)$, is finitely generated.

For all $s,t \in S\backslash L$, $x,y \in L$, $a \in K$ we have $\beta(s) \centerdot \beta(t) = s \centerdot t = \beta(st)$; $\beta(s) \centerdot \beta(x) = s \centerdot (ax/a) = asx/a = \beta(sx)$; and $\beta(x) \centerdot \beta(y) = \alpha(x) \centerdot \alpha(y) = \alpha(xy) = \beta(xy)$, since α is a homomorphism. Thus β is a homomorphism.

Let φ be a homomorphism of S into a complete semigroup T of index $\leqq m$, and g be the idempotent of the archimedean component C of T which contains $\varphi(A)$. By Lemma 4.2, $\varphi(K) \subseteq gC = H_g$ and $\varphi(L) \subseteq gT$. Hence $\varphi(a)$ is a unit of gT for every $a \in K$. The universal property of $K^{-1}L$ (Proposition II.1.3), applied to the restriction $L \longrightarrow gT$ of φ, begets a unique homomorphism $\psi : K^{-1}L \longrightarrow gT$ such that $\psi(\alpha(x)) = \varphi(x)$ for every $x \in L$; namely $\psi(x/a) = \varphi(x)\,\varphi(a)^{-1}$ for all $x \in L$ and $a \in K$, where $\varphi(a)^{-1}$ is the inverse in H_g. Then the mapping $\chi : S' \longrightarrow T$ defined for all $s \in S\backslash L$, $x \in L$, $a \in K$ by

$$\chi(s) = \varphi(s),\ \chi(x/a) = \psi(x/a) = \varphi(x)\,\varphi(a)^{-1}$$

is a homomorphism, and is the only homomorphism such that $\chi \circ \beta = \varphi$. □

Lemma 4.4. *When C is an archimedean component of S, the archimedean* *component C' of S' which contains $\beta(C)$ is as follows: if $C \nleqq A$, then $C' = C$; if $C \leqq A$, then $C' = (C \backslash L) \cup K^{-1}(C \cap L)$. Also $Y(\beta) : Y(S) \longrightarrow Y(S')$ is an isomorphism. If S has index $\leqq m$, then S' has index $\leqq m$.*

Proof. Let $\pi : S \longrightarrow Y(S)$ and $\pi' : S' \longrightarrow Y(S')$ be the projections. The homomorphism $\beta : S \longrightarrow S'$ induces a homomorphism $\delta = Y(\beta) : Y(S) \longrightarrow Y(S')$ such that $\delta \circ \pi = \pi' \circ \beta$; δ sends an archimedean component $C \in Y(S)$ of S to the archimedean component C' of S' which contains $\beta(C)$.

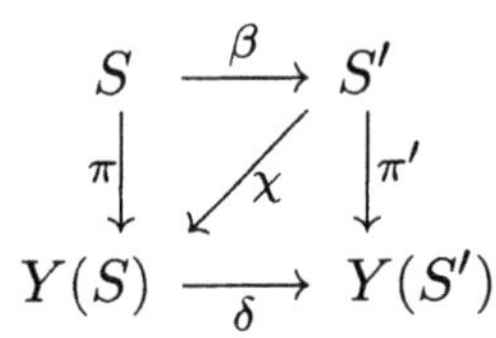

If $x/a \in K^{-1}L$, then, in $K^{-1}L$, $\alpha(a)^{-1} \mathcal{H}\ 1$, $x/a = \alpha(x)\,\alpha(a)^{-1} \mathcal{H}\ \alpha(x)$, and $x/a \mathcal{N} \alpha(x)$. Therefore every archimedean component of S' contains an element of $\beta(S)$ and δ is surjective. To prove that δ is injective we use the universal property of S' in Lemma 4.3: since $Y(S)$ is a complete semigroup of index $1 \leqq m$, there is a unique homomorphism $\chi : S' \longrightarrow Y(S)$ such that $\pi = \chi \circ \beta$; namely,

$$\chi(s) = \pi(s),\ \chi(x/a) = \pi(x)\,\pi(a)^{-1} = \pi(x) \wedge A$$

for all $s \in S \backslash L$, $x \in L$, $a \in K$. If now $s, t \in S$ and $\beta(s) \mathcal{N} \beta(t)$ in S', then $\chi(\beta(s)) \mathcal{N} \chi(\beta(t))$, which in the semilattice $Y(S)$ implies $\pi(s) = \chi(\beta(s)) = \chi(\beta(t)) = \pi(t)$ and $s \mathcal{N} t$ in S; hence δ is injective. Thus $Y(\beta) = \delta : Y(S) \longrightarrow Y(S')$ is an isomorphism.

In particular $s \mathcal{N} t$ in S if and only if $\beta(s) \mathcal{N} \beta(t)$ in S'. Since $x/a \mathcal{N} \beta(x)$ for all $x/a \in K^{-1}L$, it follows that $x/a \mathcal{N} y/b$ in S' if and only if $x \mathcal{N} y$ in S. Similarly $x/a \mathcal{N} s$ in S' if and only if $x \mathcal{N} s$ in S, for all $s \in S \backslash L$. When $C \in Y(S)$, the archimedean component C' of S' which contains $\beta(C)$ is now as follows. If $C \nleqq A$, then $C \cap L = \emptyset$, since $L \subseteq \bigcup_{B \in Y(S),\, B \leqq A} B$, and $C' = C$. If $C \leqq A$, then, for all $s \in S \backslash L$ and $x/a \in K^{-1}L$, $s \in C'$ if and only if $s \in C$, and $x/a \in C'$ if and only if $x \in C$; hence $C' = (C \backslash L) \cup K^{-1}(C \cap L)$.

Finally assume that S has index $\leqq m$, so that C^m is cancellative for every $C \in Y(S)$. If $C \nleqq A$, then $(C')^m = C^m$ is cancellative. Now let $C \leqq A$. Then $(C')^m = K^{-1}C^m$, since $C^m \subseteq L$. Also $AC \subseteq C$ and $AC^m \subseteq C^m$. If

$(x/a) \cdot (z/c) = (y/b) \cdot (z/c)$, where $x, y, z \in C^m$ and $a, b, c \in K \subseteq A$, then $xz/ac = yz/bc$, $bcdxz = acdyz$ for some $a \in K$, $bx = ay$ since bx, ay, $cdz \in AC^m \subseteq C^m$, and $x/a = y/b$. Thus $(C')^m$ is again cancellative. □

Lemma 4.5. *Let S have index $\leqq m$. Let n be the number of archimedean components of S. For every $0 \leqq k \leqq n$ there exists a homomorphism $\gamma : S \longrightarrow T$ with the following properties:*

(1) *T is finitely generated and has index $\leqq m$;*

(2) *$Y(\gamma) : Y(S) \longrightarrow Y(T)$ is an isomorphism;*

(3) *every homomorphism of S into a complete semigroup of index $\leqq m$ factors uniquely through γ;*

(4) *T has at least k idempotents.*

Proof. The proof is by induction on k. If $k = 0$ we can let $T = S$ and γ be the identity on S. Now let $k < n$ and $\gamma : S \longrightarrow T$ have properties (1) through (4). If T does not have at least $k+1$ idempotents, then there is an archimedean component A of T which contains no idempotent. Construct K and L from T as in Lemma 4.2 and $\beta : T \longrightarrow T'$ as in Lemma 4.3. By Lemmas 4.3 and 4.4, $\gamma' = \beta \circ \gamma : S \longrightarrow T'$ has properties (1), (2), and (3). Also there is for every archimedean component C' of T' a unique archimedean component C of T such that $\beta(C) \subseteq C'$. If C contains an idempotent, then so does C'. In addition the archimedean component A' which contains $\beta(A)$ contains the identity element of $K^{-1}L$. Hence T' has at least $k+1$ idempotents. □

3. We can now prove:

Theorem 4.6. *Let S be a finitely generated commutative semigroup of index $\leqq m$. If $m > 0$ is large enough, there exists a homomorphism $\gamma : S \longrightarrow T$ of S into a finitely generated complete semigroup of index $\leqq m$, such that $Y(\gamma) : Y(S) \longrightarrow Y(T)$ is an isomorphism, and every homomorphism of S into a complete semigroup of index $\leqq m$ factors uniquely through γ.*

Proof. With $k = n$, Lemma 4.5 yields a homomorphism $\gamma : S \longrightarrow T$ such that T is f.g., has index $\leqq m$, has n archimedean components by (2), has the universal property (3), and has at least n idempotents. Then T has n idempotents, every archimedean component of T contains an idempotent, and T is complete.

By Proposition 4.1 there is an embedding $\varphi : S \longrightarrow C$ of S into a complete f.g.c.s. C. Now C has finite archimedean cancellativity index, by Proposition 2.7. If m is sufficiently large (if C has index $\leqq m$), then φ factors through γ by (3), and γ is injective. □

4. It would be nice if there would exist a universal complete semigroup of S; that is, a homomorphism $\gamma : S \longrightarrow U$ of S into a complete f.g.c.s. U, with the stronger universal property that every homomorphism of S into a complete semigroup factors uniquely through γ. Alas the following example shows that this does not exist in general, even when S is archimedean.

Example 4.7. This is Example III.5.4: S is an infinite cyclic semigroup $S = \{a, a^2, \dots, a^n, \dots\}$. Then S is archimedean and has index 1. We saw that there does not exist a homomorphism $\gamma : S \longrightarrow T$ of S into a complete archimedean semigroup T, with the stronger universal property that every homomorphism of S into a complete archimedean semigroup factors uniquely through γ.

Let $m > 0$. The subsets in Lemma 4.2 are $K = L = S^m = \{a^m, a^{m+1}, \dots\}$. Then $K^{-1}L$ is an infinite cyclic group $G = \{b^n \mid n \in \mathbb{Z}\} \cong \mathbb{Z}$, with $\alpha(a^n) = b^n$ for all $n \geqq m$. Hence the semigroup S' in Lemma 4.3 is $S' = (S\backslash K) \cup G = \{a, a^2, \dots, a^{m-1}\} \cup \{b^n \mid n \in \mathbb{Z}\}$ with multiplication $a^i a^j = a^{i+j}$ if $i + j < m$, $a^i a^j = a^{i+j} = b^{i+j}$ if $i + j \geqq m$, $a^i b^j = b^{i+j}$; this is the semigroup T_m in Example III.5.4. We see that S' has index m. Also $\beta(a^i) = a^i$ if $i < m$, $\beta(a^i) = b^i$ if $i \geqq m$. By Lemma 4.3, every homomorphism of S into a complete semigroup of index $\leqq m$ factors uniquely through $\beta : S \longrightarrow S'$. Moreover S' coincides with the complete semigroup constructed in the proof of Theorem 4.6.

Assume that there is a homomorphism $\gamma : S \longrightarrow U$ of S into a complete semigroup U, with the stronger universal property that every homomorphism of S into a complete semigroup factors uniquely through γ. As in Example III.5.4, $\gamma(S)$ is archimedean and is contained in an archimedean component T of U, which contains an idempotent e. Since T is archimedean we have $\gamma(a^k) = et$ for some $k > 0$ and $t \in T$. Let $m > k$. If $\beta : S \longrightarrow S'$ factors through γ (if $\beta = \varphi \circ \gamma$ for some homomorphism $\varphi : U \longrightarrow S'$), then $\varphi(eT)$ is a subgroup of S', since eT is a subgroup of T, $\varphi(eT) \subseteq G$, and $\beta(a^k) = \varphi(\gamma(a^k)) = \varphi(et) \in G$, which does not happen when $m > k$. □

5. IRREDUCIBLE SEMIGROUPS.

In this section we determine all finitely generated commutative semigroups that are finitely subdirectly irreducible. The results are from Grillet [1977] and imply Malcev's Theorem [1958].

1. Let S be a f.g.c.s. If S is irreducible, then S is either cancellative or nil or subelementary, by Theorem 2.2.

If S is cancellative, then, by Proposition IV.1.8, S is irreducible if and only if S is either a nontrivial cyclic p-group, or isomorphic to $\mathbb{Z}$ or to a subsemigroup of $\mathbb{N}$.

If S is a nilsemigroup, then S is finite; by Propositions IV.3.6 and V.5.1, S is irreducible if and only if S is subdirectly irreducible, if and only if S is weakly irreducible and has just one minimal element.

2. We now let $S = C \cup N$ be subelementary. The cases where C or N are trivial are covered by Proposition V.5.2: $C \cup \{0\}$ is irreducible if and only if C is irreducible; $N \cup \{1\}$ is irreducible if and only if N is irreducible; and $\{0, 1\}$ is irreducible. In general:

Proposition 5.1. *When $S = C \cup N$ is subelementary, then S is (finitely) subdirectly irreducible if and only if $C^{-1}S$ is (finitely) subdirectly irreducible.*

Proof. First we show that the equality is the only congruence $\mathcal{C}$ on $C^{-1}S$ which is the equality on S. When indeed $s, t \in C^{-1}S$, then $cs, ct \in S$ for some $c \in C$; hence $s \; \mathcal{C} \; t$ implies $cs \; \mathcal{C} \; ct$, $cs = ct$, and $s = t$.

Now assume that S is irreducible. If the equality on $C^{-1}S$ is the intersection of finitely many congruences $\mathcal{C}_i$, then the equality on S is the intersection of the finitely many congruences on S induced by the congruences $\mathcal{C}_i$; therefore some $\mathcal{C}_i$ is the equality on S, and then $\mathcal{C}_i$ is the equality on $C^{-1}S$. Thus $C^{-1}S$ is irreducible. If similarly S is subdirectly irreducible, then $C^{-1}S$ is subdirectly irreducible.

To prove the converse we construct for every congruence $\mathcal{C}$ on S a congruence $\mathcal{C}^*$ on $C^{-1}S$ as follows:

$$s \; \mathcal{C}^* \; t \text{ if and only if } cs, ct \in S \text{ and } cs \; \mathcal{C} \; ct \text{ for some } c \in C^1.$$

It is immediate that $\mathcal{C}^*$ is a congruence. Also $\mathcal{C}^*$ is the equality on $C^{-1}S$ if and only if $\mathcal{C}$ is the equality on S. Moreover let $\mathcal{C} = \bigcap_{i \in I} \mathcal{C}_i$. Then $\mathcal{C} \subseteq \mathcal{C}_i$ and $\mathcal{C}^* \subseteq \mathcal{C}_i^*$ for all i, so that $\mathcal{C}^* \subseteq \bigcap_{i \in I} \mathcal{C}_i^*$. Conversely assume that $s \; \mathcal{C}_i^* \; t$ for all i. For every $i \in I$ we have $c_i s, c_i t \in S$ and $c_i s \; \mathcal{C}_i \; c_i t$ for some $c_i \in C^1$. If I is finite, then $c = \prod_{i \in I} c_i \in C^1$, $cs, ct \in S$, $cs \; \mathcal{C}_i \; ct$ for all i, and $s \; \mathcal{C}^* \; t$. Thus $\left(\bigcap_{i \in I} \mathcal{C}_i\right)^* = \bigcap_{i \in I} \mathcal{C}_i^*$, provided that I is finite.

Now assume that $C^{-1}S$ is irreducible. If the equality on S is the intersection of finitely many congruences $\mathcal{C}_i$, then by the above the equality on $C^{-1}S$ is the intersection of the finitely many congruences $\mathcal{C}_i^*$, some $\mathcal{C}_i^*$ is the equality on $C^{-1}S$, and some $\mathcal{C}_i$ is the equality on $C^{-1}S$. Thus S is irreducible.

Finally assume that $C^{-1}S$ is subdirectly irreducible. By Theorem V.5.6 and Proposition IV.1.7, $G(C)$ is either trivial or a cyclic or quasicyclic p-group. Therefore the subsemigroups of $G(C)$ are subgroups of $G(C)$, $C = G(C)$, and S is elementary; hence $S = C^{-1}S$ is subdirectly irreducible. □

Lemma 5.2. *When $S = C \cup N$ is subelementary, then N is weakly irreducible if and only if $C^{-1}N$ is weakly irreducible.*

Proof. Let M be the set of all minimal elements of N. By Proposition 3.3, $x/a \in C^{-1}S$ is minimal in $C^{-1}N$ if and only if x is minimal in N; thus the set of all minimal elements of $C^{-1}N$ is $C^{-1}M$. Recall that N is weakly irreducible if and only if its congruence

$$x \mathcal{M} y \iff (\forall u \in N^1)(ux \in M \iff uy \in M \implies ux = uy)$$

is the equality. For any $c \in C$, we have $x \in M$ if and only if $cx \in M$; hence $x \mathcal{M} y$ implies $cx \mathcal{M} cy$.

We show that $x/a \mathcal{M} y/b$ in $C^{-1}N$ if and only if $bx \mathcal{M} ay$ in N. Indeed assume $x/a \mathcal{M} y/b$ and let $u \in N^1$. If $ubx \in M$, then $ux/a = ubx/ab \in C^{-1}M$, $uy/b = ux/a \in C^{-1}M$ since $x/a \mathcal{M} y/b$, and $uay = ubx$. Similarly $uay \in M$ implies $ubx = uay$. Thus $bx \mathcal{M} ay$. Conversely assume $bx \mathcal{M} ay$ and let $u \in N$ and $c \in C$. If $(u/c)(x/a) \in C^{-1}M$, then $ubx/abc = ux/ac \in C^{-1}M$, $ubx \in M$, $uay = ubx$ since $bx \mathcal{M} ay$, and $(u/c)(y/b) = uay/abc = ubx/abc = (u/c)(x/a)$. Similarly $(u/c)(y/b) \in C^{-1}M$ implies $(u/c)(x/a) = (u/c)(y/b)$; $x/a \in C^{-1}M$ implies $bx/ab = x/a \in C^{-1}M$, $bx \in M$, $ay = bx$, and $y/b = x/a$; and $y/b \in C^{-1}M$ implies $x/a = y/b$. Thus $x/a \mathcal{M} y/b$.

If now N is weakly irreducible, so that $\mathcal{M}$ is the equality on N, then $x/a \mathcal{M} y/b$ in $C^{-1}N$ implies $bx \mathcal{M} ay$ in N, $bx = ay$, and $x/a = y/b$; hence $C^{-1}N$ is weakly irreducible. If conversely $C^{-1}N$ is weakly irreducible, then $x \mathcal{M} y$ in N implies, for any $c \in C$, $cx \mathcal{M} cy$ in N, $x/c \mathcal{M} y/c$ in $C^{-1}N$, $x/c = y/c$, and $x = y$; thus N is weakly irreducible. □

3. We call a subelementary semigroup $S = C \cup N$ **homogeneous** when $C^{-1}S$ is homogeneous; equivalently, if $cx = dx \neq 0$ implies $c = d$ when $c, d \in C$. Combining Proposition 5.1 and results from Section V.5 (Propositions V.5.1 and V.5.2, Theorem V.5.6) we obtain:

Theorem 5.3. *A finitely generated commutative semigroup is finitely subdi-*

rectly irreducible if and only if it is one of the following:

a nontrivial cyclic p-group; an infinite cyclic group; isomorphic to a subsemigroup of $\mathbb{N}$;

a finite nilsemigroup on which $\mathcal{P}$ is the equality (equivalently, which is weakly irreducible with one minimal element);

$\{0,1\}$; a finitely subdirectly irreducible cancellative semigroup with a zero element adjoined; a finite nilmonoid on which $\mathcal{P}$ is the equality; a homogeneous subelementary semigroup $S = C \cup N$ in which C is finitely subdirectly irreducible, N is weakly irreducible, and the minimal elements of N form an orbit.

Proof. This follows from Propositions V.5.1 and V.5.2 except in the case where $S = C \cup N$ is subelementary with C and N nontrivial. Then S is irreducible if and only if $C^{-1}S$ is irreducible (Proposition 5.1); if and only if $G(C)$ is irreducible, $C^{-1}N$ is weakly irreducible, the minimal elements of $C^{-1}N$ form an orbit, and $C^{-1}S$ is homogeneous (Theorem V.5.6); if and only if C is irreducible (Proposition IV.1.8), the minimal elements of N form an orbit (Proposition 3.3), N is weakly irreducible (Lemma 5.2), and S is homogeneous. □

4. From Theorem 5.3 we can deduce Malcev's Theorem that f.g.c.s. that are subdirectly irreducible are finite (Malcev [1958]); other proofs were given by Carlisle [1971] and Lallement [1971]. Since $\mathbb{Z}$ and subsemigroups of $\mathbb{N}$ are not subdirectly irreducible, a cancellative f.g.c.s. is subdirectly irreducible if and only if it is a nontrivial cyclic p-group. Hence a f.g.c.s. is subdirectly irreducible if and only if it is one of the following:

a nontrivial cyclic p-group;

a finite nilsemigroup on which $\mathcal{P}$ is the equality (equivalently, which is weakly irreducible with one minimal element);

$\{0,1\}$; a nontrivial cyclic p-group with a zero element adjoined; a finite nilmonoid on which $\mathcal{P}$ is the equality; a homogeneous elementary semigroup $S = C \cup N$ in which C is a nontrivial cyclic p-group, N is weakly irreducible, and the minimal elements of N form an orbit.

All these semigroups are finite; in the last case S is elementary with finitely many finite orbits. Thus:

Corollary 5.4 (Malcev). *Every finitely generated c.s. which is subdirectly irreducible is finite.*

6. ARCHIMEDEAN SEMIGROUPS.

This section studies finitely generated archimedean semigroups. Most of the results are from MacAlister & O'Carroll [1970].

1. By Proposition 2.4, a finitely generated archimedean semigroup S has a cancellative power S^m and its universal group $G(S) \cong G(S^m)$ is finitely generated.

Proposition 6.1. *A complete archimedean semigroup S is f.g. if and only if its kernel K is f.g. and S/K is finite.*

Proof. If S is f.g., then the f.g. nilsemigroup S/K is finite and $K = G(K) \cong G(S)$ is f.g. (as an abelian group). If conversely K is f.g. as an abelian group, then K is f.g. as a c.s. (by the generators of K and their inverses); if moreover S/K is finite, then S is f.g. (by $S \backslash K$ and the generators of K). □

2. We now assume that S has no idempotent. Proposition II.5.5 provides a smallest power cancellative congruence $\mathfrak{T}$ on S, for which

$$x \;\mathfrak{T}\; y \text{ if and only if } x^n = y^n \text{ for some } n > 0.$$

If S is archimedean without idempotent, then $T = S/\mathfrak{T}$ is archimedean without idempotent and cancellative (Proposition III.4.1). Also $x \neq xy$ for all $x, y \in S$ (Corollary III.4.2).

Proposition 6.2. *A finitely generated archimedean semigroup without idempotent is power joined.*

Proof. In the above $T = S/\mathfrak{T}$ is f.g. archimedean without idempotent, power cancellative, and cancellative. Since T is archimedean, T has only one face (T itself) and has dimension 1 by Proposition II.6.7. Thus $G(T)$ has rank 1 and S is power joined by Proposition III.6.3. □

Proposition 6.2 was proved by Tamura [1968], Levin [1968] and, in case S is cancellative, by Petrich [1964]. Chrislock [1969] showed that an N-semigroup is f.g. if and only if its structure group is finite. F.g. N-semigroups were also investigated by Higgins [1966]. Higgins & Tamura [1973] showed that an N-semigroup S is f.g. if and only if $G(S)$ is f.g. of rank 1.

Corollary 6.3. *When S is a f.g. archimedean semigroup without idempotent, then $S/\mathfrak{T}$ is isomorphic to a subsemigroup of $\mathbb{N}^+$.*

Proof. By Corollary III.6.4, $S/\mathfrak{T}$ is isomorphic to a subsemigroup Q of $\mathbb{Q}^+$. But Q is f.g. like S; if p_1/q_1, p_2/q_2, ..., p_n/q_n generate Q, then every element of Q is an integer multiple of $1/q_1 \cdots q_n$; hence Q is isomorphic to a subsemigroup of $\mathbb{N}^+$. □

3. To sharpen Corollary 6.3 we use McAlister & O'Carroll's congruence $\mathcal{M}_a$, defined for any element a of S by

$x \,\mathcal{M}_a\, y$ if and only if $x = y$, or $x = au$, $y = av$ for some $u, v \in S^1$ such that $a^p u = a^q v$ for some $p, q > 0$

The proof of Lemma 6.4 below shows that $\mathcal{M}_a$ is the smallest congruence $\mathcal{C}$ on S such that $a \,\mathcal{C}\, a^2$.

Lemma 6.4. *$\mathcal{M}_a$ is a congruence on S; if S is archimedean without idempotent, S^m is cancellative, and $a \in S^m$, then $\mathcal{M}_a \cap \mathcal{T}$ is the equality on S.*

Proof. If $x = au$, $y = av$, where $u, v \in S^1$ and $a^p u = a^q v$, and $y = aw$, $z = at$, where $w, t \in S^1$ and $a^r w = a^s t$, then $av = aw$ and $a^{p+r} u = a^{q+r} v = a^{q+r} w = a^{q+s} w$. Hence $\mathcal{M}_a$ is transitive; then $\mathcal{M}_a$ is clearly a congruence.

Now let $a \in S^m$, where S^m is cancellative. Assume that $x \,\mathcal{M}_a\, y$, with $x = au$, $y = av$, where $u, v \in S^1$ and $a^p u = a^q v$, and that $x \,\mathcal{T}\, y$, so that $x^n = y^n$. We may arrange that $n \geqq m$. Then $a^n u^n = a^n v^n$ and $a^{np} u^n = a^{nq} v^n = a^{nq} u^n$. Since $x \neq xy$ for all $x, y \in S$ we have $np = nq$ and $p = q$. Then $a^p u = a^q v$ implies $x = au = av = y$, since either $p = q = 1$ or a, $a^{p-1} u$, $a^{p-1} v \in S^m$ and S^m is cancellative. Thus $\mathcal{M}_a \cap \mathcal{T}$ is the equality on S. □

Lemma 6.5. *When S is a f.g. archimedean without idempotent, then $S/\mathcal{M}_a$ is finite.*

Proof. $M = S/\mathcal{M}_a$ is f.g. and power joined like S, and has an idempotent since $a \,\mathcal{M}_a\, a^2$. Since M is power joined, every element of M has an idempotent power and has only finitely many distinct powers. Then the elements of M are products of powers of finitely many generators and M is finite. □

Now S^m is cancellative for some m and $\mathcal{T} \cap \mathcal{M}_a$ is the equality on S when $a \in S^m$ by Lemma 6.4; hence S is a subdirect product of $S/\mathcal{T}$, which by Proposition 6.2 is isomorphic to a subsemigroup of $\mathbb{N}^+$, and $S/\mathcal{M}_a$, which by Lemma 6.5 is finite and archimedean. This proves:

Proposition 6.6. *A f.g. archimedean semigroup without idempotent is a subdirect product of a finite archimedean semigroup and a subsemigroup of $\mathbb{N}^+$.*

If conversely F is a finite archimedean semigroup, then F and $F \times \mathbb{N}^+$ are power joined, and every subsemigroup of $F \times \mathbb{N}^+$ is power joined, and is archimedean without idempotent. Higgins [1966], [1969S] proved Proposition

6.6 in the case of N-semigroups (in which case S is f.g. if and only if it is a subdirect product of a finite abelian group and a subsemigroup of $\mathbb{N}^+$).

Proposition 6.7. *When* $m, n > 0$,

$$S = \langle a, b \mid a^m = b^n \rangle$$

is an N-semigroup with two generators, and, up to isomorphism, every N-semigroup with two generators can be constructed in this fashion.

Proof. Let F be the free c.s. on $\{a, b\}$. The congruence $\mathcal{C}$ on F generated by (a^m, b^n) consists of all pairs $(a^{i+km} b^j, a^i b^{j+kn})$ and $(a^i b^{j+kn}, a^{i+km} b^j)$ with $i, j, k \geqq 0$, $i + j > 0$, and coincides with the congruence induced by the subgroup $\langle a^m b^{-n} \rangle$ of $G(F)$. Hence $S = \langle a, b \mid a^m = b^n \rangle$ is cancellative by Proposition II.5.1. Also S has no idempotent, since $\mathcal{C}$ contains no pair $(a^i b^j, a^{2i} b^{2j})$. Finally $(a^i b^j)^n = a^{ni+mj}$; therefore S is power joined, and is an N-semigroup.

Conversely let S be an N-semigroup with two generators a and b. Then S is power joined by Proposition 6.2 and $a^r = b^s$ for some $r, s > 0$. By Proposition III.6.6, S is a subdirect product of $S/\mathcal{T}$ and a torsion group G, namely its structure group $G = G(S)/A$, where $A \cong \mathbb{Z}$ is the subgroup generated by a. Now G is generated by $b + A$ and is a finite cyclic group. Let m, n be the least $r, s > 0$ such that $a^r = b^s$. Then $a^m = b^n$ and G has order n. If $a^i b^j = a^k b^l$, with, say, $j \leqq l$, then $l = j + tn$ for some $t \geqq 0$, since G has order n, and then $a^i b^j = a^k b^{j+tn} = a^{k+tm} b^j$ and $i = k + tm$, since S is cancellative and a has infinite order. Thus $S \cong \langle a, b \mid a^m = b^n \rangle$. □

Proposition 6.7 is due to Petrich [1964]; the above proof is from McAlister & O'Carroll [1970]. More generally, Hall [1971] constructed all cancellative c.s. with two generators.

7. THE A.C.C. ON SUBSEMIGROUPS.

This section looks at commutative semigroups whose subsemigroups satisfy the ascending chain condition. These semigroups were studied by MacAlister & O'Carroll [1970], Shevrin & Ovsyannikov [1990], and Rosales & García-García[1999].

1. We call a c.s. S **strongly noetherian** when its subsemigroups satisfy the ascending chain condition; equivalently, when every subsemigroup of S is finitely generated. A strongly noetherian c.s. is f.g.; hence its ideals and congruences also

satisfy the ascending chain condition. Rosales & García-García[1999] call these semigroups **hereditarily finitely generated**.

Example 2.5 shows a f.g.c.s. is not necessarily strongly noetherian. So does Example 1.5, which will be used below, in which S is the free c.s. on $\{x,y\}$ but contains a subsemigroup $\{x^m y^n \mid m,n > 0\}$ which is not f.g.

The following results are from MacAlister & O'Carroll [1970].

Proposition 7.1. *A f.g.c.s. is strongly noetherian if and only if its archimedean components are strongly noetherian.*

Proof. Let S be a f.g.c.s. If S is strongly noetherian, then every archimedean component of S is strongly noetherian. Conversely assume that every archimedean component of S is strongly noetherian. Let T be a subsemigroup of S. Since S is f.g., $Y(S)$ is finite and T is the union of the finitely many subsemigroups $T \cap A$, $A \in Y(S)$. Every $T \cap A \subseteq A$ is f.g.; therefore T is f.g. □

Proposition 7.2. *Every f.g. archimedean semigroup without idempotent is strongly noetherian.*

Proof. Let S be a f.g. archimedean semigroup without idempotent and T be a nonempty subsemigroup of S. By Proposition 6.2, S is power joined and so is T. Also S^m is cancellative for some $m > 0$; since S is archimedean, S^m contains an element a of T.

Since $x \neq xy$ for every $x,y \in S$, we have $a \in T \backslash Ta$. Let $t \in T$. Since T is power joined we have $a^k = t^l$ for some $k,l > 0$. Then $t \notin Ta^k$, otherwise $a^k = a^k s$ for some $s \in S$. Therefore there is a greatest $n \geqq 0$ such that $t \in Ta^n$ ($= T$ if $n = 0$). Then $t \notin Ta^{n+1}$ and $t = ua^n$ for some $u \in T \backslash Ta$. Thus T is generated by $T \backslash Ta$.

Let $s,t \in T \backslash Ta$. Assume that $s \mathcal{M}_a t$, with $s = au$, $y = av$, where $u,v \in S^1$ and $a^p u = a^q v$ for some $p,q > 0$. Let $r = \min(p,q)$ and $p = r + i$, $q = r + j$, so that $i,j \geqq 0$, one of i, j is 0, and $a^{r+i} u = a^{r+j} v$. If $r = 1$, then $a^i s = a^{r+i} u = a^{r+j} v = a^j t$ (with, say, $a^i s = s$ if $i = 0$). If $r > 1$, then a^{r-1}, $a^{i+1} u$, $a^{j+1} v \in S^m$ and $a^{r+i} u = a^{r+j} v$ implies $a^i s = a^{i+1} u = a^{j+1} v = a^j t$, since S^m is cancellative. In either case $a^i s = a^j t$. This contradicts $s,t \notin Ta$ if only one of i and j is 0. Therefore $i = j = 0$ and $s = t$. Thus $\mathcal{M}_a$ is the equality on $T \backslash Ta$, and $T \backslash Ta$ is finite by Lemma 6.5. □

3. We now turn to archimedean semigroups with idempotents.

Recall that a f.g. abelian group A is the direct sum $A = F \oplus G$ of a free abelian group F and a finite abelian group G. The rank of A is the rank of F.

Thus a f.g. abelian group has rank 1 if and only if it is isomorphic to $\mathbb{Z} \oplus G$ for some finite abelian group G; a f.g. abelian group has rank at most 1 if and only if it is either finite or isomorphic to $\mathbb{Z} \oplus G$ for some finite abelian group G.

Lemma 7.3. *An abelian group is strongly noetherian (as a commutative semigroup) if and only if it is finitely generated of rank at most* 1.

Proof. An abelian group A which is strongly noetherian (as a c.s.) is f.g. as a c.s. and f.g. as an abelian group. Hence A is the direct sum $A = F \oplus G$ of a free abelian group F and a finite abelian group G. If F has rank 2 or more, then F has a subsemigroup which is free with two generators, which we saw in Example 1.5 contains a subsemigroup which is not f.g. Therefore A has rank at most 1.

Conversely let A be a f.g. abelian group of rank at most 1. If A is finite, then A is strongly noetherian. Otherwise $A \cong \mathbb{Z} \oplus G$, where G is finite. Then $\mathbb{N}^+ \times G \subseteq A$ is power joined and f.g. (by $\{1\} \times G$), and is strongly noetherian by Proposition 7.2. Then every subsemigroup T of $\mathbb{Z} \oplus G$ is a union

$$T = \big(T \cap (\{0\} \times G)\big) \cup \big(T \cap (\mathbb{N}^+ \times G)\big) \cup \big(T \cap (-\mathbb{N}^+ \times G)\big)$$

in which the first term is finite and the remaining two terms are isomorphic to subsemigroups of $\mathbb{N}^+ \times G$ and are f.g.; therefore T is f.g. Thus A is strongly noetherian. □

Proposition 7.4. *Let S be an archimedean semigroup S with an idempotent, and K be the kernel of S. Then S is strongly noetherian if and only if S/K is finite and K is finitely generated of rank at most* 1.

Proof. If S is strongly noetherian, then S/K is finite and K is f.g. by Proposition 6.1 and has rank at most 1 by Lemma 7.3. Conversely assume that S/K is finite and K is finitely generated of rank at most 1. Every subsemigroup T of S is a union

$$T = \big(T \cap (S \backslash K)\big) \cup \big(T \cap K\big)$$

in which the first term is finite and the second term is a f.g. subsemigroup by Lemma 7.3; therefore T is f.g. □

4. The next results, due to Rosales & García-García[1999], are global characterizations of strongly noetherian semigroups.

Proposition 7.5. *A cancellative commutative semigroup S is strongly noetherian if and only if it finitely generated and $G(S)$ has rank at most* 1.

Proof. If S is strongly noetherian, then S is f.g. and has a least archimedean component C which is f.g. and an ideal of S. If C does not contain an idempotent, then C is power joined by Proposition 6.2 and $G(C)$ is f.g. of rank 1 by Proposition III.6.3. If C contains an idempotent, then $G(C)$, which is isomorphic to the kernel of C, is f.g. of rank at most 1 by Proposition 7.4. In either case $G(S) \cong G(C)$ is f.g. of rank at most 1.

If conversely S is f.g. and $G(S)$ has rank at most 1, then $G(S)$, which is f.g., is strongly noetherian by Lemma 7.3, and so is $S \subseteq G(S)$. □

Proposition 7.5 appears as Proposition 3.1 in Rosales & García-García[1999], with a somewhat different statement and a different proof.

The following example shows that Proposition 7.5 does not extend to all semigroups.

Example 7.6. Let

$$S = \langle\, x, y, z \mid xz = yz = z \,\rangle.$$

S is the disjoint union $S = F \cup C$ of $F = \langle x, y \rangle$, which is free on $\{x, y\}$, and $C = \langle z \rangle \cong \mathbb{N}^+$. S is not strongly noetherian since F, as in Example 1.5, is not strongly noetherian. But C is an ideal of S and $G(S) \cong G(C) \cong \mathbb{Z}$ has rank 1. □

Proposition 7.7. *For a commutative semigroup S with a finite generating subset X the following conditions are equivalent:*

(1) *S is strongly noetherian;*

(2) *the universal group of $\langle a, b \rangle \subseteq S$ has rank at most 1 for every $a, b \in S$;*

(3) *the universal group of $\langle x, y \rangle \subseteq S$ has rank at most 1 for every $x, y \in X$;*

(4) *for every $x \neq y$ in X a relation $x^p y^q = x^r y$ with $q > 1$ holds in S.*

This is Theorem 1.8 of Rosales & García-García[1999], to which the reader is referred for a proof. Condition (3) holds if and only if x, y satisfy a nontrivial relation $x^p y^q = x^r y^s$. In a strongly noetherian semigroup one may assume that this relation has the form $x^p y^q = x^r y$ with $q > 1$: indeed the subsemigroup

$$T = \{x^i y^j \mid i \geqq 0, j > 0\}$$

of S must be finitely generated; then some $x^r y$ is not a generator and is a product of two or more generators.

Chapter VII.

SUBCOMPLETE SEMIGROUPS.

Subcomplete semigroups are a natural generalization of finitely generated commutative semigroups. This chapter contains their main properties. Most notable are Ponizovsky families, with transport a number of properties of finite semigroups to all subcomplete semigroups. This brings additional properties of finitely generated commutative semigroups and reveals a deep similarity to their finite relatives.

1. COMPLETIONS.

In this section we define completions and subcomplete semigroups, and prove their first properties. The results are from Grillet [2001S].

1. A **completion** of a commutative semigroup S is a complete semigroup T of which S is a subsemigroup such that every archimedean component of T contains an element of S; a **sharp** completion of S is a complete semigroup T of which S is a subsemigroup such that every archimedean component of T contains a unique archimedean component of S.

The embedding of a cancellative semigroup into its group of fractions shows that sharp completions are not necessarily the "best" completions.

Other examples of completions include the embedding of a cancellative semigroup into its group of fractions; the embedding of an archimedean semigroup with a nonempty cancellative ideal into a complete archimedean semigroup (Proposition III.4.4); the embedding of a separative semigroup into a semilattice of abelian groups (Proposition IV.2.1); the embedding of a subelementary semigroup into an elementary semigroup (Proposition VI.2.1). The Completion Theorem VI.3.6 implies that every f.g.c.s. has a sharp completion.

Most of these examples have very nice universal properties. However, Example VI.4.7 shows that a c.s. S need not have a minimal completion T in the sense that every embedding of S into a complete semigroup extends uniquely to T.

Completions are related to Shevrin's problem 3.58 [1989] of embedding into epigroups. Recall that a c.s. S is an epigroup when every archimedean component of S contains an idempotent. Thus, complete semigroups are epigroups; conversely, a commutative epigroup S is complete if and only if there also exists for every $a \in S$ a least idempotent e of S^1 such that $ea = a$. For instance, commutative epigroups with finitely many idempotents are complete.

Complete semigroups have properties, such as Theorems IV.4.3 and V.3.4, which are not shared by all epigroups. Hence the author feels that completions are more valuable than embeddings into epigroups.

2. A **subcomplete** semigroup is a commutative semigroup which has a completion. For instance, an archimedean semigroup is subcomplete if and only if it is subcomplete as defined in Section III.4. Example III.4.3 shows that not every c.s. is subcomplete. By Theorem VI.3.6,

Proposition 1.1. *Every finitely generated c.s. has a finitely generated sharp completion and is subcomplete.*

Every archimedean component of a subcomplete semigroup is contained in a complete archimedean semigroup and has a nonempty cancellative ideal by Proposition III.4.1. Corollary IV.4.6 provides a stronger necessary condition:

Proposition 1.2. *A c.s. which can be embedded into a complete semigroup is a subdirect product of nilsemigroups, cancellative c.s., and subelementary semigroups.*

Proof. By Corollary IV.4.6 a complete semigroup T is a subdirect product of a nilsemigroup T_1, a group T_0, and elementary semigroups $T_i = G_i \cup N_i$. A subsemigroup S of T is a subdirect product of subsemigroups S_i of the semigroups T_i; $S_1 \subseteq T_1$ is a nilsemigroup, $S_0 \subseteq T_0$ is cancellative, and $S_i \subseteq G_i \cup N_i$ is a nilsemigroup if $S_i \subseteq N_i$, cancellative if $S_i \subseteq G_i$, or subelementary if S_i intersects both G_i and N_i. □

The converse of Proposition 1.2 does not hold:

Example 1.3. Let S be Example III.4.3, which consists of all positive powers of $s_0, s_1, \dots, s_p, \dots$, with $s_p^n = s_q^n$ whenever $n > p, q$ and $s_p^m s_q^n = s_r^{m+n}$, where $r = \max(p,q)$. This nefarious semigroup has no nonempty cancellative ideal and cannot be embedded into a complete semigroup. Nevertheless S is a subdirect product of nilsemigroups. Indeed let

$$I_k = \{ s_p^n \mid p \geqq k \text{ or } n \geqq k \}.$$

Then I_k is an ideal of S and S/I_k is a nilsemigroup since $x^k \in I_k$ for all

$x \in S$. Also $\bigcap_{k \geqq 0} I_k = \emptyset$, since $s_p^n \notin I_k$ when $k > n,p$. Hence the equality on S is the intersection of the Rees congruences of the ideals I_k, and S is a subdirect product of the nilsemigroups S/I_k. □

3. Nevertheless Proposition 1.2 has a partial converse, which is similar to Proposition VI.3.1 but more general:

Proposition 1.4. *For a c.s. S the following conditions are equivalent:*

(1) *S can be embedded into a complete semigroup with finitely many idempotents;*

(2) *S has a completion with finitely many idempotents;*

(3) *S is a subdirect product of finitely many nilsemigroups, cancellative semigroups, and subelementary semigroups.*

Proof. (1) $\Longrightarrow$ (2). Let S be a subsemigroup of a complete semigroup T with finitely many idempotents. Let U be the union of all the archimedean components of T which intersect S. U is a subsemigroup of T since $\mathcal{N}$ is a congruence. Every archimedean component of U is an archimedean component of T and contains an idempotent. Also U, like T, has finitely many idempotents. Hence U is complete.

(2) $\Longrightarrow$ (3). By Theorem IV.4.3 and Proposition IV.4.5, a completion T of S which has finitely many idempotents is a subdirect product of a group, a nilsemigroup, and finitely many elementary semigroups. Then, as in the proof of Proposition 1.2, S is a subdirect product of finitely many nilsemigroups, cancellative semigroups, and subelementary semigroups.

(3) $\Longrightarrow$ (1). Let S be a subdirect product of finitely many nilsemigroups N_j, cancellative semigroups C_k, and subelementary semigroups $S_l = C_l \cup N_l$. Then S can be embedded into the direct product T of the finitely many nilsemigroups $T_j = N_j' = N_j$, abelian groups $T_k = G_k = G(C_k)$, and elementary semigroups $T_l = C_l^{-1} S_l = G_l \cup N_l'$. An archimedean component of T is a (finite) direct product of subsemigroups $A_i \subseteq T_i$, where $A_i = G_i$ or $A_i = N_i'$, for every i: indeed these semigroups are archimedean and constitute a partition of T. Hence every archimedean component of T contains an idempotent. Also T has finitely many idempotents; hence T is complete. □

Proposition 1.4 requires some finiteness condition, since, in general, a subsemigroup of a complete semigroup T need not have a completion $U \subseteq T$:

Example 1.5. Let T be the commutative monoid

$$T = \{e_1, e_2, \dots, e_n, \dots, e, a, f\}$$

in which $e_1 \geqq e_2 \geqq \cdots \geqq e_n \geqq \cdots \geqq e \geqq f$ are idempotents and

$$e_n a = ea = a, \quad a^2 = af = f.$$

Associativity is readily verified. The archimedean components of T are $\{a, f\}$, $\{e\}$, and every $\{e_n\}$; all contain idempotents. Also there is for every $x \in T$ a least idempotent g of T such that $gx = x$: if x is idempotent then $g = x$, otherwise $x = a$ and $g = e$. Thus T is complete. Next let

$$S = \{e_1, e_2, \ldots, e_n, \ldots, a, f\} = T \setminus \{e\}.$$

Then S is a submonoid of T. However, S (seen before as Example IV.4.2) is not complete, since there is no least idempotent g of T such that $ga = a$. Moreover T is not a completion of S, since the archimedean component $\{e\}$ of T does not intersect S. Thus S has no completion $S \subseteq U \subseteq T$. □

4. When S has an identity element, a **monoid completion** of S is a completion T of S in which the identity element of S is also the identity element of T.

Proposition 1.6. *A c.s. S has a completion if and only if S^1 has a monoid completion. Every f.g.c.m. has a sharp f.g. monoid completion.*

Proof. Assume that S does not have an identity element. If T is a completion of S, then $T \cup \{1\}$ (with an identity element adjoined even if T already has one) is a monoid completion of S^1, which is f.g. if T is f.g. Conversely let T be a monoid completion of S^1. If the group of units U of T does not intersect S (if $U \cap S^1 = \{1\}$), then $T \backslash U$ is a completion of S; otherwise T is a completion of S. Thus S has a completion if and only if S^1 has a monoid completion.

Now let S have an identity element u. Let T be a completion of S. Then $S \subseteq uT$. We show that the c.m. uT is a monoid completion of S.

First, the idempotents of uT coincide with the idempotents $e \leqq u$ of T. If $x \in uT$ and e is the least idempotent of T^1 such that $ex = x$, then $e \leqq u$ and e is the least idempotent of uT such that $ex = x$.

If $a, b \in uT$ and $a \leqq_{\mathcal{N}} b$ in T, then $a^n = bt$ for some $n > 0$ and $t \in T$, $a^n = a^n u = b(tu)$ with $tu \in uT$, and $a \leqq_{\mathcal{N}} b$ in uT. Conversely $a \leqq_{\mathcal{N}} b$ in uT implies $a \leqq_{\mathcal{N}} b$ in T. Hence the archimedean components of uT are the intersections of uT and the archimedean components of T. In particular every archimedean component of uT contains an element of S. For every $x \in uT$ there is an idempotent e of T such that $x \,\mathcal{N}\, e$ in T; then $x = ux \,\mathcal{A}\, ue$ in T and $x \,\mathcal{N}\, ue$ in uT. Thus every archimedean component of uT contains an idempotent, and uT is a monoid completion of S.

If S is f.g., then we can arrange that T has finitely many generators $t_1, \cdots, t_n$ and that every archimedean component of T contains one and only one archimedean component of S (Proposition 1.1). Then uT is f.g., by u and $ut_1, \cdots, ut_n$, and every archimedean component of uT contains one and only one archimedean component of S, so that uT is a sharp monoid completion of S. □

2. PONIZOVSKY FAMILIES.

Ponizovsky families are a basic property of finitely generated commutative semigroups and, more generally, of subcomplete semigroups. The results in this section are from Grillet [2001S].

1. When S has a completion T, the congruence $\mathcal{N}$ on T induces a congruence $\mathcal{B}$ on S; by definition, $a \mathcal{B} b$ in S if and only if $a \mathcal{N} b$ in T.

Proposition 2.1. *When T is a completion of S, then $\mathcal{B}$ is a semilattice congruence on S, in particular $\mathcal{N} \subseteq \mathcal{B}$, and $S/\mathcal{B} \cong Y(T)$. If T is a sharp completion of S, then $\mathcal{B} = \mathcal{N}$ and $Y(S) \cong Y(T)$.*

Proof. $\mathcal{B}$ is the congruence induced by the homomorphism $S \xrightarrow{\subseteq} T \longrightarrow Y(T)$, which is surjective since the archimedean components of T all intersect S. Hence $S/\mathcal{B} \cong Y(T)$. The isomorphism sends an $\mathcal{B}$-class to the archimedean component of T that contains it. In particular $\mathcal{B}$ is a semilattice congruence and contains the least semilattice congruence $\mathcal{N}$ on S.

Let T be a sharp completion of S, so that an archimedean component of T contains exactly one archimedean component of S. When $a, b \in S$, then $a \mathcal{N} b$ in T if and only if $a \mathcal{N} b$ in S; hence $\mathcal{B} = \mathcal{N}$ in S and $Y(S) = S/\mathcal{B} \cong Y(T)$; the isomorphism sends an archimedean component of S to the archimedean component of T that contains it. □

If T is a sharp completion of S, then $S/\mathcal{B} = Y(S)$. In general, $S/\mathcal{B} \cong Y(T)$ is a homomorphic image of $Y(S) = S/\mathcal{N}$, since $\mathcal{N} \subseteq \mathcal{B}$. Proposition 1.1 implies an isomorphism $S/\mathcal{B} \longrightarrow E(T)$ which sends $A \in S/\mathcal{B}$ to the idempotent of the archimedean component of T which contains A. We denote the $\mathcal{B}$-class of $x \in S$ by $B(x)$.

2. When S has a completion T, the partial Ponizovsky factors of T induce another interesting partition of S. For technical reasons we assume that S is a monoid and that T is a monoid completion of S; by Proposition 1.6, this entails no loss of generality.

Proposition 2.2. *Let S be a c.m. and T be a monoid completion of S. For* *every $\mathcal{B}$-class A of S, let $P_A = S \cap P^*_{e(A)}$, where $e(A)$ is the idempotent of the archimedean component of T which contains A. Then*

(P1) $\mathcal{P} = (P_A)_{A \in S/\mathcal{B}}$ *is a partition of* S

and, for every $\mathcal{B}$-class A:

(P2) $L_A = \bigcup (P_B \mid B \in S/\mathcal{B},\, B \leqq A)$ *is an ideal of* S*;*

(P3) $P_A \subseteq \bigcup (B \in S/\mathcal{B} \mid B \leqq A)$ *and* $K_A = P_A \cap A \neq \varnothing$*;*

(P4) *if $A \geqq B$ in $S/\mathcal{B}$, $a \in L_A$, and $k \in K_A$, then $ka \in P_B$ if and only if $a \in P_B$;*

(P5) *K_A is cancellative in L_A (if $k \in K_A$, $a, b \in L_A$, and $ka = kb$, then $a = b$).*

(P6) $(\forall a \in A)(\exists n > 0)\ a^n \in K_A$.

Moreover $K_A = S \cap H_e$ and $L_A = S \cap eT$, where $e = e(A)$ as above.

Proof. (P1) By Proposition IV.4.4, the partial Ponizovsky factors of T constitute a partition $(P^*_e)_{e \in E(T)}$ of T. Since $A \longmapsto e(A)$ is bijective, the sets P_A are pairwise disjoint and cover S. Let $A \in S/\mathcal{B}$ and $e = e(A)$.

The archimedean component of e in T contains some $s \in A$, and then $s^n \in H_e$ in T for some $n > 0$, by Proposition VI.1.2, and $s^n \in H_e \cap S \subseteq P^*_e \cap S$. Therefore $P_A \neq \varnothing$.

(P2) In T,

$$\bigcup (P^*_f \mid f \in E(T),\, f \leqq e) \;=\; \bigcup (fT \mid f \in E(T),\, f \leqq e) \;=\; eT$$

is an ideal; therefore

$$L_A \;=\; \bigcup (P_B \mid B \in S/\mathcal{B},\, B \leqq A) \;=\; S \cap eT$$

is an ideal of S.

(P3) Let C be the archimedean component of T which contains A and e. Then $A = S \cap C$ and

$$K_A \;=\; P_A \cap A \;=\; S \cap P^*_e \cap C \;=\; S \cap H_e\,.$$

The proof of (P1) shows that $S \cap H_e \neq \varnothing$. When $x \in S \cap P^*_e$ and $k \in S \cap H_e$, then $x \leqq_{\mathcal{N}} k$ in T and $B(x) \leqq B(k) = A$ in $S/\mathcal{B}$; hence P_A is contained in $\bigcup (B \in S/\mathcal{B} \mid B \leqq A)$.

(P4) Let $A \geqq B$ in $S/\mathcal{B}$, $a \in L_A$, and $k \in K_A$. Then $e = e(A) \geqq e(B) = f$, $a \in eT$, and $k \in H_e$ has a group inverse $l \in H_e$ in T. If $ka \in P_B$ in S, then

$ka \in P_f^*$ and $\epsilon(ka) = f$; hence $fka = ka$, $fa = fea = fkla = kla = ea = a$, and $ga = a$ implies $gka = ka$ and $g \geqq f$, so that $f = \epsilon(a)$ and $a \in P_f^* \cap S = P_B$. If conversely $a \in P_B$, then $\epsilon(a) = f$; hence $fa = a$, $fka = ka$, and $gka = ka$ implies $gea = gkla = kla = a$ and $g \geqq ge \geqq f$, so that $\epsilon(ka) = f$ and $ka \in P_B$.

(P5) Let $e = e(A)$. If $k \in K_A$, $a, b \in L_A \subseteq eT$, and $ka = kb$, then, in T, $k \in H_e$, k has a group inverse $l \in H_e$, and $ka = kb$ implies $a = kla = klb = b$.

(P6) When $a \in A$, then $a^n \in H_e$ in T for some $n > 0$, by Proposition VI.1.2, and $a^n \in S \cap H_e = K_A$. □

If T is a sharp completion of S, then every $\mathcal{B}$-class A is archimedean and (P6) follows from (P1) through (P5). Indeed let $a \in A$. Since $K_A \neq \emptyset$ by (P3) there exists $k \in K_A$. Then $k \in A$, $a^n = kt$ for some $n > 0$ and $t \in A$, and $a^n \in K_A$ since K_A is an ideal of A by Lemma 2.3 below.

A **Ponizovsky family** of a commutative monoid S, relative to a semilattice congruence $\mathcal{B}$ on S, is a family $\mathcal{P} = (P_A)_{A \in S/\mathcal{B}}$ with properties (P1) through (P6) in Proposition 2.2; if $\mathcal{B} = \mathcal{N}$, then $\mathcal{P}$ is a **sharp** Ponizovsky family. When T is a monoid completion of S, the family $\left(P_{e(A)}^* \cap S\right)_{A \in S/\mathcal{B}}$ in Proposition 2.2 is the Ponizovsky family of S **induced by** T.

A Ponizovsky family of a commutative semigroup $S \neq S^1$ is a Ponizovsky family of S^1.

Lemma 2.3. *Let S be a c.m. When $\mathcal{P}$ is a Ponizovsky family:*

(P7) $L_A \subseteq \bigcup\left(B \in S/\mathcal{B} \mid B \leqq A\right)$ *and* $K_A = L_A \cap A$;

(P8) $L_A L_B \subseteq L_{A \wedge B}$ *and* $K_A K_B \subseteq K_{A \wedge B}$;

(P9) $1 \in K_{B(1)}$ *and* $L_{B(1)} = S$.

Moreover, $\mathcal{P}$ is a sharp Ponizovsky family if and only if every K_A is archimedean.

Proof. (P7) If $x \in L_A$, then $x \in P_B$ for some $B \leqq A$ and $C = B(x) \leqq B$ by (P3); hence $L_A \subseteq \bigcup\left(C \in S/\mathcal{B} \mid C \leqq A\right)$. Next, $K_A = P_A \cap A \subseteq L_A \cap A$. If conversely $x \in L_A \cap A$, then $x \in P_B \cap A$ for some $B \leqq A$ in $S/\mathcal{B}$, and $B = A$ since $A = B(x) \leqq B$ by (P3); therefore $x \in P_A \cap A = K_A$.

(P8) $A \wedge B$ is the product of A and B in the semilattice $S/\mathcal{B}$. Let $a \in L_A$ and $b \in L_B$. By (P1), $ab \in P_C$ for some $C \in S/\mathcal{B}$. By (P2), $ab \in L_A$, so $C \leqq A$. Similarly $ab \in L_B$ and $C \leqq B$. Hence $C \leqq A \wedge B$ and $ab \in L_{A \wedge B}$. Then $K_A K_B \subseteq L_A L_B \subseteq L_{A \wedge B}$, $K_A K_B \subseteq AB \subseteq A \wedge B$, and $K_A K_B \subseteq L_{A \wedge B} \cap (A \wedge B) = K_{A \wedge B}$ by (P7).

(P9) We have $1 \in P_A$ for some $A \in S/\mathcal{B}$ by (P1); then $1 \in B$ for some $B \leqq A$, by (P3); since $U = B(1)$ is the greatest element of $S/\mathcal{B}$ this implies $B = A = U$ and $1 \in P_U \cap U = K_U$. By (P1),

$$L_U = \bigcup(P_B \mid B \in S/\mathcal{B},\, B \leqq U) = \bigcup(P_B \mid B \in S/\mathcal{B}) = S.$$

If finally $\mathcal{P}$ is a sharp Ponizovsky family, then every $\mathcal{B}$-class A is archimedean; if $k, l \in K_A$, then $k^n = lt$ for some $n > 0$ and $t \in A$, $k^{n+1} = l(kt)$ with $kt \in K_A$, since K_A is an ideal of A by (P7), and K_A is archimedean. If conversely every K_A is archimedean, then $a, b \in A$ implies a^m, $b^n \in K_A$ by (P6) and $a \mathcal{N} a^m \mathcal{N} b^n \mathcal{N} b$; thus every $\mathcal{B}$-class A is archimedean, $\mathcal{B} = \mathcal{N}$, and $\mathcal{P}$ is a sharp Ponizovsky family. □

3. We now look at some particular cases.

Proposition 2.4. *A complete c.m. has only one Ponizovsky family, namely its partition into partial Ponizovsky factors.*

Proof. If S is complete, then $T = S$ is a monoid completion of S and $(P^*_{e(A)})_{A \in Y(S)}$ is a sharp Ponizovsky family, by Proposition 2.2. Conversely let $\mathcal{P} = (P_A)_{A \in S/\mathcal{B}}$ be a Ponizovsky family relative to a semilattice congruence $\mathcal{B}$.

We show that $\mathcal{B} = \mathcal{N}$. Let $e, f \in E(S)$. Assume that $e \mathcal{B} f$ and let $A = B(e) = B(f)$. Then $e = e^n \in K_A$ by (P6), $f \in K_A$, and $e = f$, since K_A is cancellative by (P5), (P7). Thus a $\mathcal{B}$-class contains only one idempotent of S. Since S is complete, a $\mathcal{B}$-class contains only one archimedean component of S, and $\mathcal{B} = \mathcal{N}$.

Let $e \in E(S)$ and $A = A(e)$. Then $H_e \subseteq L_A \cap A = K_A$. If conversely $a \in K_A$, then $ea \in L_A$, $ea = e(ea)$, and $a = ea \in A \cap eS = H_e$ by (P5). Thus $K_A = H_e$.

Next, $eS \subseteq L_A$, since $e \in H_e \subseteq K_A$. If conversely $a \in L_A$, then as above $ea \in L_A$, $ea = e(ea)$, and $a = ea \in eS$ by (P5). Thus $L_A = eS$.

Finally $P^*_e \subseteq eS = L_A$. Let $a \in L_A$. If $f < e$ in $E(S)$, then $a \in fS$ implies $a \in L_{A(f)}$, $a \in P_B$ for some $B \leqq A(f) < A$, and $a \notin P_A$ by (P1). If conversely $a \notin P_A$, then $a \in P_B \subseteq L_B$ for some $B < A$ in $Y(S)$, and $a \in fS$ for some $f < e$ in $E(S)$. Hence

$$P_A = L_A \setminus \bigcup(P_B \mid B \in S/\mathcal{B},\, B < A) = eS \setminus \left(\bigcup_{f<e} fS\right) = P^*_e.$$

Thus $\mathcal{P}$ is the partition of S into its partial Ponizovsky factors. □

On the other hand, a c.m. may have many Ponizovsky families:

Example 2.5. Let $S = \{x^n \mid n \geqq 0\}$ be the free c.m. with one generator x (cf. Example VI.3.7). There are two archimedean components, $U = \{1\}$ and $C = \{x^n \mid n > 0\}$. For any given integer $m > 0$, let

$$P_U = \{x^n \mid n < m\} \text{ and } P_C = \{x^n \mid n \geqq m\};$$

then $K_U = \{1\}$, $L_U = S$, and $K_C = L_C = P_C$. Then $\mathcal{P}$ is a partition of S; L_U and L_C are ideals of S; $P_U \subseteq U \cup C$, $P_C \subseteq C$, and K_U, K_C are nonempty; (P4) ($ka \in P_B$ if and only if $a \in P_B$ when $a \in L_A$ and $k \in K_A$) is trivial when $A = U$ (then $k = 1$) and when $A = C$ (then $B = A$); (P5) holds since S is cancellative; and (P6) holds since $a^m \in K_C$ for all $a \neq 1$ in S. Thus $\mathcal{P}$ is a sharp Ponizovsky family. Different values of m yield different Ponizovsky families of S.

Proposition 2.6. *Every subelementary monoid $S = C \cup N$ has a Ponizovsky family in which $S/\mathcal{B} = \{C, N\}$, $P_C = S\backslash 0$, $K_C = C$, $L_C = S$, and $P_N = K_N = L_N = \{0\}$.*

Proof. $T = C^{-1}S$ is a monoid completion of S with two idempotents 1 and 0 and two partial Ponizovsky factors $P_1^* = T\backslash\{0\}$ and $P_0^* = \{0\}$; this yields two $\mathcal{B}$-classes, C and N, and a Ponizovsky family in which $P_C = S\backslash 0$, $L_C = S$, $K_C = C$, and $P_N = L_N = K_N = \{0\}$. □

When S is subelementary, the Ponizovsky family in Proposition 2.6 is the **standard Ponizovsky family** of S.

By Proposition 1.6, every f.g.c.s. has a Ponizovsky family. In fact:

Proposition 2.7. *When S is finitely generated and m is sufficiently large, then S has a sharp Ponizovsky family in which $A^m \subseteq K_A$ for all $A \in Y(S)$.*

Proof. If m is sufficiently large, then S^1 has a sharp completion T in which $A^m \subseteq H_{e(A)}$ for all $A \in Y(S^1)$, by Lemma VI.4.2. □

3. ANOTHER COMPLETION THEOREM.

In this section we prove that every Ponizovsky family is induced by a completion, and that there is a smallest such completion (Grillet [2001S]).

1. The main result constructs a very tight completion.

Theorem 3.1. *A commutative monoid S has a completion if and only if it has a Ponizovsky family $\mathcal{P}$, and then S has a monoid completion in which every*

$\mathcal{H}$-class intersects S, and which induces the given Ponizovsky family.

One such completion is the monoid $\widehat{S}_{\mathcal{P}}$ of all fractions a/k with $a \in L_A$ and $k \in K_A$ for some $A \in S/\mathcal{B}$, in which $(a/k)(b/l) = ab/kl$ and $a/k = b/l$ if and only if $a,b \in P_C$ for some $C \in S/\mathcal{B}$ and $al = bk$.

Like the completion in Theorem VI.3.6, $\widehat{S}_{\mathcal{P}}$ consists of fractions, but it is constructed in one step and much more simply. The proof of Theorem 3.1 is given below. The Theorem implies:

Corollary 3.2. *A c.m. is subcomplete if and only if it has a Ponizovsky family.*

Corollary 3.3. *A family of subsets of a c.m. S is a Ponizovsky family if and only if it is induced by the partition of a monoid completion of S into its partial Ponizovsky factors.*

Theorem 3.1 is related to the following result of Tretjakova [1993]: a c.s. S is embeddable into an epigroup if and only if it embeddable into a commutative epigroup, if and only if it contains a subset K such that every element of S has a power in K and $ak^2 = bkl$, $akl = bl^2$ implies $ak = bl$, for all $a,b \in S^1$ and $k,l \in K$. In Theorem 3.1, $K = \bigcup_{A \in S/\mathcal{B}} K_A$ has these properties, by (P5) and (P6).

The **index** of an epigroup S is the least $m > 0$ such that a^m belongs to a subgroup for every $a \in S$ (or is infinite); for a commutative epigroup, this equals the archimedean cancellativity index. Tretjakova [1993] also proved that a c.s. S is embeddable into an epigroup of index $\leqq m$ if and only if it embeddable into a commutative epigroup of index $\leqq m$, if and only if $ax^{2m} = bx^m y^m$, $ax^m y^m = by^{2m}$ implies $ax^m = by^m$, for all $a,b,x,y \in S^1$. This amounts to using $K = \bigcup_{A \in Y(S)} A^m$, as in Proposition 2.7.

2. When a c.m. S has a completion, then S has a monoid completion by Proposition 1.6 and has a Ponizovsky family by Proposition 2.2. We now assume that S has a Ponizovsky family $\mathcal{P}$ relative to some semilattice congruence $\mathcal{B}$, and construct $\widehat{S}_{\mathcal{P}}$ as in Theorem 3.1. Let

$$\mathbb{T} = \{(a,k) \mid a \in L_A,\ k \in K_A \text{ for some } A \in S/\mathcal{B}\}.$$

$\mathbb{T}$ is closed under componentwise multiplication: if $a \in L_A$, $k \in K_A$, $b \in L_B$, $l \in K_B$, then $ab \in L_{A \wedge B}$ and $kl \in K_{A \wedge B}$ by (P8), and $(ab, kl) \in \mathbb{T}$. Thus $\mathbb{T}$ is a c.s.; $\mathbb{T}$ has an identity element, since $(1,1) \in \mathbb{T}$ by (P9).

Lemma 3.4. *The relation*

$$(a,k) \equiv (b,l) \text{ if and only if } a,b \in P_C \text{ for some } C \in S/\mathcal{B} \text{ and } al = bk$$

is a congruence on $\mathbb{T}$.

Proof. $\equiv$ is reflexive and symmetric. If $(a,k) \equiv (b,l) \equiv (c,m)$, with, say, $a \in L_A$, $k \in K_A$, $b \in L_B$, $l \in K_B$, $c \in L_C$, $m \in K_C$, $a,b \in P_D$, and $b,c \in P_E$, then $D = E \leqq A,B,C$ by (P1), $al = bk$, $bm = cl$, $alm = bkm = ckl$, and $am = ck$ since $am, ck \in P_D \subseteq L_B$ by (P4) and $l \in K_B$ is cancellative in L_B by (P5). Hence $\equiv$ is an equivalence relation.

Let $(a,k) \equiv (b,l)$, with, say, $a \in L_A$, $k \in K_A$, $b \in L_B$, $l \in K_B$; then $a,b \in P_D$ for some $D \in S/\mathcal{B}$, $D \leqq A,B$, and $al = bk$. Let $(c,m) \in \mathbb{T}$, with $c \in L_C$, $m \in K_C$. Then $ac, bc \in L_{C \wedge D}$ by (P8). Let $ac \in P_E$, so that $E \leqq D \wedge C$. By (P4), $acl \in P_E$ since $ac \in L_{C \wedge D} \subseteq L_B$; $bck = acl \in P_E$; and $bc \in P_E$, since $bc \in L_{C \wedge D} \subseteq L_A$. Finally $aclm = bckm$. Thus $(ac, km) \equiv (bc, lm)$. Hence $\equiv$ is a congruence on $\mathbb{T}$. $\square$

3. $\widehat{S}_{\mathcal{P}}$ is the quotient monoid $T = \mathbb{T}/\equiv$. We denote the $\equiv$-class of $(a,k) \in \mathbb{T}$ by a/k; thus a/k is defined when $a \in L_A$ and $k \in K_A$ for some $A \in S/\mathcal{B}$, and

$$a/k = b/l \text{ if and only if } a,b \in P_C \text{ for some } C \in S/\mathcal{B} \text{ and } al = bk.$$

In particular $a/1$ is defined for all $a \in S = L_U$ by (P9) (where $U = B(1)$) and $a/1 = b/1$ if and only if $a = b$. Hence $\iota : a \longmapsto a/1$ is an injective homomorphism of S into T. It is convenient to identify $a \in S$ and $a/1 \in T$, so that S is a submonoid of T.

Lemma 3.5. *In* $T = \widehat{S}_{\mathcal{P}}$*:* $a \,\mathcal{H}\, a/k$ *in* T *for every* $a/k \in T$*; in particular* $a \,\mathcal{N}\, a/k$ *in* T *for every* $a/k \in T$*; and* $a/k \,\mathcal{N}\, b/l$ *in* T *if and only if* $a \,\mathcal{B}\, b$ *in* S.

Proof. Let $a/k \in T$, with, say, $a \in L_A$, $k \in K_A$, and $a \in P_B$, where $B \leqq A$. In T, $(k/1)(a/k) = ak/k = a/1$, since $a, ak \in P_B$ by (P4) and $(ak)\,1 = ak$. Similarly $(k/k^2)(a/1) = ak/k^2 = a/k$, since $a, ak \in P_B$ and $(ak)\,k = ak^2$. Thus $a \,\mathcal{H}\, a/k$. In particular $a \,\mathcal{N}\, a/k$.

Next we show that $a/k = b/l$ implies $a \,\mathcal{B}\, b$. Let $a \in L_A$, $k \in K_A$, $b \in L_B$, $l \in K_B$, and $a,b \in P_C$, so that $C \leqq A,B$. Then $B(a)$, $B(b) \leqq C \leqq A,B$ by (P3),

$$B(a) = B(a) \wedge B(l) = B(al) = B(bk) = B(b) \wedge B(k) = B(b),$$

and $a \,\mathcal{B}\, b$.

If now $a \,\mathcal{N}\, b$ holds in T, then $a^n = b(c/m) = bc/m$ for some $n > 0$ and

$c/m \in T$, $a^n \mathcal{B}\, bc$ in S, and $B(a) \leqq B(b)$; similarly $B(b) \leqq B(a)$ and $a \mathcal{B} b$ in S. To prove the converse we note that $k \mathcal{H} l$ in T for all $k, l \in K_A$: indeed $k(l/k) = kl/k = l/1 = l$, since $kl, l \in K_A$ by (P8), and similarly $l(k/l) = k$. If now $a \mathcal{B} b$ in S, then a^m, $b^n \in K_A$ for some $m, n > 0$ by (P6), $a^m \mathcal{H} b^n$ in T, and $a \mathcal{N} b$ in T. □

By Lemma 3.5, every $\mathcal{H}$-class of T intersects S; hence every archimedean component of T intersects S.

Lemma 3.6. *In $T = \widehat{S}_{\mathcal{P}}$, $e_A = k/k$ is idempotent whenever $k \in K_A$. Every idempotent of T is of this form; moreover $e_A \leqq e_B$ in T if and only if $A \leqq B$ in $S/\mathcal{B}$.*

Let $a \in L_A$ and $k \in K_A$. When $a \in P_B$, then e_B is the least idempotent e of T such that $e(a/k) = a/k$. Moreover $a/k = (a/1)(k/k^2)$, and k/k^2 is the inverse of k in H_{e_A}.

Proof. First $k/k = l/l$ for every $k, l \in K_A$, so that $e_A = k/k$ depends only on A, and $e_A = k/k = k^2/k^2 = (k/k)^2$ is idempotent.

If a/k is idempotent, with, say, $a \in L_A$, $k \in K_A$, and $a \in P_B$, then $C = B(a) \leqq B \leqq A$ and $a/k = a^n/k^n$ for every $n > 0$. Since $a^n \in K_C \in P_C$ for some $n > 0$ by (P6), this implies $a \in P_C \cap C = K_C$. Then $a^2/k^2 = a/k$ yields $ka^2 = k^2 a$, $a^2 = ka$ by (P5) since $a^2, ka \in L_A$, and $a/k = a/a = e_C$. Thus every idempotent of T has the form e_A for some $A \in S/\mathcal{B}$.

If $A \leqq B$ in $S/\mathcal{B}$, then $(k/k)(l/l) = kl/kl = k/k$ for all $k \in K_A$, $l \in K_B$, since $kl \in K_A$ by (P8). If conversely $k \in K_A$, $l \in K_B$, and $kl/kl = (k/k)(l/l) = k/k$, then $kl \in K_A$ and $A = A \wedge B \leqq B$ by (P8). Thus $e_A \leqq e_B$ in T if and only if $A \leqq B$ in $S/\mathcal{B}$.

Let $a/k \in T$, where $a \in P_B$, $a \in L_A$, $B \leqq A$, and $k \in K_A$. Let $l \in K_B$. Then $la \in P_B$ by (P4) and $(l/l)(a/k) = la/lk = a/k$. If conversely $(m/m)(a/k) = a/k$, where $m \in K_C$, then $am \in P_B$, $B \leqq C$ since $am \in L_C$ by (P2), and $l/l \leqq m/m$ in T. Thus $e_B = l/l$ is the least idempotent e of T such that $e(a/k) = a/k$. Finally, $(a/1)(k/k^2) = ak/k^2 = a/k$, since $a, ak \in P_B$ and $(ak)k = ak^2$, and $(k/1)(k/k^2) = k^2/k^2 = e_A$, so that k/k^2 is the inverse of k in H_{e_A}. □

We can now complete the proof of Theorem 3.1. By Lemmas 3.5, 3.6, every archimedean component of T contains an $\mathcal{B}$-class A of S, contains some $a \in A$, contains $k = a^n \in K_A$ by (P6), and contains the idempotent k/k. Moreover there is for every $x = a/k \in T$ a least idempotent e of T such that $ex = x$, by Lemma 3.6. Thus T is complete.

By Lemma 3.5 the congruence induced on S by the congruence $\mathcal{N}$ on T is the given semilattice congruence $\mathcal{B}$. When $e = e_C$, Lemma 3.6 implies that $\epsilon(a/k) = e$ if and only if $a \in P_C$; hence $a/k \in P_e^*$ in T if and only if $a \in P_C$, $P_C = P_e^* \cap S$, and $\mathcal{P}$ is the Ponizovsky family induced by T. □

4. We illustrate the construction of $\widehat{S}_{\mathcal{P}}$ with two examples.

Proposition 3.7. *Let $S = C \cup N$ be a subelementary monoid. If $\mathcal{P}$ is the standard Ponizovsky family of S, then $\widehat{S}_{\mathcal{P}} = C^{-1}S$.*

Proof. In the standard Ponizovsky family $\mathcal{P}$, $P_C = S\backslash 0$, $L_C = S$, $K_C = C$, and $P_N = L_N = K_N = \{0\}$ (Proposition 2.6). By definition, $T = \widehat{S}_{\mathcal{P}}$ consists of all fractions a/k where either $a \in S\backslash 0$ and $k \in C$, or $a = k = 0$, with $(a/k)(b/l) = ab/kl$ and $a/k = b/l$ if and only if $al = bk$ and either $a,b \in S\backslash 0$ or $a = b = 0$. The fraction $0/0 \in T$ is the zero element of T and may be identified with the zero element $0/k$ of $C^{-1}S$; then $T = C^{-1}S$. □

Example 3.8. Let $S = \{x^n \mid n \geqq 0\}$ be the free c.m. with one generator x. As in Example 2.5, let $m > 0$ and

$$P_U = \{x^n \mid n < m\} \text{ and } P_C = \{x^n \mid n \geqq m\};$$

then $\mathcal{P}$ is a sharp Ponizovsky family and $K_U = \{1\}$, $L_U = S$, $K_C = L_C = P_C$.

$\widehat{S}_{\mathcal{P}}$ consists of all fractions a/k where either $a \in S$, $k = 1$ or $a,k \in C$, with $a/k = b/l$ if and only if $al = bk$ and either $a,b \in P_U$ or $a,b \in K_C$. The fractions k/l with $k,l \in K_C$ constitute the group of fractions of K_C, which is an infinite cyclic group $\{y^n \mid n \in \mathbb{Z}\}$, where $x^n/x^p = y^{n-p}$ whenever $n,p \geqq m$. We see that $x^n/1 = y^n$ for all $n \geqq m$, but $x^n = x^n/1$ equals no other element of $\widehat{S}_{\mathcal{P}}$ when $n < m$. Thus

$$\widehat{S}_{\mathcal{P}} = \{x^n \mid 0 \leqq n < m\} \cup \{y^n \mid n \in \mathbb{Z}\},$$

where $x^i y^j = y^{i+j}$ and $x^n \in S$ is identified with $y^n \in \widehat{S}_{\mathcal{P}}$ if $n \geqq m$. Thus $\widehat{S}_{\mathcal{P}}$ is the completion of index m constructed in Example VI.3.7. □

5. Finally we show that $\widehat{S}_{\mathcal{P}}$ is the smallest monoid completion of S which induces $\mathcal{P}$:

Proposition 3.9. *If T is a monoid completion of S which induces the Ponizovsky family $\mathcal{P}$, then there exists a unique homomorphism of $\widehat{S}_{\mathcal{P}}$ into T*

which is the identity on S, and it is injective.

Proof. Since T induces $\mathcal{P}$, $\mathcal{B}$ is induced by the congruence $\mathcal{N}$ on T, there is for every $\mathcal{B}$-class A an idempotent $e(A)$ in the archimedean component of T which contains A, and $A \longmapsto e(A)$ is an isomorphism of $S/\mathcal{B} \cong E(T)$. When $e = e(A)$, then $K_A \subseteq H_e$ and $L_A \subseteq eT$, by Proposition 2.2; let $k^{-1} \in T$ denote the inverse of $k \in K_A$ in H_e.

We show that $a/k = b/l$ in $\widehat{S}_{\mathcal{P}}$ if and only if $ak^{-1} = bl^{-1}$ in T. Let a/k, $b/l \in T$, with $a \in L_A$, $k \in K_A$, $b \in L_B$, and $l \in K_B$. Assume that $a/k = b/l$, so that $al = bk$ and $a,b \in P_C$ for some $C \in S/\mathcal{B}$. Then $C \leqq A,B$ and $e = e(A)$, $f = e(B)$, $g = e(C)$ satisfy $g \leqq e,f$. Also $\epsilon(a) = \epsilon(b) = g$ in T, since $a,b \in P_C = S \cap P_g^*$; hence $ga = a$, $gb = b$, $ea = ega = ga = a$, and similarly $fa = a$, $eb = b$, and $fb = b$. Now $k \in H_e$, $l \in H_f$, and

$$ak^{-1} = afk^{-1} = all^{-1}k^{-1} = bkk^{-1}l^{-1} = bel^{-1} = bl^{-1}.$$

If conversely $ak^{-1} = bl^{-1}$ in T, then $a = ae \ \mathcal{H} \ ak^{-1} = bl^{-1} \ \mathcal{H} \ bf = b$ and $\epsilon(a) = \epsilon(b)$. If $g = \epsilon(a) = \epsilon(b)$, then $g \leqq e,f$, since $ea = fa = a$, and $a,b \in S \cap P_g^* = P_C$ for some $C \in S/\mathcal{B}$. Then $a/k = b/l$, since

$$al = ael = alkk^{-1} = blkl^{-1} = bfk = bk.$$

By the above a mapping $\varphi : \widehat{S}_{\mathcal{P}} \longrightarrow T$ is well defined by: $\varphi(a/k) = ak^{-1}$, and it is injective. Moreover φ is the identity on S, since $1^{-1} = 1$ in T and $\varphi(s) = \varphi(s/1) = s$ for all $s \in S = L_{B(1)}$ by (P9). Let a/k, $b/l \in \widehat{S}_{\mathcal{P}}$, where $a \in L_A$, $k \in K_A$, $b \in L_B$, and $l \in K_B$. By (P8), $ab \in L_C$ and $kl \in K_C$, where $C = A \wedge B$. Let $e = e(A)$, $f = e(B)$, and $g = e(C)$. Then $g = ef$, $kl \in H_e H_f \subseteq H_g$, $k^{-1} l^{-1} \in H_g$, $(kl)(k^{-1} l^{-1}) = ef = g$, and $k^{-1} l^{-1} = (kl)^{-1}$. Hence

$$\varphi\big((a/k)(b/l)\big) = ab\,(kl)^{-1} = (ak^{-1})(bl^{-1}) = \varphi(a/k)\,\varphi(b/l).$$

Thus φ is a homomorphism.

Conversely let $\psi : \widehat{S}_{\mathcal{P}} \longrightarrow T$ be a homomorphism which is the identity on S. For every $A \in S/\mathcal{B}$, ψ sends the archimedean component of $\widehat{S}_{\mathcal{P}}$ which contains A into the archimedean component of T which contains A, and sends the idempotent of the former to the idempotent of the latter. Thus $\psi(e_A) = e(A)$. By Lemma 3.6, k/k^2 is the inverse of $k \in K_A$ in H_{e_A}, so that $\psi(k/k^2) = \psi(k)^{-1} = k^{-1}$, and

$$\psi(a/k) = \psi((a/1)(k/k^2)) = \psi(a/1)\,\psi(k/k^2) = ak^{-1}.$$

Thus $\psi = \varphi$. □

4. PROPERTIES.

In this section we use Ponizovsky families to transfer properties of complete semigroups to subcomplete semigroups; this includes explicit subdirect decompositions and a congruence which is similar to but larger than $\mathcal{H}$.

1. By Proposition VI.2.2, a subcomplete semigroup S is a subdirect product of nilsemigroups, cancellative semigroups, and subelementary semigroups. With a Ponizovsky family we can construct an explicit decomposition. In the complete c.m. $\widehat{S}_{\mathcal{P}}$, the projection $\widehat{S}_{\mathcal{P}} \longrightarrow P_e$ induces a congruence on $\widehat{S}_{\mathcal{P}}$; the intersection of these congruences is the equality on $\widehat{S}_{\mathcal{P}}$, since the projections $\widehat{S}_{\mathcal{P}} \longrightarrow P_e$ separate the elements of $\widehat{S}_{\mathcal{P}}$. Restriction to S yields:

Proposition 4.1. *Let S be a c.m. and $\mathcal{P}$ be a Ponizovsky family on S. When $C \in S/\mathcal{B}$ and $m \in K_C$, the congruence on S induced by the projection $\widehat{S}_{\mathcal{P}} \longrightarrow P_e$ is the relation*

$$a \,\mathcal{P}_C\, b \text{ if and only if either } am = bm \in P_C, \text{ or } am, bm \notin P_C;$$

hence $\mathcal{P}_C$ is a congruence and does not depend on the choice of m; $S/\mathcal{P}_C$ is cancellative if C is the least $\mathcal{B}$-class of S, otherwise $S/\mathcal{P}_C$ is subelementary; and $\bigcap_{C \in S/\mathcal{B}} \mathcal{P}_C$ is the equality on S.

Proof. Let $T = \widehat{S}_{\mathcal{P}}$ and $e \in E(T)$. By definition, the projection $\pi_e : T \longrightarrow P_e$ takes $x \in T$ to ex if $ex \in P_e^*$, to 0 if $ex \in \bigcup_{f<e} fT$; since $ex \in eT$, $ex \in \bigcup_{f<e} fT = \bigcup_{f<e} P_f^*$ if and only if $ex \notin P_e^*$. Hence $\pi_e(x) = \pi_e(y)$ if and only if either $ex = ey \in P_e^*$, or $ex, ey \notin P_e^*$.

Let $m \in K_C$ and $e = e_C = m/m \in E(T)$. When $a, b \in S$, then $ea = eb$ in T if and only if $am/m = (m/m)(a/1) = (m/m)(b/1) = bm/m$, if and only if $am, bm \in P_D$ for some $D \in S/\mathcal{B}$ and $am^2 = bm^2$; since $am, bm \in L_C$ we have $D \leqq C$ and $am^2 = bm^2$ implies $am = bm$ by (P5). Thus $ea = eb$ if and only if $am = bm$. Moreover $ea, eb \in P_e^*$ if and only if $D = C$ in the above, by Lemma 3.6. Thus $\pi_e(a) = \pi_e(b)$ in T if and only if $a \,\mathcal{P}_C\, b$.

It follows that $\mathcal{P}_C$ is a congruence and does not depend on m. Furthermore

$S/\mathcal{P}_C$ is isomorphic to a subsemigroup of P_e, which intersects H_e since $\pi_e(k) \in H_e$ when $k \in K_C$. Hence $S/\mathcal{P}_C$ is cancellative if C is the least $\mathcal{B}$-class of S, otherwise $S/\mathcal{P}_C$ is subelementary. Finally the intersection of all $\mathcal{P}_C$ is the equality on S, since the projections π_e separate the elements of T. □

2. The partial Ponizovsky factors of $\widehat{S}_{\mathcal{P}}$ already induce subsets P_A of S with similar properties. We now look at the congruence on S induced by the congruence $\mathcal{H}$ on $\widehat{S}_{\mathcal{P}}$. First we show:

Lemma 4.2. *Let $e = e_C$. In $\widehat{S}_{\mathcal{P}}$, $g \in H_e$ if and only if $g = k/l$ for some $k,l \in K_C$; hence $H_e = G(K_C)$.*

Proof. If $k,l \in K_C$, then $(k/l)(l/k) = kl/kl = e_C$ and $e_C(k/l) = kk/kl = k/l$; hence $k/l \in H_e$.

Conversely let $a/k \in H_e$, with, say, $a \in L_A$, $k \in K_A$. Let $m \in K_C$. Then $am/km = (m/m)(a/k) = a/k$ and $ab/kl = (a/k)(b/l) = m/m$ for some $b/l \in \widehat{S}_{\mathcal{P}}$. By Lemma 3.5, $am \;\mathcal{B}\; a$ and $ab \;\mathcal{B}\; m$, so that $a \;\mathcal{B}\; m$ and $a \in C$. Hence $C \leqq A$ by (P4). Then $a \in P_C \cap C = K_C$ and $a/k = am/km$, with am, $km \in K_C$ by (P8).

When $k,l,m,n \in K_C$, then $k/l = m/n$ if and only if $kn = lm$. Thus $H_e = K_C^{-1}K_C = G(K_C)$. □

Proposition 4.3. *Let S be a c.m. with a Ponizovsky family $\mathcal{P}$. The relation*

$$a \;\mathcal{K}\; b \text{ if and only if } a,b \in P_A \text{ and } ak = bl, \text{ for some } A \in S/\mathcal{B} \text{ and } k,l \in K_A$$

is a congruence on S; $a \;\mathcal{K}\; b$ in S if and only if $a \;\mathcal{H}\; b$ in $\widehat{S}_{\mathcal{P}}$; and $S/\mathcal{K} \cong \widehat{S}_{\mathcal{P}}/\mathcal{H}$ is complete groupfree. If S is finitely generated, then $S/\mathcal{K}$ is finite.

Proof. Let $a,b \in S$, $a \in P_A$. By Lemma V.2.2, $a \;\mathcal{H}\; b$ in $T = \widehat{S}_{\mathcal{P}}$ if and only if $b = ga$ for some $g \in H_e \subseteq T$, where $e = \epsilon(a)$. By Lemmas 3.6 and 4.2, $e = e_A$ and $g = k/l$ for some $k,l \in K_A$. Now $ak \in P_A$ by (P4), so $b/1 = b = ga = (k/l)(a/1) = ak/l$ if and only if $b \in P_A$ and $ak = bl$. Thus $a \;\mathcal{H}\; b$ in T if and only if $a \;\mathcal{K}\; b$ in S. Since every $\mathcal{H}$-class of T intersects S it follows that $S/\mathcal{K} \cong T/\mathcal{H}$, and $S/\mathcal{K}$ is complete groupfree by Corollary V.1.6.

If S is f.g., then $S/\mathcal{K}$ is f.g. and the Ponizovsky factors of $S/\mathcal{K}$, which are nilmonoids, are f.g. and therefore finite. Hence $S/\mathcal{K}$ is a subdirect product of finitely many finite monoids and is finite. □

For example every K_A is a $\mathcal{K}$-class: if $a,b \in K_A$, then $a \;\mathcal{K}\; b$, since $a,b \in P_A$ and $ab = ba$; if conversely $a \in K_A$ and $a \;\mathcal{K}\; b$, then $a \;\mathcal{B}\; b$ by Lemma 3.5 and

$b \in A \cap P_A = K_A$.

$\mathcal{K}$ is similar to the orbit congruence of a subelementary semigroup. In fact:

Proposition 4.4. *If* $S = S^1 = C \cup N$ *is subelementary, then, relative to the standard Ponizovsky family,* $\mathcal{K}$ *is the orbit congruence on* S.

Proof. $\widehat{S}_{\mathcal{P}} = C^{-1}S$, by Proposition 3.7. By Proposition 4.3, $\mathcal{K}$ is induced by Green's relation $\mathcal{H}$ on $C^{-1}S$ and is the orbit congruence on S. □

$\mathcal{K}$ is also similar to $\mathcal{H}$, as the next section will show. Subelementary semigroups show that it may be larger than $\mathcal{H}$. For another example:

Example 4.5. Let $S = \{x^n \mid n \geqq 0\}$ be the free c.m. with one generator x, and $\mathcal{P}$ consist of $P_U = \{x^n \mid n < m\}$ and $P_C = \{x^n \mid n \geqq m\}$, where $m > 0$, so that $K_U = \{1\}$, $L_U = S$, and $K_C = L_C = P_C$ (as in Example 2.5). Then $\mathcal{H}$ is the equality on S, whereas $\mathcal{K}$ is the Rees congruence of L_C. □

5. SCHÜTZENBERGER FUNCTORS.

Continuing Section 4 we construct from the congruence $\mathcal{K}$ on S an expanded Schützenberger-like functor. This actually provides a direct construction of the Schützenberger functor of the complete semigroup $\widehat{S}_{\mathcal{P}}$, which may in turn be used to reconstruct $\widehat{S}_{\mathcal{P}}$ as a group coextension of $\widehat{S}_{\mathcal{P}}/\mathcal{H} \cong S/\mathcal{K}$.

1. When K is a $\mathcal{K}$-class,

$$\mathrm{St}(K) = \{a \in S \mid aK \subseteq K\} = \{a \in S \mid ac \in K \text{ for some } c \in K\},$$

since $\mathcal{K}$ is a congruence, and $\mathrm{St}(K)$ is a submonoid of S. Also:

Lemma 5.1. $\mathrm{St}(K) = \mathrm{St}(K_C)$ *for every* $\mathcal{K}$*-class* $K \subseteq P_C$.

Proof. Let $c \in K$ and $k \in K_C$. If $a \in \mathrm{St}(K_C)$, then $ak = l \in K_C$. Hence $ack = cl$. Also $ac \in L_C$, $ack = cl \in P_C$ by (P4), and $ac \in P_C$ by (P4); hence $ac \,\mathcal{K}\, c$ and $a \in \mathrm{St}(K)$.

Conversely let $a \in \mathrm{St}(K)$ and $c \in K$, so that $ac \in K$. Let $A = B(a) \in S/\mathcal{B}$. By (P6), $a^n \in K_A$ for some $n > 0$. In $\widehat{S}_{\mathcal{P}}$ we have $ac \,\mathcal{H}\, c$ by Proposition 4.3, $a^n c \,\mathcal{H}\, c$ since $\mathcal{H}$ is a congruence, and $a^n \,\mathcal{H}\, e_A$ by Lemma 4.2; hence $e_A c \,\mathcal{H}\, c$, $c = e_A c$, $e_A \geqq \epsilon(c)$, and $A \geqq C$ by Lemma 3.6. Hence $k \in K_C$ implies $ak \in A \wedge C = C$, $ak \in C \cap L_C = K_C$, and $a \in \mathrm{St}(K_C)$. □

For every $a \in \mathrm{St}(K)$ a mapping $\sigma_K(a) : K \longrightarrow K$ is well defined by:

$$\sigma_K(a)(c) = ac \quad \text{for all } c \in K.$$

Proposition 5.2. *Let S be a c.m. with a Ponizovsky family $\mathcal{P}$. For every $\mathcal{K}$-class K,*

$$\Sigma(K) \;=\; \{\sigma_K(a) \mid a \in \mathrm{St}(K)\}$$

is a cancellative c.m. of injective transformations of K; $\sigma_K : \mathrm{St}(K) \longrightarrow \Sigma(K)$ is a homomorphism; and $\Sigma(K)$ acts simply on K (if $\sigma, \tau \in \Sigma(K)$ and $\sigma c = \tau c$ for some $c \in K$, then $\sigma = \tau$).

Proof. $\sigma_K(1)$ is the identity on K and $\sigma_K(ab) = \sigma_K(a) \circ \sigma_K(b)$ for all $a, b \in \mathrm{St}(K)$; hence σ_K is a homomorphism of monoids and $\Sigma(K)$ is a c.m.

Let $K \subseteq P_C$. If $a \in \mathrm{St}(K)$ and $ac = ad$, with $c, d \in K \subseteq P_C$, then, for any $k \in K_C$, we have $akc = akd$ and $c = d$ by (P5), since $ak \in K_C$ by Lemma 5.1. Hence every $\sigma_K(a) \in \Sigma(K)$ is injective. Therefore $\Sigma(K)$ is cancellative.

Finally let $a, b \in \mathrm{St}(K)$ and $ac = bc$ for some $c \in K$. Let $d \in K$, so that $d \in P_C$ and $ck = dl$ for some $k, l \in K_C$. Then $adl = ack = bck = bdl$ and $ad = bd$ by (P5), since $ad, bd \in L_C$ and $l \in K_C$. Hence $\Sigma(K)$ acts simply on K. □

$\Sigma(K)$ is the **Schützenberger monoid** of the $\mathcal{K}$-class K. It is, in general, larger than the Schützenberger groups of the $\mathcal{H}$-classes $H \subseteq K$, but not transitive or a group.

Since $\Sigma(K)$ is cancellative it has a group of fractions $\Gamma(K) = G(\Sigma(K))$; this is the **Schützenberger group** of K.

Example 5.3. As in Examples 2.5 and 4.5, let $S = \{x^n \mid n \geqq 0\}$ be the free c.m. with one generator x, and $\mathcal{P}$ consist of $P_U = \{x^n \mid n < m\}$ and $P_C = \{x^n \mid n \geqq m\}$, where $m > 0$, so that $K_U = \{1\}$, $L_U = S$, and $K_C = L_C = P_C$. Then $\mathrm{St}(K_C) = S$, σ_{K_C} is injective since S is cancellative, and $\Sigma(K_C) \cong S$ is larger than the Schützenberger groups of the $\mathcal{H}$-classes $H \subseteq K_C$, which are trivial; but $\Sigma(K_C)$ is not a group. When $j \geqq m$, so that $x^j \in K_C$, we have $\sigma_{K_C}(x^i)(x^j) = x^i x^j = x^{i+j}$; $\Sigma(K_C)$ is not transitive since there is no $\sigma \in \Sigma(K_C)$ such that, say, $\sigma\, x^{m+1} = x^m$. Similarly $\Gamma(K_C) \cong G(S)$ is an infinite cyclic group; on the other hand the Schützenberger groups of the $\mathcal{H}$-classes of S are all trivial. □

Example 5.3 shows that the action of $\Sigma(K)$ on K does not in general extend to an action of $\Gamma(K)$ on K. A partial action of $\Gamma(K)$ on K is defined below.

2. The Schützenberger groups of the $\mathcal{K}$-classes of S are naturally related to the Schützenberger groups of $\widehat{S}_{\mathcal{P}}$.

Proposition 5.4. *Let S be a c.m. with a Ponizovsky family $\mathcal{P}$. Let H be an $\mathcal{H}$-class of $\widehat{S}_{\mathcal{P}}$ and $K = H \cap S$. Then $\Gamma(K) \cong \Gamma(H)$.*

By Proposition 4.3 every $\mathcal{K}$-class of S has the form $H \cap S$ for some $\mathcal{H}$-class H of $\widehat{S}_{\mathcal{P}}$.

Proof. Let $T = \widehat{S}_{\mathcal{P}}$. Recall that the Schützenberger group of an $\mathcal{H}$-class H of T is $\Gamma(H) = \{\gamma_H(t) \mid t \in \mathrm{St}(H)\}$, where $\mathrm{St}(H) = \{t \in T \mid tH \subseteq H\}$ and $\gamma_H(t)$ is well defined by: $\gamma_H(t)(h) = th$ for all $h \in H$; it is a simply transitive group of permutations of H. Proposition 5.4 follows from a more detailed result:

Lemma 5.5. *Let H be an $\mathcal{H}$-class of $\widehat{S}_{\mathcal{P}}$ and $K = H \cap S$. Then $\mathrm{St}(K) \subseteq \mathrm{St}(H)$ and there is an isomorphism $\theta_K : \Gamma(K) \cong \Gamma(H)$, which is the unique homomorphism such that $\theta_K \circ \sigma_K = \gamma_K$; if $\sigma \in \Sigma(K)$, then $\theta_K(\sigma)$ is the unique $\gamma \in \Gamma(H)$ which extends σ.*

Proof. For any $c \in K$, $a/k \in \mathrm{St}(H)$ (in $T = \widehat{S}_{\mathcal{P}}$) is equivalent to $(a/k)\,c\ \mathcal{H}\ c$, $ac\ \mathcal{H}\ c$ in T, $ac\ \mathcal{K}\ c$ in S, and $a \in \mathrm{St}(K)$ (in S), by Lemma 3.5 and Proposition 4.3. In particular $\mathrm{St}(K) \subseteq \mathrm{St}(H)$.

When $a, b \in \mathrm{St}(K)$, then $\sigma_K(a) = \sigma_K(b)$ if and only if, for any $c \in K$, $ac = bc$ and $\gamma_H(a) = \gamma_H(b)$. Therefore there is a unique monoid homomorphism $\iota : \Sigma(K) \longrightarrow \Gamma(H)$ such that $\iota(\sigma_K(a)) = \gamma_H(a)$ for all $a \in \mathrm{St}(K)$. Moreover, ι is injective. Therefore ι extends to an injective homomorphism $\theta = \theta_K$: $\Gamma(K) \longrightarrow \Gamma(H)$, defined by $\theta(\sigma/\tau) = \iota(\sigma) \circ \iota(\tau)^{-1}$ for all $\sigma, \tau \in \Sigma(K)$. The square

$$\begin{array}{ccc} \mathrm{St}(K) & \xrightarrow{\subseteq} & \mathrm{St}(H) \\ {\scriptstyle \sigma_K}\downarrow & & \downarrow{\scriptstyle \gamma_H} \\ \Gamma(K) & \xrightarrow[\theta_K]{} & \Gamma(H) \end{array}$$

commutes; since $\Gamma(K)$ is generated (as a group) by $\mathrm{Im}\,\sigma_K$, θ is the only homomorphism with this property. For every $a \in \mathrm{St}(K)$, $\theta(\sigma_K(a)) = \gamma_H(a)$ extends $\sigma_K(a)$, since $\gamma_H(a)(c) = ac = \sigma_K(a)(c)$ for every $c \in K \subseteq H$; it is the unique $\gamma \in \Gamma(H)$ which extends $\sigma_K(a)$ since $\Gamma(H)$ is simply transitive.

It remains to show that θ is surjective. Let $\gamma \in \Gamma(H)$, so that $\gamma c\ \mathcal{H}\ c$ in T for any $c \in K = H \cap S$. By Lemma V.2.2, $\gamma c = gc$ for some $g \in H_e$, where

$e = \epsilon(c)$; by Lemma 3.6, $e = e_C$, where $c \in P_C$, and then $g = m/n$ for some $m, n \in K_C$, by Lemma 4.2. Thus $\gamma c = (m/n)c$ and $\gamma = \gamma_H(m/n)$ for some $m, n \in K_C$. In T, $m/n = m(n/n^2)$ and n/n^2 is the inverse of n in H_e. Now $H_e \subseteq \mathrm{St}(H)$, since $\mathcal{H}$ is a congruence on T, and the restriction of γ_H to H_e is a group homomorphism; hence

$$\gamma = \gamma_H(m/n) = \gamma_H(m)\,\gamma_H(n)^{-1} = \theta\big(\sigma_K(m)/\sigma_K(n)\big).$$

Thus θ_K is surjective. □

A partial action of $\Gamma(K)$ on K can be defined as follows: if $\gamma = \sigma/\tau \in \Gamma(K)$, where $\sigma, \tau \in \Sigma(K)$, and $c \in K$, then $\gamma \,.\, c$ is defined if and only if there exists $d \in K$ such that $\tau d = \sigma c$, and then $\gamma \,.\, c = d$. That this is well defined follows from Lemma 5.5:

Proposition 5.6. *Let H be an $\mathcal{H}$-class of $\widehat{S}_{\mathcal{P}}$ and $K = H \cap S$. Up to the isomorphism $\Gamma(K) \cong \Gamma(H)$, the partial action of $\Gamma(K)$ on K is the restriction to K of the action of $\Gamma(H)$ on K. For every $a, b \in K$ there is a unique $\gamma \in \Gamma(K)$ such that $\gamma \,.\, a = b$.*

Proof. Denote by $\gamma \longrightarrow \overline{\gamma}$ the isomorphism $\theta = \theta_K : \Gamma(K) \longrightarrow \Gamma(H)$ in Lemma 5.5. When $\gamma = \sigma/\tau \in \Gamma(K)$, where $\sigma, \tau \in \Sigma(K)$, then $\overline{\gamma} = \overline{\tau}^{-1} \circ \overline{\sigma}$, and σ, τ are the restrictions to K of $\overline{\sigma}$ and $\overline{\tau}$. If $c, d \in K$ and $\tau d = \sigma c$, then $\overline{\tau} d = \overline{\sigma} c$ and $\overline{\gamma} c = d$; thus, if $\gamma \,.\, c$ is defined, then $\overline{\gamma} c \in K$ and $\gamma \,.\, c = \overline{\gamma} c$. If conversely $d = \overline{\gamma} c \in K$, then, whenever $\gamma = \sigma/\tau \in \Gamma(K)$ (with $\sigma, \tau \in \Sigma(K)$), $\overline{\tau} d = \overline{\sigma} c$, $\tau d = \sigma c$, and $\gamma \,.\, c$ is defined, and then $\gamma \,.\, c = \overline{\gamma} c$. In this sense the partial action of γ on K is exactly the restriction to K of the action of $\overline{\gamma}$.

For every $a, b \in K$ there is a unique $\overline{\gamma} \in \Gamma(H)$ such that $\overline{\gamma} a = b$; therefore there is a unique $\gamma \in \Gamma(K)$ such that $\gamma \,.\, a = b$. □

3. In S, the Schützenberger monoids and groups of $\mathcal{K}$-classes organize themselves into functors. By Proposition 4.3, $S/\mathcal{K} \cong \widehat{S}_{\mathcal{P}}/\mathcal{H}$ is partially ordered by $\leqq_{\mathcal{H}}$; in other words, $K \geqq L$ in $S/\mathcal{K}$ if and only if $KM = L$ in $S/\mathcal{K}$ for some $M \in S/\mathcal{K}$, if and only if $Ks \subseteq L$ for some $s \in S$.

Proposition 5.7. *Let S be a c.m. with a Ponizovsky family $\mathcal{P}$. When $K \geqq L$ in $S/\mathcal{K}$, then $\mathrm{St}(K) \subseteq \mathrm{St}(L)$ and there exists a unique monoid homomorphism $\sigma_L^K : \Sigma(K) \longrightarrow \Sigma(L)$ such that the square*

$$\begin{array}{ccc} \mathrm{St}(K) & \xrightarrow{\subseteq} & \mathrm{St}(L) \\ {\scriptstyle \sigma_K}\downarrow & & \downarrow{\scriptstyle \sigma_L} \\ \Sigma(K) & \xrightarrow[\sigma_L^K]{} & \Sigma(L) \end{array}$$

commutes. The c.m. $\Sigma(K)$ *and homomorphisms* σ^K_L *constitute a monoid valued functor* $\mathbf{\Sigma}$ *on* $S/\mathcal{K}$.

As in Section V.2, $\mathbf{\Sigma} = (\Sigma, \sigma)$ is actually a thin functor on $H(S/\mathcal{K})$, in which $\Sigma(K)$ is the Schützenberger monoid of K and $\sigma_{K,T} = \sigma^K_{KT}$ is provided by Proposition 5.7; $\mathbf{\Sigma}$ is the **Schützenberger monoid functor** of S relative to $\mathcal{P}$.

Proof. We have $Ks \subseteq L$ for some $s \in S$. Hence $aK \subseteq K$ implies $aKs \subseteq Ks$ and $ad \in L$ for some $d \in L$, and $U(K) \subseteq U(L)$. Also $\sigma_K(a) = \sigma_K(b)$ in $U(K)$ implies, for any $c \in K$, $ac = bc$, $acs = bcs$, and $\sigma_L(a) = \sigma_L(b)$, since $cs \in L$. Therefore there is a unique monoid homomorphism $\sigma^K_L : \Sigma(K) \longrightarrow \Sigma(L)$ such that $\sigma^K_L(\sigma_K(a)) = \sigma_L(a)$ for all $a \in U(K)$. Since σ^K_L is unique with this property, σ^K_K is the identity on $\Sigma(K)$, and $\sigma^L_M \circ \sigma^K_L = \sigma^K_M$ whenever $K \geqq L \geqq M$ in $S/\mathcal{K}$. $\square$

When $K \geqq L$, the monoid homomorphism $\sigma^K_L : \Sigma(K) \longrightarrow \Sigma(L)$ induces a homomorphism $\gamma^K_L : \Gamma(K) \longrightarrow \Gamma(L)$ of abelian groups such that the square

$$\begin{array}{ccc} \Sigma(K) & \xrightarrow{\sigma^K_L} & \Sigma(L) \\ \subseteq\downarrow & & \downarrow\subseteq \\ \Gamma(K) & \xrightarrow[\gamma^K_L]{} & \Gamma(L) \end{array}$$

commutes. Since γ^K_L is unique with this property, γ^K_K is the identity on $\Gamma(K)$, and $\gamma^L_M \circ \gamma^K_L = \gamma^K_M$ whenever $K \geqq L \geqq M$ in $S/\mathcal{K}$. Hence:

Proposition 5.8. *The groups* $\Gamma(K)$ *and homomorphisms* γ^K_L *constitute an abelian group valued functor* $\mathbf{\Gamma}$ *on* $S/\mathcal{K}$.

$\mathbf{\Gamma}$ is the **extended Schützenberger functor** of S relative to $\mathcal{P}$ and is actually a thin abelian group valued functor on $H(S/\mathcal{K})$.

Proposition 5.9. *Let* S *be a c.m. with a Ponizovsky family* $\mathcal{P}$. *Up to the isomorphism* $S/\mathcal{K} \cong \widehat{S}_{\mathcal{P}}/\mathcal{H}$, *the extended Schützenberger functor of* S *relative to* $\mathcal{P}$ *is isomorphic to the Schützenberger functor of* $\widehat{S}_{\mathcal{P}}$.

Proof. Let $T = \widehat{S}_{\mathcal{P}}$. The isomorphism $S/\mathcal{K} \cong \widehat{S}_{\mathcal{P}}/\mathcal{H}$ in Proposition 4.3 sends a $\mathcal{K}$-class K of S to the $\mathcal{H}$-class H of T which contains K.

Let H and J be $\mathcal{H}$-classes of T and $K = H \cap S$, $L = J \cap S$. Since

$S/\mathcal{K} \cong T/\mathcal{H}$, we have $H \geqq J$ in $T/\mathcal{H}$ if and only if $K \geqq L$ in $S/\mathcal{K}$; then $Ks \subseteq L$ for some $s \in S$, and $Hs \subseteq J$. By Lemma V.2.1, the inclusion $\mathrm{St}\,(H) \subseteq \mathrm{St}\,(J)$ induces a homomorphism $\gamma_J^H : \Gamma(H) \longrightarrow \Gamma(J)$ of abelian groups such that the square

$$\begin{array}{ccc} \mathrm{St}\,(H) & \xrightarrow{\gamma_H} & \Gamma(H) \\ {\scriptstyle\subseteq}\downarrow & & \downarrow{\scriptstyle\gamma_J^H} \\ \mathrm{St}\,(L) & \xrightarrow[\gamma_J]{} & \Gamma(J) \end{array}$$

commutes. Now Lemma 5.5 provides isomorphisms $\theta_K : \Gamma(K) \longrightarrow \Gamma(H)$ and $\theta_L : \Gamma(L) \longrightarrow \Gamma(J)$. We show that the square

$$\begin{array}{ccc} \Gamma(K) & \xrightarrow{\theta_K} & \Gamma(H) \\ {\scriptstyle\gamma_L^K}\downarrow & & \downarrow{\scriptstyle\gamma_J^H} \\ \Gamma(L) & \xrightarrow[\theta_L]{} & \Gamma(J) \end{array}$$

commutes. For every $a \in \mathrm{St}\,(K)$ we have

$$\gamma_J^H\left(\theta_K(\sigma_K(a))\right) = \gamma_J^H\left(\gamma_H(a)\right) = \gamma_J(a) = \theta_L\left(\sigma_L(a)\right) = \theta_L\left(\gamma_L^K\left(\sigma_K(a)\right)\right);$$

since $\mathrm{Im}\,\sigma_K$ generates Γ_K, this implies $\gamma_J^H \circ \theta_K = \theta_L \circ \gamma_L^K$. Thus, up to the isomorphism $S/\mathcal{K} \cong \widehat{S}_{\mathcal{P}}/\mathcal{H}$, θ is a natural isomorphism from the extended Schützenberger functor of S to the Schützenberger functor of T. □

Chapter VIII.

OTHER RESULTS.

Far too much worthwhile work has been done on commutative semigroups to be exposed in some detail in one book. This chapter is a brief survey of topics which are not covered by other chapters. As noted in the Preface, some subjects have been omitted: partially ordered semigroups; varieties and pseudovarieties; factorization theory; and semigroup rings.

1. EXAMPLES.

Commutative semigroups arise in various and sundry ways.

1. Finite semigroups with small numbers of elements have been an early source of abundant examples. Lists of all distinct semigroups with n elements have been computed for $n \leqq 8$, incidentally providing all c.s. with $n \leqq 8$ elements; c.s. with $n = 9$ elements have been computed separately (Grillet [1996N]). Early lists were compiled by hand (Tamura [1953], $n \leqq 3$; Tamura [1954], $n = 4$; Tetsuya, Hashimoto, Akazawa, Shibata, Inui, & Tamura [1955], $n = 5$). Other early hand-drawn lists include all group-free c.s. of order $n \leqq 6$ (Sagastume Berra [1957]) and all archimedean semigroups of order $n \leqq 8$ (Tamura, Dehara, Iwata, Saito, & Tsukomo [1960]); see also Takahashi [1985]. Computers were soon brought to bear on the problem (Forsythe [1955], $n = 4$; Motzkin & Selfridge [1956], $n = 5$; Plemmons [1966], $n = 6$; Jürgensen & Wick [1976], $n = 7$; Sato, Yama, & Tokizawa [1991], $n = 8$), using backtrack algorithms (Jürgensen, Pallas, & Wick [1976]; see also Grillet [1996N]). Actual tables of semigroups with $n \leqq 6$ elements were published by Plemmons [1966] and Jürgensen [1990].

These computations suggest the **Kleitman-Rothschild-Spencer conjecture**, which states that, on a statistical basis, almost all finite semigroups are nilpotent of index $\leqq 3$ ($N^3 = 0$). This conjecture is a statement in Kleitman, Rothschild, & Spencer [1976]; there is considerable evidence that it is true, but gaps in the original proof have remained unfilled. There is some evidence that the commutative version of the conjecture is also true (almost all finite c.s. are nilpotent of index $\leqq 3$).

2. C.s. also arise from number systems. Additive subsemigroups of $\mathbb{N}$ and their ties to algebraic geometry have been mentioned in Section II.4. On a lighter note, Sasaki [1988] showed that they are determined up to isomorphism by their power semigroups; Ch'üan & Lee [1982] showed that the identity on $\mathbb{N}$ is the only permutation of $\mathbb{N}$ under which all direct images and inverse images of subsemigroups are subsemigroups. Various other explicit c.s. operations on $\mathbb{N}$ and $\mathbb{Z}$ have been constructed e.g. by Fraenkel, Porta, & Stolarsky [1989]. Hippisley-Coz [1992] studied n-point compactifications of $\mathbb{N}$. See also Painter [1966].

Integers modulo n constitute a c.s. $\mathbb{Z}_n$ under multiplication. Parízek [1957] showed that $\mathbb{Z}_n$ is Clifford if and only if n is a product of distinct primes. Hewitt & Zuckerman [1960] and Parízek & Schwarz [1961] found all its characters. Painter [1967] studied its groups of units. Lenzi [1980] studied which of its $\mathcal{H}$-classes is 0-cancellative in $\mathbb{Z}_n/\mathcal{H}$. Abrgan [1984] studied its prime and semiprime ideals. Dyadchenko & Shokuev [1985] determined its idempotents. Jones & Ligh [1980] determined all ring additions on $\mathbb{Z}_n$.

Free multiplicative semigroups of real numbers > 1 have been studied for density properties by Čudakov & Pavlyučuk [1951], Bredihin [1953], [1954], [1958A], [1958B], Čudakov & Bredihin [1956], Wegmann [1966], and Cibul'skite [1970]; many of these results extend to real valued homomorphisms of countably generated free c.s.

Closed intervals in $\mathbb{R}$ form c.s. under additions and under multiplication. These semigroups were studied by Ratschek [1972] and Mitrović [1976], [1977] with an eye toward numerical analysis.

3. Free commutative semigroups are seen in Chapters VI, IX, and XIII; their finitely generated subsemigroups have ties to algebraic geometry and have been mentioned in Section II.7. In what follows F denotes a free c.s. Tamura & Sasaki [1963] and Sasaki [1964] studied subsemigroups of F assuming that F is countably generated. Decompositions of F (finding subsets $(A_i)_{i\in I}$ of F such that every element of F can be written uniquely as a product $\prod_{i\in I} a_i$ with $a_i \in A_i$) were investigated by de Bruijn [1956], Hansen [1969], Niven [1971], Marcus [1974], and Lepetit [1973]. (Hansen [1969] studied similar decompositions of $\mathbb{Z}_n$.) Grillet [1969F] gave a characterization, using the Riesz interpolation property, and showed that the fixed points of any set of endomorphisms of F constitute a free c.s.; another interesting characterization, in terms of semigroup algebras, was given by Bruns & Gubeladze [1999]. Bredihin [1960] studied the asymptotic behavior of $\nu(x)$, the number of elements of norm $\leqq x$,

for multiplicative norms on F such that $\nu(x)$ is always finite. The retractions of F were studied by Stambolieva [1973] when F is finitely generated. The maximal size of independent systems of equations in F was studied by Karhumäki & Plandowski [1996]. Commutative languages (sets of subsets of F, where F is finitely generated) were considered by Perrot [1965], Eilenberg & Schützenberger [1969], and Butzbach [1972].

4. Other examples are provided by commutative subsemigroups of noncommutative semigroups, such as full transformation semigroups (Rankin, Reis, & Wang [1976]). C.s. of polynomials under composition were studied by Lidl & Müller [1986], Nöbauer [1986], [1987], Turnwald [1986], Eigenthaler, Nöbauer, & Wiesenbauer [1987], and Eigenthaler & Woracek [1997].

5. Characterizing multiplicative semigroups of commutative rings remains an open problem. The class of c.s. with this property is not axiomatizable (Kogalowski [1961]) and its finitely generated members are finite (Isbell [1959], Anderson & Stickles [2000]). Aull [1967] observed that a semigroup ideal which is directed upward under $\leqq_{\mathcal{H}}$ must be a ring ideal. Satyanarayana [1973] and Isbell [2000] obtained various necessary properties.

6. As noted in the preface, commutative semigroups turn up in divers places besides algebraic geometry. Examples include the monoid of equivalence classes of Boolean lattices (Dobbertin [1982], [1983]); the c.s. of family graphs (Farrell [1992]); certain continuous lattices (Wehrung [1998D); distributive lattices (Berg [1985]); semigroups of circulant matrices (Schwarz [1977], Guričan [1984], Huang [1996]); semigroups of subvarieties of varieties of associative algebras (Gonchigdorzh & Mal'tsev [1982]); bilinear spaces over a scheme, under orthogonal sum (Knebusch [1969]); the semigroup of generalized divisors on certain schemes (Hartshorne [1994]); the semigroup of stable equivalence classes of algebraic tori (Čistov [1978]); the semigroup of Newton polyhedra in rings of formal power series (Lipkovski [1988]); the branch multiplicity semigroup of a formal power series (Kulkarni [1990]); semigroups of convergent power series (Groza [1995]). No doubt a few others have been omitted from this list.

Bazzoni & Salce [1996] showed that the class semigroup of a valuation domain is a Clifford semigroup; domains with this property were studied by Zanardo & Zannier [1994]. Kunze [1978] studied the class semigroups of ordered local rings. Shirota [1952] showed that a locally compact Hausdorff space is determined up to homeomorphism by its multiplicative semigroup of continuous real valued functions; this result was generalized by Thành [1992].

2. PRODUCTS AND SUBSETS.

This section surveys properties of commutative semigroups related to products, subsemigroups, ideals, and other subsets.

1. A number of interesting combinatorial properties have been established for c.s. Some have applications to graph theory (Puš [1991]). Burgess [1969] proved that, in an archimedean semigroup with n elements, every sequence of n elements contains a subsequence whose product is idempotent. De Luca & Restivo [1984] proved (for semigroups in general) that a finitely generated semigroup S is finite if and only if there exists an integer $m > 0$ such that every sequence $s_1, \dots, s_m$ of m elements of S satisfies $s_1 \cdots s_j = (s_1 \cdots s_j)(s_i \cdots s_j)$ for some $1 \leqq i \leqq j \leqq m$. See also Niculescu [1995], [1996].

Khovansky [1992] showed that $|A^n|$ has polynomial growth in n when S is commutative and $A \subseteq S$ is finite. Nathanson [2000] generalized this to $\left|B\, A_1^{\ n_1} \cdots A_r^{\ n_r}\right|$. In a finitely generated c.s. S, an element a which is infinitely divisible ($(\forall n > 0)(\exists x \in S)\, x^n = a$) is idempotent (Clark, Holland, & Székely [1998]). Hindman & Woan [1993] studied systems of linear equations $AX = BX$ in c.s., with coefficients in $\mathbb{N}$.

2. C.s. with various additional elementary properties have been considered (for instance, archimedean semigroups and separative semigroups). Tamura [1971] studied c.s. in which $xyz \in \{xy, xz, yz\}$ for all x, y, z. Hora & Kimura [1972] considered c.s. in which $(\forall x)(\exists y)\, x^m y^p = x^n y^q$ (where $m, n, p, q \geqq 0$ are given). Blagojević [1982] investigated c.s. in which $(\forall x)(\exists y)\, x = xy^2$, $y = yx^2$. Cancellativity properties are mentioned in Chapter II. Other properties were studied by Galbiati & Veronesi [1986].

The **Riesz interpolation property** states that $ab = cd$ implies $a = xy$, $b = zt$, $c = xz$, $d = yt$ for some x, y, z, t. C.s. with this property (also called **refinement** semigroups) were studied by Dobbertin [1983], [1984], and Shortt & Rao [1989]; see also Goodearl [1986] and (for cancellative c.s.) Grillet [1970C], [1970I], and Caillot & Wehrung [2000]. Embeddings into refinement semigroups were constructed by Wehrung [1998E].

Krob [1987] defined a **complete** c.s. as a set S with commutative and associative product mappings $\prod_{i \in I} S \longrightarrow S$, one for every set I.

3. The subsemigroups of a semigroup constitute a complete lattice $\Sigma(S)$. According to the lucid and well organized survey by Shevrin & Ovsyannikov [1983], the study of this lattice consists of three principal areas: lattice isomor-

phisms, and determinacy (when is S determined up to isomorphism by $\Sigma(S)$); properties of $\Sigma(S)$; semigroups with certain types of lattices. Early work relevant to commutative semigroups includes the following. Semigroups S such that $\Sigma(S)$ is distributive, complemented, or a chain, were characterized by Shevrin [1961]; distributive, by Ego [1961]; modular or semimodular, by Ego [1962A]; semidistributive, by Shiryaev [1985]. Ego [1962B], [1962C], [1963] also considered various weakened forms of modularity. Laplaza [1974] showed that a c.s. whose subsemigroup lattice has finite length must be finite. Kacman [1979] studied c.s. whose subsemigroup lattice is self-dual (more generally, pairs of c.s. with anti-isomorphic subsemigroup lattices). See also Morel [1979]. Repnitsky & Kacman [1988] and Repnitsky [1994] determined semigroups whose lattice of subsemigroups satisfies a nontrivial identity. Shevrin [1966] showed that, in a number of classes of commutative semigroups, subsemigroup lattices can be characterized by finitely many axioms. Isomorphisms and determinacy were studied by Mogiljanskaja [1972], Baransky [1975], Ovsyannikov [1977A], [1977B], [1981], Liberman [1980], and Borisov [1982], [1984], [1986], [1992]. Ovsyannikov [1977A] showed that every c.s. can be embedded into a c.s. which is determined by subsemigroup lattice. Jürgensen & Wick [1973] gave an efficient algorithm which computes $\Sigma(S)$ when S is a finite c.s.

Subsemigroups give rise to other structures. The lower semilattice of subsemigroups was considered by Evseev [1965]. The subsemigroups of a semigroup and their isomorphisms constitute an inverse semigroup; the determinacy problem for this inverse semigroup was studied by Libih [1974] for certain commutative Clifford semigroups. The subsemigroups of a c.s. also constitute a c.s. under multiplication; this was studied by Mogiljanskaja [1971]. See also Etayo [1965].

Various concepts of rank have been defined for a c.s. S. Shevrin [1963] defined the rank of S as the smallest r such that every finitely generated subsemigroup of S is generated by at most r elements; he showed that S has finite rank if and only if its archimedean components have finite rank. Okniński [1988] defined rank by means of 0-cancellative homomorphic images, equivalently, by means of free subsemigroups; this is useful in studying the Krull dimension of semigroup rings.

4. Evseev [1991] defined a **closing element** c by the property that every subset which contains c is a subsemigroup, and showed that such exists in a c.s. S if and only if every partition of S has a class which is a subsemigroup.

A number of conditions on subsemigroups have been considered. Commutative semigroups all of whose subsemigroups are finitely generated have been seen in Section VI.7. Jensen & Miller [1968] showed than the only infinite c.s., all of whose proper subsemigroups are finite, are the groups $\mathbb{Z}(p^\infty)$. Gluskin [1971], [1973], [1976] studied dense extensions for subsemigroups and for normal subsemigroups. Nordahl [1973] found all c.s. whose every proper subsemigroup is power joined. Schein [1975] studied cancellative c.s. in which every subsemigroup is dense in its idealizer. Morel [1979] constructed locally cyclic semigroups. Hmelnitsky [1985] studied c.s. whose every proper subsemigroup is different from its idealizer, and c.s. whose every proper subsemigroup appears in an ideal series. Krivenko [1985] studied c.s. whose every cyclic subsemigroup is normal. Aucoin [1999] characterized c.s. in which all ideals of every subsemigroup are induced by ideals of the semigroup.

The possibility of direct summands was explored by Shershin & Moore [1965]. Hora [1973] studied more general decompositions $S = AB$ where A and B are proper subsemigroups. Hule [1976] and Toffalori [1980] considered analogues of purity. Poyatos [1968] derived a formula for the length of maximal chains of unitary subsemigroups of a finite c.s. Chen [1988] studied semigroups in which every subsemigroup is unitary. Páles [1989] studied subsemigroups admitting a pair of commuting endomorphisms. See also Šimon [1979].

5. We now turn to ideals. Fruitful similarity with ideals of commutative rings, particularly in cancellative c.s., has led to Noether-like intersections of primary ideals (Ward & Dilworth [1939], Aubert [1962], MacKenzie [1954], Johnson [1986], Zhang & Qi [1997]) or irreducible ideals (Lesieur [1955]), Krull decompositions (Perel [1957]), a hull-kernel topology for prime ideals (Kist [1963]), various radicals (Aubert [1962], Satyanarayana [1975], Lal [1972R], Roĭz & Schein [1978]), integrally closed semigroups (Maury [1958], [1959]), Dedekind semigroups (Aubert [1962], Dorofeeva [1972], Mannepalli & Satyanarayana [1974], Jain [1982], [1983], [1989]), hereditary and semihereditary semigroups (Dorofeeva [1972]), and other topics (Perić [1966], Murata [1980], Abrgan [1984], Rachunek [1987]). Properties of commutative rings, domains, and valuation rings were systematically extended to cancellative and power cancellative c.s. by Matsuda [1997], [1998A], [1999], Tanabe & Matsuda [1999], Kanemitsu & Bansho [1999]; see also Matsuda [1998B], Kanemitsu, Okabe, & Matsuda [1999], Hirabuki & Matsuda [1999], Kitsuneda, Matsuda, & Sugatani [1999].

Commutative semigroups provide a fine setting for abstract ideal theory, as carried out by Lorenzen [1939] in cancellative c.s. (later generalized in Lorenzen [1949], [1952]) and by Aubert [1962] in arbitrary c.s. Aubert's theory of x-ideals includes a Krull-Stone theorem, a hull-kernel topology for maximal ideals, intersections of primary ideals, and analogues of Dedekind rings; it has been continued by Holme [1966], Skula [1975], [1976], Porubský [1977], and Karásek

[1985].

See also the survey of ideal theory by Anderson & Johnson [1984].

6. Ideal extensions of commutative semigroups of various kinds were studied by Heuer & Miller [1966] and Heuer [1971], for cancellative semigroups; Yamada [1965], [1968] and McNeil [1971] for null semigroups; Petrich [1973E] for Clifford semigroups; Arendt & Stuth [1972] for 0-free nilsemigroups. For more general results, see Putcha & Weissglass [1973], Hildebrant [1984], Takahashi [1984A], [1985]. Ideal extensions and dense ideal extensions were studied for commutative semigroups by Gluskin [1971], [1973], [1976D], [1976E], [1979], [1983I].

7. The ideals of a c.s. S form a complete distributive lattice which was first studied by Aubert [1953] and was characterized by Anderson & Johnson [1984]. This lattice is trivial ($= \{\emptyset, S\}$) if and only if S is a group. Porubský [1977], [1978] showed that it is a chain if and only if every ideal of S is prime, if and only if every ideal of S is primary, if and only if every ideal of S is semiprimary. These semigroups were studied by Lesohin [1964G], Satyanarayana [1971A], [1972], [1978], Lal [1975], Jain [1988], [1989], and Geroldinger [1996]. Megyesi & Pollák [1968], [1977] studied and constructed semigroups in which every one-sided ideal is principal; when a c.s. S has this property, then $S/\mathcal{H}$ is a co-well ordered semilattice of infinite cyclic semigroups and cyclic nilsemigroups; Jain [1988] related this property to valuations on the universal group.

An ideal I is **categorical** when $xyz \in I$ implies $xy \in I$ or $yz \in I$. (If $\mathcal{C}$ is a small category, then the zero element of the semigroup of morphisms $\mathcal{C} \cup \{0\}$ is an ideal with this property.) Commutative semigroups in which every ideal is categorical were studied by McMorris & Satyanarayana [1972], Monzo [1973]

Other conditions have been considered: every primary ideal is prime (Satyanarayana [1971B], Lal [1972P]); every nonzero ideal is prime (Satyanarayana [1971B]); $A = B(A : B)$ for all ideals A, B (Mannepalli [1976A], [1976B]); every ideal is finitely generated (Satyanarayana [1977]); every ideal is a retract (Tully [1969]); every ideal is a finitely generated disjunctive semigroup (Johnson & McMorris [1977]).

8. Other subsets of c.s. have been studied. Rational sets were defined by Eilenberg & Schützenberger [1969]. Iwanik & Plonka [1975] studied linearly independent subsets (= which satisfy only trivial relations).

The subsets of a commutative semigroup S constitute a commutative semigroup (under multiplication of subsets), the **power** semigroup or **global** of S (Tamura & Shafer [1967]). Power semigroups of various c.s. have been studied by Tamura & Shafer [1967] (abelian groups and chains), Szimtenings [1970] (cancellative c.s.), Mogiljanskaja [1972] (determinacy), Byrd, Lloyd, Pedersen, & Stepp [1977] (finite cyclic groups), Pedersen & Sizer [1978] (torsion free abelian groups), Tamura [1984C] (chains), [1984Z], [1985] ($\mathbb{Z}$), Spake [1986A], [1986F], [1988Z] ($\mathbb{Z}$), [1988Q] ($\mathbb{Q}$), Sasaki [1988] (determinacy), Sasaki, Spake, & Tamura [1987] ($\mathbb{Q}$), Spake & Hanlon [1993] ($\mathbb{Z}$). Subsemigroups of the power semigroup can be used to construct semigroups of quotients (Toca [1977]).

3. HOMOMORPHISMS AND CONGRUENCES.

This section surveys properties of commutative semigroups related to homomorphisms and congruences.

1. We begin with homomorphisms into particular c.s.; homomorphisms into semilattices are an example.

Homomorphisms into $\mathbb{N}$ were used by Abellanas [1965] and Grillet [1970E] to study cancellative, power cancellative, and reduced c.s. Homomorphisms into $\mathbb{Q}^+$ were used similarly by Abellanas [1965], and in Section III.6 to study power joined semigroups.

Homomorphisms into $\mathbb{R}^+$ or $\mathbb{R}^+ \cup \{\infty\}$ have been studied for existence (Kaufman [1966B], [1967], Kobayashi [1977C], [1977E], Kobayashi & Tamura [1977]) or for extension from subsemigroups (Kaufman [1966A], Kranz [1974], [1979], Putcha & Tamura [1976], Kobayashi & Tamura [1977]). Krause [1989] used homomorphisms into $\mathbb{R}^+$ in factorization theory. Homomorphisms of N-semigroups into $\mathbb{R}^+$ were studied by Kobayashi [1974] and Tamura [1982] and were used by Tamura [1974B] to construct N-semigroups and by Kobayashi [1978] to characterize their Rees quotients. Homomorphisms that preserve a given compatible preorder (which could be $\leqq_{\mathcal{H}}$) were considered by Putcha [1976], Kobayashi [1980], and Plappert [1995]. Kominek [1990] studied subadditive mappings to $\mathbb{R}$.

Homomorphisms into groups have been studied by Dubreil [1941] and many of his disciples. Putcha [1971] determined c.s. S such that $\operatorname{Im}\varphi$ is a subgroup of G for every group G and homomorphism φ of S into G.

Representations are homomorphisms into more explicit semigroups. Matrix representations of finite c.s. were studied by Ponizovsky [1970], who determined, in terms of Ponizovsky factors, when there are only finitely many inequivalent irreducible representations. Increasingly refined representations by products in various categories were studied by Trnkova [1975A], [1975B], [1976], [1978], [1981] and Adámek & Koubek [1977]. A commutative Clifford semigroup S has faithful representations by partial bijections of sets; the minimal size of such a set was determined for finite S by Byleveld & Easdown [1992]. C.s. were also represented by compact convex subsets (Ratschek & Schröder [1977]), by endomorphisms of boolean algebras (Nasirov [1980]), by products of complete 0-dimensional metric spaces (Vinarek [1982]), by continuous linear operators on Hilbert spaces (Ressel & Ricker [1998]), and by subsets of simpler c.s. (Lau [1979], Gould & Iskra [1984R], Gould, Iskra, & Pálfy [1986]).

2. When A and B are c.s., we saw that $\mathrm{Hom}(A,B)$ is a c.s. under the pointwise operation. The semigroup $\mathrm{Hom}(A,B)$ was studied by Brameret [1962] and, when A and B are commutative Clifford semigroups, by Fulp [1966], Anderson [1979A], and Kupcov [1979H], [1980], [1981H].

Endomorphisms of a commutative semigroup constitute a monoid under composition, a commutative semigroup under the pointwise operation, and a semiring under both. The monoid of endomorphisms of commutative semigroups was studied by Lesohin [1966H] and described, in terms of archimedean decomposition, by Nakajima [1977] (see also Manukjanc [1974] for commutative Clifford semigroups). Kupcov [1979E] showed that a commutative Clifford semigroup has only the obvious endomorphisms $x \longmapsto x^n$ if and only if it is a finite cyclic group. Kublanovsky [1983] showed that every semigroup with at least 4 elements has at least 4 endomorphisms, and that there are only finitely many nonisimorphic finite c.s. with exactly 4 endomorphisms.

The determination problem for endomorphisms consists in detecting c.s. S such that $\mathrm{End}\,(S) \cong \mathrm{End}\,(T)$ implies $S \cong T$ when T is limited to some class of semigroups. This problem was considered by Trepetin [1971], [1974] and Boguta [1976]. If for instance S and T are nilpotent and S has only homogeneous defining relations, then $\mathrm{End}\,(S) \cong \mathrm{End}\,(T)$ (as semirings) implies $S \cong T$ (Trepetin [1971]).

Bredihin [1976] proposed a generalization of endomorphisms and automorphisms. Kozlov [1976] studied rigid cells. Kim & Neumann [1991] studied the operation $x * y = (\alpha x)(\beta y)$, where α and β are commuting endomorphisms.

3. Given commutative semigroups A, B, and C, a **bihomomorphism** of $A \times B$ into C is a mapping $\beta : A \times B \longrightarrow C$ such that $\beta(aa',b) = \beta(a,b)\,\beta(a',b)$ and $\beta(a,bb') = \beta(a,b)\,\beta(a,bb')$ for all $a,a' \in A$ and $b,b' \in B$. **Bicharacters** are bihomomorphisms into the multiplicative submonoid $\mathbb{S} \cup \{0\}$ of $\mathbb{C}$. Bihomomorphisms have been called **bilinear** mappings and other names but they are not gay. Under pointwise operation they constitute a commutative semi-

group BiHom (A,B,C). The latter is in fact a semigroup of homomorphisms: indeed bihomomorphisms of $A \times B$ into C may be regarded as homomorphisms of A into Hom(B,C) and as homomorphisms of B into Hom(A,C). Since Head [1967] and Grillet [1969T] showed that commutative semigroups have a tensor product, there is also a natural isomorphism BiHom $(A,B,C) \cong$ Hom$(A \otimes B, C)$; for instance, bicharacters of A and B are characters of $A \otimes B$.

Lesohin gave conditions under which BiHom (A,B,C) is nonempty [1967D] (see also Kupcov [1981B], Popyrin [1988]), finite [1967E], has a nonidempotent element [1969], or is finitely approximable [1976]. Lesohin also studied the particular case of semiring multiplications on a cancellative semigroup or a group with zero [1964M] and the structure of BiHom (A,B,C) in the case of bicharacters [1971B]. Manukjanc [1976A], [1976B] studied multiplications on certain commutative Clifford semigroups and on certain c.s. with cyclic archimedean components. Bicharacters and bicharacter semigroups were investigated by Lesohin [1969], [1971B] and Popyrin [1986]; see also the survey by Lesohin & Popyrin [1987].

Any bihomomorphism $\beta : A \times B \longrightarrow C$ leads to separability questions: does β separate the elements of A (if $a \neq a'$ in A, is $\beta(a,b) \neq \beta(a',b)$ for some $b \in B$)? and if so, is A maximal with this property, given B and C? Under the names of **regularity** and **completeness**, these questions were investigated by Lesohin [1958], [1961], [1963M], [1963S], [1964G], [1967E], [1967S], [1969] and Popyrin [1984], [1985], [1986]. Lesohin also studied duality relative to a bihomomorphism $A \times B \longrightarrow C$ [1966D] and gave conditions under which the natural homomorphism $A \longrightarrow$ Hom(Hom$(A,C),C)$ is an isomorphism for all or some commutative semigroups C [1968D], [1980] or under which $\beta : A \times B \longrightarrow C$ induces isomorphisms $A \longrightarrow$ Hom(B,C) and $B \longrightarrow$ Hom(A,C) (Lesohin & Kupcov [1980], Lesohin [1992]); weaker conditions were considered by Lesohin [1980], Lesohin & Kupcov [1980], Popyrin [1985], Kupcov & Lesohin [1986], Ermolina [1988], and Bogacheva [1988], [1989], [1990A], [1990B]. See also the survey by Lesohin & Popyrin [1987].

4. Other mappings have also been considered. Putcha [1976] found good use for positive mappings (mappings φ into a partially ordered set, such that $\varphi(xy) \geqq \varphi(x), \varphi(y)$ for all x, y). Kowol & Mitsch [1976] studied monomial mappings $x \longmapsto ax^n$ of a c.s.; Tichy [1979], [1981] studied polynomial mappings $a x_1^{n_1} x_2^{n_2} \cdots x_k^{n_k}$. Hule [1976] showed that a system of polynomial equations which has a solution in a larger c.s. does not necessarily have more than one solution in some larger c.s.; in which respect c.s. differ from groups and abelian groups. See also Grossman [1982].

5. We now turn to congruences which, on a c.s., have some similarity to ideals. In fact, an ideal theory of congruences, complete with Noether primary intersections, was developed by Drbohlav [1963], [1964] and Hoehnke [1964]. Hoehnke [1965] also defined the radical of a commutative semigroup as the congruence $s \mathcal{R} t$ if and only if $s^n = t^n$ for all sufficiently large $n > 0$; equivalently, $s - t$ is nilpotent in $\mathbb{Z}[S]$. Arendt [1975] studied a Jacobson like radical, the intersection of all maximal modular congruences.

Conditions on congruences have been considered since Lyapin [1950S] determined c.s. without proper nontrivial congruences (they are cyclic groups of prime order or have at most two elements) and c.s. on which all congruences are Rees congruences [1950R] (revisited by Aucoin [1995]). Jensen & Miller [1968] characterized c.s. whose every proper homomorphic image is finite; these are precisely the subsemigroups of $\mathbb{Z}(p^\infty) \cup \{0\}$. Tamura [1969] and Schein [1969] determined all c.s. whose congruences form a chain; these are precisely the subgroups of $\mathbb{Z}(p^\infty)$ and the nilsemigroups and nilmonoids that are totally ordered by divisibility. Hamilton [1975] determined all c.s. S whose congruences commute: either congruences on S form a chain, or S is an elementary semigroup whose monoid of orbits is totally ordered by divisibility. Schein [1981] determined all c.s. that are isomorphic to all their nontrivial quotients.

Group congruences have attracted attention since Dubreil [1941]. Tamura & Hamilton [1972] showed that a minimal group congruence $\mathcal{G}$ on a c.s. is a smallest group congruence (then $S/\mathcal{G}$ is the universal group of S and $S \longrightarrow G(S)$ is surjective); a shorter proof was given by Woan [1975]. Szasz [1976] used congruences similar to principal equivalences (also considered by Tamura & Hamilton [1971]) to study semilattices.

6. A special place belongs to the difficult study of congruence lattices. Hamilton & Nordahl [1978] characterized c.s. whose lattice of congruences is boolean. Hamilton [1978] characterized cancellative c.s. whose lattice of congruences is modular (which consist of abelian groups and of the positive or nonnegative parts of subgroups of $\mathbb{Q}$) or distributive (which consist of locally cyclic abelian groups and locally cyclic abelian groups and of the positive or nonnegative parts of subgroups of $\mathbb{Q}$). Tamura & Etterbeek [1966] proved distributivity for all locally cyclic semigroups. The survey by Mitsch [1983] has a section on commutative semigroups; some errors are pointed out in its review MR 84d:20064. Eberhart [1982] studied c.s. whose lattice of congruences is simple; these are either finite p-groups or nilsemigroups. Hamilton [1982] studied archimedean semigroups whose lattice of congruences is distributive, and separative c.s. whose lattice of congruences is modular. Trueman [1983] studied direct products of finite cyclic

semigroups, whose lattice of congruences is always upper semimodular, but is lower semimodular only for groups. Trueman [1985] also gave examples, including free c.s. with at least two generators, whose lattice of congruences is neither upper semimodular nor lower semimodular.

Affine completeness is the property that every congruence preserving mapping is polynomial; this was studied by Kaarli, Márki, & Schmidt [1985] for semilattices and by Kaarli & Márki [1992], [1995] for commutative Clifford semigroups.

7. A **tolerance** relation on a semigroup S is, like a congruence, a reflexive and symmetric subsemigroup of $S \times S$, but is not necessarily transitive. They were introduced by Zelinka [1975] who showed that all tolerances on a group are in fact congruences [1975], and that groups are the only periodic c.s. with this property [1977]. Tolerance lattices of c.s. were studied by Pondělíček [1983], [1985A], [1985B], [1988A], [1988B] for various lattice properties; Pondělíček [1988C] also characterized c.s. in which every singly generated tolerance relation is a congruence. In a similar vein, the reflexive and symmetric binary relations $\mathcal{T}$ on S such that $a \mathcal{T} b$ implies $xay \mathcal{T} xby$ for all $x, y \in S^1$ constitute a complete distributive lattice; Sitnikov [1983] characterized c.s. for which this lattice is boolean. See also Chandran [1987] and Pondělíček [1988D].

Other relations that have been studied include positive preorders (Putcha [1973]) and compatible partial or total order relations, which have been investigated extensively and deserve a survey of their own.

4. OTHER TOPICS.

"Other topics" include extensions, certain constructions, coextensions, categories, S-sets, and miscellany.

1. Extensions (other than ideal extensions) have been studied by a number of authors. Gluskin studied **dense** extensions (such that homomorphisms that are injective on S are injective on E), **normal** extensions (such that S is unitary in E in the sense that $ab \in S$ implies $b \in S$ when $a \in S$ and $b \in E$), and other kinds [1971], [1973], [1976D], [1976E], [1979], [1983I]. A cancellative semigroup S has a maximal dense normal extension if and only if $ab^n \in S$ implies $b \in S$ when $a \in S$, $n > 0$, and $b \in G(S)$; equivalently, S is its own complete integral closure (Gluskin [1985]).

Embedding into $\mathbb{Z}_n$ was studied by Parker [1954] and Burgess [1968].

A **Stone** semigroup is a c.m. S with a zero element but no other nilpotent element, in which for every annihilator ideal $A = \mathrm{Ann}(T) = \{x \in S \mid xT = 0\}$ there is an isomorphism $S \cong A \times \mathrm{Ann}(A)$ given by $x \longmapsto (ex, fx)$, where $e \in A$, $f \in \mathrm{Ann}(A)$ are idempotents. Keimel [1971] showed that every c.m. with a zero element but no other nilpotent element has a minimal embedding into a Stone semigroup.

2. A **tensor product** for commutative semigroups was developed independently by Head [1967A], Grillet [1969], and Nikolaev [1971]. Head [1967A] showed that tensor products yield the universal semilattice of S (as $S \otimes \{1\}$), the universal Clifford semigroup of S (as $S \otimes \mathbb{Z}$), and the universal uniquely divisible semigroup of S (as $S \otimes \mathbb{Q}^+$). Tensor products of Clifford semigroups were studied by Head [1967B], Nikolaev [1971], Kuroki [1973], and Anderson [1974]; their structure was finally elucidated by Anderson & Kimura [1978], [1979]. Anderson [1997], [1998] also determined tensor products of Clifford semigroups generated by partially ordered sets of groups. The archimedean components of tensor products were considered by Head [1968] and determined by Anderson [1978]. Other results include functor properties (Head [1967D], Grillet [1969]), kernels (Head [1967C]), maximal subgroups (Head [1969], Fulp [1970]), faces (Head [1969], Grillet [1969]), universal cancellative semigroup (Kuroki [1971]), and universal group (Head & Kuroki [1974], Li [1986]). Flatness was considered by Grillet [1970I], Bulman-Fleming & McDowell [1980], Head & Anderson [1980], and Galanová [1982].

Cancellative c.s. and other classes of c.s. also have tensor products; see Grillet [1970T], Li [1986], [1987], Galanová [1987], Li & Zhang [1993]. For cancellative c.s. and their tensor product, Zhang, Li, & Wang [1993] showed that $G(S) \cong \mathbb{Z} \otimes S$; Wehrung [1996] showed that $S \otimes T$ has the Riesz interpolation property if S and T do. Fraser & Albert [1984] applied the tensor product of lattices to semilattices.

3. Semigroups and their homomorphisms constitute a category, and similarly for commutative semigroups.

Semigroups resemble rings in that monomorphisms of semigroups are injective, but epimorphisms of semigroups are not necessarily surjective; this also holds for commutative semigroups. The **dominion** of a subsemigroup is the intersection of all the equalizers that contain it. This was defined and described by Isbell [1965] for semigroups in general; the same construction applies to c.s. (Howie & Isbell [1967]); the original proofs were improved by Stenström [1971], Philip [1974], Storrer [1976], and especially Higgins [1990]. Howie & Isbell [1967] also studied **saturated** c.s. (= which cannot be epimorphically embedded in a larger semigroup), showing, among other results, that finite c.s. are saturated (so that epimorphisms of finite c.s. are surjective). Dominions are useful in studies of free products with amalgamation (Howie [1968], Imaoka [1976], Shoji [1990]). Khan [1982], Hsieh [1982], Higgins [1983], [1985], and the survey by Higgins [1984] contain interesting examples; for instance Hsieh [1982] showed that commutative semigroups which satisfy an identity of a certain kind are saturated.

A **projective** commutative semigroup P is usually defined by the property that every homomorphism $P \longrightarrow S$ can be factored through every surjective homomorphism $T \longrightarrow S$. Free commutative semigroups have this property. A result of Grillet [1969F] implies that a commutative semigroup is projective if and only if it is free. Projectives have been determined in the smaller category of commutative Clifford semigroups (Bulman-Fleming & McDowell [1980]).

An **injective** commutative semigroup I can be defined by the property that every homomorphism $S \longrightarrow I$ can be extended to a homomorphism $T \longrightarrow I$ whenever S is a subsemigroup of T. Schein [1981I] showed that all such semigroups are trivial (thereby correcting earlier results by Schein [1976] and Gluskin [1976E]). The search for injectives has been more fruitful in smaller categories. Hancock [1960] showed that injective cancellative c.s. coincide with injective abelian groups (thereby correcting an earlier result of Wiegandt [1958]); this was rediscovered by Li & Liu [1992]. Injective separative semigroups coincide with injective (commutative) Clifford semigroups and have been determined by Schein [1974].

A number of other categories were considered by Takahashi [1987].

4. Rings make a good living acting on abelian groups. Actions of semigroups on sets (which can be regarded as representations by transformations) have been studied fairly extensively; few of these results, however, apply specifically to commutative semigroups. Kil'p [1973] showed that every principal ideal of a c.s. S is projective if and only if S is separative. Murty [1983] extended noetherian decompositions to S-sets. Poyatos [1988] devised an archimedean decomposition for S-sets. Khovansky [1995] studied orbits of free c.s. See also Bogomolov & Mustafin [1989]; Ahsan, Khan, Shabir & Takahashi [1991].

Semimodules are commutative semigroups on which a semiring acts in a suitable manner (the word has also been applied to plain commutative semigroups). Their archimedean decompositions were studied by Poyatos [1985]; their endomorphisms, by Wang [1988].

5. We conclude with a number of topics which do not fit easily in any of the above.

Jezek & Kepka [1975] used uniquely 2-divisible semigroups to construct certain (non-associative) groupoids. Cho [1990] used c.s. to construct certain medial algebras. Tamura [1977S] studied c.s. with a least archimedean component that contains no idempotent. Batbedat [1978A] [1978B] used commutative semigroups to construct schemes. Measures on c.s. have been studied e.g. by Ebanks [1979] and Bisgaard [1998]. Toader [1990] studied various kinds of infinite sequences of elements of c.s.

A **holoid**, also called **naturally totally ordered** c.s., is a c.s. on which Green's preorder $\leqq_{\mathcal{H}}$ is a total order relation. Then the ideals form a chain. Holoids have long been studied as totally ordered c.s. Markov [1995] used them for abstract interval arithmetic. Borisov [1992] characterized c.s. which can be embedded into holoids.

Commutative semigroups have been applied to other areas of Mathematics.

Seitz & Blickle [1974] and Seitz, Blickle & Grega [1975] found that certain types of semilattices are useful in Chemical Engineering. Duske [1976] studied commutative automata and there are a number of papers on commutative Thue systems (see e.g. Simmons [1976], Kapur & Narendran [1983], [1985], Narendran, O'Dunlaing, & Rolletschek [1985], Huỳnh [1986], Narendran & O'Dunlaing [1989], Yap [1991]). Commutative semigroups have been used to solve functional equations (see e.g. Brillouet-Belluot [1996] and Taylor [1999]). See also Warner [1960], Lal [1967], Speed [1968], Moszner [1979], Thron [1981], Mushtaq & Yusuf [1987], [1988], and Kisielewicz & Newrly [1993].

Chapter IX.

NILSEMIGROUPS.

This chapter gives a construction of finite nilmonoids, which is due to the author [1991N], [2001N] and expands earlier ideas of Arendt & Stuth [1970A]; a shorter account is given in Grillet [1995]. Unlike previous constructions for these semigroups, this is a global construction with a very geometric character, in which nilmonoids are obtained as quotient of free commutative monoids by suitable congruences. It accounts well for various structural features of nilmonoids, such as the greatest pure congruence and the universal group of $N\backslash 0$. A more general construction was given by Grillet [2001N] and applied to fully invariant congruences in [2001F].

Finite commutative nilsemigroups are very abundant. The construction in this chapter quickly generates large numbers of nilsemigroups but without enough control to, say, prove the commutative Kleitman-Rothschild-Spencer conjecture (that, on a statistical basis, almost all finite c.s. are nilpotent of index 3). Finding a more powerful construction would be a major advance.

Inductive constructions of nilsemigroups were devised by Yamada [1964], [1965], [1968], Yamada & Tamura [1969], and John [1973]. Tamura [1966], [1968] constructed nilsemigroups from trees. Some special classes have also been constructed by various methods. Tree-ordered nilsemigroups were constructed by Tully [1966], [1974] and Grillet [1991T]. Rees quotients of N-semigroups were characterized by Kobayashi [1978], [1982] and Clarke [1980], [1984]. Locally cyclic nilsemigroups were constructed by Buzashi & Vishnyakova [1986].

Due perhaps to lack of structural information, commutative nilsemigroups have not been studied extensively, in spite of their importance. Ideal extensions of 0-free nilsemigroups were studied by Arendt & Stuth [1970B]. The nilpotency indices of the elements of a nilsemigroup form a set of integers which was investigated by Tamura [1982N]. Stratification in nilsemigroups was studied by the author [1995S] and used in an unsuccessful attempt to prove the commutative Kleitman-Rothschild-Spencer conjecture [1995N].

1. FREE COMMUTATIVE MONOIDS.

Constructing congruences requires a number of additional properties of free c.m. (= commutative monoids). This section also contains Rédei's general description of congruences by kernel functions [1956].

In what follows, we denote by F or by F_X the free c.m. on a set X, and by G or G_X the free abelian group on the same set. We write both additively. As in Chapter I we write the elements of F as linear combinations $a = \sum_{x \in X} a_x x$ with nonnegative integer coefficients a_x that are almost all 0; we write the elements of G as linear combinations $\sum_{x \in X} a_x x$ with integer coefficients that are almost all 0.

1. A **face** of a c.s. S is a nonempty subsemigroup T of S such that $xy \in T$ implies $x \in T$ and $y \in T$; equivalently, the complement $T = S \backslash P$ of a proper prime ideal P of S.

Proposition 1.1. *The faces of the free c.m. on a set X are the free c.m. generated by subsets of X.*

Proof. Let Y be a subset of X. Then $a = \sum_{x \in X} a_x x \in F$ is in F_Y if and only if $a_x = 0$ for all $x \in X \backslash Y$. Hence F_Y is a submonoid of F. In fact F_Y is a face of F: if $a, b \in F$ and $a + b \in F_Y$, then $a_x + b_x = 0$ for all $x \in X \backslash Y$, $a_x = b_x = 0$ for all $x \in X \backslash Y$, and $a, b \in F_Y$.

Conversely let T be a face of F. Then $0 + t \in T$ for some $t \in T$, since $T \neq \emptyset$, and $0 \in T$. Then $Y = X \cap T \subseteq X$ and $F_Y \subseteq T$. If conversely $a = \sum_{x \in X} a_x x \in T$, then $a_x > 0$ implies $a - x \in F$, $a = x + (a - x) \in T$ and $x \in T \cap X = Y$; therefore $a \in F_Y$. Thus $T = F_Y$. □

By Lemma III.1.1, $a \mathcal{N} b$ holds in any c.s. S if and only if a and b belong to the same prime ideals of S, if and only if a and b belong to the same faces of S. Hence Proposition 1.1 yields the archimedean components of F:

Proposition 1.2. *$Y(F_X)$ is isomorphic to the semilattice of all finite subsets of X under union; the archimedean component of F_X that corresponds to $Y \subseteq X$ is $\{a = \sum_{x \in Y} a_x x \in F_Y \mid a_x > 0$ for all $x \in Y\}$.*

Proof. Let $a = \sum_{x \in X} a_x x \in F$ and $S(a) = \{x \in X \mid a_x > 0\}$. For every subset Y of X we have $a \in F_Y$ if and only if $S(a) \subseteq Y$. Hence that a and $b \in F$ belong to the same faces if and only if $S(a) = S(b)$. □

$Y(F_X)$ is the **free semilattice** on the set X; every mapping of X into a semilattice Y extends uniquely to a homomorphism of $Y(F_X)$ into Y.

When $A = F_Y$ is a face of F, with $Y \subseteq X$, then $A' = F_{X\backslash Y}$ is the **complementary** face of F_Y. The corresponding abelian groups G_Y and $G_{X\backslash Y}$ provide a direct sum decomposition $G = G_Y \oplus G_{X\backslash Y}$ of G. There is a very similar decomposition of F:

Proposition 1.3. *Let $A = F_Y$ be a face of F_X. Every element a of F_X can be written uniquely in the form $a = p_A a + p'_A a$ with $p_A a \in A$, $p'_A a \in A' = F_{X\backslash Y}$; namely, $p_A a = \sum_{x\in Y} a_x x$ and $p'_A a = \sum_{x\in X\backslash Y} a_x x$.*

p_A and p'_A are the **projections** of a on the faces A and A'. It follows from Proposition 1.3 that $F \cong A \times A'$.

2. There is a natural partial order on G: when $a = \sum_{x\in X} a_x x$ and $b = \sum_{x\in X} b_x x \in G$, then

$$a \leqq b \text{ if and only if } a_x \leqq b_x \text{ for all } x, \text{ if and only if } b - a \in F_X.$$

If $a, b \in F$, then $a \leqq_{\mathcal{H}} b$ in F if and only if $a \geqq b$.

Proposition 1.4. *G_X is a lattice ordered abelian group.*

Proof. Suprema and infima in G are pointwise: for all $a = \sum_{x\in X} a_x x$ and $b = \sum_{x\in X} b_x x \in G$,

$$a \vee b = \textstyle\sum_{x\in X} \max(a_x, b_x)\, x \quad\text{and}\quad a \wedge b = \sum_{x\in X} \min(a_x, b_x)\, x.$$

The equalities

$$(a \vee b) + c = (a+c) \vee (b+c), \quad (a \wedge b) + c = (a+c) \wedge (b+c)$$

hold for all $a, b, c \in G$. □

We see that F is a sublattice of G. Note the following useful properties:

Lemma 1.5. *For all $a, b, c \in G_X$:*

$$\begin{aligned} (a \vee b) + c &= (a+c) \vee (b+c), & (a \wedge b) + c &= (a+c) \wedge (b+c) \\ -(a \wedge b) &= (-a) \vee (-b), & -(a \vee b) &= (-a) \wedge (-b), \\ c - (a \wedge b) &= (c-a) \vee (c-b), & c - (a \vee b) &= (c-a) \wedge (c-b). \end{aligned}$$

The **positive** and **negative** parts of $a \in G$ are:

$$a^+ = a \vee 0 \quad\text{and}\quad a^- = (-a) \vee 0.$$

By Lemma 1.5,

Lemma 1.6. *For all $a \in G_X$, $a^+, a^- \geqq 0$, $a^+ \wedge a^- = 0$, and $a = a^+ - a^-$.*

3. We complete this section with the description of all congruences on F, due to Rédei [1956], which arises from the following observations.

Lemma 1.7. *Let $\mathcal{C}$ be a congruence on F_X. Let*

$$R = R(\mathcal{C}) = \{r \in G_X \mid r = a - b \text{ for some } a,b \in F_X \text{ such that } a \ \mathcal{C} \ b\}.$$

For every $r \in R$, let

$$f(r) = f_{\mathcal{C}}(r) = \{c \in F \mid c + r^+ \ \mathcal{C} \ c + r^-\}.$$

Then R is a subgroup of G_X; every $f(r)$ is a nonempty ideal of F_X; and

$$a \ \mathcal{C} \ b \text{ if and only if } a - b \in R \text{ and } a \wedge b \in f(a - b).$$

Proof. Since $\mathcal{C}$ is a congruence it is immediate that R is a subgroup of G and that $f(r)$ is an ideal of F.

Let $a,b \in F$ and $r = a - b \in G$. By Lemma 1.5,

$$\begin{aligned} a - r^+ &= a - ((a-b) \vee 0) = b \wedge a, \quad \text{and} \\ b - r^- &= b - ((b-a) \vee 0) = a \wedge b. \end{aligned}$$

Hence $a \ \mathcal{C} \ b$ implies $r = a - b \in R$, $(a \wedge b) + r^+ = a \ \mathcal{C} \ b = (a \wedge b) + r^-$, and $a \wedge b \in f(r)$. In particular, $f(r) \neq \emptyset$ when $r \in R$. If conversely $r = a - b \in R$ and $a \wedge b \in f(r)$, then $a = (a \wedge b) + r^+ \ \mathcal{C} \ (a \wedge b) + r^- = b$. □

In Lemma 1.7, R is the **Rédei group** of $\mathcal{C}$ and f is the **kernel function** of $\mathcal{C}$; f is a mapping of R into the set of all non-empty ideals of F. By Lemma 1.7, $\mathcal{C}$ is completely determined by its Rédei group and kernel function.

Lemma 1.8. *The kernel function of a congruence $\mathcal{C}$ on F has the following properties:*

(1) $f(0) = F$;

(2) $f(-r) = f(r)$ *for all* $r \in R$;

(3) $(r^+ + f(r)) \cap (s^+ + f(s)) \subseteq (r \vee s) + f(r - s)$, *for all* $r,s \in R$.

Proof. Properties (1) and (2) are immediate since $\mathcal{C}$ is reflexive and symmetric. Property (3) expresses the transitivity of $\mathcal{C}$. Let $c \in (r^+ + f(r)) \cap (s^+ + f(s))$. Then $c - r^+$, $c - s^+ \in F$, $c - r = c - r^+ + r^- \in F$, and $c - s \in F$. By definition of $f(r)$, $c = c - r^+ + r^+ \ \mathcal{C} \ c - r^+ + r^- = c - r$. Similarly $c \ \mathcal{C} \ c - s$. Hence $c - r \ \mathcal{C} \ c - s$ and

$$c - (r \vee s) = (c - s) \wedge (c - r) \in f(r - s)$$

by Lemma 1.7, which proves (3). □

Theorem 1.9. *Let F_X be the free commutative monoid on a set X, R be a subgroup of the free abelian group on X, and f assign to each $r \in R$ a non-empty ideal $f(r)$ of F_X, so that properties* (1), (2), *and* (3) *in Lemma 1.8 hold. Then the binary relation $\mathcal{C}$ on F_X defined by*

$$a \,\mathcal{C}\, b \iff a - b \in R \text{ and } a \wedge b \in f(a-b) \tag{C}$$

is a congruence on F_X with Rédei group R and kernel function f. Moreover every congruence on F_X can be constructed in this fashion.

Proof. It follows from Lemmas 1.7 and 1.8 that every congruence on F can be constructed as in the statement. Now let R and f have properties (1), (2), and (3); define $\mathcal{C}$ by (C).

$\mathcal{C}$ is reflexive by (1) and symmetric by (2). Now assume $a \,\mathcal{C}\, b$ and $b \,\mathcal{C}\, c$, so that $r = b - a \in R$, $s = b - c \in R$, $a \wedge b \in f(r) = f(-r)$, and $b \wedge c \in f(s)$. Then $a - c = s - r \in R$. As in the proof of Lemma 1.7, $b - r^+ = b - ((b-a) \vee 0) = a \wedge b$ and $b - s^+ = b - ((b-c) \vee 0) = c \wedge b$, so that $b \in (r^+ + f(r)) \cap (s^+ + f(s))$. By (3), $b \in (r \vee s) + f(r-s)$. But $r \vee s = (b-a) \vee (b-c) = b - (a \wedge c)$. Hence $a \wedge c = b - (r \vee s) \in f(r-s) = f(a-c)$, and $a \,\mathcal{C}\, c$. Thus $\mathcal{C}$ is transitive.

If $a \,\mathcal{C}\, b$, then, for all $c \in F$, $(a+c) - (b+c) = a - b \in R$,

$$(a+c) \wedge (b+c) = (a \wedge b) + c \in f(a-b) + c \subseteq f(a-b)$$

since $f(a-b)$ is an ideal of F, and $a + c \;\mathcal{C}\; b + c$. Thus $\mathcal{C}$ is a congruence.

By definition $a \,\mathcal{C}\, b$ implies $a - b \in R$. Conversely let $r \in R$. Since $f(r) \neq \emptyset$ there exists $c \in f(r)$, and for any such c we have $a = c + r^+ \;\mathcal{C}\; c + r^- = b$, since $a - b = r^+ - r^- = r$ and $a \wedge b = c + (r^+ \wedge r^-) = c \in f(r)$. This shows that $R(\mathcal{C}) = R$, and that $f(r) \subseteq f_{\mathcal{C}}(r)$ for all $r \in R$. If conversely $c \in f_{\mathcal{C}}(r)$, then $c + r^+ \;\mathcal{C}\; c + r^-$ and $c = (c + r^+) \wedge (c + r^-) \in f((c + r^+) - (c + r^-)) = f(r)$. Thus $f = f_{\mathcal{C}}$ is the kernel function of $\mathcal{C}$. □

For an example we calculate the kernel function of a Rees congruence. First define

$$S : a = \{x \in F \mid a + x \in S\}$$

for any subset S of F and $a \in F$. We see that S is an ideal of F ($x \in S$ implies $x + y \in S$ for all $y \in F$) if and only if $S \subseteq S : a$ for all $a \in F$.

Let $\mathcal{I}$ be the Rees congruence of a nonempty ideal I of F. Let $a \in I$. For any $r \in G$ we have $a + r^+$, $a + r^- \in I$ and $r = (a + r^+) - (a + r^-) \in R$. Thus the Rédei group of $\mathcal{I}$ is G. If $r \neq 0$, then $r^+ \neq r^-$ and $c \in f(r)$ if and only if $c+r^+ \ \mathcal{J} \ c+r^-$, if and only if $c+r^+, \ c+r^- \in I$. Thus the kernel function of $\mathcal{J}$ is

$$f(r) \ = \ (I : r^+) \cap (I : r^-)$$

for all $r \in G\backslash 0$, with $f(0) = G$.

Much more information on kernel functions is given in Rédei [1956]. This includes a classification of congruences by various ranks, and the first proof of Rédei's Theorem that congruences on a free f.g.c.s. satisfy the ascending chain condition.

2. THE ZERO CLASS.

A **nilmonoid** congruence is a congruence whose quotient semigroup is a nilmonoid. The first step in the construction of these congruences determines the equivalence classes which yield the identity and zero elements of the quotient.

1. In a commutative nilsemigroup N, Green's preorder $\leqq_{\mathcal{H}}$ is a partial order relation on N, which we denote by just $\leqq$, and $xy < x$ whenever $x \neq 0$ (Proposition IV.3.1). These properties extend to any commutative nilmonoid $N^1 = N \cup \{1\}$.

When S is a monoid we abuse the language by calling a surjective homomorphism $\pi : F_X \longrightarrow S$ a **presentation** of S; then $\pi(X)$ generates S and an actual presentation can be obtained by finding generating relations for the congruence induced by π. Conversely every generating subset X of S gives rise to a presentation $\pi : F_X \longrightarrow S$ of S.

For finite nilmonoids there exists a most economical presentation. An element a of a semigroup S is **irreducible** when $a = xy$ implies $x = a$ or $y = a$; if S is a monoid we also require $a \neq 1$.

Proposition 2.1. *Let N^1 be a finite commutative nilmonoid. An element a of N^1 is irreducible if and only if a is maximal in N, if and only if $a \in N\backslash N^2$. Moreover, N^1 is generated (as a monoid) by its irreducible elements, and therefore has a smallest generating subset $N\backslash N^2$.*

Proof. Let $a \in N$. If a is not irreducible in N^1, then $a = xy$ for some $x,y \in N^1$, $x,y \neq a$, and $a < x \neq 1$ is not maximal. If a is not maximal, then $a < b \in N$, $a = bc$ for some $c \in N$, and $a \in N^2$. If finally $a \in N^2$, then $a = bc$ for some $b,c \in N$, which in the nilsemigroup N implies $a < b,c$ and

$b, c \neq a$; thus a is not irreducible. This proves the first part of the statement.

An element of $N \backslash N^2$ cannot be written as a product of other elements of N and must belong to every generating subset of N^1. It remains to show that $N \backslash N^2$ generates N^1. This is proved by noetherian induction. If N^1 is not generated (as a monoid) by its irreducible elements, then N^1 has an element which is not a product of irreducible elements and has an element m which is maximal with this property. This includes empty products and one term products, so that $m \neq 1$ and m is not irreducible. Therefore $m \in N^2$ and $m = xy$ for some $x, y \in N$, $x, y \neq m$. Then $m < x, y$; by the choice of m, x and y are products of irreducible elements of N, and then so is $m = xy$, which is the required contradiction. □

When N is finite, the set $X = N \backslash N^2$ provides the most economical homomorphism $\pi : F_X \longrightarrow N^1$; we refer to this as the **standard presentation** of N^1. The results in Grillet [1991N] are stated in terms of this standard presentation. This restriction is unnecessary; in fact we do not even assume that F is finitely generated in what follows.

2. A surjective homomorphism π of F onto a nilmonoid N^1 sends $0 \in F$ onto $1 \in N^1$ and generally sends $a = \sum_{x \in X} a_x x \in F$ to $\pi(a) = \prod_{x \in X} \pi(x)^{a_x} \in N^1$. A congruence $\mathcal{C}$ on $F = F_X$ is a **nilmonoid** congruence on F when $F/\mathcal{C}$ is a nilmonoid; equivalently, when $\mathcal{C}$ is induced by a surjective homomorphism π of F onto a nilmonoid N^1. Then $F/\mathcal{C} \cong N^1$; the isomorphism sends the $\mathcal{C}$-class C_a of $a \in F$ to $\pi(a) \in N^1$. If N^1 is finite, then $\mathcal{C}$ is a **finite nilmonoid** congruence. If F is finitely generated (if X is finite), then every nilmonoid congruence on F is a finite nilmonoid congruence.

The **identity class** of a nilmonoid congruence $\mathcal{C}$ is the $\mathcal{C}$-class $U = C_0$ of $0 \in F$; if $\mathcal{C}$ is induced by $\pi : F \longrightarrow N^1$, then $U = \pi^{-1}(1)$. We call a nilmonoid congruence **reduced** when its identity class is trivial.

Proposition 2.2. *When $\mathcal{C}$ is a nilmonoid congruence on F_X, then the identity class U of $\mathcal{C}$ is a face of F_X; the restriction $\mathcal{C}'_U$ of $\mathcal{C}$ to U' is a reduced nilmonoid congruence; $a \;\mathcal{C}\; b$ if and only if $p'_U a \;\mathcal{C}'_U\; p'_U b$; and $F/\mathcal{C} \cong U'/\mathcal{C}'_U$.*

Proof. $\{1\}$ is a face of N^1, since $xy \neq 1$ when $x \in N$ or $y \in N$. Hence $U = \pi^{-1}(1)$ is a face of F. Moreover, $\pi(a) = \pi(p'_U a)$ for all $a \in F$, since $\pi(c) = 1$ for all $c \in U$; hence $a \;\mathcal{C}\; p'_U a$ and $a \;\mathcal{C}\; b$ if and only if $p'_U a \;\mathcal{C}'_U\; p'_U b$. In particular, $F/\mathcal{C} \cong U'/\mathcal{C}'_U$, and $\mathcal{C}'_U$ is reduced, since $a \;\mathcal{C}'_U\; 0$ with $a \in U'$

implies $a \in U \cap U'$ and $a = 0$. □

Thus we may assume when constructing nilmonoids that $\mathcal{C}$ is reduced.

3. An ideal J of F is a **nilmonoid ideal** when the Rees quotient F/J is a nilmonoid, in particular $J \neq F$ is a proper ideal. An ideal J of F is **cofinite** when its complement $F \backslash J$ is finite.

The **zero class** of a nilmonoid congruence $\mathcal{C}$ on F is the zero element J of $F/\mathcal{C}$; if $\mathcal{C}$ is induced by $\pi : F \longrightarrow N^1$, then $J = \pi^{-1}(0)$.

Proposition 2.3. *Let $\mathcal{C}$ be a reduced nilmonoid congruence on F_X. The zero class J of $\mathcal{C}$ is a nilmonoid ideal of F_X; if F_X is finitely generated, then J is a proper cofinite ideal of F_X.*

Proof. $J = \pi^{-1}(0)$ is a nonempty ideal of F, since $\{0\}$ is an ideal of N^1, and $J \neq F$, since $0 \notin J$.

For every $x \in X$ we have $\pi(x) \in N$, since $\mathcal{C}$ is reduced, $\pi(x)^{k_x} = 0$ in N for some $k_x > 0$, and $k_x x \in J$. Let $a = \sum_{x \in X} a_x x \in F$. If $a \neq 0$, then $a_x > 0$ for some x and $k_x a \in J$. Hence F/J is a nilmonoid.

Since $a \in J$ whenever $a_x \geqq k_x$ for some x, we have

$$F \backslash J \subseteq \{a \in F \mid a_x < k_x \text{ for all } x\}.$$

If X is finite, then so is $F \backslash J$. □

4. Sometimes the ideal J suffices to construct N^1. A commutative nilmonoid is 0-**free** when it is isomorphic to the Rees quotient of a free commutative monoid by a nilmonoid ideal. 0-free nilmonoids are also called **unique factorization** nilmonoids, due to the following result:

Proposition 2.4. *A finite commutative nilmonoid N^1 is 0-free if and only if there exists a subset S of N such that every nonzero element of N^1 can be written uniquely as a product of nonnegative powers of distinct elements of S.*

Proof. Let $N^1 = F/J$ be 0-free and $\pi : F \longrightarrow N^1$ be the projection. Every element of F can be written uniquely as a linear combination $a = \sum_{x \in X} a_x x$ of elements of X. If $a_x > 0$ for some $x \in J$, then $a \in J$; hence every element of $F \backslash J$ can be written uniquely as a linear combination $a = \sum_{x \in Y} a_x x$ of elements of $Y = X \backslash J$ and every nonzero element of N^1 can be written uniquely as a product of nonnegative powers $\pi(a) = \prod_{x \in Y} \pi(x)^{a_x}$ of elements of $S = \pi(Y)$.

Conversely assume that there exists a subset X of N such that every nonzero element of N^1 can be written uniquely as a product of nonnegative powers of distinct elements of X. Let $\pi : F = F_X \longrightarrow N^1$ send $\sum_{x \in X} a_x x \in F$ to $\prod_{x \in X} x^{a_x}$; let $J = \pi^{-1}(0)$. The hypothesis states that π is injective on $F \backslash J$, so that N^1 is isomorphic to the Rees quotient F/J. □

A homomorphism φ between nilsemigroups or nilmonoids is **pure** in case $\varphi^{-1}(0) = \{0\}$; equivalently, in case $\ker \varphi$ is a pure congruence.

Corollary 2.5. *Every (finite) commutative nilmonoid is a pure homomorphic image of a (finite) 0-free nilmonoid.*

Proof. By Proposition 2.2 a commutative nilmonoid N^1 has a presentation $\pi : F \longrightarrow N^1$ for which $\mathcal{C} = \ker \pi$ is reduced. Then the zero class $J = \pi^{-1}(0)$ of $\mathcal{C}$ is a nilmonoid ideal of F by Proposition 2.3, and F/J is a 0-free nilmonoid. If N^1 is finite, then we may use the standard presentation, and then J is a proper cofinite ideal of F and F/J is finite.

Since the Rees congruence of J is contained in $\mathcal{C}$, $\pi : F \longrightarrow N^1$ factors through the projection $\rho : F \longrightarrow F/J$: there is a surjective homomorphism $\varphi : F/J \longrightarrow N^1$ such that $\pi = \varphi \circ \rho$. If $a \in F \backslash J$, then $\pi(a) \neq 0$ and $\varphi(a) = \pi(a) \neq 0$; thus φ is pure. □

Corollary 2.5 is due to Arendt & Stuth [1970A].

5. When J is an ideal of F, a J-**congruence** is a congruence on F modulo which J is a class.

Proposition 2.6. *When J is a nilmonoid ideal of a free c.m. F, then every J-congruence is a reduced nilmonoid congruence. Conversely every commutative nilmonoid is isomorphic to $F/\mathcal{C}$ for some free c.m. F, nilmonoid ideal J of F, and J-congruence $\mathcal{C}$.*

Proof. As before a commutative nilmonoid N^1 has a presentation $\pi : F \longrightarrow N^1$ such that $\mathcal{C} = \ker \pi$ is reduced. Then the zero class $J = \pi^{-1}(0)$ of $\mathcal{C}$ is a nilmonoid ideal of F by Proposition 2.3, and $\mathcal{C}$ is a J-congruence. Conversely let J be a nilmonoid ideal of F and $\mathcal{C}$ be a J-congruence on F. Since J is a nonempty ideal, the $\mathcal{C}$-class J is a zero element of $F/\mathcal{C}$. When $a \in F$, $a \neq 0$, then $na \in J$ for some $n > 0$, since F/J is a nilmonoid; hence $F/\mathcal{C}$ is a nilmonoid. Moreover $\mathcal{C}$ is reduced: if $a \neq 0$ and $a \,\mathcal{C}\, 0$, then $na \,\mathcal{C}\, 0$ for all $n > 0$ and $0 \in J$, a contradiction since the nilmonoid ideal J is proper. □

Proposition 2.7. *A proper cofinite ideal J of a free c.m. F is a nilmonoid ideal of F, and every J-congruence is a finite reduced nilmonoid congruence. Conversely every finite commutative nilmonoid is isomorphic to $F/\mathcal{C}$ for some free c.m. F_X, proper cofinite ideal J of F_X, and J-congruence $\mathcal{C}$; moreover* *it may be assumed that* F_X *is finitely generated and that* $\{x\}$ *is a* $\mathcal{C}$*-class for every* $x \in X$, *and then* F_X, J, *and* $\mathcal{C}$ *are unique up to isomorphism.*

Proof. A proper cofinite ideal J of F is a nilmonoid ideal: if $a \in F$, $a \neq 0$, then a, $2a$, ..., na, ... are all distinct and cannot all be contained in the finite set $F \backslash J$; hence $na \in J$ for some $n > 0$, and F/J is a nilmonoid. By Proposition 2.6, every J-congruence $\mathcal{C}$ is a reduced nilmonoid congruence; moreover $F/\mathcal{C}$ is finite, since a $\mathcal{C}$-class is either contained in $F \backslash J$ or equal to J.

Conversely every finite commutative nilmonoid N^1 has a standard presentation $N^1 \cong F_X/\mathcal{C}$. Then $X = N \backslash N^2$ is finite and the zero class J of $\mathcal{C}$ is a proper cofinite ideal of F_X by Proposition 2.3. Also the projection $\pi : F \longrightarrow N^1$ is injective on X and $\{x\}$ is a $\mathcal{C}$-class for every $x \in X$.

Let F_X be any finitely generated free c.m., J be a proper cofinite ideal of F_X, and $\mathcal{C}$ be a J-congruence such that $F_X/\mathcal{C} \cong N^1$. If every $\{x\}$ is a $\mathcal{C}$-class, then π is injective on X and $N \backslash N^2$ contains every $\pi(x)$, since equalities $\pi(x) = \pi(a)\pi(b)$ with $a, b \neq 0$ are not possible. Since $\pi(X)$ generates N^1 it follows from Proposition 2.1 that $\pi(X) = N \backslash N^2$. Thus π induces a bijection $X \longrightarrow N \backslash N^2$ and an isomorphism $F_X \longrightarrow F_{N \backslash N^2}$, which takes F_X and $\mathcal{C}$ to the standard presentation of N^1. □

3. CORNER POINTS.

This section investigates how J-congruences are determined by the ideal J.

1. The Rees congruence of a nonempty ideal I is the smallest congruence for which I is an equivalence class. The largest such congruence was first described by Teissier [1951].

Proposition 3.1. *Let* I *be a nonempty ideal of a c.s.* S. *There is a largest congruence* $\mathcal{C}$ *on* S *such that* I *is a* $\mathcal{C}$*-class; for all* $a, b \in S$,

$$a \;\mathcal{C}\; b \quad \textit{if and only if} \quad I : a = I : b.$$

Proof. Recall that $I : a = \{x \in S^1 \mid ax \in I\} \subseteq S^1$. $\mathcal{C}$ is a congruence since $I : a = I : b$ implies $I : ac = (I : a) : c = (I : b) : c = I : bc$. Also $a \in I$ if and only if $I : a = S^1$, since I is an ideal; therefore I is a $\mathcal{C}$-class.

Conversely let $\mathcal{E}$ be a congruence on S such that I is an $\mathcal{E}$-class. Let $a \;\mathcal{E}\; b$. For every $x \in S^1$, ax and bx are in the same $\mathcal{E}$-class; hence $ax \in I$ if and only if $bx \in I$, and $I : a = I : b$. Thus $\mathcal{E} \subseteq \mathcal{C}$. □

Proposition 3.1 provides a greatest J-congruence $\mathcal{D}$ (not to be confused with Green's relation $\mathcal{D}$) for every ideal J of a free c.m. F.

Proposition 3.2. *Let J be a nilmonoid ideal of F and $\mathcal{C}$ be a J-congruence on F. The greatest pure congruence on $F/\mathcal{C}$ is the congruence induced by the greatest J-congruence $\mathcal{D}$.*

Proof. Let $\mathcal{P} = \pi(\mathcal{D})$ be the congruence on $F/\mathcal{C}$ induced by $\mathcal{D} \supseteq \mathcal{C}$ (where $\pi : F \longrightarrow F/\mathcal{C}$ is the projection); then $\pi^{-1}(\mathcal{P}) = \mathcal{D}$ (Proposition I.2.6). Let $j \in J$. $\mathcal{P}$ is a pure congruence since $\mathcal{D}$ and $\mathcal{C}$ are J-congruences: indeed $\pi(a)\ \mathcal{P}\ \pi(j) = 0$ if and only if $a\ \mathcal{D}\ j$, if and only if $a \in J$, if and only if $\pi(a) = 0$. If $\mathcal{A}$ is a pure congruence on $F/\mathcal{C}$, then $\pi^{-1}(\mathcal{A})$ is a J-congruence on F, $\pi^{-1}(\mathcal{A}) \subseteq \mathcal{D} = \pi^{-1}(\mathcal{P})$, and $\mathcal{A} \subseteq \mathcal{P}$. □

By Propositions IV.3.6 and V.5.2, a finite commutative nilmonoid $F/\mathcal{C}$ is subdirectly irreducible if and only if the greatest pure congruence on $F/\mathcal{C}$ is the equality. By Propositions 3.2 this is equivalent to $\mathcal{C} = \mathcal{D}$. Thus finite subdirectly irreducible nilmonoids are determined by proper cofinite ideals.

2. The next result gives a sharper description of the greatest J-congruence $\mathcal{D}$ when J is a proper cofinite ideal of F. Then $H = F \backslash J$ is finite and the set $K = K(J)$ of all maximal elements of H is nonempty; in fact, for every $a \in H$ there exists some $k \in K$ such that $a \leqq k$. The elements of K are the **corner points** of H, and also of J and of every J-congruence. The corner points determine

$$H = \{a \in F \mid a \leqq k \text{ for some } k \in K\},$$

$J = F \backslash H$, and the greatest J-congruence:

Proposition 3.3. *Let J be a proper cofinite ideal of F. The greatest J-congruence $\mathcal{D}$ on F is given by*

$$a\ \mathcal{D}\ b \iff J : a = J : b \iff H : a = H : b \iff K : a = K : b,$$

where $H = F \backslash J$ and K is the set of all maximal elements of H.

In particular, K is a $\mathcal{D}$-class, the $\mathcal{D}$-class of all $a \in F$ such that $K : a = \{0\}$.

Proof. Since H is the complement of J, $H : a$ is the complement of $J : a$; hence $J : a = J : b$ if and only if $H : a = H : b$.

Assume $H : a = H : b$. If $a + t \in K$, so that $a + t$ is a maximal element of H, then $b + t \in H$ and $b + t$ is maximal in H, otherwise $b + t + u \in H$ for some $u > 0$, $a + t + u \in H$, and $a + t$ is not maximal in H; hence $b + t \in K$. Similarly $b + t \in K$ implies $a + t \in K$. Thus $K : a = K : b$. Conversely assume

$K:a = K:b$. If $a+t \in H$, then since H is finite we have $a+t \leqq k$ for some $k \in K$, $a+t+u \in K$ for some $u \in F$, $b+t+u \in K$, and $b+t \in H$, since $b+t \leqq b+t+u$. Similarly $b+t \in H$ implies $a+t \in H$. Thus $H:a = H:b$. □

Corollary 3.4. *If $|K| = 1$, then the Rees congruence of J is the only J-congruence.*

Proof. If $K = \{k\}$, then $H = \{t \in F \mid t \leqq k\}$ and $K:a = \{k-a\}$ if $a \in H$, otherwise $K:a = \emptyset$. Therefore $\mathcal{D}$ is the Rees congruence of J, and this the only J-congruence. □

3. We illustrate the above with some examples.

Example 3.5. Let $X = \{x,y\}$ and J be the ideal generated by $4x$ and $3y$. Then $a = a_x x + a_y y \in F$ is in J when $a_x \geqq 4$ or $a_y \geqq 3$, and $a = a_x x + a_y y \in H = F \backslash J$ when $a_x \leqq 3$ and $a_y \leqq 2$. Thus $H = H(k)$, where $k = 3x + 2y$, and $K = \{k\}$. By Corollary 2.8, $\mathcal{D}$ is the Rees congruence of J; $F/\mathcal{D}$ is the 0-free nilmonoid $\langle\, x,y \mid x^4 = y^3 = 0 \,\rangle$. □

When $X = \{x,y\}$, an element $a = a_x x + a_y y$ of F can be represented by a point (a_x, a_y) with nonnegative integer coordinates. In the examples we draw only the elements of H. Thus example 3.5 is pictured by:

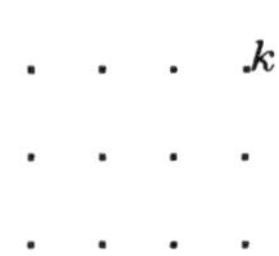

Example 3.5

Example 3.6. Let $X = \{x,y\}$ and J be the ideal generated by $8x$, $5x+4y$, and $6y$. Then $a = a_x x + a_y y \in F$ is in J when $a_x \geqq 8$, or $a_y \geqq 6$, or $a_x \geqq 5$ and $a_y \geqq 4$; and $a \in H = F \backslash J$ when $a_x \leqq 7$, $a_y \leqq 5$, and ($a_x \leqq 4$ or $a_y \leqq 3$).

Thus H has two maximal elements $k = 4x + 5y$ and $l = 7x + 3y$. We see that $a = 2x + 4y \;\mathcal{D}\; 5x + 2y = b$: indeed $a \leqq k$, $a \nleqq l$, $b \nleqq k$, $b \leqq l$, and $k - a = l - b$, so that $K:a = \{k-a\} = K:b$. But neither $a - x \;\mathcal{D}\; b - x$ nor $a - y \;\mathcal{D}\; b - y$ hold. Moreover, $x + 5y = a - x + y \;\mathcal{D}\; b - x + y = 4x + 3y$ does not hold: indeed $x + 5y \leqq k$, $x + 5y \nleqq l$, $4x + 3y \leqq k$, and $4x + 3y \leqq l$, so that $|K:(x+5y)| = 1$ but $|K:(4x+3y)| = 2$. Similarly, $3x + 3y =$

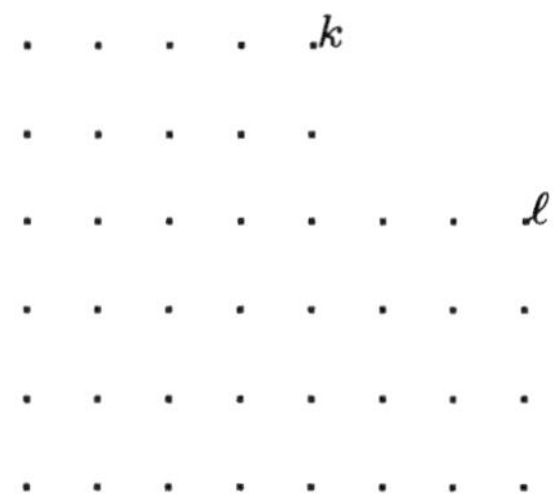

Example 3.6

$a - y + x \ \mathcal{D} \ b - y + x = 6x + y$ does not hold. Hence $\mathcal{D}$ is generated (as a J-congruence) by $(2x + 4y,\ 5x + 2y)$ and $F/\mathcal{D}$ is the commutative nilmonoid

$$F/\mathcal{D} \ \cong \ \langle\, x, y \mid \ x^8 = x^5 y^4 = y^6 = 0,\ x^2 y^4 = x^5 y^2 \,\rangle. \ \square$$

Example 3.7

Example 3.7. Let $X = \{x, y\}$ and J be the ideal of all $a \in F$ such that $\|a\| \geqq 8$, where $\|a_x x + a_y y\| = i + j$. Then $a \in H$ if and only if $\|a\| \leqq 7$, $a \in K$ if and only if $\|a\| = 7$, and there are 8 corner points. When $a \in H$ we have $H : a = \{ t \in F \mid \|t\| \leqq 7 - \|a\| \}$; hence $a \ \mathcal{D} \ b$ if and only if either $\|a\| = \|b\|$ or $a, b \in J$, and $F/\mathcal{D}$ is the cyclic nilmonoid $\langle\, x \mid x^8 = 0 \,\rangle$. $\square$

4. NESTS.

This section constructs nilmonoid congruences in terms of nests of equivalence relations, as in Grillet [2001N], and brings the chapter's main result.

1. Let J be a proper cofinite ideal of a free c.m. F. Proposition 3.3 shows that the corner points of a J-congruence $\mathcal{C}$ put very strict upper limits on the size of $\mathcal{C}$. If indeed $a \;\mathcal{C}\; b$ and $a,b \notin J$, then $a \leqq m$ for some $m \in K$ and $K : a = K : b$ implies that there exists $t \in F$ such that $a+t$, $b+t \in K$; in other words (a,b) has the form $(k-t,\, l-t)$ for some $k,l \in K$ and $t \in F$ and is translated down from a pair (k,l) of corner points.

Thus a J-congruence $\mathcal{C}$ is determined by: (1) K (which determines H and J); (2) which pairs of corner points are in $\mathcal{C}$; and (3) how far in $\mathcal{C}$ can each pair of corner points be translated down. This is the main idea behind the present construction of nilmonoid congruences.

2. Let J be a proper cofinite ideal of a free c.m. F. The **support** of a J-congruence $\mathcal{C}$ is the restriction of $\mathcal{C}$ to its set K of corner points; it specifies which pairs of corner points are in $\mathcal{C}$.

Proposition 4.1. *When J is a proper cofinite ideal of F, every equivalence relation on $K(J)$ is the support of some J-congruence on F.*

Proof. Given an equivalence relation $\mathcal{E}$ on K, define $\mathcal{C}$ by:

$$a \;\mathcal{C}\; b \text{ if and only if } a=b, \text{ or } a,b \in K,\ a \;\mathcal{E}\; b, \text{ or } a,b \in J.$$

$\mathcal{C}$ is a congruence, since $a,b \in K$ implies either $a+c=a$, $b+c=b$, or $a+c$, $b+c \in J$. We see that $\mathcal{C}$ is a J-congruence with support $\mathcal{E}$. □

The support of $\mathcal{C}$ determines the universal group of $(F/\mathcal{C})\backslash 0$. This in turn determines partial homomorphisms into abelian groups. The following result was also anticipated by Arendt & Stuth [1970B].

Proposition 4.2. *Let J be a proper cofinite ideal of F_X disjoint from X and $\mathcal{C}$ be a reduced J-congruence on F_X with support $\mathcal{E}$. The universal group of $(F_X/\mathcal{C})\backslash 0$ is isomorphic to G_X/R, where R is the subgroup of G_X generated by all $k-l$ with $k \;\mathcal{E}\; l$.*

R is similar to the Rédei group of $\mathcal{C}$ except that we only use elements of K to construct R. Note that $N\backslash 0$ and $N^1\backslash 0$ have the same universal abelian group, and that when the finite nilsemigroup N is given it can always be arranged that J be disjoint from X, by Proposition 2.7.

Proof. We use Proposition III.3.4. Let $N^1 = F/\mathcal{C}$ and $\pi : F \longrightarrow N^1$ be the projection. We have $\pi(X) \subseteq N$ since $\mathcal{C}$ is reduced (Proposition 2.7). Let $\mathcal{C}' = \{(a,b) \in H \times H \mid a \;\mathcal{C}\; b\}$ be the restriction of $\mathcal{C}$ to H. As a monoid with zero, N^1 is generated by X subject to all relations $a=b$ ($(a,b) \in \mathcal{C}'$) and

$t = 0$ $(t \in J)$. We have $a = b \neq 0$ in N^1 for all $(a,b) \in \mathcal{C}'$. Also $x \neq 0$ in N^1 for all $x \in X$, since $X \cap J = \emptyset$. By Proposition III.3.4, $G(N^1\backslash 0)$ is isomorphic to the abelian group generated by X subject to all relations $a = b$ $((a,b) \in \mathcal{C}')$. Thus $G(N^1\backslash 0) \cong G/S$, where S is the subgroup of G generated by all $a - b$ with $(a,b) \in \mathcal{C}'$.

When $(a,b) \in \mathcal{C}'$, then $a \mathrel{\mathcal{C}} b$, $a,b \in H$, $a \leqq k$ for some $k \in K$, and $a = k - s$ for some $s \in F$. Then $s \in K : a$; by Proposition 3.3, $s \in K : b$ and $l = b + s \in K$. Also $k \mathrel{\mathcal{C}} l$, since $a \mathrel{\mathcal{C}} b$, and $k \mathrel{\mathcal{E}} l$. Thus $a - b = k - l \in R$. Therefore $S \subseteq R$. If conversely $k,l \in K$ and $k \mathrel{\mathcal{E}} l$, then $k \mathrel{\mathcal{C}} l$, $(k,l) \in \mathcal{C}'$, and $k - l \in S$. Therefore $R \subseteq S$. Thus $R = S$. $\square$

Example 4.3. Let N^1 be the Volkov nilmonoid (Example III.3.6)

$$N = \langle\, x,y \mid x^3 = x^2 y = x y^2 = y^4 = 0,\ x^2 = xy = y^3 \,\rangle.$$

In the standard presentation $N^1 \cong F_X/\mathcal{C}$, $X = \{x,y\}$ and the zero class of $\mathcal{C}$ is the ideal J of F_X generated by $3x$, $2x + y$, $x + 2y$, and $4y$. Thus H consists of 0, x, $2x$, y, $x + y$, $2y$, and $3y$; and $K = \{k,l,m\}$, where $k = 3y$, $l = x + y$, and $m = 2x$.

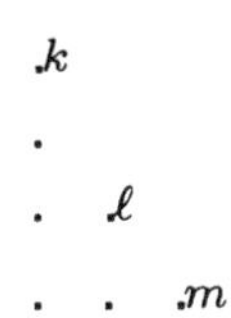

Example 4.3

The presentation $N^1 = F_X/\mathcal{C}$ shows that the support of $\mathcal{C}$ is the universal equivalence relation on K. By Proposition 4.2, $G(N^1\backslash 0)$ is isomorphic to G_X/R, where R is the subgroup of G_X generated by $k - l = 2y - x$, $k - m = 3y - 2x$, and $l - m = y - x$. Then $y = (2y - x) - (y - x) \in R$, $x = y - (y - x) \in R$, $R = G_X$, and $G(N^1\backslash 0)$ is trivial (as in Example IV.3.6).

We also have $K : 0 = K$;

$$\begin{aligned}
K : x &= \{l - x, m - x\} = \{x, 2y\};\\
K : y &= \{k - y, l - y\} = \{x + y, 2y\};\\
K : (x + y) &= \{l - x - y\} = \{y\};\\
K : 2y &= \{k - 2y, l - 2y\} = \{x,y\};
\end{aligned}$$

and $K : k = K : l = K : m = \{0\}$. Hence $\mathcal{D} = \mathcal{C}$, confirming that the Volkov nilmonoid is subdirectly irreducible, as noted after Corollary IV.3.7. □

3. We call a subset H of F a **coideal** when its complement $F \backslash H$ is an ideal; equivalently, when $x \leqq y \in H$ in F implies $x \in H$. These subsets are called **filters** in Grillet [1991N] and have also been called **order ideals**, **lower sections**, and other possibly insulting names.

As noted above, a J-congruence $\mathcal{C}$ is determined by: (1) its corner points (which determine J); (2) which pairs of corner points are in $\mathcal{C}$; and (3) how far in $\mathcal{C}$ can each pair of corner points be translated down. Classifying (3) by pairs of corner points yields the coideal sets in Grillet [1991N]. The author has come to prefer classification by translations, as in Grillet [2001N], even though [1991N] has additional properties of coideal sets.

Let $\mathcal{C}$ be a J-congruence, where J is a proper cofinite ideal of a free c.m. F. We saw that every pair (a,b) of elements of H with $a \;\mathcal{C}\; b$ has the form $(k-t,\, l-t)$ for some $k,l \in K$ and $t \in F$; actually, $t \in H$, since $t \leqq k,l$. For every $t \in H$ define an equivalence relation $\mathcal{C}_t$ on K by:

$$k \;\mathcal{C}_t\; l \text{ if and only if } k = l, \text{ or } k,l \geqq t \text{ and } k-t \;\mathcal{C}\; l-t.$$

Thus $\mathcal{C}_t$ specifies which pairs of corner points can be translated down by t (within $\mathcal{C}$). For instance $\mathcal{C}_0$ is the support of $\mathcal{C}$, but $\mathcal{C}_t$ is the equality on K whenever $t \in K$. In general:

Lemma 4.4. *When $\mathcal{C}$ is a J-congruence, then $a \;\mathcal{C}\; b$ if and only if either $a,b \in J$, or $a,b \in H$ and $a+t \;\mathcal{C}_t\; b+t$ for some $t \in H$.*

The family $N(\mathcal{C}) = (\mathcal{C}_t)_{t \in H}$ is the **nest** of equivalence relations of the J-congruence $\mathcal{C}$. Similarly $N(\mathcal{D}) = (\mathcal{D}_t)_{t \in H}$ is the nest of equivalence relations of the greatest J-congruence $\mathcal{D}$.

Lemma 4.5. *When $\mathcal{C}$ is a J-congruence, then every $\mathcal{C}_t$ is an equivalence relation on K and, for every $t,u \in H$:*

(1) *$t \leqq u$ implies $\mathcal{C}_u \leqq \mathcal{C}_t$;*

(2) *$\mathcal{C}_t \subseteq \mathcal{D}_t$;*

(3) *if $k,l,m,n \in K$ and $k - l = m - n$, then, for all $t \in H$,*

$$k \;\mathcal{C}_{p+t}\; l \quad \textit{if and only if} \quad m \;\mathcal{C}_{q+t}\; n,$$

where $p = k - k \wedge m$ and $q = m - k \wedge m$.

The proof shows that $k,l \geqq p+t$ if and only if $m,n \geqq q+t$ in (3).

Proof. $\mathcal{C}_t$ is an equivalence relation like $\mathcal{C}$.

(1) If $t \leqq u$, $k \mathcal{C}_u l$, and $k \neq l$, then $k,l \geqq u \geqq t$, $k-u \ \mathcal{C} \ l-u$,

$$k-t = k-u+(u-t) \ \mathcal{C} \ l-u+(u-t) = l-t$$

since $\mathcal{C}$ is a congruence, and $k \ \mathcal{C}_t \ l$.

(2) follows from $\mathcal{C} \subseteq \mathcal{D}$.

(3) Let k,l,m,n,p,q be as in (3). Then $k-p = k \wedge m = m-q$ and $l-p = n-q$ (since $k-l = m-n$). Hence $p+t \leqq k,l$ if and only if $q+t \leqq m,n$, and then $k-p-t \ \mathcal{C} \ l-p-t$ if and only if $m-q-t \ \mathcal{C} \ n-q-t$. Thus $k \ \mathcal{C}_{p+t} \ l$ if and only if $m \ \mathcal{C}_{q+t} \ n$. □

4. A **coherent nest** of equivalence relations on K is a family $(\mathcal{C}_t)_{t \in H}$ of equivalence relations on K with properties (1), (2), and (3) in Lemma 4.5.

Theorem 4.6. *Let F be a free commutative monoid, J be a proper cofinite ideal of F, and $(\mathcal{C}_t)_{t \in H}$ be a coherent nest of equivalence relations on $K(J)$. Then the binary relation $\mathcal{C}$ on F defined by: $a \ \mathcal{C} \ b$ if and only if*

$$a,b \in J, \text{ or } a,b \in H \text{ and } a+t \ \mathcal{C}_t \ b+t \text{ for some } t \in H$$

is a J-congruence on F whose nest of equivalence relations is $(\mathcal{C}_t)_{t \in H}$. Moreover, every J-congruence on F can be constructed in that fashion.

Proof. By Lemma 4.5, every J-congruence on F can be constructed as in the statement. For the converse we prove:

Lemma 4.7. *Let $(\mathcal{C}_t)_{t \in H}$ be a coherent nest of equivalence relations on $K = K(J)$. If $t,u \in H$, $b+u \ \mathcal{C}_u \ c+u$, and $b+t \in K$, then $c+t \in K$ and $b+t \ \mathcal{C}_t \ c+t$.*

Proof. Let $t,u \in H$, $b+u \ \mathcal{C}_u \ c+u$, and $b+t \in K$. Then $b+u \ \mathcal{D}_u \ c+u$ by (2) and $b \ \mathcal{D} \ c$, by Lemma 4.5 applied to $\mathcal{D}$. Hence $b+t \in K$ implies $c+t \in K$. We may now assume that $b \neq c$. Then $b+t$, $c+t$, $b+u$, $c+u \in K$ and $(b+t)-(c+t) = (b+u)-(c+u) \neq 0$; by (3),

$$b+t \ \mathcal{C}_{p+s} \ c+t \iff b+u \ \mathcal{C}_{q+s} \ c+u$$

for every $s \in H$, where $p = (b+t)-(b+t) \wedge (b+u) = t - t \wedge u$ and $q = (b+u)-(b+t) \wedge (b+u) = u - t \wedge u$. With $s = t \wedge u$, $b+u \ \mathcal{C}_u \ c+u$ then yields $b+t \ \mathcal{C}_t \ c+t$. □

Now let $(\mathcal{C}_t)_{t \in H}$ be a coherent nest of equivalence relations on $K = K(J)$. The relation $\mathcal{C}$ in the statement is reflexive and symmetric, since every $\mathcal{C}_t$ is an equivalence relation. Now assume $a \ \mathcal{C} \ b \ \mathcal{C} \ c$. We may assume that $a,b,c \in H$ (otherwise $a \ \mathcal{C} \ c$ already holds). Then $a+t \ \mathcal{C}_t \ b+t$ and $b+u \ \mathcal{C}_u \ c+u$ for some $t, u \in H$, $b + t \;\mathcal{C}_t\; c + t$ by Lemma 4.7, $a + t \;\mathcal{C}_t\; c + t$, and $a \;\mathcal{C}\; c$. Thus $\mathcal{C}$ is transitive.

Assume $a \;\mathcal{C}\; b$ and let $c \in F$. If $a + c \in J$, then $b + c \in J$ (since $\mathcal{C} \subseteq \mathcal{D}$ by (2)) and $a + c \;\mathcal{C}\; b + c$. Now assume $a + c$, $b + c \in H$. Then $a, b \in H$ and $a+t \;\mathcal{C}_t\; b+t$ for some $t \in H$. Since $b+c \in H$ we also have $b+c \leqq k$ for some $k \in K$ and $b + c + u \in K$ for some $u \in H$. Then $a + c + u \;\mathcal{C}_{c+u}\; b + c + u$ by Lemma 4.7, $a + c + u \;\mathcal{C}_u\; b + c + u$ by (1), and $a + c \;\mathcal{C}\; b + c$. Thus $\mathcal{C}$ is a J-congruence.

Let $k, l \in K$, $t \in H$, $t \leqq k, l$. Assume $k - t \;\mathcal{C}\; l - t$. Then $k - t + u \;\mathcal{C}_u\; l - t + u$ for some $u \in H$. By Lemma 4.7,

$$k = k - t + t \;\mathcal{C}_{p+s}\; l - t + t = l \iff k - t + u \;\mathcal{C}_{q+s}\; l - t + u$$

for every $s \in H$, where $p = k - \bigl(k \wedge (k - t + u)\bigr)$ and $q = (k - t + u) - \bigl(k \wedge (k - t + u)\bigr)$. Now $k - t \leqq k \wedge (k - t + u)$,

$$u \;=\; (k - t + u) - (k - t) \;\geqq\; (k - t + u) - \bigl(k \wedge (k - t + u)\bigr) \;=\; q,$$

and $u - q = \bigl(k \wedge (k - t + u)\bigr) - (k - t) = t - p$. With $s = t - p = u - q$, $k - t + u \;\mathcal{C}_u\; l - t + u$ and (3) yield $k \;\mathcal{C}_t\; l$. If conversely $k \;\mathcal{C}_t\; l$, then $k - t \;\mathcal{C}\; l - t$ by definition of $\mathcal{C}$. Thus $(\mathcal{C}_t)_{t \in H}$ is the nest of equivalence relations of $\mathcal{C}$. □

5. Theorem 4.6 readily yields generators of any J-congruence $\mathcal{C}$, and a presentation of the corresponding nilmonoid $F/\mathcal{C}$.

Proposition 4.8. *When J is a proper cofinite ideal of F, a J-congruence $\mathcal{C}$ on F is generated by all pairs $(r, r + x)$ where $x \in X$ and r is a minimal element of J, and all pairs $(k - s, l - s)$ where $k, l \in K(J)$, $k \neq l$, and $s \notin J$ is maximal such that $s \leqq k, l$ and $k \;\mathcal{C}_s\; l$.*

Consequently the quotient $N^1 = F/\mathcal{C}$ of F by a reduced J-congruence $\mathcal{C}$ is generated (as a c.m. with zero) by X subject to all relations $r = 0$, where r is a minimal element of J, and all relations $k - s = l - s$ where $k, l \in K(J)$, $k \neq l$, and s is maximal such that $k \;\mathcal{C}_s\; l$. This is a finite presentation, by Dickson's Theorem (Proposition VI.1.5).

6. Finally we find the kernel function of a J-congruence $\mathcal{C}$. Let

$$f_J(g) \;=\; (J : g^+) \cap (J : g^-);$$

f_J is the kernel function of the Rees congruence of J. For all $t \in H$, let

$$f_t(g) \;=\; \{\, (k \wedge l) - t \mid k, l \in K,\ t \leqq k, l,\ k \;\mathcal{C}_t\; l, \text{ and } k - l = g \,\}.$$

For instance, $f_0(0) = K$.

Proposition 4.9. *In Theorem* 4.6, *the Rédei group of* $\mathcal{C}$ *is* $G = G(F)$ *and the kernel function of* $\mathcal{C}$ *is given for all* $g \in G$ *by*

$$f(g) \;=\; f_J(g) \cup \textstyle\bigcup_{t \in H} f_t(g).$$

Proof. By definition, $c \in f(g)$ if and only if $c + g^+ \;\mathcal{C}\; c + g^-$, if and only if either $c + g^+, c + g^- \in J$, or $c + g^+ + t \;\mathcal{C}_t\; c + g^- + t$ for some $t \in H$. In the first case, $c + g^+, c + g^- \in J$ if and only if $c \in f_J(g)$. In the second case, $k = c + g^+ + t$ and $l = c + g^- + t$ satisfy $k, l \in K$, $t \leqq k, l$, $k \;\mathcal{C}_t\; l$, and $k - l = g$; hence $c = (c + g^+) \wedge (c + g^-) = (k - t) \wedge (l - t) = (k \wedge l) - t \in f_t(g)$. Conversely assume $c \in f_t(g)$, so that $c = (k \wedge l) - t$, where $k, l \in K$, $t \leqq k, l$, $k \;\mathcal{C}_t\; l$, and $k - l = g$; then $g^+ = (k - l) \vee (k - k) = k - (k \wedge l)$, $g^- = (l - k) \vee (l - l) = l - (k \wedge l)$, $c + g^+ = k - t$, $c + g^- = l - t$, $c + g^+ + t \;\mathcal{C}_t\; c + g^- + t$ for some $t \in H$, and $c \in f(g)$. □

5. EXAMPLES.

In this section we look at some examples, which indicate that conditions (1) through (3) in Lemma 4.5 allow large numbers of J-congruences even when $\mathcal{D}$ is relatively small.

Example 5.1. This is Example 3.6, in which $X = \{x, y\}$ and J is the ideal generated by $8x$, $5x + 4y$, and $6y$. We saw that $K = \{k, l\}$, where $k = 4x + 5y$ and $l = 7x + 3y$, and that $\mathcal{D}$ is generated as a J-congruence by (a, b), where $a = 2x + 4y$ and $b = 5x + 2y$.

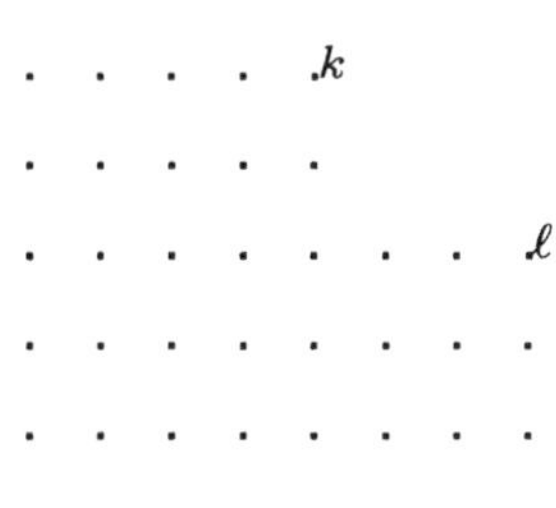

Example 5.1

There are but two equivalence relations on K, the equality $\mathcal{E}$ and the universal relation $\mathcal{U}$. We have $k\ \mathcal{D}_t\ l$ if and only if $t \leqq k,l$ and $k-t\ \mathcal{D}\ l-t$; hence $\mathcal{D}_0 = \mathcal{D}_x = \mathcal{D}_{2x} = \mathcal{D}_y = \mathcal{D}_{x+y} = \mathcal{D}_{2x+y} = \mathcal{U}$, since $k-(2x+y) = a$ and $l-(2x+y) = b$, and $\mathcal{D}_t = \mathcal{E}$ for all other $t \in H$.

Conditions (1), (2), and (3) in Lemma 4.5 are as follows. By (1), $C = \{t \in H \mid \mathcal{C}_t = \mathcal{U}\}$ is a coideal of F, which by (2) is contained in $D = \{0,\ x,\ 2x,\ y,\ x+y,\ 2x+y\}$. Condition (3) is trivial. Hence there is one coherent nest of equivalence relations on K for every coideal $C \subseteq D$. There are ten such coideals: $\varnothing$, $\{0\}$, $\{0,x\}$, $\{0,x,2x\}$, $\{0,y\}$, $\{0,x,y\}$, $\{0,x,2x,y\}$, $\{0,x,y,x+y\}$, $\{0,x,2x,y,x+y\}$, and $D_{k,l}$. This yields 10 J-congruences, ranging from the Rees congruence ($C = \varnothing$) to $\mathcal{D}$ ($C = D$). □

Example 5.2. Let $X = \{x,y\}$ and J be the ideal generated by $8x$, $6x+2y$, $3x+4y$, and $5y$. Then $K = \{k,l,m\}$, where $k = 2x+4y$, $l = 5x+3y$, $m = 7x+y$. There are five equivalence relations on K: the equality $\mathcal{E}$; the universal relation $\mathcal{U}$; the equivalence relation $\mathcal{K}$ whose classes are $\{k\}$ and $\{l,m\}$; the equivalence relation $\mathcal{L}$ whose classes are $\{l\}$ and $\{k,m\}$; and the equivalence relation $\mathcal{M}$ whose classes are $\{m\}$ and $\{k,l\}$.

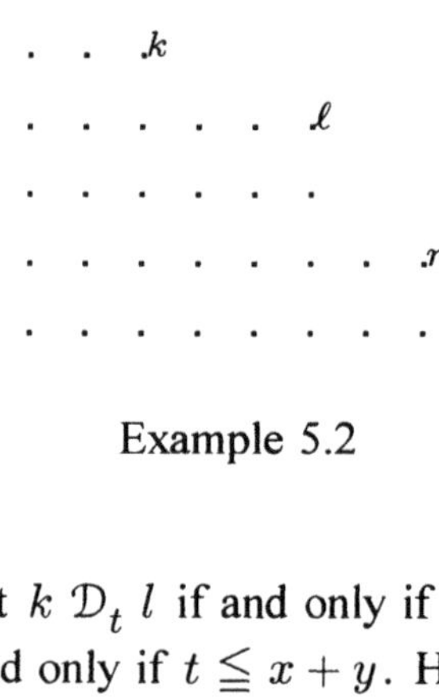

Example 5.2

It is readily verified that $k\ \mathcal{D}_t\ l$ if and only if $t \leqq 2x$, $k\ \mathcal{D}_t\ m$ if and only if $t \leqq x$, and $l\ \mathcal{D}_t\ m$ if and only if $t \leqq x+y$. Hence $N(\mathcal{D})$ is specified by

$$\begin{array}{ccccc} \mathcal{D}_0 & \mathcal{D}_x & \mathcal{D}_y & \mathcal{D}_{x+y} & \mathcal{D}_{2x} \\ \mathcal{U} & \mathcal{U} & \mathcal{K} & \mathcal{K} & \mathcal{M} \end{array}$$

with $\mathcal{D}_t = \mathcal{E}$ for all other $t \in H$.

Since condition (3) is again trivial, a coherent nest on K consists of equivalence relations $(\mathcal{C}_t)_{t\in H}$ such that $t \leqq u$ implies $\mathcal{C}_u \subseteq \mathcal{C}_t$, and $\mathcal{C}_t \subseteq \mathcal{D}_t$ for all t. In particular $\mathcal{C}_t = \mathcal{E}$ unless $t = 0,\ x,\ y,\ x+y$, or $2x$. The resulting gaggle of coherent nests can be classified by $\mathcal{C}_0$.

If $\mathcal{C}_0 = \mathcal{E}$, then $\mathcal{C}_t = \mathcal{E}$ for all t; $\mathcal{C}$ is the Rees congruence of J.

If $\mathcal{C}_0 = \mathcal{K}$, then $\mathcal{C}_t \subseteq \mathcal{K} \cap \mathcal{D}_t$ for all t; hence $\mathcal{C}_{2x} = \mathcal{E}$ and $\mathcal{C}_t$ is either $\mathcal{E}$ or $\mathcal{K}$ for all t. Thus $\mathcal{C}$ is determined by the coideal $C = \{t \in H \mid \mathcal{C}_t = \mathcal{K}\}$. Note that $0 \in C \subseteq \{0, x, y, x+y\}$. There are five such coideals: $\{0\}$, $\{0, x\}$, $\{0, y\}$, $\{0, x, y\}$, $\{0, x, y, x+y\}$. This yields five nests:

$\mathcal{C}_0$	$\mathcal{C}_x$	$\mathcal{C}_y$	$\mathcal{C}_{x+y}$	$\mathcal{C}_{2x}$
$\mathcal{K}$	$\mathcal{E}$	$\mathcal{E}$	$\mathcal{E}$	$\mathcal{E}$
$\mathcal{K}$	$\mathcal{K}$	$\mathcal{E}$	$\mathcal{E}$	$\mathcal{E}$
$\mathcal{K}$	$\mathcal{E}$	$\mathcal{K}$	$\mathcal{E}$	$\mathcal{E}$
$\mathcal{K}$	$\mathcal{K}$	$\mathcal{K}$	$\mathcal{E}$	$\mathcal{E}$
$\mathcal{K}$	$\mathcal{K}$	$\mathcal{K}$	$\mathcal{K}$	$\mathcal{E}$

If $\mathcal{C}_0 = \mathcal{L}$, then, similarly, $\mathcal{C}_y = \mathcal{C}_{x+y} = \mathcal{C}_{2x} = \mathcal{E}$ and $\mathcal{C}_x$ is either $\mathcal{E}$ or $\mathcal{L}$. This yields two nests:

$\mathcal{C}_0$	$\mathcal{C}_x$	$\mathcal{C}_y$	$\mathcal{C}_{x+y}$	$\mathcal{C}_{2x}$
$\mathcal{L}$	$\mathcal{E}$	$\mathcal{E}$	$\mathcal{E}$	$\mathcal{E}$
$\mathcal{L}$	$\mathcal{L}$	$\mathcal{E}$	$\mathcal{E}$	$\mathcal{E}$

If $\mathcal{C}_0 = \mathcal{M}$, then $\mathcal{C}_y = \mathcal{C}_{x+y} = \mathcal{E}$ and $\mathcal{C}_t$ is either $\mathcal{E}$ or $\mathcal{M}$ for all t. Thus $\mathcal{C}$ is determined by the nonempty coideal $C = \{t \in H \mid \mathcal{C}_t = \mathcal{M}\} \subseteq \{0, x, 2x\}$. This yields three nests:

$\mathcal{C}_0$	$\mathcal{C}_x$	$\mathcal{C}_y$	$\mathcal{C}_{x+y}$	$\mathcal{C}_{2x}$
$\mathcal{M}$	$\mathcal{E}$	$\mathcal{E}$	$\mathcal{E}$	$\mathcal{E}$
$\mathcal{M}$	$\mathcal{M}$	$\mathcal{E}$	$\mathcal{E}$	$\mathcal{E}$
$\mathcal{M}$	$\mathcal{M}$	$\mathcal{E}$	$\mathcal{E}$	$\mathcal{M}$

Nests with $\mathcal{C}_0 = \mathcal{U}$ can be classified by $\mathcal{C}_x$. If $\mathcal{C}_x = \mathcal{E}$, then $\mathcal{C}_{x+y} = \mathcal{C}_{2x} = \mathcal{E}$ and $\mathcal{C}_y \subseteq \mathcal{K}$; this yields two nests:

$\mathcal{C}_0$	$\mathcal{C}_x$	$\mathcal{C}_y$	$\mathcal{C}_{x+y}$	$\mathcal{C}_{2x}$
$\mathcal{U}$	$\mathcal{E}$	$\mathcal{E}$	$\mathcal{E}$	$\mathcal{E}$
$\mathcal{U}$	$\mathcal{E}$	$\mathcal{K}$	$\mathcal{E}$	$\mathcal{E}$

If $\mathcal{C}_x = \mathcal{K}$, then $\mathcal{C}_{2x} = \mathcal{E}$ and $\mathcal{C}_t \subseteq \mathcal{K}$ for all $t \neq 0$. Thus $\mathcal{C}$ is determined by the coideal $C = \{t \in H \mid \mathcal{C}_t \supseteq \mathcal{K}\}$, which satisfies $\{0, x\} \subseteq C \subseteq \{0, x, y, x+y\}$. This yields four nests:

$$\begin{array}{ccccc} \mathcal{C}_0 & \mathcal{C}_x & \mathcal{C}_y & \mathcal{C}_{x+y} & \mathcal{C}_{2x} \\ \mathcal{U} & \mathcal{K} & \mathcal{E} & \mathcal{E} & \mathcal{E} \\ \mathcal{U} & \mathcal{K} & \mathcal{K} & \mathcal{E} & \mathcal{E} \\ \mathcal{U} & \mathcal{K} & \mathcal{E} & \mathcal{K} & \mathcal{E} \\ \mathcal{U} & \mathcal{K} & \mathcal{K} & \mathcal{K} & \mathcal{E} \end{array}$$

If $\mathcal{C}_x = \mathcal{L}$, then $\mathcal{C}_{x+y} = \mathcal{C}_{2x} = \mathcal{E}$ and $\mathcal{C}_y \subseteq \mathcal{K}$; this yields two nests:

$$\begin{array}{ccccc} \mathcal{C}_0 & \mathcal{C}_x & \mathcal{C}_y & \mathcal{C}_{x+y} & \mathcal{C}_{2x} \\ \mathcal{U} & \mathcal{L} & \mathcal{E} & \mathcal{E} & \mathcal{E} \\ \mathcal{U} & \mathcal{L} & \mathcal{K} & \mathcal{E} & \mathcal{E} \end{array}$$

If $\mathcal{C}_x = \mathcal{M}$, then $\mathcal{C}_{x+y} = \mathcal{E}$, $\mathcal{C}_y \subseteq \mathcal{K}$, $\mathcal{C}_{x+y} = \mathcal{E}$ (since $\mathcal{C}_{x+y} \subseteq \mathcal{C}_x \cap \mathcal{C}_y$), and $\mathcal{C}_{2x} \subseteq \mathcal{M}$. This yields four nests:

$$\begin{array}{ccccc} \mathcal{C}_0 & \mathcal{C}_x & \mathcal{C}_y & \mathcal{C}_{x+y} & \mathcal{C}_{2x} \\ \mathcal{U} & \mathcal{M} & \mathcal{E} & \mathcal{E} & \mathcal{E} \\ \mathcal{U} & \mathcal{M} & \mathcal{K} & \mathcal{E} & \mathcal{E} \\ \mathcal{U} & \mathcal{M} & \mathcal{E} & \mathcal{E} & \mathcal{M} \\ \mathcal{U} & \mathcal{M} & \mathcal{K} & \mathcal{E} & \mathcal{M} \end{array}$$

If finally $\mathcal{C}_x = \mathcal{U}$, then $\mathcal{C}_{x+y} \subseteq \mathcal{C}_y \subseteq \mathcal{K}$ and $\mathcal{C}_{2x} \subseteq \mathcal{M}$. This yields six nests:

$$\begin{array}{ccccc} \mathcal{C}_0 & \mathcal{C}_x & \mathcal{C}_y & \mathcal{C}_{x+y} & \mathcal{C}_{2x} \\ \mathcal{U} & \mathcal{U} & \mathcal{E} & \mathcal{E} & \mathcal{E} \\ \mathcal{U} & \mathcal{U} & \mathcal{K} & \mathcal{E} & \mathcal{E} \\ \mathcal{U} & \mathcal{U} & \mathcal{K} & \mathcal{K} & \mathcal{E} \\ \mathcal{U} & \mathcal{U} & \mathcal{E} & \mathcal{E} & \mathcal{M} \\ \mathcal{U} & \mathcal{U} & \mathcal{K} & \mathcal{E} & \mathcal{M} \\ \mathcal{U} & \mathcal{U} & \mathcal{K} & \mathcal{K} & \mathcal{M} \end{array}$$

This example has $1+5+2+3+(2+4+2+4+6) = 29$ J-congruences. □

Example 5.3. In this example condition (3) is not trivial. Let $X = \{x,y\}$ and J be the ideal generated by $5x$, $3x+y$, $x+3y$, and $5y$. Then $K = \{k,l,m\}$, where $k = 4y$, $l = 2x+2y$, $m = 4x$. We see that $k-l = l-m$. As in the previous example there are five equivalence relations on K: the equality $\mathcal{E}$; the universal relation $\mathcal{U}$; the equivalence relation $\mathcal{K}$ whose classes are $\{k\}$ and $\{l,m\}$; the equivalence relation $\mathcal{L}$ whose classes are $\{l\}$ and $\{k,m\}$; and the equivalence relation $\mathcal{M}$ whose classes are $\{m\}$ and $\{k,l\}$.

It is readily seen that $k \;\mathcal{D}_t\; l$ if and only if $t \leqq 2y$; $k \;\mathcal{D}_t\; m$ if and only if $t = 0$; and $l \;\mathcal{D}_t\; m$ if and only if $t \leqq 2x$. Hence $N(\mathcal{D})$ is specified by

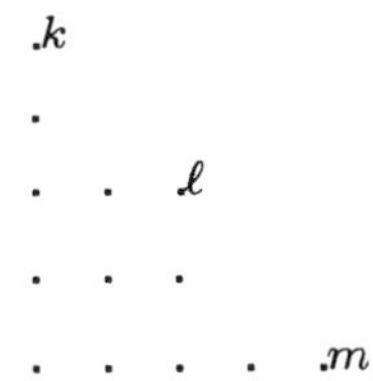

Example 5.3

$$\begin{array}{ccccc} \mathcal{D}_0 & \mathcal{D}_x & \mathcal{D}_{2x} & \mathcal{D}_y & \mathcal{D}_{2y} \\ \mathcal{U} & \mathcal{K} & \mathcal{K} & \mathcal{M} & \mathcal{M} \end{array}$$

with $\mathcal{D}_t = \mathcal{E}$ for all other $t \in H$.

Let $(\mathcal{C}_t)_{t\in H}$ be a coherent nest of equivalence relations on K. Conditions (1) and (2) state that $\mathcal{C}_{2x} \subseteq \mathcal{C}_x \subseteq \mathcal{C}_0$, $\mathcal{C}_x \subseteq \mathcal{K}$, $\mathcal{C}_{2y} \subseteq \mathcal{C}_y \subseteq \mathcal{C}_0$, $\mathcal{C}_y \subseteq \mathcal{M}$, and $\mathcal{C}_t = \mathcal{E}$ for all $t \neq 0,\ x,\ 2x,\ y,\ 2y$.

Since $k - l = l - m$, with $k - k \wedge l = 2y$, $l - k \wedge l = 2x$, condition (3) reads: $k\ \mathcal{C}_{2y+t}\ l$ if and only if $l\ \mathcal{C}_{2x+t}\ m$, for all $t \in H$; equivalently, $\mathcal{C}_{2y} = \mathcal{M}$ if and only if $\mathcal{C}_{2x} = \mathcal{K}$.

As before, the resulting nests can be classified by $\mathcal{C}_0$. If $\mathcal{C}_0 = \mathcal{E}$, then $\mathcal{C}_t = \mathcal{E}$ for all t; $\mathcal{C}$ is the Rees congruence of J.

If $\mathcal{C}_0 = \mathcal{K}$, then $\mathcal{C}_t \subseteq \mathcal{K} \cap \mathcal{D}_t$ for all t; hence $\mathcal{C}_y = \mathcal{C}_{2y} = \mathcal{E}$, $\mathcal{C}_{2x} = \mathcal{E}$ by (3), and $\mathcal{C}_x \subseteq \mathcal{K}$. This yields two nests:

$$\begin{array}{ccccc} \mathcal{C}_0 & \mathcal{C}_x & \mathcal{C}_{2x} & \mathcal{C}_y & \mathcal{C}_{2y} \\ \mathcal{K} & \mathcal{E} & \mathcal{E} & \mathcal{E} & \mathcal{E} \\ \mathcal{K} & \mathcal{K} & \mathcal{E} & \mathcal{E} & \mathcal{E} \end{array}$$

If $\mathcal{C}_0 = \mathcal{L}$, then $\mathcal{C}_t \subseteq \mathcal{K} \cap \mathcal{D}_t$ for all t; hence $\mathcal{C}_t = \mathcal{E}$ for all $t \neq 0$. This yields one nest.

If $\mathcal{C}_0 = \mathcal{M}$, then $\mathcal{C}_x = \mathcal{C}_{2x} = \mathcal{E}$, $\mathcal{C}_{2y} = \mathcal{E}$ by (3), and $\mathcal{C}_y \subseteq \mathcal{M}$. This yields two nests:

$$\begin{array}{ccccc} \mathcal{C}_0 & \mathcal{C}_x & \mathcal{C}_{2x} & \mathcal{C}_y & \mathcal{C}_{2y} \\ \mathcal{M} & \mathcal{E} & \mathcal{E} & \mathcal{E} & \mathcal{E} \\ \mathcal{M} & \mathcal{E} & \mathcal{E} & \mathcal{M} & \mathcal{E} \end{array}$$

Nests with $\mathcal{C}_0 = \mathcal{U}$ can be classified by $\mathcal{C}_x \subseteq \mathcal{K}$ and $\mathcal{C}_y \subseteq \mathcal{M}$. If $\mathcal{C}_x = \mathcal{C}_y = \mathcal{E}$, then $\mathcal{C}_t = \mathcal{E}$ for all $t \neq 0$. If $\mathcal{C}_x = \mathcal{K}$ and $\mathcal{C}_y = \mathcal{E}$, then $\mathcal{C}_{2y} = \mathcal{E}$ and $\mathcal{C}_{2x} = \mathcal{E}$ by (3). If $\mathcal{C}_x = \mathcal{E}$ and $\mathcal{C}_y = \mathcal{M}$, then $\mathcal{C}_{2x} = \mathcal{E}$ and $\mathcal{C}_{2y} = \mathcal{E}$ by

(3). If $\mathcal{C}_x = \mathcal{K}$ and $\mathcal{C}_y = \mathcal{M}$, then, by (3), either $\mathcal{C}_{2x} = \mathcal{C}_{2y} = \mathcal{E}$ or $\mathcal{C}_{2x} = \mathcal{K}$, $\mathcal{C}_{2y} = \mathcal{M}$. This yields five nests:

$\mathcal{C}_0$	$\mathcal{C}_x$	$\mathcal{C}_{2x}$	$\mathcal{C}_y$	$\mathcal{C}_{2y}$
$\mathcal{U}$	$\mathcal{E}$	$\mathcal{E}$	$\mathcal{E}$	$\mathcal{E}$
$\mathcal{U}$	$\mathcal{K}$	$\mathcal{E}$	$\mathcal{M}$	$\mathcal{E}$
$\mathcal{U}$	$\mathcal{E}$	$\mathcal{E}$	$\mathcal{M}$	$\mathcal{E}$
$\mathcal{U}$	$\mathcal{K}$	$\mathcal{E}$	$\mathcal{M}$	$\mathcal{E}$
$\mathcal{U}$	$\mathcal{K}$	$\mathcal{K}$	$\mathcal{M}$	$\mathcal{M}$

This example has $1 + 2 + 1 + 2 + 5 = 11$ J-congruences. □

In each of these examples there is for every pair k, l of corner points some $h \in H$ such that $k \; \mathcal{D}_t \; l$ if and only if $t \leqq h$. This is not true in general but holds whenever F has two generators (Grillet [1991N]).

Chapter X.

GROUP-FREE SEMIGROUPS.

This chapter contains a global construction of complete group-free semigroups, as quotients of free commutative monoids; finite group-free semigroups are a particular case. This construction bypasses the difficulties, noted earlier, in reassembling archimedean components and Ponizovsky factors, and accounts well for the main structural features of these semigroups (idempotents, $\mathcal{H}$-classes, archimedean components, and Ponizovsky factors); its relation to extension groups will be noted in Chapter XIII. It was first obtained by the author in the case of finite congruences [1993], and generalized to complete group-free congruences in [2001C]. The class of partially free complete semigroups arises as an application.

A different construction of finite group-free semigroups (by induction) was devised by John [1973].

For brevity's sake we often abbreviate "complete group-free" as **c.g.f.** and "complete group-free monoid" as **c.g.f.m.** in what follows.

1. SEMILATTICE CONGRUENCES.

In this section we show that semilattice congruences on a free commutative monoid F are determined by certain sets of faces of F. This preliminary result expresses a basic property of semilattice congruences, that their classes run parallel to faces of F. The results, due to the author [2001C], extend to any commutative monoid S but will be used later only when S is free.

1. Let $\mathcal{Y}$ be a semilattice congruence on a free c.m. F. The **direction face** of $a \in F$ is the set

$$D_{\mathcal{Y}}(a) = D(a) = \{t \in F \mid a+t \ \mathcal{Y} \ a\}.$$

Thus $t \in D(a)$ if and only if, in the semilattice $(F/\mathcal{Y}, \wedge)$, $Y_{a+t} = Y_a \wedge Y_t = Y_a$, if and only if $Y_t \geqq Y_a$. The $\mathcal{Y}$-class Y_a of a extends "in the direction of" $D(a)$ in the sense that is contains $a+t$ for every $t \in D(a)$.

Lemma 1.1. *Let $\mathcal{Y}$ be a semilattice congruence on F. Every direction face is a union of $\mathcal{Y}$-classes and a face of F. Moreover, $D_{\mathcal{Y}}(a)$ is the smallest direction face which contains a.*

Proof. We saw that $t \in D(a)$ if and only if $Y_t \geqq Y_a$ in the semilattice $F/\mathcal{Y}$. Hence $D(a)$ is a union of $\mathcal{Y}$-classes; $D(a) \neq \varnothing$, since $a \in D(a)$; and $D(a)$ is a face of F since, in $F/\mathcal{Y}$, $s \wedge t \geqq Y_a$ if and only if $s \geqq Y_a$ and $t \geqq Y_a$. If $a \in D(b)$, then $Y_a \geqq Y_b$, $Y_c \geqq Y_a$ implies $Y_c \geqq Y_b$, and $D(a) \subseteq D(b)$. □

2. When $\mathcal{Y}$ is a semilattice congruence on a free c.m. F, the **direction set** of $\mathcal{Y}$ is the set

$$D(\mathcal{Y}) = \{ D_{\mathcal{Y}}(a) \mid a \in F \}$$

of all the direction faces.

Proposition 1.2. *Let $\mathcal{Y}$ be a semilattice congruence on a free c.m. F. The direction set $D(\mathcal{Y})$, partially ordered by inclusion, is an upper semilattice, in which $D_{\mathcal{Y}}(a) \vee D_{\mathcal{Y}}(b) = D_{\mathcal{Y}}(a+b)$.*

Proof. For every $a, b \in F$ we have $ab \in D(ab)$, $a, b \in D(ab)$, since $D(ab)$ is a face, and $D(a)$, $D(b) \subseteq D(ab)$ by Lemma 1.1. If moreover $D(a)$, $D(b) \subseteq D(c)$, then similarly $a, b \in D(c)$, $ab \in D(c)$, and $D(ab) \subseteq D(c)$. Thus $D(ab) = D(a) \vee D(b)$ in $D(\mathcal{Y})$. □

Proposition 1.3. *A semilattice congruence $\mathcal{Y}$ on a free c.m. F is determined as follows by its direction set $\mathcal{D} = D(\mathcal{Y})$. For every $C \in \mathcal{D}$,*

$$\begin{aligned} C^* &= \{ a \in F \mid D(a) = C \} \\ &= \{ c \in C \mid c \notin B \text{ when } B \in \mathcal{D} \text{ and } B \subsetneqq C \} \end{aligned}$$

is a $\mathcal{Y}$-class; every $\mathcal{Y}$-class can be constructed in this fashion; and $C \longmapsto C^*$ is an isomorphism $(D(\mathcal{Y}), \vee) \cong F/\mathcal{Y}$.

Proof. Let $C^* = \{ c \in C \mid c \notin B$ when $B \in \mathcal{D}$ and $B \subsetneqq C \}$. When $a \in C^*$, then $D(a) \subseteq C$ by Lemma 1.1, and $D(a) \subsetneqq C$ is not possible since $a \in D(a)$; hence $D(a) = C$. If conversely $D(a) = C$, then $a \in C$, and $a \notin B$ when $B \in \mathcal{D}$ and $B \subsetneqq C$, otherwise $C = D(a) \subseteq B$ by Lemma 1.1. Thus $C^* = \{ a \in F \mid D(a) = C \}$.

By Lemma 1.1, $D(a) \subseteq D(b)$ if and only if $a \in D(b)$, if and only if $Y_a \leqq Y_b$ in $F/\mathcal{Y}$. Hence $D(a) = D(b)$ if and only if $Y_a = Y_b$, and C^* is the $\mathcal{Y}$-class of all $a \in F$ such that $D(a) = C$; then $C \longmapsto C^*$ is an isomorphism $(D(\mathcal{Y}), \vee) \cong F/\mathcal{Y}$ by Proposition I.2.4. □

Corollary 1.4. *$C^* \geqq D^*$ in $F/\mathcal{Y}$ if and only if $C \subseteq D$; and $C^* + D^* \subseteq$*

$(C \vee D)^*$.

Proof. If $c \in C^*$ and $d \in D^*$, then $D(c) = C$, $D(d) = D$, $D(c+d) = D(c) \vee D(d) = C \vee D$, and $c + d \in (C \vee D)^*$. □

3. Direction sets of semilattice congruences can be characterized as follows.

Proposition 1.5. *Let F be a free c.m. A set $\mathcal{D}$ of faces of F is the direction set of a semilattice congruence on F if and only if*

(D1) *for every $a \in F$ there is a smallest $C \in \mathcal{D}$ such that $a \in C$;*

(D2) *no element of $\mathcal{D}$ is the union of smaller elements of $\mathcal{D}$.*

Then there is only one semilattice congruence $\mathcal{Y}$ on F whose direction set is $\mathcal{D}$; when $a, b \in F$, then $a \mathcal{Y} b$ if and only if $(\forall C \in \mathcal{D})(a \in C \iff b \in C)$.

Proof. When $\mathcal{D} = D(\mathcal{Y})$, then $D(a)$ is the smallest $C \in \mathcal{D}$ which contains a, by Lemma 1.1, and when $C = D(a)$ then a does not belong to any direction face $B \subsetneqq C$, since C is the smallest direction face which contains a. Thus (D1) and (D2) hold.

Conversely assume that (D1) and (D2) hold. Define $\mathcal{Y}$ as in the statement. By (D1), $a \mathcal{Y} b$ if and only if the smallest $C \in \mathcal{D}$ which contains a coincides with the smallest $C \in \mathcal{D}$ which contains b. Hence $\mathcal{Y}$ is an equivalence relation. If $a \mathcal{Y} b$ and $c \in F$, then, for every $C \in \mathcal{D}$, $a + c \in C$ if and only if $a \in C$ and $c \in C$ (since C is a face), if and only if $b \in C$ and $c \in C$, if and only if $b + c \in C$; hence $a + c \;\mathcal{Y}\; b + c$. Moreover $a \mathcal{Y} 2a$. Thus $\mathcal{Y}$ is a semilattice congruence. We show that $D(\mathcal{Y}) = \mathcal{D}$.

Let $C \in \mathcal{D}$. By (D2) there exists $c \in C \setminus \bigcup (B \in \mathcal{D} \text{ and } B \subsetneqq C)$. By (D1) there is a smallest $D \in \mathcal{D}$ which contains c. Then $D \subseteq C$ and the choice of c rules out $D \subsetneqq C$. Therefore C is the smallest $D \in \mathcal{D}$ which contains c. We show that $C = D_{\mathcal{Y}}(c)$. When $x \in C$, then $x \in C$ and $c + x \notin B$ when $B \in \mathcal{D}$ and $B \subsetneqq C$, since $c \notin B$. Hence C is the smallest $D \in \mathcal{D}$ which contains $c + x$. Therefore $c + x \;\mathcal{Y}\; c$ and $x \in D_{\mathcal{Y}}(c)$. Conversely $x \in D_{\mathcal{Y}}(c)$ implies $c \;\mathcal{Y}\; c + x$, $c + x \in C$, and $x \in C$. Thus $C = D_{\mathcal{Y}}(c) \in D(\mathcal{Y})$.

Conversely let $a \in F$. By (D1) there is a smallest $C \in \mathcal{D}$ which contains a. Then $a \notin B$ when $B \in \mathcal{D}$ and $B \subsetneqq C$. By the above $D_{\mathcal{Y}}(a) = C \in \mathcal{D}$. Thus $D(\mathcal{Y}) = \mathcal{D}$; by Proposition 1.3, $\mathcal{Y}$ is the only semilattice congruence with this property. □

A **direction set** on a free c.m. F is the direction set of a semilattice congruence on F; equivalently, a set of faces of S with properties (D1) and (D2).

Proposition 1.6. *When F is finitely generated, then every direction set on*

F is finite.

Proof. A direction set $\mathcal{D}$ on F is the direction set $\mathcal{D} = D(\mathcal{Y})$ of a semilattice congruence $\mathcal{Y}$ on F; if F is finitely generated, then $F/\mathcal{Y}$ is finitely generated, $F/\mathcal{Y}$ is finite, and $\mathcal{D} \cong F/\mathcal{Y}$ is finite by Proposition 1.3. □

4. Finite direction sets have a simpler characterization.

Proposition 1.7. *Let F be a free c.m. A finite set of faces is a direction set of F if and only if it contains F and is closed under intersections.*

Proof. Let $\mathcal{Y}$ be a semilattice congruence whose direction set $\mathcal{D}$ is finite. For every nonempty subset S of F we show that there is a smallest $C \in \mathcal{D}$ which contains S. Since $\mathcal{D}$ is a finite upper semilattice, the finite set $\{ D(s) \mid s \in S\}$ has a least upper bound D in $\mathcal{D}$. Then $s \in D(s) \subseteq D$ and $S \subseteq D$. If conversely $S \subseteq C \in \mathcal{D}$, then $s \in C$ for all $s \in S$, $D(s) \subseteq C$ by Lemma 1.1, and $D \subseteq C$. Thus D is the smallest $C \in \mathcal{D}$ which contains S.

In particular F is contained in some $C \in \mathcal{D}$ and $F \in \mathcal{D}$. Let $S = \bigcap_{i \in I} D_i$ be an intersection of elements of $\mathcal{D}$. Then $0 \in D_i$ for all i and $0 \in S \neq \emptyset$. By the above there is a smallest $C \in \mathcal{D}$ which contains S. Then C is contained in every D_i and $C = S$. Thus $\mathcal{D}$ is closed under intersections.

Conversely let $\mathcal{D}$ be a finite set of faces of F which contains F and is closed under intersections. For every $a \in F$ the intersection of all $D \in \mathcal{D}$ which contain a is the smallest $C \in \mathcal{D}$ which contain a; thus (D1) holds. Next we show that a face C of F is not the union $C = B_1 \cup \cdots \cup B_n$ of finitely many faces $B_1, \dots, B_n \subsetneq C$: if indeed $b_1 \in C \backslash B_1, \dots, b_n \in C \backslash B_n$, then $b_1 + \cdots + b_n \in C$ but $b_1 + \cdots + b_n \notin B_1, \dots, b_1 + \cdots + b_n \notin B_n$. Therefore no $C \in \mathcal{D}$ is the union of the finitely many faces $B \in \mathcal{D}$ such that $B \subsetneq C$, and (D2) holds. □

5. The following examples shows that not every direction set contains F or is closed under intersections, and that, in general, not every subset of F is contained in a smallest direction face.

Example 1.8. Let $F = F_X$ be an infinitely generated free c.m. The *support* of $a = \sum_{x \in X} a_x x \in F$ is the finite set

$$S(a) = \{x \in X \mid a_x \neq 0\},$$

The smallest semilattice congruence $\mathcal{N}$ on F is given by:

$$a \,\mathcal{N}\, b \text{ if and only if } S(a) = S(b).$$

Then $t \in D(a)$ if and only if $S(a) = S(a+t)$, if and only if $S(t) \subseteq S(a)$; hence $D(a)$ is the face generated (as a submonoid of F) by $S(a)$. Thus the direction faces of $\mathcal{C}$ are the finitely generated faces. The direction set of $\mathcal{N}$ is closed under intersections but is not finite and does not contain F. No infinitely generated face is contained in a direction face, let alone a smallest direction face. □

Example 1.9. Let S be the semilattice $\{e_1, e_2, e_n, \dots, f_1, f_2, f_n, \dots, r_1, r_2, r_n, \dots, s_1, s_2, s_n, \dots, 1\}$ in which: 1 is the greatest element; $e_m = e_{m+1} \wedge r_m$, $f_{n+1} = f_n \wedge s_n$, $e_m \leqq f_n$ for all m, n; but no two of $r_1, r_2, r_n, \dots, s_1, s_2, s_n, \dots$ are comparable. We see that S is not a complete semilattice.

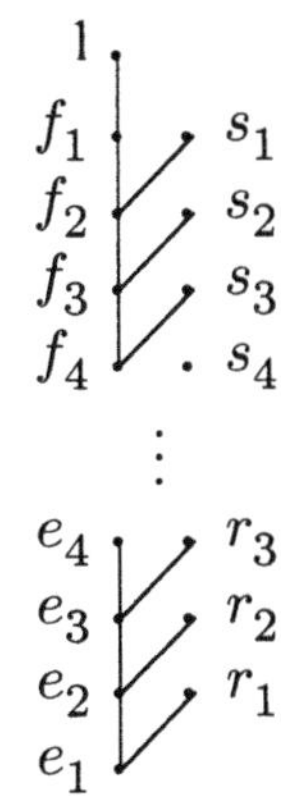

Let F be the free c.m. with an infinite basis $x_1, x_2, x_n, \dots, y_1, y_2, y_n, \dots, z_1, z_2, z_n, \dots, w_1, w_2, w_n, \dots$. Let $\pi : F \longrightarrow S$ be the homomorphism such that $\pi x_n = e_n$, $\pi y_n = f_n$, $\pi z_n = r_n$, $\pi w_n = s_n$ for all n. The congruence $\mathcal{Y}$ induced by π is a semilattice congruence on F, and $F/\mathcal{Y} \cong S$.

In S the upper sections $U(s) = \{u \in S \mid u \geqq s\}$ are: $U(1) = \{1\}$; $U(r_n) = \{r_n, 1\}$; $U(s_n) = \{s_n, 1\}$;

$U(f_n) = \{1, f_1, f_2, \dots, f_n, s_1, s_2, \dots, s_{n-1}\}$; and $U(e_n) = \{1, e_n, e_{n+1}, \dots, f_1, f_2, \dots, f_n, \dots, r_n, r_{n+1}, \dots, s_1, \dots, s_n, \dots\}$.

In F, $t \in D(a)$ if and only if $\pi(t) \geqq \pi(a)$. Hence $D(a) = \pi^{-1}(U(\pi a))$ and the direction faces of $\mathcal{Y}$ are: $D(0) = \{0\}$;

$D(z_n)$ is the face generated by $\{z_n\}$;

$D(w_n)$ is the face generated by $\{w_n\}$;

$D(y_n)$ is the face generated by $\{y_1, y_2, \dots, y_n, w_1, w_2, \dots, w_{n-1}\}$;

$D(x_n)$ is the face generated by $\{x_n, x_{n+1}, \dots, y_1, y_2, \dots, y_n, \dots, z_n, z_{n+1}, \dots, w_1, \dots, w_n, \dots\}$.

These are all the direction faces. Then $\bigcap_{n>0} D(x_n)$ is the face generated by $\{y_1, y_2, \dots, y_n, \dots, w_1, w_2, \dots, w_n, \dots\}$ and is not a direction face; thus the direction set of $\mathcal{Y}$ is not closed under intersections. Also there does not exist a smallest direction face which contains $\bigcap_{n>0} D(x_n)$ or $\{y_1, y_2, \dots, y_n, \dots, w_1, w_2, \dots, w_n, \dots\}$. □

2. DIRECTION SETS.

We now begin in earnest the study of complete group-free congruences on a free commutative monoid F. In Section 1 we saw that semilattice congruences on F have classes that run parallel to faces of F. This property extends to every complete group-free congruence; the resulting faces constitute a direction set as defined in Section 1, and determine the archimedean components of the quotient monoid.

1. First we note some properties of complete group-free monoids. When S is c.g.f., all subgroups of S are trivial; $\mathcal{H}$ is the equality on S; and Green's preorder $\leqq_{\mathcal{H}}$ is a partial order relation on S (Proposition V.2.3), which we denote by $\leqq$.

Proposition 2.1. *Let S be a complete group-free monoid. Every element of S has an idempotent power. If S is finitely generated, then S is finite.*

Proof. Since S is complete, every element of S has a power in a subgroup of S and has an idempotent power. In particular an element of S has only finitely many distinct powers. If S is finitely generated, then every element of S is a product of powers of finitely many generators, and S is finite. □

By Proposition 2.1 we can arrange that S is finite simply by letting F be finitely generated. In this case there is a standard presentation, just as in the case of nilmonoids. Recall that an element a of a monoid S is **irreducible** when $a \neq 1$, and $a = bc$ implies $b = a$ or $c = a$.

Proposition 2.2. *A finite group-free monoid is generated (as a monoid) by its irreducible elements, and therefore has a smallest generating subset.*

Proof. An irreducible element of a monoid S is not the product of other elements of S and must therefore be included in every generating subset of S.

We prove by noetherian induction on a that every element a of a finite group-free monoid S is a (possibly empty) product of irreducible elements of S. If this is not true then S has an element m which is not a product of irreducible elements of S and is maximal (under $\leqq$) with this property. In particular, m is not an empty or one-term product of irreducible elements, so $m \neq 1$ and m is not irreducible. Therefore $m = bc$ for some $b, c \in S$, $b, c \neq m$. Then $b, c > m$; by the choice of m, b and c are products of irreducible elements of S; hence so is $m = bc$, which is the required contradiction. □

By Proposition 2.2, a finite group-free monoid has a **standard presentation** $\pi : F \longrightarrow S$, where F is free on the set X of all irreducible elements of S and π is the identity on X.

Lemma 2.3. *Let S be a complete group-free monoid. For every $s \in S$,*

$$U(s) \;=\; \{u \in S \mid us = s\}$$

is a face of S.

$U(s)$ is the stabilizer of the trivial $\mathcal{H}$-class $\{s\}$.

Proof. $U(s)$ is a submonoid of S since $1s = s$, and $us = vs = s$ implies $uvs = s$. If conversely $uvs = s$, then $s = uvs \leqq us \leqq s$ and $us = s$; similarly $vs = s$. Thus $U(s)$ is a face of S. □

2. In what follows, F is a free commutative monoid; $\mathcal{C}$ is a **complete group-free** congruence on F (a congruence $\mathcal{C}$ on F such that the quotient monoid $F/\mathcal{C}$ is complete group-free; equivalently, the congruence induced by a surjective homomorphism of F onto a c.g.f.m.); S is a c.g.f.m. and $\pi : F \longrightarrow S$ is a surjective homomorphism which induces $\mathcal{C}$.

The direction of the $\mathcal{C}$-class $C(a)$ of $a \in F$ is measured as in Section 1 by its **direction face**, which is the set

$$D_{\mathcal{C}}(a) \;=\; D(a) \;=\; \{t \in F \mid a + t \;\mathcal{C}\; a\};$$

$C(a)$ extends "in the direction of" $D(a)$ in the sense that is contains $a + t$ for every $t \in D(a)$. Since $\mathcal{C} = \ker \pi$, we have $t \in D(a)$ if and only if $\pi(t) \in U(\pi(a))$.

Lemma 2.4. *$D_{\mathcal{C}}(a)$ is a face of F; moreover, $a \;\mathcal{C}\; b$ implies $D_{\mathcal{C}}(a) = D_{\mathcal{C}}(b)$, and $a \leqq b$ implies $D_{\mathcal{C}}(a) \subseteq D_{\mathcal{C}}(b)$.*

Proof. $D(a) = \pi^{-1}\big(U(\pi(a))\big)$ is a face of F since $U(\pi(a))$ is a face of S. If $a \;\mathcal{C}\; b$, then $a + t \;\mathcal{C}\; b + t$ and $a + t \;\mathcal{C}\; a$ if and only if $b + t \;\mathcal{C}\; b$; thus $D(a) = D(b)$. If $a \leqq b$, then $b = a + c$, and $a + t \;\mathcal{C}\; a$ implies $b + t = a + t + c \;\mathcal{C}\; a + c = b$; thus $D(a) \subseteq D(b)$. □

3. By Proposition I.2.6, the smallest semilattice congruence $\mathcal{N}$ on S induces a semilattice congruence $\mathcal{Y} = \pi^{-1}(\mathcal{N})$ on F, which is the smallest semilattice congruence which contains $\mathcal{C}$. Moreover, $F/\mathcal{Y} \cong S/\mathcal{N}$; the isomorphism sends the $\mathcal{Y}$-class Y_a of $a \in F$ to the archimedean component $A(\pi(a))$ of $\pi(a)$.

The main result of this section is that $\mathcal{C}$ and $\mathcal{Y}$ have the same set of direction faces.

Lemma 2.5. $D_{\mathcal{Y}}(a) = \{t \in F \mid \pi(t) \geqq_{\mathcal{N}} \pi(a)\} = \{t \in F \mid na \ \mathcal{C} \ t+u$ *for some* $n > 0$ *and* $u \in F\}$.

Proof. $t \in D_{\mathcal{Y}}(a)$ is equivalent to: $Y_t \geqq Y_a$ in $F/\mathcal{Y}$; $A(\pi(t)) \geqq A(\pi(a))$ in $S/\mathcal{N}$; $\pi(t) \geqq_{\mathcal{N}} \pi(a)$ in S; $\pi(a)^n = \pi(t)\, s$ for some $n > 0$ and some $s \in S$; and $na \ \mathcal{C} \ t+u$ for some $n > 0$ and some $u \in F$. □

By Lemma 2.5, $D_{\mathcal{C}}(a) \subseteq D_{\mathcal{Y}}(a)$. This may be a strict inclusion.

Example 2.6. Let $F = \mathbb{N}$ be free on one generator and $\mathcal{C}$ be the Rees congruence of the ideal $J = \{m,\, m+1,\, \dots\}$, where $m > 1$. Then $F/\mathcal{C}$ is a cyclic nilmonoid and is complete group-free. For every $t \in F$ we have $n1 = t+u$ for some $n = t+1 > 0$ and $u = 1 \in F$; hence $D_{\mathcal{Y}}(1) = F$. On the other hand, $\{1\}$ is a $\mathcal{C}$-class, since $m > 1$, and $D_{\mathcal{C}}(1) = \{0\}$. □

Nevertheless $\mathcal{C}$ and $\mathcal{Y}$ have the same direction faces. This follows from certain properties of c.g.f.m.

Lemma 2.7. *Let* S *be a c.g.f.m. If* $e \in S$ *is idempotent, then* $e \leqq_{\mathcal{N}} s$ *if and only if* $es = e$, *if and only if* $s \in U(e)$; *if* $e = \epsilon(s)$, *then* $U(s) = U(e)$.

Proof. If $e \leqq_{\mathcal{N}} s$, then $e = e^n = st \leqq_{\mathcal{H}} s$, $e = e^2 \leqq_{\mathcal{H}} es \leqq_{\mathcal{H}} e$, $es \ \mathcal{H} \ e$, $es = e$ since $\mathcal{H}$ is the equality on S, and $s \in U(e)$; conversely $se = e$ implies $e = es \leqq_{\mathcal{N}} s$.

Now let $e = \epsilon(s)$. Then $es = s$, and $ue = e$ implies $us = ues = es = s$. Conversely assume $us = s$. Since S is complete group-free, $u^n = f$ is idempotent for some $n > 0$, and then $f \leqq_{\mathcal{H}} u$ and $fu = f$ by the above; also $fs = u^n s = s$, $e \leqq f$, and $uf = f$ yields $ue = e$. Thus $U(s) = U(e)$. □

Proposition 2.8. *When* $\mathcal{C}$ *is a c.g.f. congruence on* F, *then* $D_{\mathcal{C}}(a) = D_{\mathcal{Y}}(a)$ *whenever* $\pi(a)$ *is idempotent, and* $\{D_{\mathcal{C}}(a) \mid a \in F\} = \{D_{\mathcal{Y}}(a) \mid a \in F\}$.

Proof. When $\pi(a) = e$ is idempotent, then $s \geqq_{\mathcal{N}} e$ if and only if $s \in U(e)$, by Lemma 2.7. Now $t \in \mathcal{D}_{\mathcal{C}}(a)$ if and only if $\pi(t) \in U(e)$, and $t \in \mathcal{D}_{\mathcal{Y}}(a)$ if and only if $a+t \ \mathcal{Y} \ a$, if and only if $\pi(a)\,\pi(t) \ \mathcal{N} \ \pi(a)$, if and only if $\pi(t) \leqq_{\mathcal{N}} e$; hence $D_{\mathcal{C}}(a) = D_{\mathcal{Y}}(a)$.

Now take any $a \in F$. Since S is c.g.f., the archimedean component of $\pi(a)$ contains an idempotent $\pi(c)$. Then $a \ \mathcal{Y} \ c$. By Lemma 2.4 and the above, $D_{\mathcal{Y}}(a) = D_{\mathcal{Y}}(c) = D_{\mathcal{C}}(c)$. There also exists $\pi(d) = \epsilon(\pi(a))$; then $U(\pi(a)) = U(\pi(d))$ by Lemma 2.7, and the above yields $D_{\mathcal{C}}(a) = D_{\mathcal{C}}(d) = D_{\mathcal{Y}}(d)$. Thus

$\{ D_{\mathcal{C}}(a) \mid a \in F \} = \{ D_{\mathcal{Y}}(a) \mid a \in F \}$. □

The proof of Proposition 2.8 shows that $D_{\mathcal{Y}}(a)$ depends only on the archimedean component of $\pi(a)$, whereas $D_{\mathcal{C}}(a)$ depends only on $\epsilon(\pi(a))$ and on the Ponizovsky factor of $\pi(a)$. In either case the direction faces are the direction faces of idempotents.

4. The **direction set** of a complete group-free congruence $\mathcal{C}$ on F is the set $D(\mathcal{C}) = \{ D_{\mathcal{C}}(a) \mid a \in F \}$ of its direction faces. By Proposition 2.8, $D(\mathcal{C})$ is also the direction set of the semilattice congruence $\mathcal{Y} = \pi^{-1}(\mathcal{N})$ and is a direction set as defined in Section 1. Propositions 1.2 and 1.3 yield:

Proposition 2.9. *Let $\mathcal{C}$ be the c.g.f. congruence on F induced by a surjective homomorphism $\pi : F \longrightarrow S$. Then $\mathcal{D} = D(\mathcal{C})$ is a direction set and $\mathcal{D}$, partially ordered by inclusion, is an upper semilattice. For every $C \in \mathcal{D}$ let*

$$C^* = \{ c \in C \mid c \notin B \text{ when } B \in \mathcal{D} \text{ and } B \subsetneqq C \};$$

Then $C^* = \{ a \in F \mid D_{\mathcal{Y}}(a) = C \}$; C^* is a union of $\mathcal{C}$-classes; $\pi(C^*)$ is an archimedean component of S; every archimedean component of S can be constructed in this fashion; and $C \longmapsto \pi(C^*)$ is an isomorphism $(\mathcal{D}, \vee) \cong S/\mathcal{N}$.

Corollary 2.10. *Let $A \in \mathcal{D}$ and $a \in A^*$. Then $\pi(na)$ is idempotent for some $n > 0$, and then $na \in A^*$ and $D(na) = A$.*

Proof. By Proposition 2.9, $D_{\mathcal{Y}}(a) = A$ and $na \in A^*$ for all $n > 0$, since $\pi(A^*)$ is an archimedean component of S. Also $\pi(na)$ is idempotent for some $n > 0$, since S is c.g.f.; then $D_{\mathcal{C}}(na) = D_{\mathcal{Y}}(na) = A$ by Proposition 2.8. □

5. We complete this section with two examples, which will be seen again in later sections. When $X = \{x, y\}$ we represent every element $a = a_x x + a_y y$ of F by a point (a_x, a_y) with nonnegative integer coordinates. F has four faces: $\{0\}$, $A = F_{\{x\}} = \{ a \in F \mid a_y = 0 \}$, $B = F_{\{y\}} = \{ a \in F \mid a_x = 0 \}$, and F. Straight lines in the figures indicate the partition of F into $\mathcal{C}$-classes.

Example A. As shown $\mathcal{C}$ is the nilmonoid congruence with identity class $B = \{ a \in F \mid a_x = 0 \}$, zero class $\{ a \in F \mid a_x \geqq 3 \}$, and two other classes $\{ a \in F \mid a_x = 1 \}$ and $\{ a \in F \mid a_x = 2 \}$. Then S is the cyclic nilmonoid $\langle x \mid x^3 = 0 \rangle$. We see that $D(0) = D(x) = D(2x) = B$ and $D(3x) = F$. Thus $D(\mathcal{C}) = \{B, F\}$. □

Example B. This example is more typical. The congruence $\mathcal{C}$ is shown below. Slanted lines indicate two more $\mathcal{C}$-classes, $\{ 2x + 6y,\ 5x + 4y \}$ and $\{ 3x + 6y,\ 6x + 4y \}$. We see that $D(0) = \{0\} = D(\{ 2x + 6y,\ 5x + 4y \})$,

$D(9x) = D(7x+2y) = F_{\{x\}} = A$, and $D(7y) = D(4x+5y) = D(7x+4y) = F$. Thus $D(\mathcal{C}) = \{0, A, F\}$. The archimedean components of $F/\mathcal{C}$ in Example B are shown by the thicker lines below.

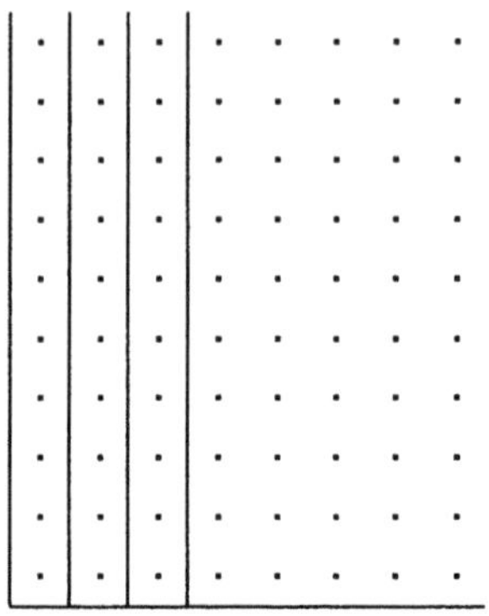

Example A

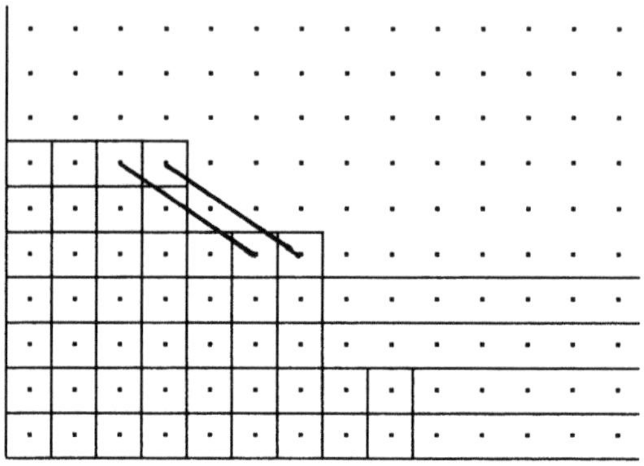

Example B

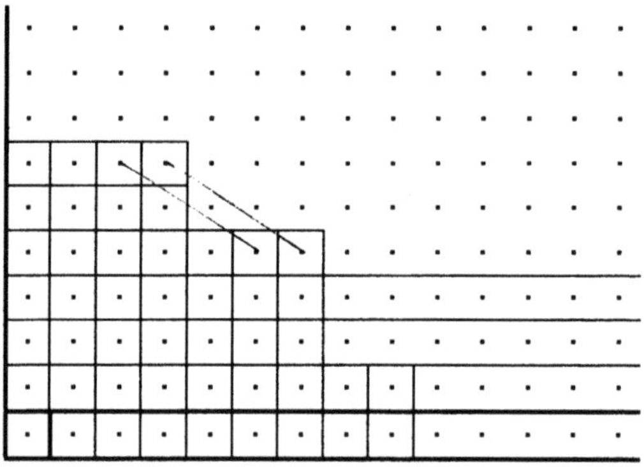

Example B: archimedean components

3. EXTENT CELLS.

In this section we construct the extent cells of a complete group-free congruence and show how they determine the idempotents and Ponizovsky factors of the quotient monoid.

1. As before F is a free commutative monoid and $\mathcal{C}$ is the complete group-free congruence on F induced by a surjective homomorphism $\pi : F \longrightarrow S$, where S is a complete group-free monoid; $\mathcal{D} = D(\mathcal{C})$ is the direction set of $\mathcal{C}$.

The **extent cells** of $\mathcal{C}$ are the sets

$$E_A = \{a \in F \mid D_{\mathcal{C}}(a) = A\},$$

one for every $A \in \mathcal{D}$. The family $\mathcal{E} = (E_A)_{A \in \mathcal{D}}$ is the **extent cell family** of $\mathcal{C}$; it is a partition of F. (The similar sets $A^* = \{a \in F \mid D_{\mathcal{Y}}(a) = A\}$ yield the archimedean components of S in Proposition 2.9.)

The extent cells in Examples A and B (introduced in Section 2) are shown by thick lines:

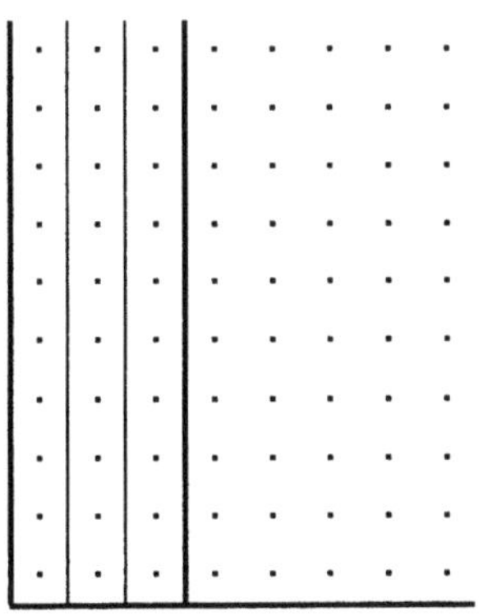

Example A: extent cells

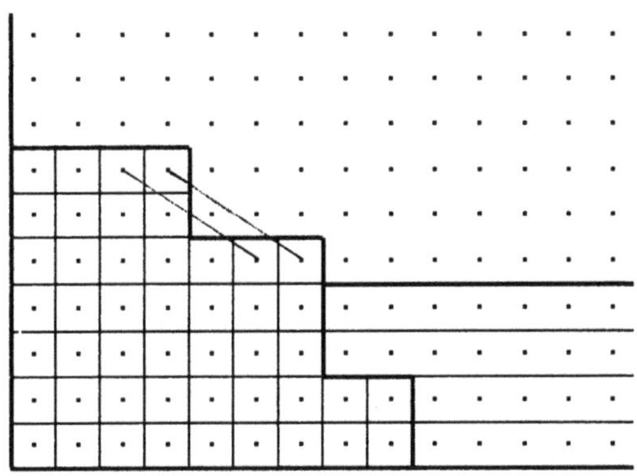

Example B: extent cells

2. The extent cells of $\mathcal{C}$ determine the idempotents of S:

Proposition 3.1. *For every $A \in D(\mathcal{C})$, $I_A = A \cap E_A \subseteq A^*$ is a $\mathcal{C}$-class and an ideal of A; $e_A = \pi(I_A)$ is idempotent; this yields every idempotent of S.*

Proof. For any $a \in A^*$, Corollary 2.10 shows that $na \in A^*$ and $D(na) = A$ for some $n > 0$. Thus $I_A \neq \emptyset$, in fact I_A intersects A^*.

If $a, b \in I_A = A \cap E_A$, then $a, b \in A = D(a) = D(b)$ and $a + b \;\mathcal{C}\; a$, $a + b \;\mathcal{C}\; b$. Hence I_A is a $\mathcal{C}$-class; then $I_A \subseteq A^*$, since A^* is a union of $\mathcal{C}$-classes. If similarly $a \in I_A$ and $c \in A$, then $c \in D(a)$, $a + c \;\mathcal{C}\; a$, $D(a + c) = A$ by Lemma 2.4, and $a + c \in I_A$; thus I_A is an ideal of A.

Let $a \in F$. Then $a \in E_A$ for some $A = D(a) \in \mathcal{D}$, and $\pi(a)$ is idempotent if and only if $a + a \;\mathcal{C}\; a$, if and only if $a \in D(a) = A$. Thus $\pi(a)$ is idempotent if and only if $a \in I_A$ for some $A \in \mathcal{D}$. Hence $\pi(I_A)$ is idempotent, and every idempotent of S is obtained in this fashion. □

Corollary 3.2. *$\pi(a) \geqq_{\mathcal{H}} e_A$ if and only if $a \in A$.*

Proof. If $a \in A$, then $a+c \in I_A$ for any $c \in I_A$ and $\pi(a) \geqq_{\mathcal{H}} \pi(a+c) = e_A$. If conversely $\pi(a) \geqq_{\mathcal{H}} e_A$, then $a+t \in I_A \subseteq A$ for some $t \in F$ and $a \in A$. □

Proposition 3.3. *Let $\mathcal{C}$ be the c.g.f. congruence on F induced by a surjective homomorphism $\pi : F \longrightarrow S$, and $\mathcal{D}$ be the direction set of $\mathcal{C}$. Then $e_A \leqq e_B$ in $E(S)$ if and only if $B \subseteq A$ in $\mathcal{D}$; hence $E(S) \cong (\mathcal{D}, \vee)$.*

Proof. Let $a \in I_A$, $b \in I_B$, so that $\pi(a) = e_A$, $\pi(b) = e_B$. If $e_A \leqq e_B$, then $e_A e_B = e_A$ and $a+b \;\mathcal{C}\; a$; hence $b+t \;\mathcal{C}\; b$ implies $a+t \;\mathcal{C}\; a+b+t \;\mathcal{C}\; a+b \;\mathcal{C}\; a$, and $B = D(b) \subseteq D(a) = A$. If conversely $B \subseteq A$, then $b \in A$, $a+b \in I_A$ since I_A is an ideal of A, and $e_A e_B = \pi(a+b) = \pi(a) = e_A$. □

The isomorphism $E(S) \cong (\mathcal{D}, \vee)$ also follows from Proposition 2.9, since S is complete and $E(S) \cong Y(S)$.

As noted after Proposition 2.8, the extent cells of $\mathcal{C}$ also determine the Ponizovsky factors of S:

Proposition 3.4. *Let $\mathcal{C}$ be the c.g.f. congruence on F induced by a surjective homomorphism $\pi : F \longrightarrow S$. Then $a \in E_A$ if and only if $e_A = \epsilon(\pi(a))$; E_A is a union of $\mathcal{C}$-classes; $\pi(E_A)$ is the partial Ponizovsky factor of e_A in S.*

Proof. Let $a \in F$ and $c \in I_A$, so that $\pi(c) = e_A$. If $e_A = \epsilon(\pi(a))$, then $U(\pi(c)) = U(\pi(a))$ by Lemma 2.7, $D(c) = D(a) = A$, and $a \in E_A$. Conversely let $a \in E_A$. Then $a+c \;\mathcal{C}\; a$, since $c \in A = D(a)$, and $e_A \pi(a) = \pi(a)$. If $e_B \pi(a) = \pi(a)$, then $e_B = \pi(b)$ for some $b \in I_B$, $a+b \;\mathcal{C}\; a$, $D(b) \subseteq D(a+b) = D(a)$ by Lemma 2.4, $B \subseteq A$, and $e_A \leqq e_B$ by Proposition 3.3. Thus $e_A = \epsilon(\pi(a))$. Thus $a \in E_A$ if and only if $e_A = \epsilon(\pi(a))$. □

In Example B the partial Ponizovsky factors of $F/\mathcal{C}$ are shown below by the thicker lines; the idempotent classes are clear.

3. Extent cells also enjoy the following properties:

Lemma 3.5. *Let $\mathcal{C}$ be a c.g.f. congruence on F with direction set $\mathcal{D}$. The extent cells of $\mathcal{C}$ constitute a partition of F with the following properties:*

(E1) *if $a \in E_A$, $b \in E_B$, and $a \leqq b$, then $A \subseteq B$;*

(E2) $(\forall a \in A^*)(\exists n > 0)\; na \in E_A$*;*

(E3) *for every $A \in \mathcal{D}$, the projections $p'_A E_B$ with $A \subseteq B \in \mathcal{D}$ form a*

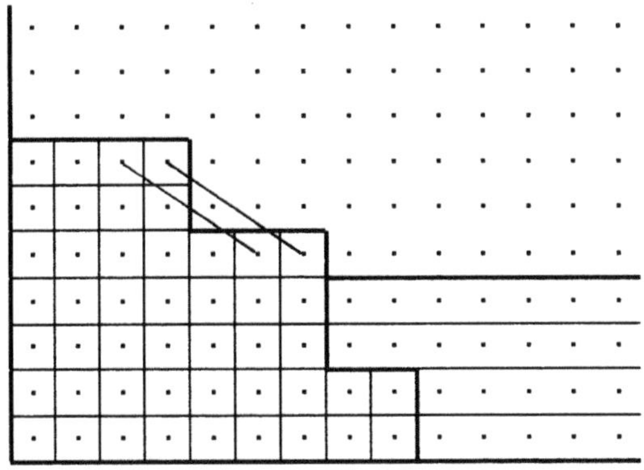

Example B: partial Ponizovsky factors

partition of A';

(E4) *every* E_A *is a union of* $\mathcal{C}$*-classes;*

(E5) *if* $t \in A$, $a, b \in E_A$, *and* $a + t \ \mathcal{C} \ b + t$, *then* $a \ \mathcal{C} \ b$;

(E6) *if* $a, b \in E_A$ *and* $p'_A a = p'_A b$, *then* $a \ \mathcal{C} \ b$.

(E7) $E_A + A \subseteq E_A$.

Proof. This can be deduced from Proposition 3.4, as in the next chapter, but we give a direct proof.

(E1). If $a \in E_A$, $b \in E_B$, and $a \leqq b$, then $A = D(a) \subseteq D(b) = B$ by Lemma 2.6.

(E2) follows from Corollary 2.10.

(E3). Let $t \in A'$. For any $a \in A^*$ we have $na \in E_A$ for some $n > 0$ by (E2) and $na \in A$. Let $c = na + t$ and $C = D(c)$. Then $p'_A c = t$ and $A \subseteq C$ by (E1) since $c \geqq na$ and $D(na) = A$. Thus $t \in p'_A E_C$ and the projections $p'_A E_B$ with $A \subseteq B \in \mathcal{D}$ cover A'.

Assume $t \in p'_A E_B$, where $A \subseteq B$, so that $t = p'_A b$ for some $b \in E_B$. Now $b = p_A b + t$, $c = na + t$ with $p_A b \in A \subseteq C = D(c)$, $na \in A \subseteq B = D(b)$; hence $b \ \mathcal{C} \ b + na = c + p_A b \ \mathcal{C} \ c$ and $B = D(b) = D(c) = C$ by Lemma 2.4. Thus and the projections $p'_A E_B$ with $A \subseteq B \in \mathcal{D}$ are pairwise disjoint.

(E4). If $a \ \mathcal{C} \ b \in E_A$, then $D(a) = D(b) = A$ by Lemma 2.4 and $a \in E_A$.

(E5). If $t \in A$, $a, b \in E_A$, and $a + t \ \mathcal{C} \ b + t$, then $A = D(a) = D(b)$ and $a \ \mathcal{C} \ a + t \ \mathcal{C} \ b + t \ \mathcal{C} \ b$.

(E6). If $a, b \in E_A$ and $p'_A a = p'_A b$, then $p_A a,\ p_A b \in A = D(a) = D(b)$ and $a \ \mathcal{C} \ a + p'_A b = b + p'_A a \ \mathcal{C} \ b$.

(E7). If $a \in E_A$ and $t \in A$, then $D(a) = A$, $a + t \ \mathcal{C} \ a$, and $a + t \in E_A$ by (E4). □

If F is f.g., then every face is contained in a direction face by Proposition 1.7 and (E3) can be strengthened as follows:

Lemma 3.6. *If F is f.g., then, for every face A of F, the projections $p'_A E_B$ with $A \subseteq B \in \mathcal{D}$ form a partition of A'.*

Proof. Let A be a face of F. For every $x \in X$, $\pi(x)^{k_x}$ is idempotent for some $k_x > 0$, and then $\pi(x)^{k_x+1} = \pi(x)^{k_x}$ by Lemma 2.7. Then $d = \sum_{x \in X \cap A} k_x x \in A$ satisfies $d \ \mathcal{C} \ 2d$. Also $d + x \ \mathcal{C} \ d$ for every $x \in X \cap A$, since $(k_x + 1)x \ \mathcal{C} \ k_x x$; hence $A \subseteq D(d)$. In fact $D(d)$ is the smallest $B \in \mathcal{D}$ which contains A: if $A \subseteq D(b)$, then $b + d \ \mathcal{C} \ b$ and $D(d) \subseteq D(b+d) = D(b)$ by Lemma 2.4.

For every $t \in A'$ we now have $a = d + t \in E_B$ for some $B \in \mathcal{D}$, $A \subseteq D(d) \subseteq D(d+t) = B$ by Lemma 2.4, and $t = p'_A(d+t) \in p'_A E_B$. Thus the projections $p'_A E_B$ with $A \subseteq B \in \mathcal{D}$ cover A'.

Assume $t \in p'_A E_C$, where $A \subseteq C \in \mathcal{D}$. As above, $d + t \in E_B$, where $A \subseteq B \in \mathcal{D}$. Also $t = p'_A c$ for some $c \in E_C$. Since $d \in A \subseteq D(c)$ and $c = p_A c + p'_A c = p_A c + t$ with $p_A c \in A \subseteq D(d+t)$, we have

$$d + t \ \mathcal{C} \ d + t + p_A c \ = \ d + c \ \mathcal{C} \ c$$

and $C = D(c) = D(d+t) = B$. Thus t belongs to only one of the sets $p'_A E_B$ with $A \subseteq B \in \mathcal{D}$. □

4. Properties (E1) (E2) (E3) characterize extent cell families.

Theorem 3.7. *Let F be a free commutative monoid and $\mathcal{D}$ be a direction set on F. A partition $\mathcal{E} = (E_A)_{A \in \mathcal{D}}$ of F is the extent cell family of a complete group-free congruence on F with direction set $\mathcal{D}$ if and only if it has properties* (E1), (E2), *and* (E3); *the smallest such congruence $\mathcal{S}$ is given by*

$$a \ \mathcal{S} \ b \text{ if and only if } a, b \in E_A \text{ and } p'_A a = p'_A b \text{ for some } A \in \mathcal{D}.$$

Proof. Conditions (E1), (E2), and (E3) are necessary by Lemma 3.5. Now assume that (E1), (E2), and (E3) hold. Define $\mathcal{S}$ as in the statement; $\mathcal{S}$ is an equivalence relation since $\mathcal{E}$ is a partition of F.

Assume $a \ \mathcal{S} \ b$, so that $a, b \in E_A$ and $p'_A a = p'_A b$ for some $A \in \mathcal{D}$, and let

$c \in F$. Then $a + c \in E_B$, $b + c \in E_C$ for some $B, C \in \mathcal{D}$, with $A \subseteq B, C$ by (E1). Moreover $p'_A(a + c) = p'_A(b + c)$; hence $B = C$ by (E3). Then $a + c$, $b + c \in E_B$ and

$$p'_B(a + c) \;=\; p'_B(p'_A(a + c)) \;=\; p'_B(p'_A(b + c)) \;=\; p'_B(b + c)$$

since $A \subseteq B$, and $a + c \;\mathcal{S}\; b + c$. Thus $\mathcal{S}$ is a congruence. To show that $\mathcal{S}$ is a complete group-free congruence we retrace the proof of Proposition 3.1 to find the idempotents of $S = F/\mathcal{S}$.

Let $I_A = A \cap E_A$. Then $I_A \neq \varnothing$ by (E2). If $a \in I_A$ and $t \in A$, then $a + t \in A$; also $a + t \in E_B$ for some $A \subseteq B \in \mathcal{D}$ by (E1), and $B = A$ by (E3) since $p'_A(a + t) = p'_A a \in p'_A E_A$. Thus I_A is an ideal of A. In particular I_A is a subsemigroup of F.

Moreover I_A is the $\mathcal{S}$-class of all $a \in E_A$ with $p'_A a = 0$. Therefore I_A is an idempotent of S. This yields every idempotent of S: if indeed an $\mathcal{S}$-class S_a is idempotent, then a, $2a \in E_A$ and $p'_A a = p'_A(2a)$ for some $A \in \mathcal{D}$, and then $p'_A a = 0$, $a \in A \cap E_A$, and $S_a = I_A$.

Property (E2) then shows that every element of S has an idempotent power; hence every archimedean component of S contains an idempotent.

Let $a \in E_A$. If $c \in I_A$, then as above $a + c \in E_B$ for some $A \subseteq B \in \mathcal{D}$ by (E1) and $B = A$ by (E3) since $p'_A(a + c) = p'_A a \in p'_A E_A$; hence $a + c \;\mathcal{S}\; a$. Conversely assume $a + d \;\mathcal{S}\; a$, where $d \in I_B$. Then $a + d \in E_A$ and $p'_A(a + d) = p'_A a$; hence $d \in A$ and $c + d \in I_A$, since I_A is an ideal of A. Therefore $I_A \leqq I_B$ in $F/\mathcal{S}$. (Note that $B \subseteq A$ by (E1).) Thus there is for every element s of S a least idempotent e of S such that $es = s$; namely, I_A, when $s = S_a$ with $a \in E_A$. Hence S is complete.

Let $e = I_C$ be an idempotent of S and $S_a \in H_e$, where $a \in E_A$. Let $c \in I_C$. Since $S_a \in H_e$ we have $a + c \;\mathcal{S}\; a$ and $a + b \;\mathcal{S}\; c$ for some $b \in F$. Then $c \leqq a + c \in E_A$ and $a \leqq a + b \in E_C$; by (E1), $C \subseteq A \subseteq C$ and $A = C$. Hence $a + b \;\mathcal{S}\; c$ yields $p'_A a + p'_A b = p'_A c = 0$, $p'_A a = 0$, $a \in A \cap E_A$, and $S_a = I_A = e$. Thus S is group-free.

Let $a \in E_A$. If $a + t \;\mathcal{S}\; a$, then $a + t \in E_A$, $p'_A a + p'_A t = p'_A a$, $p'_A t = 0$, and $t \in A$. If conversely $t \in A$, then $a + t \in E_B$ for some $A \subseteq B \in \mathcal{D}$ by (E1), in fact $B = A$ by (E3) since $p'_A(a + t) = p'_A a \in p'_A E_A$; hence $a + t \;\mathcal{S}\; a$. Thus $D_{\mathcal{S}}(a) = A$ for all $a \in E_A$. Therefore $\mathcal{D}$ is the direction set of $\mathcal{S}$ and $\mathcal{E}$ is the extent cell family of $\mathcal{S}$. By (E6), $\mathcal{S}$ is the smallest complete group-free congruence with this property. □

When $\mathcal{D}$ is a direction set on F we call $\mathcal{E} = (E_A)_{A\in\mathcal{D}}$ an **extent cell family** over $\mathcal{D}$ when it is a partition of F with properties (E1), (E2), and (E3); equivalently, when it is the extent cell family of a c.g.f. congruence on F.

5. We illustrate properties (E1), (E2), (E3) with an example.

Example 3.8. Let F be free on $X = \{x, y\}$. F has four faces: 0, $A = F_{\{x\}}$, $B = F_{\{y\}}$, and F.

By Proposition 1.7, a direction set $\mathcal{D}$ on F contains F and is closed under intersections. In particular $\mathcal{D}$ has a least element C. Then $0 \in E_C$ by (E1). If $c \in C$, then $p'_C c = 0 \in p'_C E_C$ and $c \in E_C$ by (E3); hence $C \subseteq E_C$. Then C is a $\mathcal{C}$-class, by (E6), and is the $\mathcal{C}$-class of the identity element.

Let $\mathcal{C}$ be a c.g.f. congruence on F with direction set $\mathcal{D}$. If $\mathcal{D} = \{F\}$, then $\mathcal{C}$ has only one class, by (E6). If $\mathcal{D} = \{C, F\}$, where $C = 0$, A, or B, then E_F is an ideal of F by (E7), and every $a \in F\backslash C = F^*$ has an integer multiple in E_F. Therefore $\mathcal{C}$ is a nilmonoid congruence with identity class C and zero class E_F.

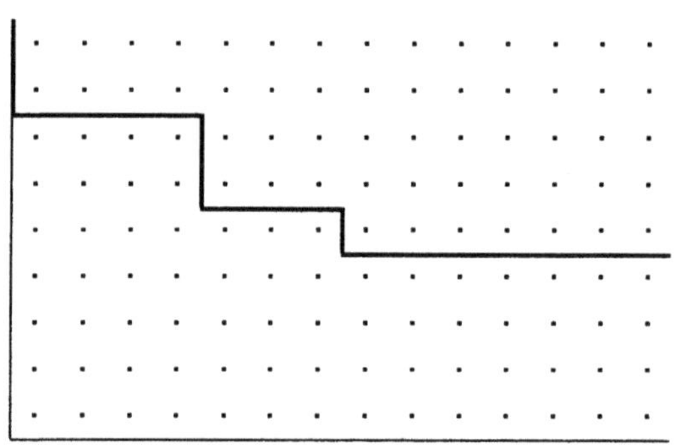

Example 3.8: E_F

Let $\mathcal{D} = \{0, A, F\}$ (the case $\mathcal{D} = \{0, B, F\}$ is similar). Since F is the only $D \in \mathcal{D}$ which contains B, it follows from Lemma 3.6 that $\{p'_B E_F\}$ is a partition of B', i.e. $p_A E_F = A$. Thus E_F contains a multiple of y and extends all the way to B. On the other hand, E_F does not extend all the way to A: indeed $\{p'_A E_A, p'_A E_F\}$ is a partition of $A' = B$ and $p'_A E_F$ is a proper ideal of B. Let $p'_A E_A = \{0, y, \dots, qy\}$. Then $a_y \geqq q+1$ for all $a = a_x x + a_y y \in E_F$, and $a_y = q+1$ for some $a \in E_F$. The first figure shows a typical shape of E_F.

Next, $E_A \cup E_F$ is an ideal of F by (E1), and then $E_0 = F\backslash(E_A \cup E_F)$.

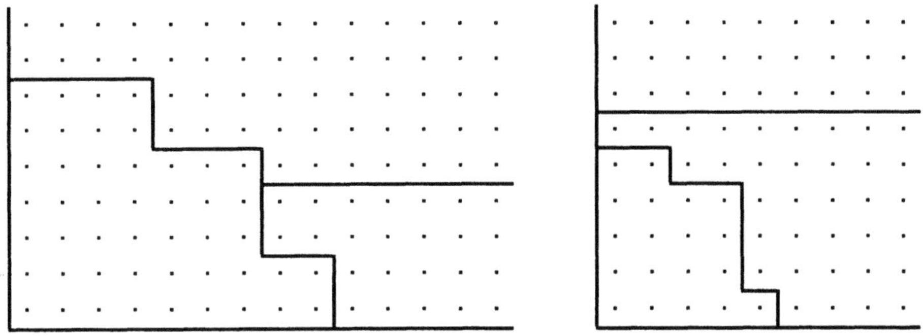

Example 3.8: E_F, E_A, and E_0

Now $a_y \leqq q$ for all $a \in E_A$, since $p'_A E_A = \{0, y, \ldots, qy\}$. Let $p \geqq 0$ be least such that $px + (q+1)y \in E_F$. Then $a_x \geqq p$ for all $a \in E_A$: if indeed $a_x < p$ for some $a \in E_A$, then $b = a_x x + (q+1)y \geqq a$ since $a_y \leqq q$, but $b \notin E_A$, $b \notin E_F$, contradicting (E1). The second figure shows typical shapes of E_F, E_A, and E_0.

The third figure then shows the $\mathcal{S}$-classes.

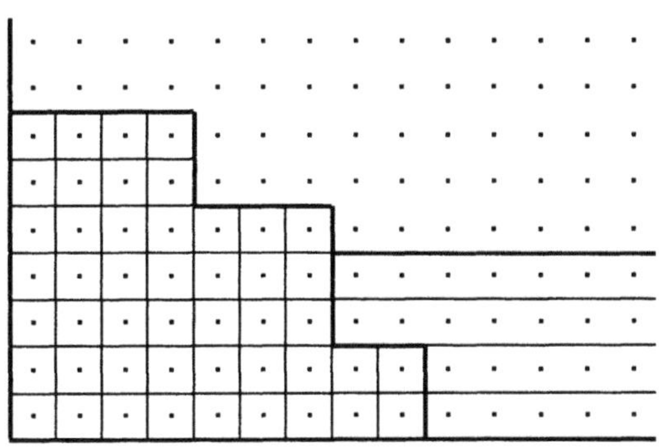

Example 3.8: $\mathcal{S}$-classes

Finally let $\mathcal{D} = \{0, A, B, F\}$. Then E_F is an ideal of F but does not extend all the way to A or B. As above let $p'_A E_A = \{0, y, \ldots, qy\}$ and $p'_B E_B = \{0, x, \ldots, rx\}$. Then $a_x \geqq r+1$ and $a_y \geqq q+1$ for every $a \in E_F$, and $a_x = r+1$, $a_y = q+1$ occur. Let $p \geqq 0$ and $s \geqq 0$ be least such that $px + (q+1)y \in E_F$ and $(r+1)x + sy \in E_F$. Then $a_x \geqq p$ for all $a \in E_A$, and $a_y \geqq s$ for all $a \in E_B$. Our last figures show typical shapes of E_F, E_A,

E_B, and E_0.

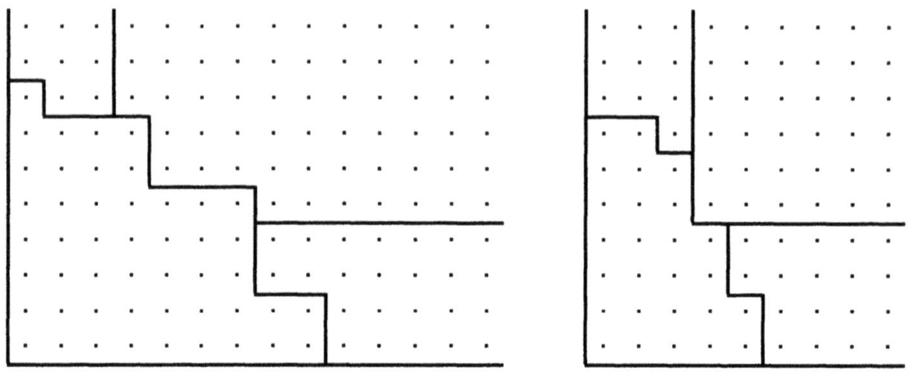

Example 3.8: E_F, E_A, E_B, and E_0

6. When $\mathcal{E} = (E_A)_{A \in \mathcal{D}}$ is an extent cell family, a **strand** is an $\mathcal{S}$-class, where $\mathcal{S}$ is the congruence in Theorem 3.7; $\mathcal{S}$ is the **strand congruence** of $\mathcal{E}$ (and of $\mathcal{C}$, when $\mathcal{E}$ is the extent cell family of $\mathcal{C}$).

Every strand is obtained by translating an ideal of a face $A \in \mathcal{D}$ along the complementary face A'. Indeed let $a \in F$, so that $a \in E_A$ for some unique $A \in \mathcal{D}$, and let $t = p'_A a \in A'$. The strand of $a \in F$ is $S_a = \{b \in E_A \mid p'_A b = t\}$. The set

$$I = \{a \in A \mid a + t \in S_a\} = \{a \in A \mid a + t \in E_A\}$$

is an ideal of A: if indeed $a \in I$ and $b \in A$, then $a + t \in E_A$, $a + t + b \in E_A$ by (E3), and $a + b \in I$. Then $S_a = I + t$ is obtained by translating the ideal I of A by $t \in A'$.

Now let $\mathcal{C}$ be a c.g.f. congruence on F and $\mathcal{E}$ be its extent cell family. Then $\mathcal{S} \subseteq \mathcal{C}$ by Theorem 3.7. Hence every $\mathcal{C}$-class is a union of strands. The last step in the construction of $\mathcal{C}$ is the assembling of strands into $\mathcal{C}$-classes, which is described in the next section.

4. TRACE CONGRUENCES.

The trace congruences of a complete group-free congruence $\mathcal{C}$ are nilmonoid congruences which determine which strands are assembled into $\mathcal{C}$-classes. Hence

$\mathcal{C}$ is determined by its direction set, extent cells, and trace congruences. This construction of $\mathcal{C}$ will be completed in the next section.

1. As before, $\mathcal{C}$ is a c.g.f. congruence on F, induced by $\pi : F \longrightarrow S$, with direction set $\mathcal{D}$ and extent cell family $\mathcal{E}$.

For every $A \in \mathcal{D}$, let

$$H_A = p'_A E_A \subseteq A' \text{ and } J_A = A' \backslash H_A .$$

Lemma 4.1. *H_A is a coideal and J_A is a nilmonoid ideal of A'. If F is f.g., then H_A is finite. Moreover $H_A + I_A \subseteq E_A$.*

Proof. By (E1), the union I of all E_B with $A \subsetneqq B \in \mathcal{D}$ is an ideal of F; hence $J_A = p'_A I$ is an ideal of A' and $H_A = A' \backslash J_A$ is a coideal of A'.

Let $h \in H_A$, $h \neq 0$. Then $h = p'_A a$ for some $a \in E_A$, $a \notin A$. Now $\pi(na)$ is idempotent for some $n > 0$, by Proposition 2.1. Then $na \in I_B$ for some $B \in \mathcal{D}$ by Proposition 3.1, $A \subseteq B$ by (E1), and $A \subsetneqq B$ since $na \notin A$. Hence $nh = p'_A(na) \in J_A$. Therefore A'/J_A is a nilmonoid. If F is f.g., then A'/J_A is f.g., A'/J_A is finite, and H_A is finite.

Finally let $h \in H_A$ and $a \in I_A$. Let $B = D(a+h)$, so that $a + h \in E_B$. Then $A = D(a) \subseteq B$ by Lemma 2.4; since $p'_A(a+h) = h \in p'_A E_A$, (E3) yields $B = A$ and $a + h \in E_A$. □

When $A \in \mathcal{D}$ the **trace congruence** of $\mathcal{C}$ on A' is the congruence $\mathcal{C}_A$ on A' defined by: $t \ \mathcal{C}_A \ u$ if and only if

$$t, u \in J_A, \text{ or } a + t \ \ \mathcal{C} \ \ a + u \text{ for some } a \in A.$$

In the above we may assume that $a \in I_A$, since I_A is an ideal of A; then $a + t$, $a + u \in E_A$ if $t, u \in H_A$, by Lemma 4.1. The **trace** of $\mathcal{C}$ is the family $(\mathcal{C}_A)_{A \in \mathcal{D}}$.

If for instance $\mathcal{C}$ is the congruence $\mathcal{S}$ in Theorem 3.7, then $\mathcal{C}_A$ is the Rees congruence of J_A. In general,

Lemma 4.2. *$\mathcal{C}_A$ is a J_A-congruence.*

Proof. First $\mathcal{C}_A$ is a congruence on A', since J_A is an ideal of A' and $\mathcal{C}$ is a congruence. Also $t \ \mathcal{C}_A \ u$ when $t, u \in J_A$. If $h \ \mathcal{C}_A \ t$ with $h \in H_A$, $t \in A'$, then $a + h \ \ \mathcal{C} \ \ a + t$ for some $a \in I_A$, $a + h \in E_A$ by Proposition 4.1, $a + t \in E_A$ by (E4), and $t = p'_A(a+t) \in H_A$. Hence J_A is a $\mathcal{C}_A$-class. □

Corollary 4.3. *If F is finitely generated, then $A'/\mathcal{C}_A$ is finite.*

2. The main properties of trace congruences are:

Proposition 4.4. *When $\mathcal{C}$ is a c.g.f. congruence on F, then $a \mathrel{\mathcal{C}} b$ if and only if there exists $A \in \mathcal{D}$ such that $a, b \in E_A$ and $p'_A a \mathrel{\mathcal{C}_A} p'_A b$.*

Proof. If $a \mathrel{\mathcal{C}} b$, then a and b are in the same extent cell E_A, so that $a' = p'_A a \in H_A$, $b' = p'_A b \in H_A$. Also $a \mathrel{\mathcal{C}} a + c$, $b \mathrel{\mathcal{C}} b + c$ for all $c \in A = D(a) = D(b)$. Hence

$$a' + p_A a + p_A b \;=\; a + p_A b \mathrel{\mathcal{C}} a \mathrel{\mathcal{C}} b \mathrel{\mathcal{C}} b + p_A a \;=\; b' + p_A b + p_A a$$

and $a' \mathrel{\mathcal{C}_A} b'$.

Conversely let $a, b \in E_A$. Assume that $a' = p'_A a \mathrel{\mathcal{C}_A} p'_A b = b'$, so that $a' + c \mathrel{\mathcal{C}} b' + c$ for some $c \in A$. We may assume that $c \in I_A$, so that $a' + c$, $b' + c \in E_A$ by Lemma 4.1. Then $a \mathrel{\mathcal{C}} a' + c$, $b \mathrel{\mathcal{C}} b' + c$ by (E6), and $a \mathrel{\mathcal{C}} b$. □

Proposition 4.5. *Let $\mathcal{C}$ be the c.g.f. congruence on F induced by a surjective homomorphism $\pi : F \longrightarrow S$. For every $A \in \mathcal{D}$, $\mathcal{C}_A$ is a reduced nilmonoid congruence on A', and $A' / \mathcal{C}_A$ is isomorphic to the Ponizovsky factor of e_A in S; the isomorphism sends the $\mathcal{C}_A$-class of $p'_A a$ to $\pi(a)$, for every $a \in E_A$.*

Proof. By Propositions 3.3 and 3.4, the Ponizovsky factor of e_A is

$$P_A \;=\; Se_A \,/ \bigcup\nolimits_{A \subsetneqq B \in \mathcal{D}} Se_B$$

and $a \in E_A$ if and only if $\pi(a)$ is in the partial Ponizovsky factor

$$P_A^* \;=\; Se_A \setminus \bigcup\nolimits_{A \subsetneqq B \in \mathcal{D}} Se_B$$

of e_A. Thus π induces a surjection $\pi : E_A \longrightarrow P_A^*$. Let $\rho : E_A \longrightarrow H_A$ be the projection $a \longmapsto p'_A a$. Then $\ker \rho \subseteq \ker \pi$ by (E6) and there is a mapping φ of A' onto P_A^* such that $\varphi(p'_A a) = \pi(a)$ for all $a \in E_A$, which is a partial homomorphism, since π and ρ are homomorphisms.

$$\begin{array}{ccc} E_A & \xrightarrow{\;\rho\;} & H_A \\ & {\scriptstyle \pi} \searrow & \downarrow {\scriptstyle \varphi} \\ & & P_A^* \end{array}$$

Moreover $\varphi(t) = \varphi(u)$ if and only if $t \mathrel{\mathcal{C}_A} u$, by Proposition 4.4.

Extend φ to $\psi : A' \longrightarrow P_A$, so that $\psi(t) = \varphi(t)$ for all $t \in H_A$ and $\varphi(t) = 0 \in P_A$ for all $t \in J_A$. Then $\ker \varphi = \mathcal{C}_A$. We show that ψ is a homomorphism; then $\ker \psi = \mathcal{C}_A$, $A' / \mathcal{C}_A \cong P_A$, and $\mathcal{C}_A$ is a nilmonoid congruence, which is reduced by Proposition 4.4 since $I_A = \rho^{-1}\{0\}$ is a $\mathcal{C}$-class.

The equality $\psi(t+u) = \psi(t)\,\psi(u)$ is clear if $t+u \in H_A$ or if $t \in J_A$ or $u \in J_A$, since φ is a partial homomorphism. Now assume $t, u \in H_A$, $t+u \in J_A$. Then $t = p'_A a$, $u = p'_A b$ with $a, b \in E_A$, $a+b \notin E_A$, and $a+b \in E_B$ for some $A \subsetneqq B \in \mathfrak{D}$ by (E2). If $c \in I_B$, so that $\pi(c) = e_B$, then $a+b+c \;\mathcal{C}\; a+b$, $\pi(a+b) = e_B\,\pi(a+b)$, and $\varphi(t)\,\varphi(u) = \pi(a)\,\pi(b) = 0$ in P_A; hence $\psi(t+u) = \psi(t)\,\psi(u)$ holds in this case also. □

Since S is a subdirect product of its Ponizovsky factors, the equality on S is the intersection of the congruences induced by the projections $S \longrightarrow Se_A \longrightarrow P_A$. Therefore:

Corollary 4.6. *$a \;\mathcal{C}\; b$ if and only if $p'_A a \;\mathcal{C}_A\; p'_A b$ for all $A \in \mathfrak{D}$.*

This can also be proved directly. If $a \;\mathcal{C}\; b$, then for any $A \in \mathfrak{D}$ and $c \in I_A$ we have $a+c \;\mathcal{C}\; b+c$, $a+c,\ b+c \in E_B$ for some $A \subseteq B \in \mathfrak{D}$, and $p'_A a = p'_A(a+c) \;\mathcal{C}_A\; p'_A(b+c) = p'_A b$. Conversely assume that $p'_A a \;\mathcal{C}_A\; p'_A b$ for all $A \in \mathfrak{D}$. We have $a \in E_A$, $b \in E_B$ for some $A, B \in \mathfrak{D}$. For any $c \in I_A$ we have $b+c \in E_C$ for some $A \subseteq C \in \mathfrak{D}$ by Lemma 2.4, and $p'_A(b+c) = p'_A b \;\mathcal{C}_A\; p'_A a \in H_A$; hence $p'_A(b+c) \in H_A$, $b+c \in E_A$ by (E3), and $B \subseteq A$. Exchanging a and b yields $A \subseteq B$, and then $a \;\mathcal{C}\; b$ by Proposition 4.4.

Since c.g.f.m. are subdirect products of nilmonoids, every c.g.f. congruence on F is an intersection of nilmonoid congruences; Corollary 4.6 provides an explicit intersection.

4. Proposition 4.4 provides a ready set of generating pairs for $\mathcal{C}$; equivalently, a presentation of $F/\mathcal{C}$. When $t \in H_A$ let

$$S_A(t) \;=\; \{a \in E_A \mid p'_A a = t\}$$

denote the strand which lies over t.

Proposition 4.7. *A c.g.f. congruence $\mathcal{C}$ on $F = F_X$ is generated by:*

(1) *one pair $(m+x, m)$ for every $A \in \mathfrak{D}$, $x \in X \cap A$, $t \in H_A$, and minimal element m of $S_A(t)$; and*

(2) *for every $A \in \mathfrak{D}$ and generator (t,u) of $\mathcal{C}_A$ with $t, u \in H_A$, one pair (m, n) in which m is a minimal element of $S_A(t)$ and n is a minimal element of $S_A(u)$.*

When F is finitely generated, then the above is a finite generating set, by Dickson's Theorem (Corollary VI.1.3). Indeed $S_A(t)$ has only finitely many minimal elements; since $\mathfrak{D}$ and every H_A are finite, (1) yields only finitely many pairs and so does (2).

Proof. Let $\mathcal{B}$ be the smallest congruence on F containing all pairs (1) and (2). Then $\mathcal{B} \subseteq \mathcal{C}$.

We show that every strand $S_A(t)$ is contained in a $\mathcal{B}$-class (i.e. $\mathcal{S} \subseteq \mathcal{B}$). Every $a \in S_A(t)$ has the form $a = m + c$ for some minimal element m of $S_A(t)$ and some $c = \sum_{x \in X \cap A} c_x x \in A$; since $m + x \ \mathcal{B} \ m$ for all $x \in X \cap A$, we have $a \ \mathcal{B} \ m$. Let m and n be any two minimal elements of $S_A(t)$; then $a = m \vee n \in S_A(t)$, since $p'_A a = p'_A m \vee p'_A n = t$, $a - m \in A = D(m)$, and $m \ \mathcal{C} \ a$; by the above, $m \ \mathcal{B} \ a \ \mathcal{B} \ n$. Hence $S_A(t)$ is contained in a single $\mathcal{B}$-class.

When $t, u \in H_A$ and (t, u) is a generator of $\mathcal{C}_A$, it follows from (2) and the above that $S_A(t)$ and $S_A(u)$ are contained in the same $\mathcal{B}$-class. Hence $S_A(t)$ and $S_A(u)$ are contained in the same $\mathcal{B}$-class whenever $t, u \in H_A$ and $t \ \mathcal{C}_A \ u$. Then Proposition 4.4 yields $\mathcal{C} \subseteq \mathcal{B}$. □

Proposition 4.4 also yields the kernel function of $\mathcal{C}$. For every $A \in \mathcal{D}$ and $g \in G$ let f'_A be the kernel function of $\mathcal{C}_A$ and

$$f_A(g) \ = \ {p'_A}^{-1}(f'_A(p'_A g)) \ = \ \{c \in F \mid p'_A c \in f'_A(p'_A g)\};$$

the proof of Proposition 4.8 below shows that f_A is the kernel function of the nilmonoid congruence induced by the projection $F \longrightarrow A' \longrightarrow A'/\mathcal{C}_A$. If F is finitely generated (more generally, if J_A is cofinite in A' for all $A \in \mathcal{D}$), then f'_A can be constructed as in Proposition IX.4.9.

Proposition 4.8. *Let $\mathcal{C}$ be a c.g.f. congruence on F. The Rédei group of $\mathcal{C}$ is $G = G(F)$. The kernel function f of $\mathcal{C}$ is given by:*

$c \in f(g)$ *if and only if* $c \in (E_A : g^+) \cap (E_A : g^-) \cap f_A(g)$ *for some* $A \in \mathcal{D}$.

Equivalently, $f(g) = \bigcap_{A \in \mathcal{D}} f_A(g)$.

Proof. For every $a \in F$ we have $a \in A^*$ for some $A \in \mathcal{D}$, $na \in E_A$ for some $n > 0$ by (E2), and $na \ \mathcal{C} \ (n+1)a$, since $a \in D(na)$, so that a belongs to the Rédei group of $\mathcal{C}$. Hence the latter is all of $G = G(F)$.

Let $c \in F$ and $g \in G$. Then $c \in f(g)$ if and only if $c + g^+ \ \mathcal{C} \ c + g^-$, if and only if $c + g^+$, $c + g^- \in E_A$ and $p'_A(c + g^+) \ \mathcal{C}_A \ p'_A(c + g^-)$ for some (unique) $A \in \mathcal{D}$, by Proposition 4.4. Now $(p'_A g)^+ = p'_A(g^+)$ and $(p'_A g)^- = p'_A(g^-)$; hence $p'_A(c + g^+) \ \mathcal{C}_A \ p'_A(c + g^-)$ if and only if $p'_A c + (p'_A g)^+ \ \mathcal{C}_A \ p'_A c + (p'_A g)^-$, if and only if $p'_A c \in f'_A(p'_A g)$. Thus $c \in f(g)$ if and only if $c \in (E_A : g^+) \cap (E_A : g^-) \cap f_A(g)$ for some $A \in \mathcal{D}$ (and then A is unique).

By Corollary 4.6, $c + g^+ \ \mathcal{C} \ c + g^-$ if and only if $p'_A(c + g^+) \ \mathcal{C}_A \ p'_A(c +$

g^{-}) for all $A \in \mathcal{D}$, if and only if (as above) $c \in f_A(g)$ for all $A \in \mathcal{D}$. $\square$

5. An example will illuminate the results in this section.

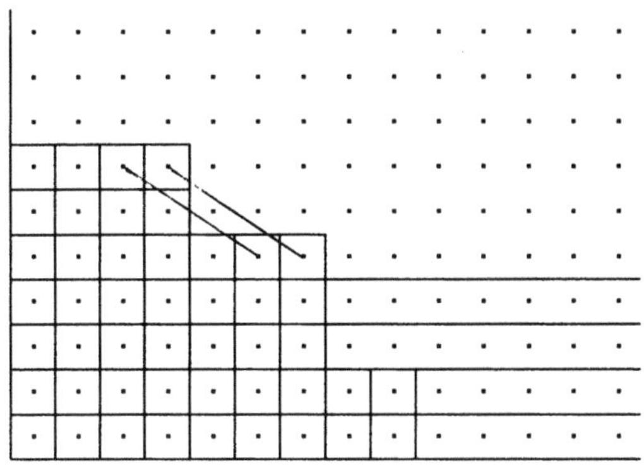

Example B

Example B. As in Section 2 let $X = \{x,y\}$, $\mathcal{D}$ consist of $0 = \{0\}$, $A = F_{\{x\}}$, and F, and $\mathcal{C}$ be the congruence in the figure. We see that $S_F(0) = E_F$ has three minimal elements $7y$, $4x + 5y$, and $7x + 4y$; $S_A(0)$ has one minimal element $9x$; $S_A(y)$ has one minimal element $9x + y$; $S_A(2y)$ has one minimal element $7x + 2y$; $S_A(3y)$ has one minimal element $7x + 3y$; the remaining strands $S_D(t)$ have $D = 0$. Hence (1) yields the following pairs:

$(x + 7y,\ 7y)$, $(8y,\ 7y)$, $(5x + 5y,\ 4x + 5y)$, $(4x + 6y,\ 4x + 5y)$, $(8x + 4y,\ 7x + 4y)$, $(7x + 5y,\ 7x + 4y)$, $(10x,\ 9x)$, $(10x + y,\ 9x + y)$, $(8x + 2y,\ 7x + 2y)$, $(8x + 3y,\ 7x + 3y)$,

some of which may be omitted; and (2) yields the single pair

$(2x + 6y,\ 5x + 4y)$.

By Proposition 4.7, $S = F/\mathcal{C}$ has the presentation (as a commutative monoid)

$$\begin{aligned} S \ = \ \langle\, a,b \mid\ & ab^7 = b^7 = b^8,\ a^5b^5 = a^4b^5 = a^4b^6, \\ & a^8b^4 = a^7b^4 = a^7b^5,\ a^{10} = a^9,\ a^8b^2 = a^9b^2,\ a^2b^6 = a^5b^4 \rangle. \end{aligned}$$

Since $S_F(0)$ is the zero element of S, it is more economical to present S as a commutative monoid with zero:

$$S \ = \ \langle\, a,b \mid b^7 = a^4b^5 = a^7b^4 = 0,\ a^{10} = a^9,\ a^8b^2 = a^9b^2,\ a^2b^6 = a^5b^4 \rangle.$$

5. MAIN RESULT.

We now complete the construction of complete group-free congruences.

1. By Proposition 4.4, a c.g.f. congruence $\mathcal{C}$ on a free c.m. F is determined by its direction set, extent cell family, and trace. The latter can be characterized as follows.

Lemma 5.1. *The trace of a c.g.f. congruence has the following properties: for every* $A \in \mathcal{D}$,

(T1) $\mathcal{C}_A$ *is a* J_A*-congruence on* A'*;*

(T2) *if* $t, u \in H_A$, $t \; \mathcal{C}_A \; u$, *and* $A \subseteq B \in \mathcal{D}$, *then* $(p'_A E_B) : t = (p'_A E_B) : u$ *in* A' *and* $p'_B(t+v) \;\; \mathcal{C}_B \;\; p'_B(u+v)$ *for all* $v \in (p'_A E_B) : t = (p'_A E_B) : u \subseteq A'$*;*

(T3) $\{0\}$ *is a* $\mathcal{C}_A$*-class.*

Proof. (T1) holds by Lemma 4.2 and (T3) holds by Proposition 4.5. Let $t, u \in H_A$, $t \; \mathcal{C}_A \; u$, and $A \subseteq B \in \mathcal{D}$. For any $a \in I_A$ we have $a + t$, $a + u \in E_A$ and $a + t \;\; \mathcal{C} \;\; a + u$ by Proposition 4.4. Let $v \in A'$. If $t + v \in p'_A E_B$, then $b + t + v \in E_B$ for some $b \in A$, $a + b + t + v \in E_B$ by (E3), $a + b + t + v \;\; \mathcal{C} \;\; a + b + u + v$, $a + b + u + v \in E_B$, and $u + v \in p'_A E_B$. Conversely $u + v \in p'_A E_B$ implies $t + v \in p'_A E_B$. Thus $(p'_A E_B) : t = (p'_A E_B) : u$ holds in A'; and in the above $t + v \in p'_A E_B$ implies $a + b + t + v \;\; \mathcal{C} \;\; a + b + u + v$, with $a + b + t + v$, $a + b + u + v \in E_B$, and

$$p'_B(t+v) \;\; = \;\; p'_B(a+b+t+v) \;\; \mathcal{C}_B \;\; p'_A(a+b+u+v) \;\; = \;\; p'_B(u+v). \; \square$$

Theorem 5.2. *Let* F *be a free commutative monoid,* $\mathcal{D}$ *be a direction set on* F, *and* $\mathcal{E}$ *be an extent cell family over* $\mathcal{D}$. *A family* $\mathcal{T} = (\mathcal{C}_A)_{A \in \mathcal{D}}$ *is the trace of a complete group-free congruence* $\mathcal{C}$ *on* F *with direction set* $\mathcal{D}$ *and extent cell family* $\mathcal{E}$, *if and only if it has properties* (T1) *and* (T2)*; then*

$$a \; \mathcal{C} \; b \text{ if and only if } a, b \in E_A \text{ and } p'_A a \;\; \mathcal{C}_A \;\; p'_A b \text{ for some } A \in \mathcal{D}.$$

Moreover, every complete group-free congruence on F *can be constructed in this fashion.*

Proof. Conditions (T1) and (T2) are necessary by Lemma 5.1, and Proposition 4.4 shows that every c.g.f. congruence on F can be constructed as in the statement.

Now let $\mathcal{D}$, $\mathcal{E}$, and $\mathcal{T}$ be given, so that (T1) and (T2) hold, and let $a \; \mathcal{C} \; b$ if and only if $a, b \in E_A$ and $p'_A a \;\; \mathcal{C}_A \;\; p'_A b$ for some $A \in \mathcal{D}$. By Theorem 3.7 there is a c.g.f. congruence $\mathcal{S}$ on F with direction set $\mathcal{D}$ and extent cell family $\mathcal{E}$;

namely, $a \ \mathcal{S} \ b$ if and only if $a, b \in E_A$ and $p'_A a = p'_A b$ for some $A \in \mathcal{D}$. We see that $\mathcal{S} \subseteq \mathcal{C}$.

$\mathcal{C}$ is an equivalence relation on F, since $\mathcal{E}$ is a partition of F and every $\mathcal{C}_A$ is an equivalence relation. Assume that $a \ \mathcal{C} \ b$ and let $t \in F$. Then $a, b \in E_A$ for some $A \in \mathcal{D}$ and $p'_A a \ \mathcal{C}_A \ p'_A b$. Also $a + t \in E_B$ and $b + t \in E_C$ for some $B, C \in \mathcal{D}$, and $A \subseteq B, C$ by (E1). Since $p'_A a + p'_A t \in p'_A E_B$, we have $p'_A b + p'_A t \in p'_A E_B$ by (T2), $B = C$ by (E3), and $b + t \in E_B$. Since $A \subseteq B$, (T2) also yields

$$p'_B a + p'_B t \ = \ p'_B(p'_A a + p'_A t) \ \mathcal{C}_B \ p'_B(p'_A b + p'_A t) \ = \ p'_B b + p'_B t,$$

and $a + t \ \mathcal{C} \ b + t$. Thus $\mathcal{C}$ is a congruence.

To prove that $\mathcal{C}$ is a c.g.f. congruence we prove:

Lemma 5.3. *If $t + v \ \mathcal{C}_A \ t$ and $t, \ t + v \in H_A$, then $v = 0$. In particular,* (T3) *holds.*

Proof. $t + v \ \mathcal{C}_A \ t$ implies

$$t \ \mathcal{C}_A \ t + v \ \mathcal{C}_A \ t + 2v \ \mathcal{C}_A \ \cdots \ \mathcal{C}_A \ t + nv$$

for all $n > 0$. Since J_A is a nilmonoid ideal, $v \neq 0$ would imply $nv \in J_A$ and $t + nv \in J_A$ for some $n > 0$, and $t \in J_A$ since $\mathcal{C}_A$ is a J_A-congruence, contradicting $t \in H_A$; therefore $v = 0$. □

Let $S = F/\mathcal{C}$. By Proposition 3.1, applied to $\mathcal{S}$, $I_A = A \cap E_A$ is an ideal of A and an $\mathcal{S}$-class. In fact I_A is a $\mathcal{C}$-class by (T3) and I_A is idempotent in S, since it is a subsemigroup of F. Conversely assume that the $\mathcal{C}$-class C_a of $a \in F$ is idempotent. Then there exists $A \in \mathcal{D}$ such that $a, \ 2a \in E_A$ and $p'_A a \ \mathcal{C}_A \ p'_A(2a)$ for all $n > 0$; by Lemma 5.3, $p'_A a = 0$, $a \in A \cap E_A$, and $C_a = I_A$. Thus every idempotent $\mathcal{C}$-class has the form I_A for some unique $A \in \mathcal{D}$. By (E2), every element of S has an idempotent power; hence every archimedean component of S contains an idempotent.

Let $a \in E_A$. If $c \in I_A$, then $a + c \in E_B$ for some $A \subseteq B \in \mathcal{D}$ by (E1) and $B = A$ by (E3), since $p'_A(a + c) = p'_A a \in p'_A E_A$; hence $a + c \ \mathcal{C} \ a$. Conversely assume $a + d \ \mathcal{C} \ a$, where $d \in I_B$. Then $a + d \in E_A$ and $p'_A(a + d) \ \mathcal{C}_A \ p'_A a$. By Lemma 5.3, $p'_A d = 0$ and $d \in A$. Hence $c + d \in I_A$, since I_A is an ideal of A, and $I_A \leqq I_B$ in S. (Note that $B \subseteq A$ by (E1).) Thus there is for every element s of S a least idempotent e of S such that $es = s$; namely, $e = I_A$, when $s = C_a$ with $a \in E_A$. Thus S is complete.

Let $e = I_C$ be an idempotent of S and $C_a \in H_e$. Let $c \in I_C$. Then

$a + c \ \mathcal{C}\ a$ and $a + b \ \mathcal{C}\ c$ for some $b \in F$. Now $c \leqq a + c \in E_A$ and $a \leqq a + b \in E_C$; by (E1), $C \subseteq A \subseteq C$ and $A = C$. Hence $a + b \ \mathcal{C}\ c$ yields $p'_A a + p'_A b \ \mathcal{C}_A\ p'_A c = 0$, $p'_A a = 0$, $a \in A \cap E_A = I_A$, and $C_a = e$. Thus S is group-free, and $\mathcal{C}$ is a c.g.f. congruence.

We show that $\mathcal{C}$ has the given direction set $\mathcal{D}$ and extent cell family $\mathcal{E}$. Let $a \in E_A$. Since $\mathcal{S}$ has direction set $\mathcal{D}$ and extent cell family $\mathcal{E}$, and $\mathcal{S} \subseteq \mathcal{C}$, we have $D_{\mathcal{C}}(a) \supseteq D_{\mathcal{S}}(a) = A$. If conversely $a + t \ \mathcal{C}\ a$, then $a + t \in E_A$, $p'_A a + p'_A t \ \mathcal{C}_A\ p'_A a$, $p'_A t = 0$ by Lemma 5.3, and $t \in A$. Thus $D_{\mathcal{C}}(a) = A$ for all $a \in E_A$. Therefore $\mathcal{C}$ has the given direction set $\mathcal{D}$, and the given extent cell family $\mathcal{E}$.

Finally we show that $\mathcal{C}$ has the given trace. Let $A \in \mathcal{D}$ and $t, u \in H_A$. If $a + t \ \mathcal{C}\ a + u$ for some $a \in A$, then $a + t$, $a + u \in E_B$ for some $A \subseteq B \in \mathcal{D}$ by (E1), $t = p'_A(a + t) \in p'_A E_B$, $B = A$ by (E3), $a + t$, $a + u \in E_A$, and $t \ \mathcal{C}_A\ u$ by definition of $\mathcal{C}$. If conversely $t \ \mathcal{C}_A\ u$, then for any $a \in I_A$ we have $a + t$, $a + u \in E_A$ by Lemma 4.1, and $a + t \ \mathcal{C}\ a + u$, since $p'_A(a + t) = t \ \mathcal{C}_A\ u = p'_A(a + u)$. Thus $\mathcal{C}_A$ is the trace congruence of $\mathcal{C}$ on A'. □

2. When F is f.g., every trace congruence $\mathcal{C}_A$ is a finite nilmonoid congruence and is determined as in Chapter IX by its zero class J_A and its nest of equivalence relations. Examples show that condition (T2) puts serious restrictions on the latter.

Example 5.4. let $X = \{x, y\}$; let $\mathcal{D}$ consist of $\{0\}$, $A = F_{\{x\}}$, and F; and let E_F, E_A, and $E_{\{0\}}$ be as in the figure (and as in Example B). Necessarily $\mathcal{C}_F$ is the equality on $F' = \{0\}$ and $\mathcal{C}_A$ is the Rees congruence of $J_A = \{4y, 5y, \cdots\}$; but $\mathcal{C}_{\{0\}}$ is more interesting. We see that $H = H_{\{0\}}$ has three maximal elements (corner points) $k = 3x + 6y$, $l = 6x + 4y$, and $m = 8x + y$. For the nest of equivalence relations $(\mathcal{D}_t)_{t \in H}$ of the largest J-congruence on F (where $J = F \backslash H$), it is readily verified that $k \ \mathcal{D}_t\ l$ if and only if $t \leqq 2x + y$; $k \ \mathcal{D}_t\ m$ if and only if $t \leqq x + y$; and $l \ \mathcal{D}_t\ m$ if and only if $t \leqq x + y$. Let $K = \{k, l, m\}$ and $\mathcal{M}$ be the equivalence relation on K with classes $\{k\}$ and $\{l, m\}$.

Since $p'_{\{0\}}$ is the identity on F, condition (T2), applied to $\mathcal{C}_{\{0\}}$, reads: if $t, u \in H$ and $t \ \mathcal{C}_{\{0\}}\ u$, then

(1) $H : t = H : u$ (since $\mathcal{C}_{\{0\}}$ is a congruence we already have $t + v \ \mathcal{C}_{\{0\}}\ u + v$ for all v);

(2) $E_A : t = E_A : u$ and $p'_A(t + v) \ \mathcal{C}_A\ p'_A(u + v)$ for all $v \in E_A : t = E_A : u$,

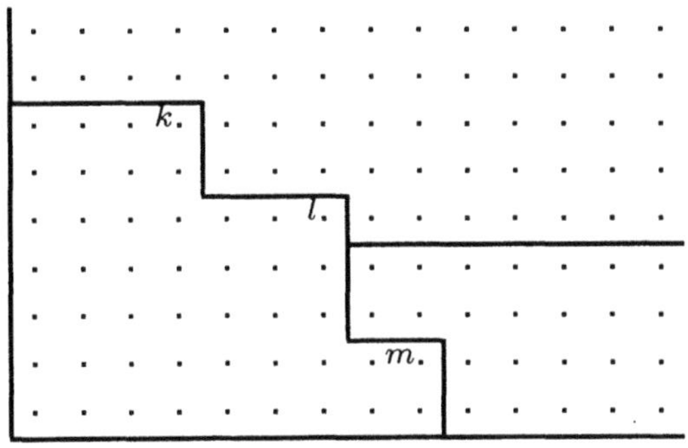

Example 5.4

equivalently $t_y = u_y$ or $t_y + v_y,\ u_y + v_y \geqq 4$ for all $v \in E_A : t = E_A : u$;

(3) $E_F : t = E_F : u$ (then $p'_F(t+v)\ \mathcal{C}_F\ p'_F(u+v)$, since $F' = 0$).

It suffices to consider (2): since $E_{\{0\}}$, E_A, E_F is a partition of F, (3) follows from (1) and (2), and we saw in Chapter IX that (1) holds for every J-congruence on F. By (2), $k\ \mathcal{C}_A\ m$ and $l\ \mathcal{C}_A\ m$ cannot hold, since $E_A : k = E_A : l = \varnothing$ and $E_A : m \neq \varnothing$. Thus the support of $\mathcal{C}_{\{0\}}$ has at most one nontrivial class $\{k, l\}$, and $C_t \subseteq \mathcal{M}$ for all $t \in H$ (whereas $\mathcal{D}_{x+y} = \mathcal{U}$).

Hence $\mathcal{C}_{\{0\}}$ and $\mathcal{C}$ are determined by the coideal $C = \{t \in H \mid \mathcal{C}_t = \mathcal{M}\}$. Now k and l satisfy (2), since $E_A : k = E_A : l = \varnothing$ and $p'_A(k+v)\ \mathcal{C}_A\ p'_A(l+v)$ holds vacuously for all $v \in E_A : k = E_A : l$. If more generally $t \in F$, $t \leqq k, l$, then $k - t$ and $l - t$ satisfy (2) if and only if $E_A : (k-t)$ and $E_A : (l-t)$ are both empty, that is, $t_y = 0$; thus $k\ \mathcal{C}_t\ l$ implies $t \leqq 2x$. Thus $C \subseteq \{0, x, 2x\}$. Hence there are four c.g.f. congruences with the given direction set and extent cells. In Example B, $C = \{0, x\}$.

3. In Example 5.4 there is a largest c.g.f. congruence with the given direction set and extent cells. This is always true:

Proposition 5.5. *Let $\mathcal{D}$ be a direction set on F and $\mathcal{E}$ be an extent cell family over $\mathcal{D}$. There exists a greatest c.g.f. congruence $\mathcal{L}$ with direction set $\mathcal{D}$ and extent cell family $\mathcal{E}$; its trace is given for every $A \in \mathcal{D}$ by: $t\ \mathcal{L}_A\ u$ if and only if $t, u \in J_A$ or $p'_A E_B : t = p'_A E_B : u$ holds in A' for every $A \subseteq B \in \mathcal{D}$.*

Proposition 5.5 generalizes Proposition IX.3.1.

Proof. We show that $\mathcal{T} = (\mathcal{L}_A)_{A \in \mathcal{D}}$ satisfies (T1) and (T2). First, $\mathcal{L}_A$ is an equivalence relation, and $t\ \mathcal{L}_A\ u$ for all $t, u \in J_A$. If $p'_A E_B : t = p'_A E_B : u$

holds in A' for every $A \subseteq B \in \mathcal{D}$, and $u \in J_A$, then $u \in p'_A E_B$ for some $A \subsetneq B \in \mathcal{D}$, $0 \in p'_A E_B : u$, $0 \in p'_A E_B : t$, $t \in p'_A E_B$, and $t \in J_A$. Thus $t \, \mathcal{L}_A \, u \in J_A$ implies $t \in J_A$, and J_A is a $\mathcal{L}_A$-class.

Assume $t \, \mathcal{L}_A \, u$ and let $v \in A'$. If $t, u \in J_A$, then $t+v$, $u+v \in J_A$ and $t+v \;\; \mathcal{L}_A \;\; u+v$. Otherwise $t, u \in H_A$, $p'_A E_B : t = p'_A E_B : u$ holds in A' for every $A \subseteq B \in \mathcal{D}$, and

$$p'_A E_B : (t+v) \;=\; (p'_A E_B : t) : v \;=\; (p'_A E_B : u) : v \;=\; p'_A E_B : (u+v)$$

holds in A' for every $A \subseteq B \in \mathcal{D}$. Hence $t+v \in J_A$ if and only if $u+v \in J_A$, and $t+v \;\; \mathcal{L}_A \;\; u+v$. Thus $\mathcal{L}_A$ is a J_A-congruence on A'.

Next we prove the following: if $A \subseteq B \subseteq C$ in $\mathcal{D}$ and $u \in A'$, then $p'_B u \in p'_B E_C$ if and only if $u \in p'_A E_C$. Indeed $u \in p'_A E_C$ implies $u+v \in E_C$ for some $v \in A$ and $p'_B u = p'_B(u+v) \in p'_B E_C$. The converse holds by (E3).

Now assume $t, u \in H_A$, $t \, \mathcal{L}_A \, u$, and $A \subseteq B \in \mathcal{D}$. Then $p'_A E_B : t = p'_A E_B : u$ holds in A'. Let $v \in p'_A E_B : t = p'_A E_B : u \subseteq A'$, so that $t+v$, $u+v \in p'_A E_B$ and $p'_B(t+v)$, $p'_B(u+v) \in p'_B E_B = H_B$. If $B \subseteq C \in \mathcal{D}$ and $w \in B' \subseteq A'$, then $p'_B(t+v)+w \in p'_B E_C$ implies $p'_B(t+v+w) \in p'_B E_C$, $t+v+w \in p'_A E_C$, $v+w \in p'_A E_C : t$, $v+w \in p'_A E_C : u$, $u+v+w \in p'_A E_C$, and $p'_B(u+v)+w = p'_B(u+v+w) \in p'_B E_C$; and conversely. Thus $p'_B E_C : p'_B(t+v) = p'_B E_C : p'_B(v+u)$, which shows $p'_B(t+v) \;\; \mathcal{L}_B \;\; p'_B(u+v)$, and proves (T2).

By Theorem 5.2 there is a c.g.f. congruence $\mathcal{L}$ on F with direction set $\mathcal{D}$, extent cell family $\mathcal{E}$, and trace $\mathcal{T} = (\mathcal{L}_A)_{A \in \mathcal{D}}$; namely, $a \, \mathcal{L} \, b$ if and only if $a, b \in E_A$ and $p'_A a \, \mathcal{L}_A \, p'_A b$ for some $A \in \mathcal{D}$. If $\mathcal{C}$ is a c.g.f. congruence on F with direction set $\mathcal{D}$ and extent cell family $\mathcal{E}$, then the trace $(\mathcal{C}_A)_{A \in \mathcal{D}}$ of $\mathcal{C}$ satisfies $\mathcal{C}_A \subseteq \mathcal{L}_A$ for all $A \in \mathcal{D}$, by Lemma 5.1, and $\mathcal{C} \subseteq \mathcal{L}$. □

Theorem 3.7 and Proposition 5.5 provide inequalities $\mathcal{S} \subseteq \mathcal{C} \subseteq \mathcal{L}$ for any c.g.f. congruence with given direction set and extent cell family.

6. PARTIALLY FREE SEMIGROUPS.

The results in Section 4 suggest a class of complete group-free semigroups, which are of interest partly because of Proposition 6.2 below, partly because of a formula in Chapter XIII for their extension groups $\mathrm{Ext}\,(S, \mathbb{G})$.

1. A **partially free** semigroup is a complete group-free semigroup S such that there is a free commutative monoid F and a surjective homomorphism $\pi : F \longrightarrow S^1$ such that the trace of $\mathcal{C} = \ker \pi$ consists of Rees congruences; equivalently, in case the strand congruence $\mathcal{S}$ of $\mathcal{C}$ is $\mathcal{C}$ itself; equivalently, every $\mathcal{C}$-class consists of a single strand.

Proposition 6.1. *A commutative nilsemigroup or nilmonoid is partially free if and only if it is 0-free.*

Proof. Let $\mathcal{C}$ be a nilmonoid congruence on a free commutative monoid F. Since a nilmonoid has just one nontrivial Ponizovsky factor, Proposition 4.5 implies that there are only two direction faces, F and the identity class U. Then E_F is the zero class J, $H_F = 0$, whereas $J_U = p'_U J$ and $H_U = U' \backslash p'_U J$. The trace of $\mathcal{C}$ consists of the trivial congruence $\mathcal{C}_F$ and the J_U-congruence $\mathcal{C}_U$. If $\mathcal{C} = \mathcal{S}$, then $F/\mathcal{C} \cong U'/\mathcal{C}_U = U'/J_U$ is a 0-free nilmonoid.

If conversely S is a 0-free nilmonoid, then there is a free commutative monoid F and a surjective homomorphism $\pi : F \longrightarrow S$ such that $\mathcal{C} = \ker \pi$ is the Rees congruence of a nilmonoid ideal J of F. Then the direction set of $\mathcal{C}$ consist of $0 = \{0\}$ and F. We see that $H_F = 0$ and $\mathcal{C}_F$ is trivial, whereas $H_0 = F \backslash J$ and $\mathcal{C}_0 = \mathcal{C}$ is the Rees congruence of J. Thus S is partially free. $\square$

Theorem 3.7 implies that partially free semigroups are not rare. Call a surjective homomorphism $\varphi : S \longrightarrow T$ **pure** in case the inverse image $\varphi^{-1}(e)$ of every idempotent e of T consists of a single element of S (necessarily an idempotent). This is consistent with the definition of a pure congruence in Chapter III. The following result generalizes the result of Arendt and Stuth [1970] (Corollary IX.2.5) that every nilsemigroup is a pure homomorphic image of a 0-free nilsemigroup.

Proposition 6.2. *Every c.g.f. semigroup (or monoid) is a pure homomorphic image of a partially free semigroup (or monoid).*

Proof. Let S be a c.g.f.m. There is a free commutative monoid F and a surjective homomorphism $\pi : F \longrightarrow S^1$. Let $\mathcal{C} = \ker \pi$ and $\mathcal{S}$ be the strand congruence of $\mathcal{C}$. Since $\mathcal{S} \subseteq \mathcal{C}$, π factors through the projection $\rho : F/\mathcal{S}$: $\pi = \varphi \circ \rho$ for some homomorphism $\varphi : F/\mathcal{S} \longrightarrow S$.

$$\begin{array}{ccc} F & \xrightarrow{\rho} & F/\mathcal{S} \\ & \searrow^{\pi} & \downarrow \varphi \\ & & S \end{array}$$

Then $F/\mathcal{S}$ is partially free and φ is surjective like π. Since $\mathcal{C}$ and $\mathcal{S}$ have the same direction set and extent cell family, they have the same idempotent classes

$I_A = E_A \cap A$ (where $A \in \mathcal{D}$). Hence φ is pure: if $t = \rho(a) \in F/\mathcal{S}$ and $\varphi(t) = e_A = \pi(I_A)$ is idempotent in S, then $\pi(a) = \pi(I_A)$, $a \in I_A$, and $t = \rho(I_A)$ is idempotent in $F/\mathcal{S}$. $\square$

2. By Proposition 4.5, the Ponizovsky factors of a partially free semigroup are 0-free. Example 6.5 below shows that this condition is not sufficient. Nevertheless, finite partially free semigroups can be characterized by their Ponizovsky factors, due to additional properties by which the irreducible elements of a partially free semigroup determine the irreducible elements of its Ponizovsky factors. Let Irr (S) denote the set of all irreducible elements of S.

Lemma 6.3. *Given a finite group-free monoid S and $e \in E(S)$, $e \neq 0$, let $Q_e = \{q \in \text{Irr}\,(S) \mid eq < e \text{ and } \epsilon(eq) = e\}$.*

(1) *If $r \in \text{Irr}\,(P_e)$, then $r = eq$ for some $q \in Q_e$.*

(2) *If $\pi : F_X \longrightarrow S$ is surjective, $x \in X$, and $e = e_A$, then $\pi(x) \in Q_e$ if and only if $\pi(x) \in \text{Irr}\,(S)$ and $x \in H_A$.*

Proof. (1). Let $\rho : S \longrightarrow eS \longrightarrow P_e = eS \,/ \bigcup_{f \in E(S),\, f<e} fS$ be the projection: $\rho(s) = es$ if $\epsilon(es) = e$ (if $es \notin fS$ when $f < e$), $\rho(s) = 0 \in P_e$ if $\epsilon(es) < e$ (if $es \in fS$ for some $f < e$). Since Irr (S) generates S, $\rho(\text{Irr}\,(S))$ generates P_e and contains every $r \in \text{Irr}\,(P_e)$: thus $r = \rho(q)$ for some $q \in \text{Irr}\,(S)$, and then $q \in Q_e$, since $\rho(q) \neq 0$ in P_e yields $\epsilon(eq) = e$, and $eq = r < e$.

(2). Assume that $\pi(x) = q \in Q_e$. Then $q \in \text{Irr}\,(S)$. Also $x \notin A$, since $x \in A$ would imply $\pi(x) \geqq e_A$ (Corollary 3.2) and $qe \geqq e$. Hence $x \in A'$. Assume that $x \notin H_A$. Then $x \in p'_A E_B$ for some $A \subsetneqq B \in \mathcal{D}$ and $x + t \in E_B$ for some $t \in A$. Hence $s = \pi(t) \in S$ and $f = e_B \in E(S)$ have the following properties: $s \geqq e$ by Corollary 3.2; $f < e$ by Corollary 3.3; and $\epsilon(qs) = f$, in particular $qs \leqq f$, by Proposition 3.4. Hence $eq \leqq sq \leqq f < e$, contradicting $\epsilon(eq) = e$ and $q \in Q_e$. Therefore $x \in H_A$.

Conversely assume that $x \in H_A$ and $q = \pi(x) \in \text{Irr}\,(S)$. Then $q \ngeqq e$ by Corollary 3.2; hence $eq < e$. Let $p \in I_A$. Then $x + p \in E_A$ by Lemma 4.1, $\pi(p) = e_A = e$, $\pi(x + p) = qe$, $\epsilon(eq) = e$ by Proposition 3.4, and $q \in Q_e$. $\square$

Theorem 6.4. *For a finite group-free commutative monoid S the following conditions are equivalent:*

(1) *S is partially free (there is a free commutative monoid F and a surjective homomorphism $\pi : F \longrightarrow S$ such that the trace of $\mathcal{C} = \ker \pi$ consists of Rees congruences);*

(2) *in the standard presentation* $\pi : F \longrightarrow S$ *of* S, *the trace of* $\mathcal{C} = \ker \pi$ *consists of Rees congruences;*

(3) *all Ponizovsky factors of* S *are 0-free and, for every* $e \in E(S)$, $e \neq 0$, *the mapping* $q \longmapsto eq$ *is a bijection of*

$$Q_e = \{q \in \operatorname{Irr}(S) \mid eq < e \text{ and } \epsilon(eq) = e\}$$

onto $\operatorname{Irr}(P_e)$.

Proof. (2) $\Longrightarrow$ (1) is clear.

(1) $\Longrightarrow$ (3). Let $\pi : F_X \longrightarrow S$ be a surjective homomorphism such that the trace of $\mathcal{C} = \ker \pi$ consists of Rees congruences. The Ponizovsky factors of S are 0-free by Proposition 4.5.

Let $e \in E(S)$, $e \neq 0$; then $e = e_A$ for some $A \in \mathcal{D}$, $A \neq X$. If $q \in Q_e$, then $q \in \operatorname{Irr}(S)$, $q = \pi(x)$ for some $x \in X$, $x \in H_A$ by Lemma 6.3, and $x + p \in E_A$ for every $p \in I_A$ by Lemma 4.1. Then x is irreducible in A' and $\pi(p) = e$. The isomorphism $\theta : A'/\mathcal{C}_A \longrightarrow P_e$ in Proposition 4.5 sends the $\mathcal{C}_A$-class of $p'_A a$ to $\pi(a)$, when $a \in E_A$, and sends the $\mathcal{C}_A$-class of $x = p'_A(x+p)$ to $\pi(x+p) = eq$. Therefore $eq \in \operatorname{Irr}(P_e)$. Thus $\widehat{e} : q \longmapsto eq$ is a mapping of Q_e into $\operatorname{Irr}(P_e)$. Now $\widehat{e}$ is surjective by Lemma 6.3. If $\widehat{e}(q) = \widehat{e}(r)$, then $r = \pi(y)$ for some $y \in X$, $y \in H_A$ by Lemma 6.3, $\pi(x+p) = eq = er = \pi(y+p)$, $x+p \;\mathcal{C}\; y+p$, $x \;\mathcal{C}_A\; y$, $x = y$ since $\mathcal{C}_A$ is the equality on H_A, and $q = r$. Thus $\widehat{e}$ is bijective.

(3) $\Longrightarrow$ (2). Assume that P_e is 0-free and that $\widehat{e} : q \longmapsto eq$ is a bijection of Q_e onto $\operatorname{Irr}(P_e)$, for every $e \in E(S)$, $e \neq 0$. Let $\pi : F_X \longrightarrow S$ be the standard presentation of S. (Thus $X = \operatorname{Irr}(S)$ and π is the identity on X.) Let $A \in \mathcal{D}$, $A \neq X$, $e = e_A$, and $\rho : A' \longrightarrow A'/\mathcal{C}_A \cong P_e$ be the projection. By Proposition 4.5, $\rho(p'_A a) = \pi(a)$ for all $a \in E_A$. If $c \in A'$ and $\rho(c)$ is irreducible in P_e, then c is irreducible in A' and $c \in X \backslash A$, $c \in H_A$. Thus $\operatorname{Irr}(P_e) \subseteq \rho(Y)$, where $Y = X \cap H_A$.

Let $p \in I_A$. When $x \in Y$, then $q = \pi(x) \in Q_e$ by Lemma 6.3 and $x + p \in E_A$ by Lemma 4.1, so that $\rho(x) = \pi(x+p) = eq = \widehat{e}(q)$. In particular $\rho(Y) \subseteq \operatorname{Irr}(P_e)$. If $y \in Y$, $\pi(y) = r$, and $\rho(x) = \rho(y)$, then $\widehat{e}(q) = \widehat{e}(r)$, $q = r$, $\pi(x) = \pi(y)$, and $x = y$, since $\pi : F \longrightarrow S$ is the standard presentation of S. Thus ρ induces a bijection of Y onto $\operatorname{Irr}(P_e)$.

Let $c = \sum_{x \in X \backslash A} c_x x \in H_A$. Since H_A is a coideal of A', $c_x > 0$ implies $x \in H_A$; hence $c = \sum_{x \in Y} c_x x$. If now $c, d \in H_A$, then $\rho(c) = \prod_{x \in Y} \rho(x)^{c_x} \neq 0$ and $\rho(d) = \prod_{x \in Y} \rho(x)^{d_x} \neq 0$. Since P_e is 0-free, a nonzero element of P_e can be written uniquely as a product of nonnegative powers of irreducible elements of P_e, and by the above can be written uniquely as a product of nonnegative powers of elements of $\rho(Y)$. Hence $\rho(c) = \rho(d) \neq 0$ implies $c_x = d_x$ for all $x \in Y$ and $c = d$. Therefore $\mathcal{C}_A = \ker \rho$ is the Rees congruence of J_A. □

3. The following example is a semigroup which is not partially free even though all its Ponizovsky factors are 0-free. This example also shows that $A' \longrightarrow A'/\mathcal{C}_A \cong P_A$ need not be the standard presentation of P_A; and that, when P_A is 0-free, $\mathcal{C}_A$ need not be the Rees congruence of J_A.

Example 6.5. Let $X = \{x,y,z\}$ and $\mathcal{D}$ consist of $0 = \{0\}$, $A = F_{\{x,y\}}$, $B = F_{\{z\}} = A'$, and F. Let $E_0 = \{0, x, 2x, y\}$; $E_A = A \backslash E_0$, which is the ideal of A generated by $3x$, $x+y$, and $2y$; $E_B = \{kz,\ x+kz,\ 2x+kz,\ y+kz \mid k > 0\}$; and $E_F = F \backslash (E_0 \cup E_A \cup E_B)$, which is the ideal of F generated by $3x+z$, $x+y+z$, and $2y+z$. It is immediate that $\mathcal{D}$ is a direction set and that $\mathcal{E}$ is an extent cell family; we see that $H_0 = E_0 = \{0, x, 2x, y\}$, $I_0 = 0$; $H_A = 0$, $I_A = E_A$; $H_B = H_0 = \{0, x, 2x, y\}$, $I_B = \{kz \mid k > 0\}$; and $H_F = 0$, $I_F = E_F$.

Let $\mathcal{C}_D = \mathcal{S}_D$ be the Rees congruence of J_D for all $D \in \mathcal{D}$, except for $\mathcal{C}_B$, for which we also have $2x \mathrel{\mathcal{C}_B} y$. We see that (T1) and (T2) hold, so Theorem 5.2 provides a congruence $\mathcal{C}$ on F with the given direction set, extent cell family, and trace, in which $a \mathrel{\mathcal{C}} b$ if and only if $a, b \in D$ and $p'_D a \mathrel{\mathcal{C}_B} p'_D b$ for some $D \in \mathcal{D}$. Let $S = F/\mathcal{C}$ and $\pi : F \longrightarrow S$ be the projection.

By Proposition 4.7, S has a presentation, as a commutative monoid with zero, whereby S is generated by x, y, z subject to: $x^3 z = xyz = y^2 z = 0$ (from E_F); $x^3 = x^4 = x^3 y$, $xy = x^2 y = xy^2$, $y^2 = y^3 = xy^2$ (from E_A); $z = z^2$ (from E_B); and $x^2 z = yz$ (from $\mathcal{C}_B$).

By Proposition 4.5, the Ponizovsky factors of S are the nilmonoids $D'/\mathcal{C}_D$. The choice of $\mathcal{C}_D$ shows that all are 0-free except perhaps for $B'/\mathcal{C}_B$. Since $2x \mathrel{\mathcal{C}_B} y$, the latter is a cyclic nilmonoid $\{1, c, c^2, 0\}$, where c is the $\mathcal{C}_B$-class of x, and is 0-free. Thus all Ponizovsky factors of S are 0-free. In particular, P_B is 0-free, even though $\mathcal{C}_B$ is not a Rees congruence, and $B' \longrightarrow B'/\mathcal{C}_B \cong P_B$ is not the standard presentation of P_B.

We have $0,\ x,\ 2x,\ y \in E_0 = H_0$; since $\mathcal{C}_0$ is the Rees congruence of J_0, $\{0\}$, $\{x\}$, $\{2x\}$ and $\{y\}$ are $\mathcal{C}$-classes. Hence $\pi(x)$, $\pi(y) \in \operatorname{Irr}(S)$. Similarly I_B is a $\mathcal{C}$-class, and $\pi(z) \in \operatorname{Irr}(S)$. Thus $\pi : F \longrightarrow S$ is the standard presentation of S. But $\mathcal{C}_B$ is not a Rees congruence and S is not partially free. □

Chapter XI.

SUBCOMPLETE SEMIGROUPS.

This chapter contains a global construction of subcomplete semigroups, as quotients of free commutative monoids; complete semigroups, finite semigroups, subelementary semigroups, and finitely generated semigroups are particular cases. In particular, this constructs all congruences on finitely generated free commutative monoids. The construction uses Ponizovsky families to generalize the results in Chapter X and relates smoothly to related structural features including archimedean components, subdirect decompositions, $\mathcal{K}$-classes, and extended Schützenberger functors. Its relationship to extension groups is less obvious and is shown in Section XIII.2. A similar construction was obtained by the author for finite congruences [1996C], then generalized to complete group-free congruences [2001C].

Subcomplete congruences are built from direction sets, extent cells, trace congruences, strand groups, and strand bases. Explicit representations of subcomplete congruences as intersections of subelementary congruences complete the chapter.

1. DIRECTION SETS.

This section defines direction sets for arbitrary congruences on a free commutative monoid.

1. In what follows, F is a free c.m. and $\mathcal{C}$ is any congruence on F; for instance, the congruence induced by a surjective homomorphism $\pi : F \longrightarrow S$.

Directionality is a property of complete group-free congruences but does not much extend to more general congruences, as later results will show. Accordingly the general definition of direction sets uses semilattice congruences rather than $\mathcal{C}$-classes. Later sections also require a definition which is not tied to archimedean components.

Proposition 1.1. *Let $\mathcal{C}$ be a congruence on a free c.m. F, For a direction set $\mathcal{D}$ on F the following conditions are equivalent:*

(1) *every* $A \in \mathcal{D}$ *is a union of* $\mathcal{C}$*-classes;*

(2) A^* *is a union of* $\mathcal{C}$*-classes for every* $A \in \mathcal{D}$*;*

(3) $\mathcal{D} = D(\mathcal{Y})$ *for some semilattice congruence* $\mathcal{Y} \supseteq \mathcal{C}$ *on* F*;*

(4) $\mathcal{D} = D(\pi^{-1}(\mathcal{B}))$ *for some semilattice congruence* $\mathcal{B}$ *on* S*.*

Proof. (1) implies (2) since $A^* = A \setminus \left(\bigcup_{A \subsetneq B \in \mathcal{D}} B\right)$. By Proposition X.1.3, the sets A^* with $A \in \mathcal{D}$ are the classes of the semilattice congruence $\mathcal{Y}$ on F such that $\mathcal{D} = D(\mathcal{Y})$; hence (2) implies that every $\mathcal{Y}$-class is a union of $\mathcal{C}$-classes, and (2) implies (3). By Lemma X.1.1, every $A \in D(\mathcal{Y})$ is a union of $\mathcal{Y}$-classes and is a union of $\mathcal{C}$-classes if $\mathcal{C} \subseteq \mathcal{Y}$; thus (3) implies (1).

(3) and (4) are equivalent by Proposition I.2.6. A semilattice congruence $\mathcal{B}$ on S induces a congruence $\mathcal{Y} = \pi^{-1}(\mathcal{B})$ on F which contains $\mathcal{C}$, under which

$$a \;\mathcal{Y}\; b \text{ if and only if } \pi(a) \;\mathcal{B}\; \pi(b);$$

$\mathcal{Y}$ is a semilattice congruence since $F/\mathcal{Y} \cong S/\mathcal{B}$. Conversely, a semilattice congruence $\mathcal{Y} \supseteq \mathcal{C}$ on F induces a congruence $\mathcal{B} = \pi(\mathcal{Y})$ on S, under which

$$\pi(a) \;\mathcal{B}\; \pi(b) \text{ if and only if } a \;\mathcal{Y}\; b;$$

$\mathcal{B}$ is a semilattice congruence on S, since $S/\mathcal{B} \cong F/\mathcal{Y}$, and $\mathcal{Y} = \pi^{-1}(\mathcal{B})$. □

A direction set $\mathcal{D}$ on F is a **direction set of** $\mathcal{C}$ when either condition holds; if $\mathcal{B}$ is known, then $\mathcal{D}$ is the direction set $\mathcal{D} = D_{\mathcal{B}}$ (actually, $D(\pi^{-1}(\mathcal{B}))$) of $\mathcal{C}$ **over** $\mathcal{B}$.

The **archimedean direction set** of $\mathcal{C}$ is its direction set $D_{\mathcal{N}}$ over the smallest semilattice congruence $\mathcal{N}$ on S. By Proposition X.2.8, the direction set of a c.g.f. congruence as defined in Chapter X is its archimedean direction set.

In general, let B_s denote the $\mathcal{B}$-class of $s \in S$. Then $B_s \leqq B_t$ in the semilattice $S/\mathcal{B}$ if and only if $B_s \wedge B_t = B_s$ in $S/\mathcal{B}$, if and only if $B_s B_t \subseteq B_s$ in S, if and only if $st \;\mathcal{B}\; s$ in S. When $\mathcal{D} = D_{\mathcal{B}}$, then $t \in D_{\mathcal{B}}(a)$ if and only if $B_{\pi(t)} \geqq B_{\pi(a)}$ in $S/\mathcal{B}$; thus

$$D_{\mathcal{B}}(a) \;=\; \{t \in F \mid a \;\mathcal{Y}\; a+t\} \;=\; \{t \in F \mid \pi(a) \;\mathcal{B}\; \pi(a+t)\}.$$

From Section X.1 we obtain the following properties. By Proposition 1.1 and Lemma X.1.1:

Proposition 1.2. *Let* $\mathcal{C}$ *be a congruence on* F *and* $\mathcal{D}$ *be a direction set of* $\mathcal{C}$*. Every* $A \in \mathcal{D}$ *is a union of* $\mathcal{C}$*-classes and a face of* F*.*

When $a, b \in F$*, then* $a \in D(b)$ *if and only if* $D(a) \subseteq D(b)$*, if and only if*

$B_{\pi(a)} \geqq B_{\pi(b)}$ *in* $S/\mathcal{B}$; *in particular,* $D(a)$ *is the smallest direction face which contains* a.

By Proposition X.1.3 and Corollary X.1.4:

Proposition 1.3. *Let* $\mathcal{C}$ *be a congruence on* F *and* $\mathcal{D} = D_{\mathcal{B}}$ *be the direction set of* $\mathcal{C}$ *over a semilattice congruence* $\mathcal{B}$ *on* $S = F/\mathcal{C}$. *Then* $\mathcal{D}$, *partially ordered by inclusion, is an upper semilattice, in which* $D(a) \vee D(b) = D(a+b)$. *For every* $C \in \mathcal{D}$ *let*

$$C^* = \{c \in C \mid c \notin B \text{ when } B \in \mathcal{D} \text{ and } B \subsetneqq C\};$$

equivalently, $C^* = \{a \in F \mid D(a) = C\}$. *Then* C^* *is a union of* $\mathcal{C}$*-classes;* $\pi(C^*)$ *is a* $\mathcal{B}$*-class of* S *(an archimedean component of* S, *if* $\mathcal{B} = \mathcal{N}$*) and every* $\mathcal{B}$*-class of* S *can be constructed in this fashion;* $C \longmapsto \pi(C^*)$ *is an isomorphism* $(\mathcal{D}, \vee) \cong S/\mathcal{B}$, *with* $\pi(C^*) \geqq \pi(D^*)$ *in* $S/\mathcal{B}$ *if and only if* $C \subseteq D$ *in* F, *and* $C^* + D^* \subseteq (C \vee D)^*$, *for all* $C, D \in \mathcal{D}$.

Corollary 1.4. *If* S *is finitely generated, then every direction set of* $\mathcal{C}$ *is finite, contains* F, *and is closed under intersections.*

Proof. If S is finitely generated, then $S/\mathcal{B}$ is finite and $(\mathcal{D}, \vee) \cong S/\mathcal{B}$ is finite. □

2. EXTENT CELLS.

The central result of this chapter is that all subcomplete congruences have extent cells.

1. Let F be a free c.m. and $\mathcal{C}$ be a subcomplete congruence on F; for instance, the congruence induced by a surjective homomorphism $\pi : F \longrightarrow S$, where S is a subcomplete c.m.

When $\mathcal{C}$ is a c.g.f. congruence we saw in Chapter X that $\mathcal{C}$ has extent cells

$$\{a \in F \mid D(a) = A\},$$

one for every face A in the direction set, which correspond to the partial Ponizovsky factors of S (Proposition X.3.4). This definition cannot be used in general, since, by Proposition 1.3, it would merely yield the $\mathcal{Y}$-classes; rather, extent cell families for $\mathcal{C}$ are obtained from Ponizovsky families of S.

Proposition 2.1. *Let* $\mathcal{C}$ *be the congruence induced on* F *by a surjective homomorphism* $\pi : F \longrightarrow S$. *When* S *is subcomplete, then there exists a* *semilattice congruence* $\mathcal{B}$ *on* S *and a Ponizovsky family* $\mathcal{P} = (P_C)_{C \in S/\mathcal{B}}$ *of* S *relative to* $\mathcal{B}$, *and then* $\mathcal{D} = D_{\mathcal{B}}$ *is a direction set of* $\mathcal{C}$ *and the family* $\mathcal{E} = \pi^{-1}(\mathcal{P}) = (E_A)_{A \in \mathcal{D}}$ *of subsets* $E_A = \pi^{-1}\big(P_{\pi(A^*)}\big)$ *of* F *has the following properties:*

(X1) E_A *is a union of* $\mathcal{C}$*-classes;*

(X2) $\mathcal{E}$ *is a partition of* F*;*

(X3) $L_A = \bigcup \big(E_B \mid B \in \mathcal{D},\ A \subseteq B\big)$ *is an ideal of* F*;*

(X4) $E_A \subseteq \bigcup \big(B^* \mid A \subseteq B \in \mathcal{D}\big)$ *and* $I_A = E_A \cap A^* \neq \emptyset$*;*

(X5) *if* $A \subseteq B$, $a \in L_A$, *and* $c \in I_A$, *then* $a + c \in E_B$ *if and only if* $a \in E_B$*;*

(X6) *if* $c \in I_A$, $a, b \in L_A$, *and* $a + c \ \mathcal{C}\ b + c$, *then* $a \ \mathcal{C}\ b$*;*

(X7) $(\forall a \in A^*)(\exists n > 0)\ na \in I_A$.

These properties imply:

(X8) $I_A = E_A \cap A = L_A \cap A$ *is a nonempty ideal of* A*;*

(X9) $L_A + L_B \subseteq L_{A \vee B}$ *and* $I_A + I_B \subseteq I_{A \vee B}$.

If S *is finitely generated and* $\mathcal{B} = \mathcal{N}$, *then in the above we may assume that*

(XM) $a_1 + \cdots + a_m \in I_A$ *for all* $a_1, \ldots, a_m \in A^*$ *and all* $A \in \mathcal{D}$, *when* m *is sufficiently large.*

Proof. By Proposition VII.2.2, S has a Ponizovsky family $\mathcal{P} = (P_C)_{C \in S/\mathcal{B}}$ relative to some semilattice congruence $\mathcal{B}$, with the following properties:

(P1) $\mathcal{P}$ is a partition of S;

(P2) $L_C = \bigcup \big(P_D \mid D \in S/\mathcal{B},\ D \leqq C\big)$ is an ideal of S;

(P3) $P_C \subseteq \bigcup \big(D \in S/\mathcal{B} \mid D \subseteq C\big)$ and $K_C = P_C \cap C \neq \emptyset$;

(P4) if $C \subseteq D$, $c \in L_C$, and $k \in K_C$, then $ck \in P_D$ if and only if $c \in P_D$;

(P5) K_C is cancellative in L_C (if $k \in K_C$, $c, d \in L_C$, and $ck = dk$, then $c = d$);

(P6) $(\forall c \in C)(\exists n > 0)\ a^n \in K_C$.

By Proposition 1.3, $A \longmapsto \pi(A^*)$ is an isomorphism $(\mathcal{D}, \vee) \cong S/\mathcal{B}$, and $\pi(A^*) \geqq \pi(B^*)$ in $S/\mathcal{B}$ if and only if $A \subseteq B$ in F. When $A \in \mathcal{D}$, then $C = \pi(A^*) \in S/\mathcal{B}$ and

$$E_A \;=\; \pi^{-1}(P_C) \;=\; \textstyle\bigcup_{s \in P_C} \pi^{-1}(s)$$

is a union of $\mathcal{C}$-classes. Then (X2)–(X7) follow from (P1)–(P6) since inverse image under π preserves unions and intersections: thus $\mathcal{E} = (E_A)_{A \in \mathcal{D}}$ is a partition of F by (P1);

$$\begin{aligned} L_A &= \bigcup (E_B \mid B \in \mathcal{D},\ A \subseteq B) \\ &= \pi^{-1} \left(\bigcup (P_D \mid D \in S/\mathcal{B},\ C \geqq D) \right) = \pi^{-1}(L_C) \end{aligned}$$

is an ideal of F by (P2);

$$E_A = \pi^{-1}(P_C) \subseteq \pi^{-1} \left(\bigcup (D \in S/\mathcal{B} \mid C \geqq D) \right) = \bigcup (B^* \mid A \subseteq B \in \mathcal{D})$$

and

$$I_A = E_A \cap A^* = \pi^{-1}(PsC \cap C) = \pi^{-1}(K_C) \neq \emptyset$$

by (P3); then (X5), (X6), and (X7) follow from (P4), (P5), and (P6).

Now let $\mathcal{D}$ be a direction set of $\mathcal{C}$ and $\mathcal{E}$ be a family with properties (X1) through (X7). If $a \in L_A$, then $a \in E_B$ for some $A \subseteq B \in \mathcal{D}$ and $a \in C^*$ for some $B \subseteq C \in \mathcal{D}$ by (X4). If $a \in L_A \cap A$, then in the above $C = D(a) \subseteq A$ by Propositions 1.3 and 1.2, so that $A = B = C$ and $a \in E_A \cap A^* = I_A$. Thus $I_A = E_A \cap A^* \subseteq E_A \cap A \subseteq L_A \cap A \subseteq I_A$. This proves (X8).

(X9). Let $a \in L_A$, $b \in L_B$. By (X2), $a + b \in E_C$ for some $C \in \mathcal{D}$. Then $A \subseteq C$, since $a + b \in L_A$ by (X3), $B \subseteq C$, since $a + b \in L_B$ by (X3), $A \vee B \subseteq C$, and $a + b \in L_{A \vee B}$. Thus $L_A + L_B \subseteq L_{A \vee B}$. Then $I_A + I_B \subseteq L_A + L_B \subseteq L_{A \vee B}$, $I_A + I_B \subseteq A^* + B^* \subseteq (A \vee B)^*$ by Proposition 1.3, and $I_A + I_B \subseteq L_{A \vee B} \cap (A \vee B)^* = I_{A \vee B}$. This proves (X9).

If finally S is finitely generated and m is sufficiently large, then S has a Ponizovsky family in which $\mathcal{B} = \mathcal{N}$ and $C^m \subseteq K_C$ for all $C \in Y(S)$ (Proposition VII.2.7); then $a_1 + \cdots + a_m \in I_A$ for all $a_1, \ldots, a_m \in A^*$ and all $A \in \mathcal{D}$. $\square$

2. When $\mathcal{C}$ is a congruence on F and $\mathcal{D}$ is a direction set of $\mathcal{C}$, an **extent cell family** of $\mathcal{C}$ **over** $\mathcal{D}$ is a family $\mathcal{E} = (E_A)_{A \in \mathcal{D}}$ with properties (X1) through (X7) in Proposition 2.1; the sets E_A are the **extent cells** of $\mathcal{C}$. We shall show that this agrees with the definition of extent cells in Chapter X when $\mathcal{C}$ is complete group-free. A **sharp** extent cell family is an extent cell family over the archimedean direction set $D_{\mathcal{N}}$.

By Proposition 2.1, a finitely generated monoid has a sharp extent cell family which also satisfies (XM). Example VII.2.5 shows that a subcomplete monoid (in fact, finitely generated) may have more than one Ponizovsky family; hence a subcomplete congruence may have more than one extent cell family.

In general:

Proposition 2.2. *Let $\mathcal{C}$ be the congruence induced on F by a surjective homomorphism $\pi : F \longrightarrow S$. Then $\mathcal{C}$ has an extent cell family $\mathcal{E}$ if and only if it is subcomplete, and then $\mathcal{E} = \pi^{-1}(\mathcal{P})$ for some Ponizovsky family $\mathcal{P}$ of S.*

Proof. Let $\mathcal{E} = (E_A)_{A\in\mathcal{D}}$ be an extent cell family of $\mathcal{C}$ over a direction set $\mathcal{D} = D_{\mathcal{B}}$ of $\mathcal{C}$, where $\mathcal{B}$ is a semilattice congruence on S. By Proposition 1.3, $A \longmapsto \pi(A^*)$ is an isomorphism $(\mathcal{D}, \vee) \cong S/\mathcal{B}$, and $\pi(A^*) \geqq \pi(B^*)$ in $S/\mathcal{B}$ if and only if $A \subseteq B$ in F. For every $\mathcal{B}$-class C let $P_C = \pi(E_A)$, where $A \in \mathcal{D}$ and $\pi(A^*) = C$. We show that $\mathcal{P} = (P_C)_{C\in S/\mathcal{B}}$ is a Ponizovsky family of S relative to $\mathcal{B}$; then S is subcomplete by Corollary VII.3.2, and $\mathcal{E} = \pi^{-1}(\mathcal{P})$, since in the above $E_A = \pi^{-1}(P_C)$ by (X1).

Properties (P1) through (P6) follow from (X2) through (X7) since every E_A is a union of $\mathcal{C}$-classes by (X1) and direct image under π preserves unions and intersections of unions of $\mathcal{C}$-classes. Thus $\mathcal{P}$ is a partition of S, since $A \longmapsto \pi(A^*)$ is a bijection and $\mathcal{E}$ is a partition of F;

$$L_C = \bigcup (P_D \mid D \in S/\mathcal{B},\ D \leqq C) = \pi(\bigcup (E_B \mid B \in \mathcal{D},\ A \subseteq B)) = \pi(L_A)$$

is an ideal of S, since L_A is an ideal of F;

$$P_C = \pi(E_A) \subseteq \pi\left(\bigcup (B^* \mid A \subseteq B \in \mathcal{D})\right) = \bigcup (D \in S/\mathcal{B} \mid D \leqq C)$$

and $K_C = P_C \cap C = \pi(E_A \cap A^*) = \pi(I_A) \neq \emptyset$; then (P4), (P5), and (P6) follow from (X5), (X6), and (X7). □

3. Propositions 2.1 and 2.2 have a number of interesting particular cases.

Proposition 2.3. *Let $\mathcal{C}$ be the congruence induced on F by a surjective homomorphism $\pi : F \longrightarrow S$. If $S = C \cup N$ is subelementary, then $\mathcal{C}$ has a direction set $\mathcal{D} = \{A, F\}$, where $A = \pi^{-1}(C)$, and an extent cell family $\mathcal{E}$ over $\mathcal{D}$ in which $E_F = \pi^{-1}(0) = Z$ is the zero class of $\mathcal{C}$ and $E_A = F\backslash Z$. Then $L_F = I_F = Z$, $L_A = F$, and $I_A = A$.*

Proof. C and N are the classes of a semilattice congruence $\mathcal{B}$ on S. By Proposition VII.2.6, S has a standard Ponizovsky family $\mathcal{P}$ over $\mathcal{B}$ in which $P_C = S\backslash 0$, $K_C = C$, $L_C = S$, and $P_N = K_N = L_N = \{0\}$. Then $\mathcal{E} = \pi^{-1}(\mathcal{P})$ is as above. □

Proposition 2.4. *Let $\mathcal{C}$ be a congruence on F and $\mathcal{E} = (E_A)_{A\in\mathcal{D}}$ be an extent cell family of $\mathcal{C}$. Then $\mathcal{E}$ is a sharp extent cell family of $\mathcal{C}$ if and only if $I_A/\mathcal{C}$ is archimedean for every $A \in \mathcal{D}$.*

Proof. Let $\mathcal{C}$ be induced by $\pi : F \longrightarrow S$. By Proposition 2.2, $\mathcal{E} = \pi^{-1}(\mathcal{P})$ for some Ponizovsky family $\mathcal{P}$ of S. By Lemma VII.2.3, $\mathcal{P}$ is a sharp Ponizovsky family if and only if every K_C is archimedean. Hence $\mathcal{E}$ is a sharp extent cell family if and only if $I_A/\mathcal{C} \cong \pi(I_A) = K_C$ is archimedean for every $A \in \mathcal{D}$, where $C = \pi(A^*)$. □

By Proposition VII.2.4, a complete monoid has only one Ponizovsky family, namely its partition into partial Ponizovsky factors. Hence:

Proposition 2.5. *A complete congruence has only one extent cell family. The extent cell family of a complete group-free congruence coincides with the extent cell family constructed in Chapter X.*

The last part of the statement follows from Proposition X.3.4.

Proposition 2.6. *Let $\mathcal{C}$ be the congruence induced on F by a surjective homomorphism $\pi : F \longrightarrow S$, and $\mathcal{E} = (E_A)_{A \in \mathcal{D}}$ be an extent cell family of $\mathcal{C}$. Then S is complete if and only if $\pi(I_A)$ is a group for every $A \in \mathcal{D}$, and then the groups $\pi(I_A)$ are the maximal subgroups of S.*

Proof. Let $\mathcal{D} = D_{\mathcal{B}}$, where $\mathcal{B}$ is a semilattice congruence on S. If S is complete, then, by Proposition VII.2.4, S has only one Ponizovsky family, namely its partition into partial Ponizovsky factors. Hence $\mathcal{B} = \mathcal{N}$ and $K_C = H_e$, where e is the idempotent in the archimedean component C of S. If now $A \in \mathcal{D}$, then $C = \pi(A^*) \in S/\mathcal{B}$ and $\pi(I_A) = K_C = H_e$.

Conversely assume that every $\pi(I_A)$ is a group. Then $\mathcal{B} = \mathcal{N}$, by Proposition 2.4. Moreover every $\mathcal{B}$-class $\pi(A^*)$ contains an idempotent, the identity element e_A of $\pi(I_A)$. When $a \in E_A$, so that $s = \pi(a) \in P_C$, we show that e_A is the least idempotent e of S such that $es = s$. Hence S is complete.

We have $e_A = \pi(c)$ for some $c \in I_A$. Then $c \;\mathcal{C}\; 2c$. If $a \in E_A$, then $a + c \;\mathcal{C}\; a + c + c$ and $a \;\mathcal{C}\; a + c$ by (X6), since a, $a + c \in L_A$ by (X3); hence $e_A s = s$. Now assume that $fs = s$, where $f \in S$ is idempotent. Let $f = \pi(d)$, where $d \in D \in \mathcal{D}$. Then $a \;\mathcal{C}\; a + d$, $a + d \in E_A$ by (X1), and $D \subseteq A$ by (X3). Hence $c + d \in I_A$ by (X9), $f e_A \in \pi(I_A)$, $f e_A = e_A$ since $f e_A$ is idempotent, and $f \geqq e_A$. □

Proposition 2.2 also implies that a congruence which is not subcomplete does not have an extent cell family. Any monoid which is not subcomplete, such as Example VII.1.3, readily provides an example.

Example 2.7. Let S be Example III.5.2 (also Example VII.1.3), which consists of all positive powers of $s_0, s_1, \dots, s_p, \dots$, with $s_p^n = s_q^n$ whenever $n > p, q$ and $s_p^m \, s_q^n = s_r^{m+n}$, where $r = \max\,(p,q)$. This villainous c.s. has no cancellative ideal and is not subcomplete.

Let F be the free c.m. with basis $e_{m,p}$ $(m,p \in \mathbb{N},\ m > 0)$, and $\mathcal{C}$ be the congruence on F generated by all pairs $(e_{m,p},\, e_{n,q})$ such that $m = n > p,q$. Then $F/\mathcal{C} \cong S^1$ is not subcomplete; by Proposition 2.2, $\mathcal{C}$ has no extent cell family. This can also be verified directly.

S^1 has two archimedean components, $\{1\}$ and S, so either $\mathcal{B} = \mathcal{N}$ or S^1 is the only $\mathcal{B}$-class. Let $\mathcal{B} = \mathcal{N}$. By Proposition 1.3, $\mathcal{D} = D_{\mathcal{N}}$ has two elements $A \subsetneqq B$, and $B = F$ by Proposition X.1.7. Assume that $\mathcal{E} = (E_C)_{C \in \mathcal{D}}$ is an extent cell family of $\mathcal{C}$ over $\mathcal{D}$. By (X3), $E_F = I_F$ is a nonempty ideal of F. Then $0 \in E_A$, otherwise E_F would contain 0 and E_A would be empty. Hence $\pi(I_F) \subseteq \pi(E_F) \subseteq S$. By (X8), I_F is a nonempty ideal of F. By (X6) the restriction of $\mathcal{C}$ to I_F is a cancellative congruence. Then $\pi(I_F)$ is a cancellative ideal of S, contradicting the choice of S.

If S^1 is the only $\mathcal{B}$-class, then similarly $D_{\mathcal{B}}$ has only one element, namely F, $E_F = F$, and $\pi(I_F)$ is a cancellative ideal of S, which again contradicts the choice of S. □

3. We now give additional properties and a simpler characterization of extent cells.

Lemma 2.8. *Every extent cell family $\mathcal{E} = (E_A)_{A \in \mathcal{D}}$ satisfies:*

(X10) $0 \in I_{D(0)}$;

(X11) *the projections $p'_A E_B$ with $A \subseteq B \in \mathcal{D}$ constitute a partition of the complementary face A';*

(X12) $A + E_A \subseteq E_A$;

(X13) *if $a \in A$, $b,c \in L_A$, and $a + b \;\mathcal{C}\; a + c$, then $b \;\mathcal{C}\; c$.*

Proof. (X10). We have $0 \in E_A$ for some $A \in \mathcal{D}$ by (X2). Then $0 \in B^*$ for some $A \subseteq B \in \mathcal{D}$ by (X4). Now $D(0)$ is the smallest direction face by Proposition 1.2, since 0 belongs to every direction face; hence $A = B = D(0)$ and $0 \in E_A \cap A^* = I_A$.

(X11). Let $t \in A'$. By (X4) there exists $c \in I_A = L_A \cap A^*$, and then $a = c + t \in L_A$ and $p'_A a = t$. Now $a \in E_B$ for some $A \subseteq B \in \mathcal{D}$, since $a \in L_A$, and $t \in p'_A E_B$.

To show that B is unique in the above, assume that $t = p'_A b$ for some $b \in L_A$. Then $p'_A(b+c) = p'_A b = t$, $b + c \in L_A$ by (X3), and $p_A(c+b) = c + p_A b \in I_A$, since I_A is an ideal of A by (X8). Hence we may assume that $p_A b \in I_A$. Then

$p_A b + a = p_A b + c + t = b + c$; by (X5), $a \in E_B$ implies $p_A b + a \in E_B$, $b + c \in E_B$, and $b \in E_B$. Thus $b \in E_B$ whenever $b \in L_A$ has $p'_A b = t$.

(X12). Let $a \in E_A$ and $c \in A$. Then $a + c \in L_A$ by (X3) and $a + c \in E_B$ for some $A \subseteq B \in \mathcal{D}$. Now $p'_A(a + c) = p'_A a \in p'_A E_B \cap p'_A E_A$. By (X11), $B = A$, and $a + c \in E_A$.

(X13). Assume that $a \in A$, $b, c \in L_A$, and $a + b \;\mathcal{C}\; a + c$. Since I_A is a nonempty ideal of A by (X8), there exists $p \in I_A$, and $a + p \in I_A$. With $b, c \in L_A$ and $a + b + p \;\mathcal{C}\; a + c + p$ this implies $b \;\mathcal{C}\; c$, by (X6). □

Proposition 2.9. *Let $\mathcal{C}$ be a congruence on F and $\mathcal{D}$ be a direction set of $\mathcal{C}$. A partition $\mathcal{E} = (E_A)_{A \in \mathcal{D}}$ of F is an extent cell family for $\mathcal{C}$ over $\mathcal{D}$ if and only if:*

(E1) *if $a \in E_A$, $b \in E_B$, and $a \leqq b$, then $A \subseteq B$;*

(E2) $(\forall a \in A^*)(\exists n > 0)\; na \in E_A$;

(E3) *for every $A \in \mathcal{D}$, the projections $p'_A E_B$ with $A \subseteq B \in \mathcal{D}$ constitute a partition of A';*

(E4) *every E_A is a union of $\mathcal{C}$-classes;*

(E5) *if $c \in A \in \mathcal{D}$, $a, b \in E_A$, and $a + c \;\mathcal{C}\; b + c$, then $a \;\mathcal{C}\; b$.*

Properties (E1)–(E5) are the same as in Lemma X.3.5.

Proof. When $\mathcal{E}$ is an extent cell family, (E1), (E2), (E3), (E4), and (E5) follow from (X3), (X7), (X11), (X1), and (X13), respectively.

Conversely let $\mathcal{E}$ be a partition of F with properties (E1)–(E5). We show that (X1)–(X7) hold. First, (X2) holds; (X1) and (X3) follow from (E4) and (E5); (X7) follows from (E2), since $a \in A^*$ implies $na \in A^*$.

(X4). Let $a \in L_A$ and $D(a) = B$. Then $a \in B^*$ and $na \in I_B$ for some $n > 0$, by (E4). Then $na \in L_A$ by (X3) and $A \subseteq B$. (X7) also implies that $I_A \neq \emptyset$, since $A^* \neq \emptyset$.

(X5). If $A \subseteq B \in \mathcal{D}$, $a \in L_A$, and $c \in I_A$, then $a + c \in L_A$; by (E3), $a + c \in E_B$ if and only if $p'_A a = p'_A(a + c) \in p'_A E_B$, if and only if $a \in E_B$.

(X6). Let $a, b \in L_A$ and $c \in I_A$. Assume that $a + c \;\mathcal{C}\; b + c$. Then $a \in E_C$ for some $A \subseteq C \in \mathcal{D}$ and $b \in E_D$ for some $A \subseteq D \in \mathcal{D}$. By (X5), $a + c \in E_C$ and $b + c \in E_D$; hence $C = D$ by (X1). Now $a + c \;\mathcal{C}\; b + c$ with $c \in A \subseteq C$ and $a, b \in E_C$, and (E5) yields $a \;\mathcal{C}\; b$. □

Corollary 2.10. *Let $\mathcal{D}$ be a direction set on a free commutative monoid F. A partition $\mathcal{E} = (E_A)_{A \in \mathcal{D}}$ of F is an extent cell family of a subcomplete* *congruence if and only if it has properties* (E1), (E2), *and* (E3). *Then* E *is an extent cell family of the complete group-free congruence* $\mathcal{S}$ *defined by*

$$a \;\mathcal{S}\; b \textit{ if and only if } a, b \in E_A \textit{ and } p'_A a = p'_A b \textit{ for some } A \in \mathcal{D}.$$

Proof. (E1)–(E3) are necessary by Proposition 2.9. The converse follows from Theorem X.3.7: if (E1)–(E3) hold, then $\mathcal{S}$ is a c.g.f. congruence with direction set $\mathcal{D}$ and extent cell family $\mathcal{E}$ (by Proposition 2.5, the extent cells of $\mathcal{S}$ defined in Chapter X coincide with the extent cells defined in this chapter). □

An **extent cell family** on F is a family $\mathcal{E} = (E_A)_{A \in \mathcal{D}}$ such that $\mathcal{D}$ is a direction set on F and (E1), (E2), (E3) hold; equivalently, an extent cell family of a subcomplete congruence on F; equivalently, the extent cell family of a c.g.f. congruence on F.

3. TRACE CONGRUENCES.

Now that extent cells are available, the construction of subcomplete congruences continues, much as in Chapter X, with trace congruences.

1. As before $\mathcal{C}$ is the subcomplete congruence on a free c.m. F induced by a surjective homomorphism $\pi : F \longrightarrow S$; $\mathcal{E} = (E_A)_{A \in \mathcal{D}}$ is an extent cell family of $\mathcal{C}$.

For every direction face $A \in \mathcal{D}$, let

$$H_A = p'_A E_A \subseteq A' \quad \text{and} \quad J_A = A' \backslash H_A = \bigcup \left(p'_A E_B \mid B \in \mathcal{D},\ A \subsetneqq B\right),$$

where the last equality follows from (E3). By Lemma X.4.1, H_A is a coideal of A' and J_A is a nilmonoid ideal of A'. If F is finitely generated, then H_A is finite. Moreover $H_A + I_A \subseteq E_A$.

Given a subcomplete congruence $\mathcal{C}$ on F, or, more generally, an extent cell family $\mathcal{E} = (E_A)_{A \in \mathcal{D}}$, the **strands** are the $\mathcal{S}$-classes, where $\mathcal{S}$ is the congruence in Corollary 2.10:

$$a \;\mathcal{S}\; b \text{ if and only if } a, b \in E_A \text{ and } p'_A a = p'_A b \text{ for some } A \in \mathcal{D}.$$

There is one strand for every $A \in \mathcal{D}$ and $t \in H_A = p'_A E_A$, namely

$$S_A(t) = \{a \in E_A \mid p'_A a = t\}.$$

Every extent cell has a partition into strands. A $\mathcal{C}$-class is contained in a single extent cell and typically intersects several strands; hence a $\mathcal{C}$-class C is determined by: (1) the strands which intersect it; (2) the partition induced on these strands by $\mathcal{C}$; and (3) which parts of various strands constitute C.

The trace of $\mathcal{C}$ specifies the strands which intersect a $\mathcal{C}$-class. For every $A \in \mathcal{D}$ the **trace congruence** of $\mathcal{C}$ on A is the binary relation $\mathcal{C}_A$ on A' defined by:

$t \; \mathcal{C}_A \; u$ if and only if either $t, u \in J_A$, or $t = p'_A a$, $u = p'_A b$ for some $a, b \in E_A$ such that $a \; \mathcal{C} \; b$.

Thus two strands $S_A(t)$ and $S_B(u)$ contain elements that are equivalent modulo $\mathcal{C}$ if and only if $A = B$ and t, u are equivalent modulo $\mathcal{C}_A$. The **trace** of $\mathcal{C}$ (relative to the extent cell family $\mathcal{E}$, which determines the strands) is the family $\mathcal{T} = (\mathcal{C}_A)_{A \in \mathcal{D}}$.

Lemma 3.1. *The trace of a subcomplete congruence has the following properties:*

(T1) $\mathcal{C}_A$ *is a* J_A*-congruence on* A'*;*

(T2) *if* $A \in \mathcal{D}$, $t, u \in H_A$, $t \; \mathcal{C}_A \; u$, *and* $A \subsetneq B \in \mathcal{D}$, *then* $p'_A E_B : t = p'_A E_B : u$ *in* A' *and* $p'_B(t + v) \; \mathcal{C}_B \; p'_B(u + v)$ *for every* $v \in p'_A E_B : t \subseteq A'$*;*

(T3) $\{0\}$ *is a* $\mathcal{C}_A$*-class;*

(T4) *when* $t, u \in H_A$, *then* $t \; \mathcal{C}_A \; u$ *if and only if* $a + t \; \mathcal{C} \; b + u$ *for some* $a, b \in I_A$.

Proof. (T4). Let $t, u \in H_A$. If $t \; \mathcal{C}_A \; u$, then $a + t \; \mathcal{C} \; b + u$ for some $a, b \in A$ such that $a + t$, $b + u \in E_A$. Since I_A is a nonempty ideal of A by (X8), adding any $c \in I_A$ to a and b yields $a + c$, $b + c \in I_A$ and $a + c + t \; \mathcal{C} \; b + c + u$. If conversely $a + t \; \mathcal{C} \; b + u$ for some $a, b \in I_A$, then $a + t$, $b + u \in E_A$ by Lemma X.4.1 and $t \; \mathcal{C}_A \; u$.

(T1) Assume $t \; \mathcal{C}_A \; u$ and $u \; \mathcal{C}_A \; v$, where $t, u, v \in A'$. If t, u, or v is in J_A, then all three are in J_A and $t \; \mathcal{C}_A \; v$. Otherwise (T4) yields $a, b, c, d \in I_A$ such that $a + t \; \mathcal{C} \; b + u$ and $c + u \; \mathcal{C} \; d + v$. Then $a + c$, $b + c$, $b + d \in I_A$; $a + c + t$, $b + c + u$, $b + d + v \in E_A$ by Lemma X.4.1; $a + c + t \; \mathcal{C} \; b + c + u \; \mathcal{C} \; b + d + v$; and $t \; \mathcal{C}_A \; v$. Thus $\mathcal{C}_A$ is an equivalence relation.

Assume $t \; \mathcal{C}_A \; u$ and let $v \in A'$. If $t, u \in J_A$, then $t + v$, $u + v \in J_A$ and $t + v \; \mathcal{C}_A \; u + v$. Now let $t, u \in H_A$, so that $t = p'_A a$, $u = p'_A b$, where $a, b \in E_A$ and $a \; \mathcal{C} \; b$. If $t + v \in J_A$, then $a + v \in L_A$, $a + v \notin E_A$ since $p'_A(a + v) = t + v \notin H_A$, $b + v \notin E_A$ since E_A is a union of $\mathcal{C}$-classes, and $u + v = p'_A(b + v) \in J_A$. Similarly $u + v \in J_A$ implies $t + v \in J_A$. In either case $t+v \;\mathcal{C}_A\; u+v$. If finally $t+v,\ u+v \notin J_A$, then $a+v,\ b+v \in E_A$ by (E3) and $a+v \;\mathcal{C}\; b+v$, and again $t+v \;\mathcal{C}_A\; u+v$. Thus $\mathcal{C}_A$ is a congruence. It is clear that J_A is a $\mathcal{C}_A$-class.

(T2) Assume $t = p'_A a$, $u = p'_A b$, where $a, b \in E_A$ and $a \;\mathcal{C}\; b$, and let $A \subsetneqq B \in \mathcal{D}$. If $t+v \in p'_A E_B \subseteq A'$, then $p'_A(a+v) \in p'_A E_B$, $a+v \in E_B$ by (E3), $b+v \in E_B$ since E_B is a union of $\mathcal{C}$-classes, and $u+v = p'_A(b+v) \in p'_A E_B$. Then $p'_B(t+v) \;\mathcal{C}_B\; p'_B(u+v)$, since $a+v \;\mathcal{C}\; b+v$. Similarly $u+v \in p'_A E_B$ implies $t+v \in p'_A E_B$. Thus $p'_A E_B : t = p'_A E_B : u$.

(T3) Let $0 \neq t \in H_A$. Then $nt \in J_A$ for some $n > 0$ since J_A is a nilmonoid ideal of A', and $0 \;\mathcal{C}_A\; t$ would imply $0 \;\mathcal{C}_A\; nt$, $0 \in J_A$ by (T1), $J_A = A'$, and $H_A = \varnothing$, contradicting $E_A \neq \varnothing$. □

In (T1) the quotient semigroup $A'/\mathcal{C}_A$ is a nilmonoid, since J_A is a nilmonoid ideal of A' by Lemma X.4.1.

2. Conditions (T1) and (T2) in Lemma 3.1 are identical to conditions (T1) and (T2) in Lemma X.5.1. Hence every family of congruences which satisfies (T1) and (T2) is the trace of a c.f.g. congruence: by Theorem X.5.2,

Proposition 3.2. *Let $\mathcal{E} = (E_A)_{A\in\mathcal{D}}$ be an extent cell family on F. A family $\mathcal{T} = (\mathcal{C}_A)_{A\in\mathcal{D}}$ of congruences is the trace of a subcomplete congruence on F with direction set $\mathcal{D}$ and extent cell family $\mathcal{E}$ if and only if it satisfies* (T1) *and* (T2), *and then it is the trace of the complete group-free congruence $\mathcal{G}$ defined by:*

$$a \;\mathcal{G}\; b \text{ if and only if } a,b \in E_A \text{ and } p'_A a \;\mathcal{C}_A\; p'_A b \text{ for some } A \in \mathcal{D};$$

in fact $\mathcal{G}$ is the largest subcomplete congruence with direction set $\mathcal{D}$, extent cell family $\mathcal{E}$, and trace $\mathcal{T}$.

The last part of the statement follows from (T4).

A **trace** for an extent cell family $\mathcal{E} = (E_A)_{A\in\mathcal{D}}$ is a family $\mathcal{T} = (\mathcal{C}_A)_{A\in\mathcal{D}}$ which enjoys properties (T1) and (T2).

3. When $\mathcal{T}$ is the trace of a subcomplete congruence $\mathcal{C}$ on F (relative to an extent cell family $\mathcal{E}$ of $\mathcal{C}$), the congruence $\mathcal{G}$ in Proposition 3.2 is the **complete group-free hull** of $\mathcal{C}$ relative to $\mathcal{E}$; we denote it by $\mathcal{C}^*$. By definition,

$$a \;\mathcal{C}^*\; b \text{ if and only if } a,b \in E_A \text{ and } p'_A a \;\mathcal{C}_A\; p'_A b \text{ for some } A \in \mathcal{D}.$$

By Proposition 3.2 and (T4) (applied to $\mathcal{C}^*$, which has the same trace as $\mathcal{C}$),

Corollary 3.3. *$\mathcal{C}^*$ is the largest congruence on F with the same direction set, extent cell family, and trace as $\mathcal{C}$, and it is a complete group-free congruence;*

moreover $a\ \mathcal{C}^*\ b$ *if and only if*

$$a,b \in E_A \text{ and } a+c\ \mathcal{C}\ b+d \text{ for some } A \in \mathcal{D} \text{ and } c,d \in I_A.$$

Since $\mathcal{C}$ is induced by the surjective homomorphism $\pi : F \longrightarrow S$, $\mathcal{C}^* \supseteq \mathcal{C}$ induces a congruence $\pi(\mathcal{C}^*)$ on S. We show that $\pi(\mathcal{C}^*)$ is none other than the congruence $\mathcal{K}$ in Proposition VII.4.3, induced by the corresponding Ponizovsky family, under which $s\ \mathcal{K}\ t$ if and only if

$$s,t \in P_C \text{ and } ks = lt \text{ for some } C \in S/\mathcal{B} \text{ and } k,l \in K_C.$$

Proposition 3.4. *Let* $\mathcal{C}$ *be the subcomplete congruence on* F *induced by a surjective homomorphism* $\pi : F \longrightarrow S$ *and* $\mathcal{E} = \pi^{-1}(\mathcal{P})$ *be an extent cell family of* $\mathcal{C}$, *where* $\mathcal{P}$ *is a Ponizovsky family of* S. *Relative to* $\mathcal{E}$, $\mathcal{C}^* = \pi^{-1}(\mathcal{K})$.

Proof. Let $\mathcal{P} = (P_C)_{C \in S/\mathcal{B}}$, where $\mathcal{B}$ is a semilattice congruence on S. When $\mathcal{E} = (E_A)_{A \in \mathcal{D}} = \pi^{-1}(\mathcal{P})$, then, as in Proposition 2.1, $\mathcal{D} = D_{\mathcal{B}}$, so that $A \longmapsto \pi(A^*)$ is an isomorphism $(\mathcal{D}, \vee) \longrightarrow S/\mathcal{B}$, and $E_A = \pi^{-1}(P_C)$, $I_A = \pi^{-1}(K_C)$ whenever $C = \pi(A^*)$. Hence $\pi(a)\ \mathcal{K}\ \pi(b)$ if and only if $a,b \in E_A$ and $a+c\ \mathcal{C}\ b+d$ for some $A \in \mathcal{D}$ and $c,d \in I_A$, if and only if $a\ \mathcal{C}^*\ b$. □

Proposition 3.5. *If* S *is finitely generated, then* $A'/\mathcal{C}_A$ *is finite and* $F/\mathcal{C}^*$ *is finite.*

Proof. If S is f.g., then $F/\mathcal{C}^* \cong S/\mathcal{K}$ is finite by Proposition VII.4.3.

Let $\mathcal{C}_A^*$ be the congruence induced by the projection $F \longrightarrow A' \longrightarrow A'/\mathcal{C}_A$; in other words, $a\ \mathcal{C}_A^*\ b$ if and only if $p'_A a\ \mathcal{C}_A\ p'_A b$. Then $F/\mathcal{C}_A^* \cong A'/\mathcal{C}_A$. Also $\mathcal{C} \subseteq \mathcal{C}_A^*$ since $a\ \mathcal{C}\ b$ implies $p'_A a \mathcal{C}_A p'_A b$. Hence the nilmonoid $A'/\mathcal{C}_A \cong F/\mathcal{C}_A^*$ is a homomorphic image of $F/\mathcal{C} \cong S$, is f.g., and is finite. □

4. STRAND GROUPS.

Strand groups constitute the next step in the construction of subcomplete congruences; they determine the partitions induced by the congruence on its strands, and the Schützenberger monoids of the quotient monoid.

1. As before $\mathcal{C}$ is the subcomplete congruence on a free c.m. F induced by a surjective homomorphism $\pi : F \longrightarrow S$; $\mathcal{E} = (E_A)_{A \in \mathcal{D}}$ is an extent cell family of $\mathcal{C}$. We denote by G the universal group of F, which is the free abelian group with the same basis as F; if A is a face of F, then G_A is the universal group

of A, which is the free abelian group with the same basis as A, and which we regard as a subgroup of G.

When $t \in H_A$, let

$$I_A(t) = \{c \in A \mid c+t \in E_A\} = \{c \in A \mid c+t \in S_A(t)\}$$

and let $\mathcal{C}_t$ be the equivalence relation on $I_A(t)$ defined by:

$$c \; \mathcal{C}_t \; d \text{ if and only if } c+t \;\; \mathcal{C} \;\; d+t.$$

Then $I_A = I_A(0)$ and every element of the strand $S_A(t) = \{a \in E_A \mid p'_A a = t\}$ can be written uniquely in the form $a = c+t$ with $c = p_A a \in I_A(t)$; and $\mathcal{C}_t$ is a copy of the restriction of $\mathcal{C}$ to $S_A(t)$.

Proposition 4.1. *Let $A \in \mathcal{D}$ and $t \in H_A$. Then $I_A(t)$ is an ideal of A; $\mathcal{C}_t$ is a cancellative congruence on $I_A(t)$;*

$$\begin{aligned} R_A(t) &= \{a-b \in G_A \mid a,b \in S_A(t) \;\; \textit{and} \;\; a \; \mathcal{C} \; b\} \\ &= \{c-d \in G_A \mid c,d \in I_A \;\; \textit{and} \;\; c+t \;\; \mathcal{C} \;\; d+t\} \end{aligned}$$

is a subgroup of G_A; when $a,b \in S_A(t)$, then $a \; \mathcal{C} \; b$ if and only if $a-b \in R_A(t)$; and $G(I_A(t)/\mathcal{C}_t) \cong G_A/R_A(t)$.

Proof. This follows from Proposition II.5.1, according to which cancellative congruences on a cancellative semigroup are induced by subgroups of its universal group.

First, if $c+t \in E_A$ and $c \leqq d \in A$, then $d+t \in L_A$ since L_A is an ideal of F, and $d+t \in E_A$ by (E3) since $p'_A(d+t) = t \in H_A = p'_A E_A$. Thus $S_A(t) + A \subseteq S_A(t)$ and $I_A(t)$ is an ideal of A. Hence $G(I_A(t)) = G(A) = G_A$, by Proposition II.2.5. Also $\mathcal{C}_t$ is a congruence on $I_A(t)$ since $\mathcal{C}$ is a congruence, and $I_A(t)/\mathcal{C}_t$ is cancellative: if $c,d,e \in I_A(t)$ and $c+e \;\; \mathcal{C}_t \;\; d+e$, then $c+e+t \;\; \mathcal{C} \;\; d+e+t$ and $c+t \;\; \mathcal{C} \;\; d+t$ by (E5).

By Proposition II.5.1,

$$R = R_A(t) = \{c-d \in G(I_A(t)) \mid c,d \in I_A(t) \text{ and } c+t \;\; \mathcal{C}_t \;\; d+t\}$$

is a subgroup of $G(I_A(t))$; $c+t \;\; \mathcal{C}_t \;\; d+t$ if and only if $c-d \in R$, for all $c,d \in I_A(t)$; and $G(I_A(t)/\mathcal{C}_t) \cong G(I_A(t))/R_A(t)$. Hence

$$R_A(t) = \{a-b \in G_A \mid a,b \in S_A(t) \text{ and } a \; \mathcal{C} \; b\};$$

R is a subgroup of G_A; when $a,b \in S_A(t)$, then $a \;\; \mathcal{C} \;\; b$ if and only if $a-b \in R$; and $G(I_A(t)/\mathcal{C}_t) \cong G_A/R_A(t)$.

Finally let $c,d \in I_A(t)$. Since $S_A(t) + A \subseteq S_A(t)$ and I_A is an ideal of A, we may add any $e \in I_A$ to c and d, to the effect that $c - d = (c+e) - (d+e)$, with $c+e$ and $d+e \in I_A$ and $c+e+t \;\mathcal{C}\; d+e+t$. If conversely $r = c-d$ with $c,d \in I_A$ and $c+t \;\mathcal{C}\; d+t$, then $c,d \in I_A(t)$ since $H_A + I_A \subseteq E_A$ (Lemma X.4.1) and $r \in R_A(t)$. Thus

$$R_A(t) \;=\; \{c-d \in G_A \mid c,d \in I_A \text{ and } c+t \;\mathcal{C}\; d+t\}. \square$$

$R_A(t)$ is similar to but smaller than the Rédei group of $\mathcal{C}$; in fact, the Rédei group of $\mathcal{C}$ is the union of all $R_A(t)$ (Proposition 5.4).

$R_A(t)$ is the **strand group** of $t \in H_A$; the family $\mathcal{R} = (R_A(t))_{A\in\mathcal{D},\, t\in H_A}$ is the **strand group family** of $\mathcal{C}$ relative to its extent cell family $\mathcal{E}$.

In Proposition 4.1 we denote the quotient $I_A(t)/\mathcal{C}_t$ by $I_A(t)/R_A(t)$. The $\mathcal{C}_t$-class of $a \in I_A(t)$ is the intersection of $I_A(t)$ and the coset $a + R_A(t)$ of $R_A(t)$ in G_A. It is convenient to view $I_A(t)/R_A(t)$ as a submonoid of $G_A/R_A(t)$; then $G\big(I_A(t)/R_A(t)\big) = G_A/R_A(t)$.

Similarly, when R is a subgroup of G_A, we denote by A/R the quotient of A by the cancellative congruence $\equiv$ under which $c \equiv d$ if and only if $c-d \in R$. The equivalence class of $c \in A$ is the intersection of A and the coset $c+R$ of c in G_A; hence we may view A/R as a submonoid of G_A/R_a. Then $G(A/R) = G_A/R$, by Proposition II.5.1.

2. **Lemma 4.2.** *The strand groups of a subcomplete congruence $\mathcal{C}$ have the following properties:*

(R1) *$R_A(t)$ is a subgroup of G_A;*

(R2) *if $t,u \in H_A$ and $t \;\mathcal{C}_A\; u$, then $R_A(t) = R_A(u)$;*

(R3) *if $A \subseteq B \in \mathcal{D}$, $t \in H_A$, $v \in A'$, and $t+v \in p'_A E_B$, then $R_A(t) \subseteq R_B(p'_B(t+v))$. (In particular, $R_A(t) \subseteq R_A(u)$ whenever $t \leqq u$ in H_A.)*

Proof. (R1) follows from Proposition 4.1.

(R2). Let $t,u \in H_A$ and $t \;\mathcal{C}_A\; u$. If $r \in R_A(t)$, then $r = a-b$, where $a,b \in I_A$ and $a+t \;\mathcal{C}\; b+t$. By (T4) we also have $c+t \;\mathcal{C}\; d+u$ for some $c,d \in I_A$. Then

$$a+d+u \;\mathcal{C}\; a+c+t \;\mathcal{C}\; b+c+t \;\mathcal{C}\; b+d+u,$$

$a+u \;\mathcal{C}\; b+u$ by (X6), and $r = a-b \in R_A(u)$. Similarly, $r \in R_A(u)$ implies $r \in R_A(t)$.

(R3). Let $A \subseteq B \in \mathcal{D}$, $t \in H_A$, $v \in A'$, $t+v \in p'_A E_B$, and $r \in R_A(t)$. Then

$r = c - d$, where $c, d \in I_A$ and $c + t \; \mathcal{C} \; d + t$. Then $c + t + v \; \mathcal{C} \; d + t + v$, with $c + t + v$, $d + t + v \in E_B$ by (E3) since $p'_A(c + t + v) = p'_A(d + t + v) = t + v \in p'_A E_B$. Hence $r \in R_B(p'_B(t + v))$. □

In the next section we shall see that properties (R1)–(R3) characterize strand group families (Corollary 5.4).

3. We illustrate properties (R1)–(R3) by two examples.

Example A. This is Example A from Chapter X: $X = \{x, y\}$, so that F has four faces, of which two, $B = F_{\{y\}}$ and F, constitute the direction set in Example A. The extent cells are

$$E_B = \{a \in F \mid a_x \leqq 2\} \text{ and } E_F = \{a \in F \mid a_x \geqq 3\};$$

hence $H_B = \{0, x, 2x\}$ and $H_F = \{0\}$. The figure shows $\mathcal{S}$. There are four strands: $S_B(0) = B$, $S_B(x)$, $S_B(2x)$, and $S_F(0) = E_F$. The only possible trace congruences are Rees congruences; hence $\mathcal{S} = \mathcal{C}^*$ for every subcomplete congruence $\mathcal{C}$ with the given extent cells; $F/\mathcal{C}^*$ is the cyclic nilmonoid $\{1, x, x^2, 0\}$.

By (R1), $R_B(0)$, $R_B(x)$, $R_B(2x)$ are subgroups of $G_B = \mathbb{Z}y$, and $R_F(0)$ is a subgroup of $G = \mathbb{Z}x \oplus \mathbb{Z}y$. (R2) is trivial since all trace congruences are Rees congruences. (R3) states that $R_B(0) \subseteq R_B(x) \subseteq R_B(2x) \subseteq R_F(0)$. □

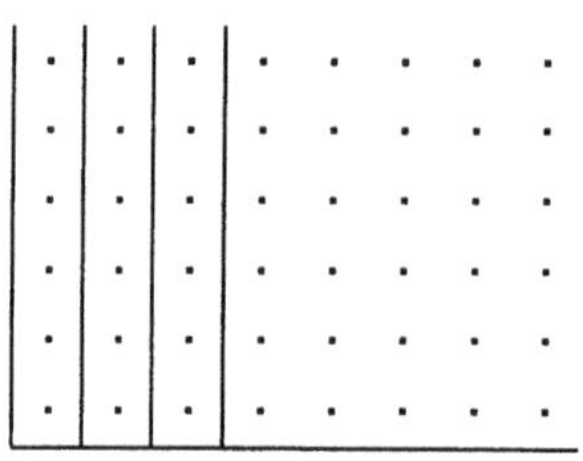

Example A

Example C

Example C. This example, like Example A, will be seen again in later sections. Let $X = \{x, y, z\}$ and $\mathcal{D}$ consists of $A = F_{\{x\}}$, $B = F_{\{x,y\}}$, and F. This example is best visualized from its projection to A' above.

We have $H_A = \{0, y, z, y + z, 2z\}$ and $E_A = \{a \in F \mid a_y y + a_z z \in H_A\}$; $E_B = \{a \in F \mid a_y \geqq 2, a_z = 0\}$ and $H_B = \{0\}$; $E_F = F \setminus (E_A \cup E_B)$ and $H_F = \{0\}$. There is a trace relative to $\mathcal{E}$ in which $k \; \mathcal{C}_A \; l$, where $k = 2z$ and $l = y + z$ are the corner points of H_A. Then $F/\mathcal{G} = \{1, y, z, yz, e, 0\}$, where 1 is the identity element, $yz = z^2$, $y^2 = e = ye = e^2$, and all other products are 0.

By (R1), $R_A(t)$ is a subgroup of $G_A = \mathbb{Z}x$ for every $t \in H_A$; $R_B(0)$ is a subgroup of $G_B = \mathbb{Z}x \oplus \mathbb{Z}y$; and $R_F(0)$ is a subgroup of $G = \mathbb{Z}x \oplus \mathbb{Z}y \oplus \mathbb{Z}z$. By (R2), $R_A(k) = R_A(l)$. By (R3), $R_A(0) \subseteq R_A(y)$, $R_A(z)$ and $R_A(y), R_A(z) \subseteq R_A(k) = R_A(l)$, $R_A(y) \subseteq R_B(0)$, and $R_A(k)$, $R_B(0) \subseteq R_F(0)$. □

4. In some cases properties stronger than (R1) hold, to the effect that the groups $R_A(0)$ (and therefore all groups $R_A(x)$) cannot be too small.

Proposition 4.3. *When $\mathcal{C}$ is a subcomplete congruence on F, $\mathcal{E}$ is a sharp extent cell family if and only if for every $A \in \mathcal{D}$ and every $a, b \in I_A$, there exist $n > 0$ and $r \in R_A(0)$ such that $na + r \geqq b$.*

Proof. By Proposition 2.4, $\mathcal{E}$ is a sharp extent cell family if and only if every $I_A/\mathcal{C}$ is archimedean. If $I_A/\mathcal{C}$ is archimedean, then for every $a, b \in I_A$ there exist $n > 0$ and $c \in I_A$ such that $na \;\mathcal{C}\; b + c$; then $r = b + c - na \in R_A(0)$ and $na + r = b + c \geqq b$. If conversely $na + r \geqq b$ for some $r \in R_A(0)$, then $na + r \in I_A$, since $c \in A$ and $c \geqq b \in I_A$ and $na + r = b + c$ for some $c \in A$, $na \;\mathcal{C}\; b + c$, and $(n+1)a \;\mathcal{C}\; b + (a + c)$ with $a + c \in I_A$; if this holds for all a and b, then $I_A/\mathcal{C}$ is archimedean. □

Proposition 4.4. *Let $\mathcal{C}$ be the congruence induced on F by a surjective homomorphism $\pi : F \longrightarrow S$, and $\mathcal{E} = (E_A)_{A \in \mathcal{D}}$ be an extent cell family of $\mathcal{C}$. Then S is complete if and only if every $I_A/R_A(0)$ is a group, if and only if for every $A \in \mathcal{D}$ and $g \in G_A$ there exists $r \in R_A(0)$ such that $g + r \in I_A$.*

Proof. By Proposition 2.6, S is complete if and only if every $I_A/\mathcal{C} = I_A/R_A(0)$ is a group. Since $I_A/R_A(0)$ is cancellative, it is a group if and only if $I_A/R_A(0) = G(I_A/R_A(0)) = G_A/R_A(0)$, if and only if every coset of $R_A(0)$ in G_A intersects I_A. □

In particular, S is complete group-free if and only if $R_A(0) = G_A$ for all $A \in \mathcal{D}$.

Proposition 4.5. *Let $\mathcal{C}$ be the congruence induced on F by a surjective homomorphism $\pi : F \longrightarrow S$. If S is finite, then $R_A(0)$ has finite index in G_A for every $A \in \mathcal{D}$.*

Proof. If S is finite, then it is complete and every coset of $R_A(0)$ in G_A intersects I_A by Proposition 4.4. These intersections are the $\mathcal{C}$-classes contained in I_A. Therefore $R_A(0)$ has only finitely many cosets in G_A. □

5. When $a \in E_A$, then A is uniquely determined by a by (X2) and it is convenient to denote G_A by G_a, $S_A(p'_A a)$ by S_a, and $R_A(p'_A a)$ by R_a. Thus

S_a is the $\mathcal{S}$-class of a and $a \mathrel{\mathcal{C}} b$ implies $R_a = R_b$, by (R2); if $a \mathrel{\mathcal{S}} b$, then $a \mathrel{\mathcal{C}} b$ if and only if $a - b \in R_a = R_b$.

Armed with this notation we show that the Schützenberger monoids and functor of S can be reconstructed from the strand groups of $\mathcal{C}$.

Proposition 4.6. *When $a \in E_A$, $\Sigma(K_{\pi(a)}) \cong A/R_a$ and $\Gamma(K_{\pi(a)}) \cong G_a/R_a$.*

Proof. As before, $\mathcal{E}$ is an extent cell family of $\mathcal{C}$, and $\mathcal{E} = \pi^{-1}(\mathcal{P})$, where $\mathcal{P}$ is a Ponizovsky family of S. By Proposition 3.4, $\mathcal{C}^* = \pi^{-1}(\mathcal{K})$.

Recall that the Schützenberger monoid $\Sigma(K)$ of a $\mathcal{K}$-class K, relative to $\mathcal{P}$, is constructed as follows (Section VII.5). Let

$$U(K) = \{u \in S \mid uK \subseteq K\}.$$

For every $u \in U(K)$ an injective mapping $\sigma_K(u) : K \longrightarrow K$ well defined by

$$\sigma_K(u)(k) = uk$$

for all $k \in K$. Then

$$\Sigma(K) = \{\sigma_K(u) \mid u \in U(K)\};$$

$\Sigma(K)$ is a commutative cancellative monoid which acts simply on K (if $\sigma, \tau \in \Sigma(K)$ and $\sigma k = \tau k$ for some $k \in K$, then $\sigma = \tau$) (Proposition VII.5.2). The Schützenberger group of K is the universal group $\Gamma(K) = G(\Sigma(K))$.

Let K be the $\mathcal{K}$-class of $\pi(a)$. Then $\pi(u) \in U(K)$ if and only if $\pi(u)\,\pi(a) \in K$, if and only if $a + u \mathrel{\mathcal{C}^*} a$. For the c.g.f. congruence $\mathcal{C}^*$ this is equivalent to $u \in D(a)$. Since $\mathcal{C}^*$ has the same direction set and extent cell family as $\mathcal{C}$ (Corollary 3.3), we have $D(a) = A$, since $a \in E_A$, and $\pi(u) \in U(K)$ if and only if $u \in A$.

If $\pi(u)$, $\pi(v) \in U(K)$, then $u, v \in A$, $a + u$, $a + v \in E_A$ by (X12), and $a + u$, $a + v \in S_a$; hence $\sigma_K(\pi(u)) = \sigma_K(\pi(v))$ if and only if $\pi(u)\,\pi(a) = \pi(v)\,\pi(a)$, if and only if $a + u \mathrel{\mathcal{C}} a + v$, if and only if $u - v \in R_a$. Therefore $\Sigma(K) \cong A/R_a$; the isomorphism sends $\sigma_K(\pi(u))$ to $\pi_a(u)$, where $\pi_a : A \longrightarrow A/R_a$ is the projection. This extends to an isomorphism $\Gamma(K) \cong G(A/R_a) = G_a/R_a$. □

Let $\pi_a : A \longrightarrow A/R_a$ be the projection. When $a \leqq b$ in F, with, say, $a \in E_A$ and $b \in E_B$, then $A \subseteq B$ by (E1) and $R_a \subseteq R_b$ by (R3); therefore $\pi_a(c) = \pi_a(d)$ implies $\pi_b(c) = \pi_b(d)$, and there is a homomorphism $\varphi^a_b : A/R_a \longrightarrow B/R_b$ which sends $\pi_a(c)$ to $\pi_b(c)$ for every $c \in A$. This in turn induces a homomorphism

$\psi^a_b : G_a/R_a = G(A/R_a) \longrightarrow G(B/R_b) = G_b/R_b$ which sends $g + R_a$ to $g + R_b$ for every $g \in G_a$.

Lemma 4.7. *Let $a \leqq b$ in F, with $a \in E_A$, $b \in E_B$, and $K = K_{\pi(a)}$, $L = K_{\pi(b)}$. The following square commutes:*

$$\begin{array}{ccc} \Sigma(K) & \xrightarrow{\cong} & A/R_a \\ {\scriptstyle \Sigma^K_L}\downarrow & & \downarrow{\scriptstyle \varphi^a_b} \\ \Sigma(L) & \xrightarrow{\cong} & B/R_b \end{array}$$

Proof. When $K \geqq L$ in $S/\mathcal{K}$, then $U(K) \subseteq U(L)$ and there is a unique homomorphism $\Sigma^K_L : \Sigma(K) \longrightarrow \Sigma(L)$ which sends $\sigma_K(u)$ to $\sigma_L(u)$ (Proposition VII.5.7). Now $a \leqq b$ implies $\pi(a) \geqq \pi(b)$ and $K \geqq L$ in $S/\mathcal{K}$. For all $u \in A$, $v \in B$, the isomorphisms $\Sigma(K) \cong A/R_a$, $\Sigma(L) \cong B/R_b$ send $\sigma_K(u)$, $\sigma_L(v)$ to $\pi_a(u)$, $\pi_b(v)$ respectively, whereas Σ^K_L sends $\sigma_K(u)$ to $\sigma_L(v)$ and φ^a_b sends $\pi_a(u)$ to $\pi_b(v)$. Hence the square above commutes. □

Let C^*_a denote the $\mathcal{C}^*$-class of $a \in F$. When $a \;\mathcal{C}^*\; b$, with $a \in E_A$, $b \in E_B$, then $A = B$, $p'_A a \;\mathcal{C}_A\; p'_A b$, and $R_a = R_b$ by (R2). Hence A/R_a and G_a/R_a depend only on C^*_a. Similarly, φ^a_b and ψ^a_b depend only on C^*_a and C^*_b.

If $a \leqq b$ in F, then $C^*_a \geqq C^*_b$ in $F/\mathcal{C}^*$. If conversely $C^*_a \geqq C^*_b$ in $F/\mathcal{C}^*$, then $C^*_b = C^*_a C^*_t = C^*_{a+t}$ for some $t \in F$, with $a + t \geqq a$. Thus $C^*_a \geqq C^*_b$ in $F/\mathcal{C}^*$ if and only if $C^*_b = C^*_c$, where $a \leqq c$ in F. Therefore the monoids A/R_a and homomorphisms φ^a_b induce a monoid valued functor on $F/\mathcal{C}^*$, the **strand monoid functor** of $\mathcal{C}$ relative to the extent cell family $\mathcal{E}$; and the groups G_a/R_a and homomorphisms ψ^a_b induce an abelian group valued functor on $F/\mathcal{C}^*$, the **strand group functor** of $\mathcal{C}$ relative to $\mathcal{E}$.

If S is complete, then $A/R_a = G_A/R_a$ for every $a \in A$, by Proposition 4.4, since $R_A(0) \subseteq R_a$; then the strand monoid functor of $\mathcal{C}$ (relative to the only extent cell family of $\mathcal{C}$) is an abelian group valued functor and coincides with the strand group functor of $\mathcal{C}$.

In general Lemma 4.7 yields:

Proposition 4.8. *Let $\mathcal{C}$ be the subcomplete congruence on F induced by a surjective homomorphism $\pi : F \longrightarrow S$ and $\mathcal{E} = \pi^{-1}(\mathcal{P})$ be an extent cell family of $\mathcal{C}$, where $\mathcal{P}$ is a Ponizovsky family of S. Up to the isomorphism $F/\mathcal{C}^* \cong S/\mathcal{K}$, the strand monoid functor of $\mathcal{C}$ (relative to $\mathcal{E}$) is isomorphic to the Schützenberger monoid functor of S (relative to $\mathcal{P}$), and the strand group* *functor of $\mathcal{C}$ is isomorphic to the extended Schützenberger functor of S. If S is complete, then, up to the isomorphism $F/\mathcal{C}^* \cong S/\mathcal{H}$, the strand group functor of $\mathcal{C}$ is isomorphic to the Schützenberger functor of S.*

5. STRAND BASES.

When $\mathcal{C}$ is a subcomplete congruence, a strand base determines when two elements of different strands lie in the same $\mathcal{C}$-class; this is the final step in the construction of $\mathcal{C}$.

1. Given an extent cell family $\mathcal{E}$ of $\mathcal{C}$, a **strand base** of $\mathcal{C}$ is a cross-section $s = \big(s_A(t)\big)_{A\in\mathcal{D},\, t\in H_A}$ of $\mathcal{S}$ such that $s_A(t) \in S_A(t)$ for all $t \in H_A$ and $s_A(t) \mathrel{\mathcal{C}} s_A(u)$ whenever $t, u \in H_A$ and $t \mathrel{\mathcal{C}_A} u$.

A cross-section of $\mathcal{S}$ with these properties always exists: for every $A \in \mathcal{D}$, we can, for one t in every $\mathcal{C}_A$-class, choose $s_A(t)$ arbitrarily in $S_A(t)$; then every other $s_A(u)$ is unique modulo $R_A(u)$, by Proposition 4.1. This construction suggests that the typical subcomplete congruence has many strand bases.

When $a \in E_A$ it is convenient to denote $s_A(p'_A a)$ by s_a; then $s_a \in S_a$, and $s_a \mathrel{\mathcal{C}} s_b$ whenever $a \mathrel{\mathcal{C}} b$ (which implies $a, b \in E_A$ and $p'_A a \mathrel{\mathcal{C}_A} p'_A b$).

A subcomplete congruence is completely determined by its extent cells, trace, strand groups, and a strand base:

Proposition 5.1. *$a \mathrel{\mathcal{C}} b$ if and only if $a, b \in E_A$, $p'_A a \mathrel{\mathcal{C}_A} p'_A b$, and $a - s_A(p'_A a) - b + s_A(p'_A b) \in R_A(p'_A a)$, for some $A \in \mathcal{D}$; equivalently, $a \mathrel{\mathcal{C}^*} b$ and $a - s_a - b + s_b \in R_a$.*

Note that $p'_A a \mathrel{\mathcal{C}_A} p'_A b$ implies $R_A(p'_A a) = R_A(p'_A b)$, by (R2).

Proof. Let $a, b \in E_A$; let $a = c + t$, $b = d + u$, where $c, d \in A$, $t = p'_A a \in H_A$, and $u = p'_A b \in H_A$. Let $s_a = q + t$ and $s_b = r + u$, where $q, r \in A$. Then $a - s_a = c + t - s_a = c - q$. Also, $c + s_b$, $q + b \in E_A$ by (X12); in fact $c + s_b$, $q + b \in S_b$.

If $a \mathrel{\mathcal{C}} b$, then $a, b \in E_A$ for some $A \in \mathcal{D}$, and in the above $t \mathrel{\mathcal{C}_A} u$ by definition of $\mathcal{C}_A$ and $s_a = s_A(t) \mathrel{\mathcal{C}} s_A(u) = s_b$. Hence

$$c + s_b \mathrel{\mathcal{C}} c + s_a = q + a \mathrel{\mathcal{C}} q + b$$

and $a - s_a - b + s_b = (c + s_b) - (q + b) \in R_b$ by Proposition 4.1, since $c + s_b$, $q + b \in S_b$.

If conversely $t \;\mathcal{C}_A\; u$ and $a - s_a - b + s_b \in R_a$, then $(c + s_b) - (q + b) = a - s_a - b + s_b \in R_b$; $c + s_b \;\mathcal{C}\; q + b$ by Proposition 4.1, since $c + s_b,\ q + b \in S_b$;

$$q + a \;=\; c + s_a \;\mathcal{C}\; c + s_b \;\mathcal{C}\; q + b;$$

and $a \;\mathcal{C}\; b$, by (E5). □

Two strand bases (for the same strand group family) are **equivalent** when they define the same congruence. By Proposition 5.1,

Proposition 5.2. *Two strand bases s and t are equivalent if and only if $a \;\mathcal{C}^*\; b$ implies $s_a - s_b - t_a + t_b \in R_a (= R_b)$.*

2. From Proposition 5.1 a set of generating pairs for $\mathcal{C}$ (a presentation of $S \cong F/\mathcal{C}$) can be set up as follows. For every $A \in \mathcal{D}$ and $t \in H_A$, let

$$I_A(t) \;=\; \{a \in A \mid a + t \in S_A(t)\} \;=\; \{a \in A \mid a + t \in E_A\}$$

and $\mathcal{R}_A(t)$ be the intersection of the Rees congruence of $I_A(t)$ and the cancellative congruence on A induced by $R_A(t)$; thus

$$a \;\mathcal{R}_A(t)\; b \text{ if and only if } a = b, \text{ or } a, b \in I_A(t) \text{ and } a - b \in R_A(t);$$

if $a, b \in I_A(t)$, then $a \;\mathcal{R}_A(t)\; b$ if and only if $a + t \;\mathcal{C}\; b + t$.

Proposition 5.3. *A subcomplete congruence $\mathcal{C}$ on $F = F_X$ is generated by:*

(1) *for every $A \in \mathcal{D}$ and $t \in H_A$, one pair $(a + t,\ b + t)$ for every generator (a, b) of $\mathcal{R}_A(t)$; and*

(2) *for every $A \in \mathcal{D}$ and $t \neq u \in H_A$ such that $t \;\mathcal{C}_A\; u$, every pair (a, b) which is minimal such that $a \in S_A(t)$, $b \in S_A(u)$, and $a - s_A(t) - b + s_A(u) \in R_A(t) = R_A(u)$.*

Proof. Let $\mathcal{B}$ be the smallest congruence on F containing all pairs (1) and (2). Then $\mathcal{B} \subseteq \mathcal{C}$.

Let $\mathcal{Q}_A(t)$ be the binary relation on A defined by:

$$a \;\mathcal{Q}_A(t)\; b \text{ if and only if } a = b, \text{ or } a, b \in I_A(t) \text{ and } a + t \;\mathcal{B}\; b + t.$$

Then $\mathcal{Q}_A(t)$ is a congruence on A, since I_A is an ideal of A and $\mathcal{B}$ is a congruence; also, $\mathcal{Q}_A(t) \subseteq \mathcal{R}_A(t)$, since $\mathcal{B} \subseteq \mathcal{C}$. On the other hand $\mathcal{Q}_A(t)$ contains every generator (a, b) of $\mathcal{R}_A(t)$, since $\mathcal{B}$ contains every pair in (1). Therefore $\mathcal{Q}_A(t) = \mathcal{R}_A(t)$. Hence $\mathcal{B}$ and $\mathcal{C}$ induce the same partition of each strand.

Now let $a \;\mathcal{C}\; b$. By Proposition 5.1 we have $a, b \in E_A$, $t \;\mathcal{C}_A\; u$, and $a - s_A(t) - b + s_A(u) = r \in R_A(t) = R_A(u)$, where $t = p'_A a \in H_A$ and $u = p'_A b \in H_A$.

By the descending chain condition in F there is a pair $(c,d) \leqq (a,b)$ which is minimal such that $c \in S_A(t)$, $d \in S_A(u)$, and $c - s_A(t) - d + s_A(u) \in R_A(t) = R_A(u)$. Then $c \mathcal{B} d$. Since $c \leqq a$ and $d \leqq b$, we have $e = d + (a - c) \in S_A(u)$ and $a \mathcal{B} e$. Also

$$a - s_A(t) - e + s_A(u) \;=\; c - s_A(t) - d + s_A(u) \;\in\; R_A(u).$$

Since $a - s_A(t) - b + s_A(u) = r \in R_A(u)$, it follows that $b - e \in R_A(u)$. Hence $b \mathcal{C} e$. Then $b \mathcal{B} e$, since $\mathcal{B}$ and $\mathcal{C}$ induce the same partition of $S_A(u)$, and $a \mathcal{B} b$. Thus $\mathcal{B} = \mathcal{C}$. □

If $\mathcal{C}$ is c.g.f., then Proposition 5.3 yields a presentation which is similar to but less economical than the presentation in Proposition X.4.7. Indeed $\mathcal{R}_A(t)$ is the Rees congruence of $I_A(t)$ and is generated by all pairs $(m + x, m)$ where m is a minimal element of $R_A(t)$ and $x \in X \cap A$; and m is minimal in $I_A(t)$ if and only if $m + t$ is minimal in $S_A(t)$. Also one can construct a strand base merely by picking one minimal element in each strand. With these choices, Proposition 5.3 yields the same pairs as Proposition X.4.7, except that we now have one pair for every (t, u) with $t \mathcal{C}_A u$, $t \neq u$, rather than one pair for every generator of $\mathcal{C}_A$.

When F is finitely generated, then Proposition 5.3 provides a finite generating set: Indeed (1) yields only finitely many pairs, since $\mathcal{R}_A(t)$ is finitely generated by Rédei's Theorem, and $\mathcal{D}$ and every H_A are finite. Similarly, pairs (a, b) which are minimal such that $a \in S_A(t)$, $b \in S_A(u)$, and $a - s_A(t) - b + s_A(u) \in R_A(t) = R_A(u)$, constitute an antichain in $F \times F$, which is finite by Dickson's Theorem (Corollary VI.1.3); since $\mathcal{D}$ and every H_A are finite, (2) yields only finitely many pairs.

3. Proposition 5.1 also yields the kernel function of $\mathcal{C}$. By Proposition X.4.8, the kernel function f^* of $\mathcal{C}^*$ is given by

$$c \in f^*(g) \text{ if and only if } c \in (E_A : g^+) \cap (E_A : g^-) \cap f_A(g) \text{ for some } A \in \mathcal{D},$$

where $f_A(g) = {p'_A}^{-1}(f'_A(p'_A g))$ and f'_A is the kernel function of $\mathcal{C}_A$. If F is finitely generated (more generally, if J_A is cofinite in A'), then f'_A can be constructed more precisely as in Proposition IX.4.9.

Proposition 5.4. *Let $\mathcal{C}$ be a c.g.f. congruence on F. The Rédei group of $\mathcal{C}$ is*

$$R \;=\; \bigcup\nolimits_{A \in \mathcal{D},\, t \in H_A} R_A(t) \;=\; \bigcup\nolimits_{A \in \mathcal{D}} R_A(0)$$

The kernel function f of $\mathcal{C}$ is given by:

$c \in f(g)$ if and only if $c \in (E_A : g^+) \cap (E_A : g^-) \cap f_A(g)$ and $p_A\big(s_A(t) - s_A(u)\big) \in p_A g + R_A(t)$ for some $A \in \mathcal{D}$, where $t = p'_A(c + g^+)$ and $u = p'_A(c + g^-)$.

Proof. Every $R_A(t)$ is contained in the Rédei group $R = \{a - b \in G \mid a \; \mathcal{C} \; b\}$ of $\mathcal{C}$, by Proposition 4.1. Conversely let $r = a - b \in R$, where $a \; \mathcal{C} \; b$. We have $a + b \in A$ for some $A \in \mathcal{D}$ by Proposition 1.2. Then $a, b \in A$, $a + c$, $b + c \in I_A$ for any $c \in I_A$, and $r = (a + c) - (b + c) \in R_A(0)$.

Let $c \in F$ and $g \in G$. Then $c \in f(g)$ if and only if $c + g^+ \; \mathcal{C} \; c + g^-$, if and only if $c + g^+$, $c + g^- \in E_A$, $p'_A(c + g^+) \; \mathcal{C}_A \; p'_A(c + g^-)$, and

$$(c + g^+) - s_A(t) - (c + g^-) + s_A(u) \in R_A(t)$$

for some unique $A \in \mathcal{D}$, by Proposition 5.1, where $t = p'_A(c + g^+)$ and $u = p'_A(c + g^-)$. As in the proof of Proposition X.4.8, $(p'_A g)^+ = p'_A(g^+)$ and $(p'_A g)^- = p'_A(g^-)$. When $c + g^+$, $c + g^- \in E_A$, then $p'_A(c + g^+) \; \mathcal{C}_A \; p'_A(c + g^-)$ if and only if $p'_A c + (p'_A g)^+ \; \mathcal{C}_A \; p'_A c + (p'_A g)^-$, if and only if $p'_A c \in f'_A(p'_A g)$. Then $t = p'_A(c + g^+) \in H_A$, $u = p'_A(c + g^-) \in H_A$, $t - u = p'_A g$, $t \; \mathcal{C}_A \; u$, $R_A(t) = R_A(u)$, and $(c + g^+) - s_A(t) - (c + g^-) + s_A(u) \in R_A(t)$ if and only if $s_A(t) - s_A(u) \in g + R_A(t)$, if and only if

$$p_A\big(s_A(t) - s_A(u)\big) \in p_A g + R_A(t),$$

since $p'_A\big(s_A(t) - s_A(u)\big) = t - u = p'_A g \in p'_A\big(g + R_A(t)\big)$ always. □

6. MAIN RESULT.

1. The final step in the construction of subcomplete congruences is the characterization of strand bases. This leads to the main result in this chapter.

Lemma 6.1. *Every strand base has the following properties.*

(S+) *if $a \; \mathcal{C}^* \; b$, then $s_a - s_b - s_{a+c} + s_{b+c} \in R_{a+c}$ for every $c \in F$.*

In particular,

(S) *if $A \subseteq B \in \mathcal{D}$, $t, u \in H_A$, $t \; \mathcal{C}_A \; u$, $v \in A'$, and $t + v \in p'_A E_B$, then $s_A(t) - s_A(u) - s_B(p) + s_B(q) \in R_B(p)$, where $p = p'_B(t + v)$ and $q = p'_B(u + v)$.*

Proof. (S+) Let $a \; \mathcal{C}^* \; b$, so that $a, b \in E_A$ and $t = p'_A a \; \mathcal{C}_A \; p'_A b = u$. By (T4), $d + t \; \mathcal{C} \; e + u$ for some $d, e \in I_A$. Then $R_a = R_b$ by (R2), $d + t$, $c + e \in E_A$

by Lemma X.4.1, and

$$(d+t) - s_a - (e+u) + s_b \in R_a$$

by Proposition 5.1. Also $c+d+t \;\mathcal{C}\; c+e+u$, $R_{a+c} = R_{b+c}$ by (R2), and

$$(d+t) - s_{a+c} - (e+u) + s_{b+c} = (c+d+t) - s_{a+c} - (c+e+u) + s_{b+c} \in R_{a+c}$$

by Proposition 5.1. Since $R_a \subseteq R_{a+c}$ by (R3), subtraction yields (S+):

$$s_a - s_b - s_{a+c} + s_{b+c} \in R_{a+c}.$$

Now assume $A \subseteq B \in \mathcal{D}$, $t, u \in H_A$, $t \;\mathcal{C}_A\; u$, $v \in A'$, and $t+v \in p'_A E_B$. By (T4), $c+t \;\mathcal{C}\; d+u$ for some $c, d \in I_A$. Then $a = c+t$, $b = d+u \in E_A$, $p'_A a = t \;\mathcal{C}_A\; u = p'_A b$, and $s_a - s_b - s_{a+v} + s_{b+v} \in R_{a+v}$, by (S+). Now $a+v \in E_B$ by (E3), since $a+v \in L_A$ and $p'_A(a+v) = t+v \in p'_A E_B$, and $b+v \in E_B$ since E_B is a union of $\mathcal{C}$-classes. Hence $s_{a+c} = s_B(p)$ and $s_{b+c} = s_B(q)$, where $p = p'_B(a+v) = p'_B(t+v)$ and $q = p'_B(b+v) = p'_B(u+v)$, just as $s_a = s_A(t)$ and $s_b = s_A(u)$. This proves (S). $\square$

2. The main result in this chapter is:

Theorem 6.2. *Let F be a free commutative monoid. Given a direction set, an extent cell family, a trace, groups $R_A(t)$, and a strand base s such that* (R1), (R2), (R3), *and* (S) *hold, define*

$$a \;\mathcal{C}\; b \text{ if and only if } a, b \in E_A,\ p'_A a \;\mathcal{C}_A\; p'_A b, \text{ and } a - s_a - b + s_b \in R_a (= R_b).$$

Then $\mathcal{C}$ is a subcomplete congruence on F with the given direction set, extent cell family, trace, strand groups, and strand base, and is the only such congruence. Conversely every subcomplete congruence on F can be constructed in this fashion.

Proof. The uniqueness and converse follow from Proposition 5.1. For the direct part let $\mathcal{D}$ be a direction set on F, $\mathcal{E} = (E_A)_{A\in\mathcal{D}}$ be an extent cell family over $\mathcal{D}$, $\mathcal{T} = (\mathcal{C}_A)_{A\in\mathcal{D}}$ be a trace for $\mathcal{E}$, $R_A(t)$ ($A \in \mathcal{D}$, $t \in H_A$) be groups such that (R1), (R2), and (R3) hold, and s be a cross section of $\mathcal{S}$ such that (S) holds. Define $\mathcal{C}$ as in the statement. As before, $a, b \in E_A$, $p'_A a \;\mathcal{C}_A\; p'_A b$ implies $R_a = R_b$, by (R2). Then $\mathcal{C}$ is contained in the congruence $\mathcal{G}$ in Proposition 3.2:

$$a \;\mathcal{G}\; b \text{ if and only if } a, b \in E_A \text{ and } p'_A a \;\mathcal{C}_A\; p'_A b \text{ for some } A \in \mathcal{D},$$

which is the largest congruence with direction set $\mathcal{D}$, extent cell family $\mathcal{E}$, and trace $\mathcal{T}$; in fact

$$a \;\mathcal{C}\; b \text{ if and only if } a \;\mathcal{G}\; b \text{ and } a - s_a - b + s_b \in R_a.$$

$\mathcal{C}$ is an equivalence relation since $a \;\mathcal{G}\; b$ implies $R_a = R_b$ by (R2).

We show that (S+) holds for $\mathcal{G}$. Let $a \;\mathcal{G}\; b$, so that $a, b \in E_A$ and $p'_A a \;\mathcal{C}_A\; p'_A b$ for some $A \in \mathcal{D}$, and let $c \in F$. Then $a + c \;\mathcal{G}\; b + c$, and $a + c$, $b + c \in E_B$, $p'_B a + p'_B b \;\mathcal{C}_B\; p'_B b + p'_B c$ for some $B \in \mathcal{D}$, and $A \subseteq B$ by (E1). Hence $t = p'_A a$, $u = p'_A b \in H_A$, $v = p'_A c \in A'$, and $t + v = p'_A(a + c) \in p'_A E_B$; by (S),

$$s_a - s_b - s_{a+c} + s_{b+c} \;=\; s_A(t) - s_A(u) - s_B(p) + s_B(q) \in R_B(p) = R_{a+c},$$

where $p = p'_B(t + v) = p'_B(a + c)$ and $q = p'_B(u + v) = p'_B(b + c)$ since $A \subseteq B$.

Now assume $a \;\mathcal{C}\; b$, so that $a \;\mathcal{G}\; b$ and $a - s_a - b + s_b \in R_a = R_b$, and let $c \in F$. Then $a + c \;\mathcal{G}\; b + c$ since $\mathcal{G}$ is a congruence, $s_a - s_b - s_{a+c} + s_{b+c} \in R_{a+c}$ by (S+), $R_a \subseteq R_{a+c}$ by (R3),

$$\begin{aligned}(a + c) &- s_{a+c} - (b + c) + s_{b+c}\\ &= (a - s_a - b + s_b) + (s_a - s_b - s_{a+c} + s_{b+c}) \in R_{a+c},\end{aligned}$$

and $a + c \;\mathcal{C}\; b + c$. Thus $\mathcal{C}$ is a congruence.

$\mathcal{D}$ is a direction set of $\mathcal{C}$: indeed $\mathcal{D}$ is a direction set of $\mathcal{G}$, so that $\mathcal{D} = D(\mathcal{Y})$ for some semilattice congruence $\mathcal{Y} \supseteq \mathcal{G} \supseteq \mathcal{C}$.

We show that $\mathcal{E}$ is an extent cell family for $\mathcal{C}$. By Proposition 3.2, $\mathcal{E}$ is an extent cell family for $\mathcal{G}$, so that (E1), (E2), and (E3) hold. Every E_A is a union of $\mathcal{C}$-classes, since every E_A is a union of $\mathcal{G}$-classes. If finally $c \in A \in \mathcal{D}$, $a, b \in E_A$, and $a + c \;\mathcal{C}\; b + c$, then $a + c$, $b + c \in E_A$ by (X12), $p'_A a = p'_A(a + c) \;\mathcal{C}_A\; p'_A(b + c) = p'_A b$, $s_{a+c} = s_a$, $s_{b+c} = s_b$,

$$(a + c) - s_{a+c} - (b + c) + s_{b+c} \;=\; a - s_a - b + s_b \in R_a = R_{a+c},$$

and $a \;\mathcal{C}\; b$. Thus (E5) holds and $\mathcal{E}$ is an extent cell family of $\mathcal{C}$, by Proposition 2.9. In particular $\mathcal{C}$ is subcomplete, by Proposition 2.2.

We show that $\mathcal{T}$ is the trace of $\mathcal{C}$ relative to $\mathcal{E}$. Let $t, u \in H_A$. If $c + t \;\mathcal{C}\; d + u$ for some $c, d \in I_A$, then $c + t$, $d + u \in E_A$ by Lemma X.4.1 and $t = p'_A(c + t) \;\mathcal{C}_A\; p'_A(d + u) = u$. Conversely assume that $t \;\mathcal{C}_A\; u$. We have $s_A(t) = q + t$ and $s_A(u) = r + u$ for some $q, r \in A$. For any $c \in I_A$, we have $a = c + q + t$, $b = c + r + u \in E_A$ by Lemma X.4.1, $s_a = q + t$, $s_b = r + u$, $a - s_a = c = b - s_b$, $a - s_a - b + s_b = 0 \in R_a$, and $c + q + t = a \;\mathcal{C}\; b = c + r + u$ with $c + q$, $c + r \in I_A$. Hence $\mathcal{C}_A$ is the trace congruence of $\mathcal{C}$, by (T4) in Lemma 3.1, and T is the trace of $\mathcal{C}$. Moreover, $\mathcal{G} = \mathcal{C}^*$.

Finally let $t \in H_A$ and $c, d \in I_A$. Then $c + t$, $d + t \in E_A$ and $c + t \;\mathcal{G}\; d + t$; hence $c + t \;\mathcal{C}\; d + t$ if and only if $c - d \in R_A(t)$. By Proposition 4.1, $R_A(t)$ is the strand group of t. Thus the strand groups of $\mathcal{C}$ are the given strand groups. Also $s_A(t)\ \mathcal{C}\ s_A(u)$ whenever $t, u \in H_A$ and $t\ \mathcal{C}_A\ u$, since $s_{s_A(t)} = s_A(t)$ and $s_A(t) - s_A(t) - s_A(u) + s_A(u) \in R_A(t)$; hence s is a strand base for $\mathcal{C}$. □

Corollary 6.3. *Let $\mathcal{E}$ be an extent cell family on F. A family $\mathcal{R} = \left(R_A(t)\right)_{A \in \mathcal{D},\, t \in H_A}$ is the strand group family of a subcomplete congruence $\mathcal{C}$ with extent cell family $\mathcal{E}$ if and only if* (R1), (R2), *and* (R3) *hold. Then a cross section of $\mathcal{S}$ is the strand base of a subcomplete congruence $\mathcal{C}$ with extent cell family $\mathcal{E}$ if and only if* (S) *holds.*

Let $\mathcal{E}$ be an extent cell family on F and $\mathcal{T}$ be a trace for $\mathcal{E}$. A family $\mathcal{R} = \left(R_A(t)\right)_{A \in \mathcal{D},\, t \in H_A}$ is a **strand group family** over $\mathcal{E}$ and $\mathcal{T}$ when it satisfies (R1), (R2), and (R3); equivalently, when it is the strand group family of a subcomplete congruence with extent cell family $\mathcal{E}$ and trace $\mathcal{T}$; then a cross section of $\mathcal{S}$ is a **strand base** for $\mathcal{R}$ when it satisfies (S); equivalently, when it is the strand base of a subcomplete congruence with extent cell family $\mathcal{E}$, trace $\mathcal{T}$, and strand group family $\mathcal{R}$.

3. Theorem 6.2 has some noteworthy particular cases.

Corollary 6.4. *In Theorem* 6.2, *$F/\mathcal{C}$ is cancellative if and only if $\mathcal{D} = \{F\}$; $F/\mathcal{C}$ is subelementary if and only if $\mathcal{D} = \{A, F\}$ for some face $A \neq F$ of F and $R_F(0) = G$.*

Proof. A cancellative monoid S has a Ponizovsky family consisting of just S. Hence a cancellative congruence $\mathcal{C}$ on F has an extent cell consisting only of $E_F = F$. If conversely $\mathcal{D} = \{F\}$, then $I_F = E_F = L_F$ and $\mathcal{C}$ is cancellative by (E5).

Now let $F/\mathcal{C}$ be subelementary. By Proposition 2.3, $\mathcal{C}$ has a direction set $\mathcal{D} = \{A, F\}$, where $A = \pi^{-1}(C)$, and an extent cell family $\mathcal{E}$ over $\mathcal{D}$ in which $E_F = I_F$ is the zero class of $\mathcal{C}$; then Proposition 4.1 yields $R_F(0) = G_F = G$.

Conversely assume that $\mathcal{D} = \{A, F\}$ for some face $A \neq F$ of F and $R_F(0) = G$. Then $L_F = I_F = E_F$, $L_A = F$, and $I_A = A$. In $F/\mathcal{C}$, $C = A/\mathcal{C} = I_A/R_A(0)$ is a cancellative submonoid; the $\mathcal{C}$-class E_F is a zero element, since $E_F = L_F$ is an ideal of F; $N = (F \backslash A)/\mathcal{C}$ is a nilsemigroup, since for every $a \in F \backslash A = F^*$ we have $na \in E_F$ for some $n > 0$ by (E2); and $F/\mathcal{C}$ is the disjoint union $F/\mathcal{C} = C \cup N$ since A and $F \backslash A$ are unions of $\mathcal{C}$-classes. □

Subelementary congruences are studied in greater detail in Section 7.

Corollary 6.5. *Let F be a finitely generated free commutative monoid.*

Given a direction set, an extent cell family, a trace, groups $R_A(x)$, *and a strand base* s *such that* (R1), (R2), (R3), *and* (S) *hold, define*

$$a\ \mathcal{C}\ b \text{ if and only if } a,b \in E_A,\ p'_A a\ \mathcal{C}_A\ p'_A b, \text{ and } a - s_a - b + s_b \in R_a (= R_b).$$

Then $\mathcal{C}$ *is a congruence on* F *with the given direction set, extent cell family, trace, strand groups, and strand base, and is the only such congruence. Conversely every congruence on* F *can be constructed in this fashion.*

This holds since finitely generated commutative semigroups are subcomplete, by Proposition VII.1.1.

Corollary 6.6. *In Theorem* 6.2, $\mathcal{E}$ *is sharp if and only if*

(R4) *for every* $A \in \mathcal{D}$ *and every* $a,b \in I_A$, *there exist* $n > 0$ *and* $r \in R_A(0)$ *such that* $na + r \geqq b$.

This follows from Proposition 4.3. We call $\mathcal{R} = \big(R_A(t)\big)_{A\in\mathcal{D},\, t\in H_A}$ **sharp** when (R4) holds.

Corollary 6.7. *In Theorem* 6.2, $F/\mathcal{C}$ *is complete if and only if*

(R4+) *every* $I_A/R_A(0)$ *is a group; equivalently, for every* $A \in \mathcal{D}$ *and* $g \in G_A$ *there exists* $r \in R_A(0)$ *such that* $g + r \in I_A$.

This follows from Proposition 4.4. We call $\mathcal{R} = \big(R_A(t)\big)_{A\in\mathcal{D},\, t\in H_A}$ **complete** when (R4+) holds.

Corollary 6.8. *In Theorem* 6.3, $F/\mathcal{C}$ *is finite if and only if* $\mathcal{D}$ *is finite,* $A'/\mathcal{C}_A$ *is finite, and* $R_A(0)$ *has finite index in* G_A, *for every* $A \in \mathcal{D}$.

Proof. These conditions are necessary by Corollary 1.4 and Propositions 3.4 and 4.5. If conversely $\mathcal{D}$ is finite, $A'/\mathcal{C}_A$ is finite, and $R_A(0)$ has finite index in G_A, for every $A \in \mathcal{D}$, then every $G_A/R_A(x)$ is finite, a strand intersects only finitely many $\mathcal{C}$-classes, there are only finitely many $\mathcal{C}$-classes contained in E_A that project to a $\mathcal{C}_A$-class, and E_A contains only finitely many $\mathcal{C}$-classes; hence $F/\mathcal{C}$ is finite. □

4. Examples will illuminate the results in this section.

Example A. In this example, $X = \{x,y\}$, $\mathcal{D} = \{B,F\}$, where $B = F_{\{y\}}$, $E_B = \{a \in F \mid a_x \leqq 2\}$, and $E_F = \{a \in F \mid a_x \geqq 3\}$. We saw that there are four strands: $S_B(0) = B$, $S_B(x)$, $S_B(2x)$, and $S_F(0) = E_F$. By Corollary 6.3, the strand groups are any subgroups $R_B(0)$, $R_B(x)$, $R_B(2x) \subseteq G_B = \mathbb{Z}y$, and $R_F(0) \subseteq G = \mathbb{Z}x \oplus \mathbb{Z}y$ such that $R_B(0) \subseteq R_B(x) \subseteq R_B(2x) \subseteq R_F(0)$.

Condition (S) is trivial; hence a strand base consists of any $s_0 \in S_B(0)$,

Example A

$s_x \in S_B(x)$, $s_{2x} \in S_B(2x)$, and $s_{3x} \in S_F(0)$. Proposition 5.2 shows that all strand bases define the same congruence $\mathcal{C}$.

For instance let $R_B(0) = R_B(x) = 6y\mathbb{Z}$, $R_B(2x) = 3y\mathbb{Z}$, and $R_F(0)$ be generated by $x - y$ and $3y$. We see that $\mathcal{R}_B(0) = \mathcal{R}_B(x)$ is generated by $(0, 6y)$ and $\mathcal{R}_B(2x)$ is generated by $(0, 3y)$; inspection shows that $\mathcal{R}_F(0)$ is generated by $(3x,\ 3x+3y)$ and $(4x,\ 3x+y)$. Proposition 5.3 yields the following generators for $\mathcal{C}$: $(0, 6y)$, $(x, 6x+y)$ (which may be omitted), $(2x,\ 2x+3y)$, $(3x,\ 3x+3y)$ (which may also be omitted), and $(4x,\ 3x+y)$.

Returning to the general case, we saw that $\mathcal{C}^* = \mathcal{S}$. By Proposition 3.4, the quotient semigroup $F/\mathcal{C}$ has four $\mathcal{K}$-classes: the cancellative monoid $B/R_B(0)$, which is isomorphic to $\mathbb{N}$ if $R_B(0) = 0$ and is a finite cyclic group otherwise; two $\mathcal{K}$-classes $S_B(x)/\mathcal{C}$ and $S_B(2x)/\mathcal{C}$, which are similar though not subsemigroups; and the cancellative semigroup $E_F/R_F(0)$. Thus $F/\mathcal{C}$ is subelementary if and only if $E_F/R_F(0)$ is trivial, if and only if $R_F(0) = G$, as in Corollary 6.4.

By Corollary 6.6, $\mathcal{E}$ is a sharp extent cell family if and only if

(R4(A)) for every $a, b \in I_A$ there exist $n > 0$ and $r \in R_A(0)$ such that $na + r \geqq b$

holds for every $A \in \mathcal{D}$. Here (R4(B)) holds since $I_B = B$ is archimedean. We show that (R4(F)) holds if and only if $r_y > 0$ for some $r = r_x x + r_y y \in R = R_F(0)$. If indeed (R4(F)) holds, then for $a = 3x$ and $b = 3x + y \in I_F = E_F$ there exist $n > 0$ and $r \in R$ such that $na + r \geqq b$, and then $r_y > 0$. If conversely $r_y > 0$ for some $r \in R$, and $a, b \in E_F$, then $a_x \geqq 3$ and $na + kr \geqq b$ when $k \geqq b_y$ and $3n \geqq b_x - kr_x$, and (R4(F)) holds.

By Corollary 6.7, $F/\mathcal{C}$ is complete if and only if

(R4+(A)) $I_A/R_A(0)$ is a group; equivalently, for every $g \in G_A$ there exists $r \in R_A(0)$ such that $g + r \in I_A$

holds for every $A \in \mathcal{D}$. Here (R4+(B)) holds if and only if $R_B(0) \neq 0$. We show that (R4+(F)) holds if and only if $R = R_F(0)$ has an element $r = r_x x + r_y y$ with positive coordinates $r_x > 0$, $r_y > 0$. If indeed (R4+(F)) holds, then $-y + r \in I_F = E_F$ for some $r \in R$, and then r has positive coordinates. If conversely $r \in R$ has positive coordinates and $g \in G$, then $kr + g \in E_F$ when $k \geqq 3 - g_x$ and $k \geqq -g_y$, and (R4+(F)) holds.

By Corollary 6.8, $F/\mathcal{C}$ is finite if and only if $G_A/R_A(0)$ is finite for all $A \in \mathcal{D}$. Since G_A is finitely generated, this happens if and only if $R_A(0)$ has the same rank as G_A, for every $A \in \mathcal{D}$. Thus $F/\mathcal{C}$ is finite if and only if $R_B(0) \neq 0$ (equivalently, $R_B(0) \subseteq \mathbb{Z}y$ has rank 1) and $R_F(0) \subseteq \mathbb{Z}x \oplus \mathbb{Z}y$ has rank 2. □

Example C. In this example, $X = \{x, y, z\}$ and $\mathcal{D}$ consists of $A = F_{\{x\}}$, $B = F_{\{x,y\}}$, and F; $H_A = \{0, y, z, y+z, 2z\}$ and $E_A = \{a \in F \mid a_y y + a_z z \in H_A\}$; $E_B = \{a \in F \mid a_y \geqq 2, a_z = 0\}$ and $H_B = \{0\}$; $E_F = F \setminus (E_A \cup E_B)$ and $H_F = \{0\}$. The figure shows the projections of the strands to A'.

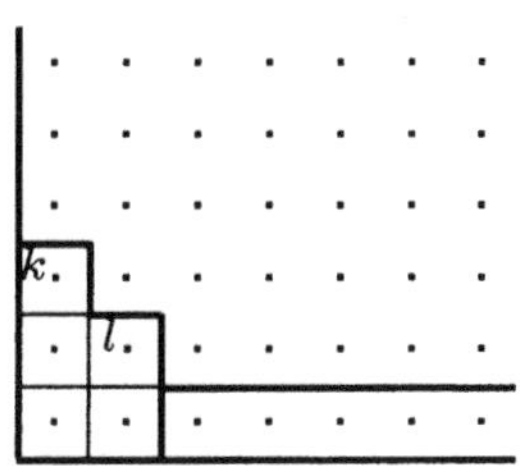

Example C

We saw that there is a trace relative to $\mathcal{E}$ in which $k \; \mathcal{C}_A \; l$, where $k = 2z$ and $l = y + z$ are the corner points of H_A. By Corollary 6.3, the strand groups are any subgroups $R_A(t) \subseteq G_A = \mathbb{Z}x$ (one for every $t \in H_A$), $R_B(0) \subseteq G_B = \mathbb{Z}x \oplus \mathbb{Z}y$, and $R_F(0) \subseteq G = \mathbb{Z}x \oplus \mathbb{Z}y \oplus \mathbb{Z}z$, such that $R_A(0) \subseteq R_A(y)$, $R_A(z)$ and $R_A(y), R_A(z) \subseteq R_A(k) = R_A(l)$, $R_A(y) \subseteq R_B(0)$, and $R_A(k), R_B(0) \subseteq R_F(0)$.

Condition (S) states that $s_A(k) - s_A(l) \in R_F(0)$. Thus a strand base consists of any $s_A(0) \in S_A(0)$, $s_A(y) \in S_A(y)$, $s_A(z) \in S_A(z)$, $s_A(k) \in S_A(k)$, $s_A(l) \in S_A(l)$, $s_B(0) \in S_B(0)$, and $s_F(0) \in S_F(0)$, such that $s_A(k) - s_A(l) \in R_F(0)$.

By Proposition 5.2, two strand bases r and s define the same congruence if and only if $r_A(k) - r_A(l) - s_A(k) + s_A(l) \in R_A(k) = R_A(l)$.

For instance let $R_A(0) = R_A(y) = R_A(z) = R_A(k) = R_A(l) = R_B(0) = 3x\mathbb{Z}$,

$R_F(0) = x\mathbb{Z}$ and $s_A(k) = x + k$, $s_A(l) = 2x + l$. Proposition 5.3 yields the following generators for $\mathcal{C}$. Pairs in (1) are $(0, 3x)$, $(y, 3x + y)$, $(z, 3x + z)$, $(k, 3x + k)$, $(l, 3x + l)$, $(2y, 2y + 3x)$, and, from $\mathcal{R}_F(0)$, $(2y + z, x + 2y + z)$, $(y + 2z, x + y + 2z)$, and $(3z, x + 3z)$; some of these may be omitted. In (2), $(t, u) = (k, l)$; when $a = mx + k \in S_A(k)$ and $b = nx + l \in S_A(l)$, then

$$a - s_A(k) - b + s_A(l) \;=\; (m - 1 - n + 2)x \;\in\; R_A(k)$$

if and only if $m \equiv n - 1$ (mod. 3); minimal such pairs are $(k, x + l)$ and $(2x + k, l)$. Thus $\mathcal{C}$ is generated by $(0, 3x)$, $(2y + z, x + 2y + z)$, $(y + 2z, x + y + 2z)$, $(3z, x + 3z)$, $(k, x + l)$, and $(2x + k, l)$.

In general, it follows from Corollary 6.6 that $\mathcal{E}$ is a sharp extent cell family if and only if (R4(D)) holds for every $D \in \mathcal{D}$. As in Example A, (R4(A)) holds since $I_A = A$ is archimedean, and (R4(B)) holds if and only if $R_B(0) \not\subseteq \mathbb{Z}y$ (if $r_x \neq 0$ for some $r \in R_B(0)$). Similarly, (R4(F)) holds if and only if $r_x > 0$ and $r_y > 0$ for some $r \in R = R_F(0)$). If indeed (R4(F)) holds, then for $a = 3z$ and $b = x + y + 3z \in I_F = E_F$ there exist $n > 0$ and $r \in R$ such that $na + r \geqq b$, and then $r_x > 0$, $r_y > 0$. If conversely $r \in R$ has $r_x > 0$, $r_y > 0$, and $a, b \in E_F$, then $a_z > 0$ and $na + kr \geqq b$ when $k \geqq b_x$, $k \geqq b_y$, and $n \geqq b_z - kr_z$.

By Corollary 6.7, $F/\mathcal{C}$ is complete if and only if (R4+(D)) holds for every $D \in \mathcal{D}$. As in Example A, (R4+(A)) holds if and only if $R_A(0) \neq 0$; (R4+(B)) holds if and only if $r_x > 0$ and $r_y > 0$ for some $r \in R = R_B(0)$); and (R4+(F)) holds if and only if $R_F(0)$ has an element with positive coordinates.

By Corollary 6.8, $F/\mathcal{C}$ is finite if and only if $R_A(0) \subseteq \mathbb{Z}x$ has rank 1 $R_B(0) \subseteq \mathbb{Z}x \oplus \mathbb{Z}y$ has rank 2, and $R_F(0) \subseteq \mathbb{Z}x \oplus \mathbb{Z}y \oplus \mathbb{Z}z$ has rank 3. □

7. SUBELEMENTARY CONGRUENCES.

Every subcomplete congruence $\mathcal{C}$ is the intersection of subelementary and cancellative congruences. We now use the results from previous sections to write $\mathcal{C}$ as an explicit intersection of subelementary and cancellative congruences, and show how these congruences give rise to the direction set, extent cells, trace, and strand groups of $\mathcal{C}$.

1. First we inspect the construction of cancellative and subelementary congruences provided by Theorem 6.2.

By Corollary 6.4, a subcomplete congruence $\mathcal{C}$ on F is subelementary if and only if it has a direction set $\mathcal{D} = \{D, F\}$, where $D \neq F$, and $R_F(0) = G$.

Let $D \neq F$ be a face of F. Then $\mathcal{D} = \{D, F\}$ is a direction set (by Proposition X.1.7). An extent cell family $\mathcal{E}$ over $\mathcal{D}$ consists of $E_F = Z$ (the zero class of $\mathcal{C}$ in Proposition 2.3) and $E_D = F \backslash Z$. Then $L_F = I_F = Z$, $L_D = F$, and $I_D = D$. (E1) is trivial when $A = D$ and states that Z is an ideal of F when $A = F$. (E2) is trivial when $A = D$ and states that

$$(\forall a \in F \backslash D)(\exists n > 0)\ na \in Z$$

when $A = F$. (E3) is trivial when $A = F$ and states that

$$H = p'_D(F \backslash Z) \text{ and } J = p'_D Z \text{ constitute a partition of } D'$$

when $A = D$. Then J is a proper nilmonoid ideal of D', by (E2), and $Z = D + J$ by (E3). If conversely J is a proper nilmonoid ideal of D' and $Z = D + J$, then (E2) and (E3) hold (with $H = D' \backslash J$).

We see that $a \mathcal{S} b$ if and only if either $a, b \in Z$, or $a, b \in F \backslash Z$ and $p'_D a = p'_D b$; thus $S_F(0) = Z$ and $S_D(t) = t + D$ for all $t \in H$.

A trace over $\mathcal{E}$ consists of $\mathcal{C}_F$, which we can ignore since $F' = \{0\}$, and $\mathcal{T} = \mathcal{C}_D$. Then (T1) is trivial when $A = F$ and states that $\mathcal{T}$ is a J-congruence on D' when $A = D$; then $\mathcal{T}$ is a nilmonoid congruence, since J is a nilmonoid ideal. (T2) is vacuous when $A = F$ and trivial when $A = D$, for then $B = F$.

A strand group family over $\mathcal{T}$ consists of groups $R(t) = R_D(t)$, one for every $t \in H$, and $R_F(0)$. To obtain subelementary congruences we let $R_F(0) = G$ (Corollary 6.4). Then (R1) is trivial when $A = F$ and states that

(R1D) $R(t)$ is a subgroup of G_D for every $t \in H$

when $A = D$. (R2) is trivial when $A = F$ and states that

(R2D) if $t, u \in H$ and $t \mathcal{T} u$, then $R(t) = R(u)$

when $A = D$. (R3) is trivial when $A = F$ or $B = F$ and states that

(R3D) if $t, u \in H$ and $t \leqq u$, then $R(t) \subseteq R(u)$

when $A = B = D$.

A strand base s consists of $s_F(0) \in Z$ and $s(t) = s_D(t) \in t + D$, one for every $t \in H$. We may identify s with the family $(s(t))_{t \in H}$. Condition (S) is trivial when $A = F$ or $B = F$; when $A = B = F$, (S) states that

(SD) if $t, u \in H$, $t \mathcal{T} u$, $v \in D'$, and $t + v \in H$, then $s(t) - s(u) - s(t + v) + s(u + v) \in R(t + v)$;

in the above we have $t+v \mathcal{T} u+v$ and $u+v \in H$, since $\mathcal{T}$ is J-congruence; then $R(t+v) = R(u+v)$ by (R2D). Corollary 6.4 now yields:

Proposition 7.1. *Let:* $D \neq F$ *be a face of* F; J *be a proper nilmonoid ideal of* D'; $Z = D + J$ *and* $H = D' \backslash J$; $\mathcal{T}$ *be a* J*-congruence on* D'; $R = (R(t))_{t \in H}$ *be a family of groups such that* (R1D), (R2D), *and* (R3D) *hold; and* $s = (s(t))_{t \in H}$ *satisfy* (SD) *and* $s(t) \in t + D$ *for all* $t \in H$. *Define*

$a \mathcal{C} b$ if and only if either $a, b \in Z$, or $a, b \in F \backslash Z$, $p'_D a \mathcal{T} p'_D b$, and $a - s(p'_D a) - b + s(p'_D b) \in R(p'_D a) = R(p'_D b)$.

Then $\mathcal{C}$ *is a subelementary congruence on* F *with direction set* $\{D, F\}$, *extent cell family* $\{F \backslash Z, Z\}$, *trace* $\mathcal{T}$, *strand groups* $\mathcal{R}$, *and strand base* s, *and is the only such congruence.*

Conversely every subelementary congruence $\mathcal{C}$ *on* F *can be constructed in this fashion, with* $D = \pi^{-1} C$ *and* $Z = \pi^{-1}(0)$ *if* $\mathcal{C}$ *is induced by a surjective homomorphism* $\pi : F \longrightarrow S$ *where* $S = C \cup N$ *is subelementary.*

In particular a subelementary congruence $\mathcal{C}$ has one nontrivial direction face and one nontrivial trace congruence. In Proposition 7.1 we call D the **direction face** of $\mathcal{C}$, $Z = E_F$ the **zero class** of $\mathcal{C}$, and $H = p'_D E_D = D' \backslash J$ the **trace coideal** of $\mathcal{C}$.

Proposition 7.2. *In Proposition* 7.1, $D'/\mathcal{T}$ *is isomorphic to the monoid of orbits of* S.

Proof. We have $a \mathcal{C}^* b$ if and only if $p'_D a \mathcal{T} p'_D b$. By Proposition 3.4, $D'/\mathcal{T} \cong S/\mathcal{K}$. Now $\mathcal{K}$ arises from the Ponizovsky family $\mathcal{P}$ in which $P_C = \pi(E_D) = S \backslash 0$ and $P_N = \pi(E_F) = \{0\}$, which is the standard Ponizovsky family of S, and is the orbit congruence on S, by Proposition VII.4.4. □

2. Cancellative congruences on F are similar to subelementary congruences but simpler. By Corollary 6.4, a subcomplete congruence $\mathcal{C}$ on F is cancellative if and only if it has a direction set $\mathcal{D} = \{F\}$. Then $\mathcal{C}$ has one strand $S_F(0) = F$ and one strand group $R = R_F(0)$. Theorem 6.2 yields $a \mathcal{C} b$ if and only if $a - b \in R$; this is according to Proposition II.5.1.

If in Proposition 7.1 we let $D = F$ and $J = \emptyset$, then $H = D' = \{0\}$, $Z = \emptyset$, $\mathcal{T}$ is the equality on D', $\mathcal{R}$ consists of one subgroup $R(0)$ of G, conditions (R2D), (R3D), and (SD) are trivial, and $a \mathcal{C} b$ if and only if $a - b \in R(0)$, as above. When $\mathcal{C}$ is a cancellative congruence, it is convenient to let the **direction face** of $\mathcal{C}$ be F and its **trace coideal** be $\{0\}$.

3. By Proposition VII.1.2, a subcomplete monoid S is a subdirect product

of cancellative semigroups and subelementary semigroups. From any Ponizovsky family $\mathcal{P} = (P_C)_{C \in S/\mathcal{B}}$ of S, Proposition VII.4.1 constructs an explicit subdirect decomposition of S: when $C \in S/\mathcal{B}$ and $m \in K_C$, the relation

$$a \ \mathcal{P}_C \ b \text{ if and only if either } am = bm \in P_C \text{ or } am, bm \notin P_C$$

is a congruence on S and does not depend on the choice of m; $S/\mathcal{P}_C$ is cancellative if C is the least $\mathcal{B}$-class of S, otherwise $S/\mathcal{P}_C$ is subelementary; $\bigcap_{C \in S/\mathcal{B}} \mathcal{P}_C$ is the equality on S; and S is a subdirect product of the monoids $S/\mathcal{P}_C$.

If now $\mathcal{C}$ is the subcomplete congruence on F induced by a surjective homomorphism $\pi : F \longrightarrow S$, then $\mathcal{C}$ is the intersection of congruences $\pi^{-1}(\mathcal{P}_C)$ which can be constructed explicitly from the extent cell family $\mathcal{E} = \pi^{-1}(\mathcal{P})$.

Proposition 7.3. *Let $\mathcal{C}$ be a subcomplete congruence on F with direction set $\mathcal{D}$ and extent cell family $\mathcal{E}$. For any $c \in I_A$, the relation*

$$a \ \mathcal{Q}_A \ b \textit{ if and only if either } a+c \ \mathcal{C} \ b+c \in E_A \textit{ or } a+c,\ b+c \in M_A = \bigcup_{A \subsetneqq B} E_B$$

is a congruence on F and does not depend on the choice of c; $\mathcal{Q}_A$ is cancellative if A is the greatest element of $\mathcal{D}$, otherwise $\mathcal{Q}_A$ is subelementary; and $\mathcal{C} = \bigcap_{A \in \mathcal{D}} \mathcal{Q}_A$.

If $\mathcal{D}$ has a greatest element A, then $A = F$: indeed every $a \in F$ belongs to some $B \in \mathcal{D}$ by (D1), and then $a \in B \subseteq A$. Then $M_A = \emptyset$ and $a \ \mathcal{Q}_A \ b$ if and only if $a + c \ \mathcal{C} \ b + c$, if and only if $b - a \in R_F(0)$.

Proof. By Proposition 2.2, $\mathcal{E} = \pi^{-1}(\mathcal{P})$ for some Ponizovsky family $\mathcal{P} = (P_C)_{C \in S/\mathcal{B}}$ on S; when $C = \pi(A^*)$, then $E_A = \pi^{-1}(P_C)$ and $I_A = \pi^{-1}(K_C)$. Thus $c \in I_A$ if and only if $\pi(c) \in K_C$; $a + c \in E_A$ if and only if $\pi(a)\,\pi(c) \in P_C$; and $a + c \in M_A$ if and only if $a + c \notin E_A$, if and only if $\pi(a)\,\pi(c) \notin P_C$ (since $a + c \in L_A$). Hence $a \ \mathcal{Q}_A \ b$ if and only if $\pi(a) \ \mathcal{P}_C \ \pi(b)$. Therefore $\mathcal{Q}_A$ is a congruence on F and does not depend on the choice of c; $F/\mathcal{Q}_A \cong S/\mathcal{P}_C$ is cancellative if A is the greatest element of $\mathcal{D}$, in which case C is the least element of $S/\mathcal{B}$ by Proposition 2.3, otherwise $F/\mathcal{Q}_A$ is subelementary; and $\mathcal{C} = \bigcap_{A \in \mathcal{D}} \mathcal{Q}_A$, since $\bigcap_{C \in S/\mathcal{B}} \mathcal{P}_C$ is the equality on S. □

It is instructive to verify directly that $\mathcal{Q}_A$ is cancellative if A is the greatest element of $\mathcal{D}$, subelementary otherwise. Let $\pi_A : F \longrightarrow F/\mathcal{Q}_A$ be the projection.

Lemma 7.4. *A is a union of $\mathcal{Q}_A$-classes; $\pi_A(a)$ is cancellative in $F/\mathcal{Q}_A$*

for every $a \in A$; if $A \neq F$, then the zero class of $\mathcal{Q}_A$ is $Z_A = M_A : c$; $\pi_A(b)$ is nilpotent in $F/\mathcal{Q}_A$ for every $b \in F \backslash A$.

Proof. If $a \; \mathcal{Q}_A \; b \in A$, then $b + c \in I_A$ by (X8), $a + c \in I_A$ since I_A is a union of $\mathcal{C}$-classes, and $a \in A$; hence A is a union of $\mathcal{Q}_A$-classes.

Assume that $a + u \; \mathcal{Q}_A \; b + u$, where $u \in A$. If $a + c + u \in E_B$ for some $A \subsetneqq B$, then $a + c \in E_B$ by (E3), since $p'_B(a + c + u) = p'_B(a + c)$, and $a + c$, $b + c \in M_A$. If $a + c + u \in E_A$, then $a + c \in E_A$ by (E3), since $p'_A(a + c + u) = p'_A(a + c)$, and $a + c + u \; \mathcal{C} \; b + c + u$ implies $a + c \; \mathcal{C} \; b + c$ by (X13) and $a \; \mathcal{Q}_A \; b$. Hence $\pi_A(u)$ is cancellative in $F/\mathcal{Q}_A$ for every $u \in A$.

Let $Z_A = M_A : c = \{z \in F \mid z + c \in M_A\}$. If $A \neq F$, then A is not the greatest element of $\mathcal{D}$, the ideal M_A of F is not empty, and $Z_A \neq \emptyset$. If $z, t \in Z_A$, then $z + c$, $t + c \in M_A$ and $z \; \mathcal{Q}_A \; t$. If $t \; \mathcal{Q}_A \; z \in Z_A$, then $z + c \in M_A$, $t + c \in M_A$, and $t \in Z_A$. Thus Z_A is an $\mathcal{Q}_A$-class. Also Z_A is an ideal of F, since M_A is an ideal of F by (E1). Hence Z_A is the zero element of $F/\mathcal{Q}_A$.

If finally $b \in F \backslash A$, then $b \in B^*$ for some $B \in \mathcal{D}$; $nb \in I_B$ for some $n > 0$ by (E2); $nb + c \in I_{A \vee B}$ by (X9); $nb + c \in M_A$, since $nb + c \in (A \vee B) \backslash A$ and $A \vee B \supsetneqq A$; $nb \in Z_A$; and $\pi_A(b)$ is nilpotent in $F/\mathcal{Q}_A$. □

4. We now let $\mathcal{C}$ be any subcomplete congruence and show that the direction faces, extent cells, trace, and strand groups of $\mathcal{C}$ are directly related to the direction faces, extent cells, trace congruences, and strand groups of the congruences $\mathcal{Q}_A$.

Proposition 7.5. *When $\mathcal{C}$ is a subcomplete congruence on F: the direction face of $\mathcal{Q}_A$ is A; the extent cells of $\mathcal{C}$ are the intersections of the extent cells of all $\mathcal{Q}_B$; the trace coideal of $\mathcal{Q}_A$ is H_A; the trace congruence of $\mathcal{Q}_A$ is $\mathcal{C}_A$; the strand group $R(x)$ of $\mathcal{Q}_A$ is $R_A(x)$; and s_A is a strand base of $\mathcal{Q}_A$.*

Proof. Let $A \in \mathcal{D}$ and $\pi_A : F \longrightarrow F/\mathcal{Q}_A$ be the projection. If $A \neq F$, then, by Lemma 7.4, $F/\mathcal{Q}_A = C \cup N$ is subelementary and $A = \pi_A^{-1}(C)$; hence A is the direction face of $\mathcal{Q}_A$. This also holds if $A = F$.

If $A \neq F$, then, by Lemma 7.4, $z \in F$ belongs to the zero ideal $Z = Z_A$ of $\mathcal{Q}_A$ if and only if $z + c \in M_A = \bigcup_{A \subsetneqq B \in \mathcal{D}} E_B$, if and only if $p'_A z \in J_A = \bigcup_{A \subsetneqq B \in \mathcal{D}} p'_A E_B$, by (E3); hence $J = p'_A Z = J_A$ and the trace coideal of $\mathcal{Q}_A$ is $H = A' \backslash J = H_A$. This also holds if $A = F$.

The extent cells of $\mathcal{Q}_B$ are its zero class Z_B and $F \backslash Z_B$ (or just F, if $B = F$). Let $A, B \in \mathcal{D}$ and $d \in I_B$. If $A \not\subseteq B$, then $a \in E_A$ implies $a + d \in L_{A \vee B}$ by (X9), $a + d \in M_B$ since $B \subsetneqq A \vee B$, and $a \in Z_B$; thus

$E_A \subseteq Z_B$. But if $A \subseteq B$, then $a \in E_A$ implies $a + d \in E_B$ by (E3), since $p'_A(a+d) = p'_A d \in p'_A E_B$, so that $a + d \notin M_B$ and $a \notin Z_B$; thus $E_A \subseteq F \backslash Z_B$. In particular, for any $a \in E_A$, A is the smallest $B \in \mathcal{D}$ such that $a \in F \backslash Z_B$. It follows that

$$E_A = \bigcap(Z_B \mid E_A \subseteq Z_B) \cap \bigcap(F \backslash Z_B \mid E_A \subseteq F \backslash Z_B) :$$

if a belongs to the right hand side, then A is the smallest $B \in \mathcal{D}$ such that $a \in F \backslash Z_B$ and necessarily $a \in E_A$.

Let $\mathcal{T}$ be the trace congruence of $\mathcal{Q}_A$. Let $t, u \in A'$. By (T1), (T4), $t \,\mathcal{T}\, u$ if and only if either $t, u \in J_A$ or $t, u \in H_A$ and $a + t \;\mathcal{Q}_A\; b + u$ for some $a, b \in A$. If $t, u \in H_A$ and $a + t \;\mathcal{Q}_A\; b + u$ for some $a, b \in A$, then $a + c + t \;\mathcal{C}\; b + c + u \in E_A$ and $t \,\mathcal{C}_A\, u$ by (T4), since $a + c$, $b + c \in I_A$. If conversely $t, u \in H_A$ and $t \,\mathcal{C}_A\, u$, then $a + t \;\mathcal{C}\; b + u$ for some $a, b \in I_A$, $a + c + t \;\mathcal{C}\; b + c + u \in E_A$ by Lemma 4.1, $a + t \;\mathcal{Q}_A\; b + u$, and $t \,\mathcal{T}\, u$. Thus $\mathcal{T} = \mathcal{C}_A$.

Let $t \in H_A$. By definition, $g \in R(t)$ if and only if $g = a - b$ for some $a, b \in A$ such that $a + t \;\mathcal{Q}_A\; b + t$; then $c + a + t \;\mathcal{C}\; c + b + t \in E_A$, with $c + a$, $c + b \in I_A$, and $g = (c + a) - (c + b) \in R_A(t)$. If conversely $g \in R_A(t)$, then $g = a - b$ for some $a, b \in I_A$ such that $a + t \;\mathcal{C}\; b + t$, $c + a + t \;\mathcal{C}\; c + b + t \in E_A$ by Lemma 4.1, $a + t \;\mathcal{Q}_A\; b + t$, and $g = a - b \in R(t)$. Thus $R(t) = R_A(t)$.

Finally let s be a strand base of $\mathcal{C}$. Then $s_A(t) \in S_A(t)$ for every $t \in H_A$, and $t, u \in H$, $t \,\mathcal{C}_A\, u$ implies $s_A(t) \;\mathcal{C}\; s_A(u)$ and $s_A(t) \;\mathcal{Q}_A\; s_A(u)$. Thus s_A serves as a strand base of $\mathcal{Q}_A$. □

Proposition 7.5 suggests an explanation for Theorem 6.2. As noted above, Proposition VII.4.1 represents any subcomplete congruence $\mathcal{C}$ on F as an intersection of subelementary and cancellative congruences. Proposition 7.1 can be established directly. The direction set, extent cells, trace, and strand groups of $\mathcal{C}$ then arise as in Proposition 7.5.

Chapter XII.

COMMUTATIVE SEMIGROUP COHOMOLOGY.

Commutative semigroup cohomology assigns abelian groups $H^n(S,\mathbb{G})$ to a commutative semigroup S and an abelian group valued functor $\mathbb{G}$ on S.

Other cohomology theories have been considered for commutative semigroups (see the introduction to Section 4). The theory we call commutative semigroup cohomology is of particular interest because $H^2(S,\mathbb{G})$ classifies commutative group coextensions of S by $\mathbb{G}$, that is, $H^2(S,\mathbb{G})$ coincides with the extension group $\mathrm{Ext}\,(S,\mathbb{G})$ in Chapter V; moreover, if S is complete group-free and $\mathbb{G}$ is Schützenberger, then, as we saw in Chapter V, $H^2(S,\mathbb{G})$ classifies complete commutative semigroups T with $T/\mathcal{H} \cong S$ and Schützenberger functor isomorphic to $\mathbb{G}$. These results make cohomology an important part of the structure theory of commutative semigroups.

Commutative semigroup cohomology is an instance of triple cohomology, which provides a definition in dimensions $n \geqq 3$ as well as valuable properties.

After a brief account of triple cohomology and two sections of preliminary results, this chapter defines commutative semigroup cohomology, and gives simpler definitions in low dimensions.

1. TRIPLE COHOMOLOGY.

This section gives, without proofs, the definition and main properties of triple cohomology. We follow Beck [1967] and Barr & Beck [1969] but have renumbered cohomology groups in the more traditional fashion. We assume a general knowledge of category theory and triples, from, say, MacLane [1971]; Grillet [1999] also has a short account of triples and the tripleability of varieties.

1. The minimal requirements for cohomology are: a category $\mathcal{C}$; a functor $\mathbb{V} : \mathcal{C} \longrightarrow \mathcal{C}$ (normally denoted by $\mathbb{G}$, but we use $\mathbb{G}$ for abelian group valued functors); a natural transformation $\epsilon : \mathbb{V} \longrightarrow 1_{\mathcal{C}}$; and a contravariant functor $\mathbb{A}$

from $\mathcal{C}$ to the category Ab of abelian groups and homomorphisms.

For every $0 \leqq i \leqq n$ the natural transformation $\epsilon : \mathbb{V} \longrightarrow 1_{\mathcal{C}}$ induces a natural transformation $\epsilon^{n,i} = \mathbb{V}^{n-i}\,\epsilon\,\mathbb{V}^{i} : \mathbb{V}^{n+1} \longrightarrow \mathbb{V}^{n}$. To each object C of $\mathcal{C}$ can then be assigned a complex of abelian groups

$$0 \longrightarrow \mathbb{A}\mathbb{V}C \longrightarrow \cdots \longrightarrow \mathbb{A}\mathbb{V}^{n}C \xrightarrow{\delta_n} \mathbb{A}\mathbb{V}^{n+1}C \longrightarrow \cdots$$

where $\delta_n = \sum_{0 \leqq i \leqq n} (-1)^i \mathbb{A}\,\epsilon_C^{n,i} : \mathbb{A}\mathbb{V}^n C \longrightarrow \mathbb{A}\mathbb{V}^{n+1}C$. That $\delta_{n+1} \circ \delta_n = 0$ for all $n \geqq 1$ follows by a standard argument from the equalities $\epsilon^{n,j} \circ \epsilon^{n+1,\,i} = \epsilon^{n,i} \circ \epsilon^{n+1,\,j+1}$, which hold for all $0 \leqq i \leqq j \leqq n$. The n-th cohomology group of C with coefficients in $\mathbb{A}$ is

$$H^n(C,\mathbb{A}) = \operatorname{Ker}\delta_n \,/\, \operatorname{Im}\delta_{n-1} \quad \text{if } n \geqq 2, \quad \text{and} \quad H^1(C,\mathbb{A}) = \operatorname{Ker}\delta_1.$$

(In Beck [1967], Barr & Beck [1969], these groups are H^{n-1} and H^0.)

In what follows $\mathbb{V}$ and ϵ arise from an adjunction $(\mathbb{F},\mathbb{U},\eta,\epsilon) : \mathcal{U} \longrightarrow \mathcal{C}$, where $\mathbb{F} : \mathcal{U} \longrightarrow \mathcal{C}$ is a left adjoint of $\mathbb{U} : \mathcal{C} \longrightarrow \mathcal{U}$, and $\eta : 1_{\mathcal{U}} \longrightarrow \mathbb{U}\mathbb{F}$, $\epsilon : \mathbb{F}\mathbb{U} \longrightarrow 1_{\mathcal{C}}$ are the corresponding natural transformations. Then $\mathbb{T} = \mathbb{U}\mathbb{F}$, η, and $\mu = \mathbb{U}\epsilon\mathbb{F}$ constitute a triple on $\mathcal{U}$; $\mathbb{V} = \mathbb{F}\mathbb{U}$, ϵ, and $\nu = \mathbb{F}\eta\mathbb{U}$ constitute a cotriple on $\mathcal{C}$.

If for example $\mathcal{C}$ is the category of commutative semigroups and homomorphisms, and $\mathcal{U}$ is the category of $Sets$ of sets and mappings, then the free c.s. functor $\mathbb{F} : Sets \longrightarrow \mathcal{C}$ is a left adjoint of the forgetful or underlying set functor $\mathbb{U} : \mathcal{C} \longrightarrow Sets$; η_X embeds a set X into the free c.s. on X; ϵ is described by Lemma 4.1 below. Like all varieties, $\mathcal{C}$ is tripleable over $Sets$.

2. To obtain the Beck cohomology groups of an object S of $\mathcal{C}$, one applies the above to the category $\overline{\mathcal{C}} = \mathcal{C}{\downarrow}S$ of objects over S; abelian group objects of $\overline{\mathcal{C}}$ provide coefficient functors (Beck [1967]). The details are as follows.

Recall that an **object over** S in $\mathcal{C}$ is an ordered pair (C,π) of an object C of $\mathcal{C}$ and a morphism $\pi : C \longrightarrow S$. A morphism $\gamma : (C,\pi) \longrightarrow (D,\rho)$ of objects over S is a morphism $\gamma : C \longrightarrow D$ in $\mathcal{C}$ such that $\rho \circ \gamma = \pi$.

$$\begin{array}{ccc} C & \xrightarrow{\gamma} & D \\ {\scriptstyle\pi}\downarrow & \swarrow{\scriptstyle\rho} & \\ S & & \end{array}$$

Every adjunction $(\mathbb{F},\mathbb{U},\eta,\epsilon) : \mathcal{U} \longrightarrow \mathcal{C}$ lifts to an adjunction $(\overline{\mathbb{F}},\overline{\mathbb{U}},\overline{\eta},\overline{\epsilon}) : \overline{\mathcal{U}} = \mathcal{U}{\downarrow}\mathbb{U}S \longrightarrow \mathcal{C}{\downarrow}S = \overline{\mathcal{C}}$; namely,

$$\overline{\mathbb{F}}(X,\pi) = (\mathbb{F}X,\overline{\pi}),\ \ \overline{\mathbb{U}}(C,\rho) = (\mathbb{U}C,\mathbb{U}\rho),\ \ \overline{\eta}_{(X,\pi)} = \eta_X,\ \ \overline{\epsilon}_{(C,\rho)} = \epsilon_C,$$

where $\overline{\pi} : \mathbb{F}X \longrightarrow S$ is the morphism such that $\mathbb{U}\overline{\pi} \circ \eta_X = \pi$. In particular, the cotriple $(\mathbb{V} = \mathbb{F}\mathbb{U},\ \epsilon,\ \nu = \mathbb{F}\eta\mathbb{U})$ induced by $(\mathbb{F},\mathbb{U},\eta,\epsilon)$ lifts to the cotriple $(\overline{\mathbb{V}} = \overline{\mathbb{F}}\overline{\mathbb{U}},\ \overline{\epsilon},\ \overline{\nu} = \overline{\mathbb{F}}\overline{\eta}\overline{\mathbb{U}})$ induced by $(\overline{\mathbb{F}},\overline{\mathbb{U}},\overline{\eta},\overline{\epsilon})$ on $\overline{\mathcal{C}} = \mathcal{C}{\downarrow}S$. By definition,

$$\overline{\mathbb{V}}(C,\pi) = \overline{\mathbb{F}}(\mathbb{U}C,\mathbb{U}\pi) = (\mathbb{V}C,\overline{\pi}),\text{ where } \mathbb{U}\overline{\pi} \circ \eta_{\mathbb{U}C} = \mathbb{U}\pi;$$

$\overline{\epsilon}_{(C,\pi)} = \epsilon_C$; and $\overline{\nu}_{(C,\pi)} = \overline{\mathbb{F}}\,\overline{\eta}_{(\mathbb{U}C,\mathbb{U}\pi)} = \mathbb{F}\,\eta_{\mathbb{U}C} = \nu_C$. In particular, $\overline{\epsilon}^{n,i}_{(C,\pi)} = \epsilon^{n,i}_C$. If $\mathcal{C}$ is tripleable over $\mathcal{U}$ (if the adjunction $(\mathbb{F},\mathbb{U},\eta,\epsilon)$ is tripleable), then $\overline{\mathcal{C}}$ is tripleable over $\overline{\mathcal{U}}$ ($(\overline{\mathbb{F}},\overline{\mathbb{U}},\overline{\eta},\overline{\epsilon})$ is tripleable).

3. An **abelian group object** of a category $\mathcal{C}$ is an object G of $\mathcal{C}$ such that every $\mathrm{Hom}_{\mathcal{C}}(C,G)$ is a set, together with an abelian group operation $+$ on every set $\mathrm{Hom}_{\mathcal{C}}(C,G)$, such that $\mathbb{A} = \mathrm{Hom}_{\mathcal{C}}(-,G)$ is a (contravariant) abelian group valued functor on $\mathcal{C}$; equivalently, such that

$$(g+h)\circ\gamma \;=\; (g\circ\gamma)+(h\circ\gamma)$$

for all morphisms $g,h : D \longrightarrow G$ and $\gamma : C \longrightarrow D$ of $\mathcal{C}$.

A morphism $\varphi : G \longrightarrow G'$ of abelian group objects of $\mathcal{C}$ is a morphism of $\mathcal{C}$ such that the mapping $\mathrm{Hom}_{\mathcal{C}}(C,\varphi) : \mathrm{Hom}_{\mathcal{C}}(C,G) \longrightarrow \mathrm{Hom}_{\mathcal{C}}(C,G')$ is a homomorphism of abelian groups for every object C of $\mathcal{C}$; equivalently,

$$\varphi\circ(g+h) \;=\; (\varphi\circ g)+(\varphi\circ h)$$

for all C and $g,h : C \longrightarrow G$. Then $\mathrm{Hom}_{\mathcal{C}}(-,\varphi)$ is a natural transformation of abelian group valued functors.

When $\mathcal{C}$ has finite products, abelian group objects of $\mathcal{C}$ can also be defined by means of suitable morphisms $m : G\times G \longrightarrow G$, $e : 0 \longrightarrow G$, $i : G \longrightarrow G$, as in MacLane [1971, first edition] (see also Lemma 2.2 below).

If for example G is an abelian group in the usual sense, then every $\mathrm{Hom}_{Sets}(C,G)$ is an abelian group under pointwise addition, and this makes G an abelian group object of $Sets$ since $(g+h)\circ\gamma = (g\circ\gamma)+(h\circ\gamma)$ for all $g,h : C \longrightarrow G$ and $\gamma : D \longrightarrow C$. If conversely G is an abelian group object of $Sets$, then $\mathrm{Hom}_{Sets}(\{1\},G)$ is an abelian group, and the bijection $\mathrm{Hom}_{Sets}(\{1\},G) \longrightarrow G$ is readily used to make G an abelian group so that pointwise addition on every $\mathrm{Hom}_{Sets}(C,G)$ is the given operation.

In general, an **action** $\centerdot$ of an abelian group object G of $\mathcal{C}$ on an object A of $\mathcal{C}$ assigns to every object C of $\mathcal{C}$ a group action

$$\mathrm{Hom}_{\mathcal{C}}(C,G)\times\mathrm{Hom}_{\mathcal{C}}(C,A) \longrightarrow \mathrm{Hom}_{\mathcal{C}}(C,A),\ \ (g,a) \longmapsto g\centerdot a$$

of the abelian group $\mathrm{Hom}_{\mathcal{C}}(C,G)$ on $\mathrm{Hom}_{\mathcal{C}}(C,A)$ which is natural in C; equivalently,

$$(g \centerdot a) \circ \gamma \;=\; (g \circ \gamma) \centerdot (a \circ \gamma)$$

for all $\gamma : C \longrightarrow D$, $g : D \longrightarrow G$, and $a : D \longrightarrow A$.

If for instance G is an abelian group which acts on a set X in the usual sense, then, for every set C, the abelian group $\mathrm{Hom}_{Sets}(C,G)$ acts pointwise on $\mathrm{Hom}_{Sets}(X,G)$, and this is an abelian group object action in $Sets$ since $(g \centerdot a) \circ \gamma = (g \circ \gamma) \centerdot (a \circ \gamma)$ for all $g : C \longrightarrow G$, $a : C \longrightarrow X$, and $\gamma : D \longrightarrow C$.

4. The ingredients for Beck cohomology are: an adjunction $(\mathbb{F}, \mathbb{U}, \eta, \epsilon) : \mathcal{U} \longrightarrow \mathcal{C}$; an object S of $\mathcal{C}$; and an abelian group object $\overline{G} = (G, \rho)$ of $\overline{\mathcal{C}} = \mathcal{C}{\downarrow}S$. For any object $\overline{T} = (T, \pi)$ of $\overline{\mathcal{C}}$, the **Beck cohomology** groups or **triple cohomology** groups of $\overline{T}$ with coefficients in $\overline{G}$ are its cohomology groups calculated from the cotriple $\overline{\mathbb{V}}$ above and coefficient functor $\mathbb{A} = \mathrm{Hom}_{\overline{\mathcal{C}}}(-, \overline{G})$. The triple cohomology groups of S are those of $(S, 1_S)$.

For a more detailed definition, let $\overline{T}_n = (T_n, \pi_n) = \overline{\mathbb{V}}^n \overline{T}$. Then $T_0 = T$, $\pi_0 = \pi$, and $(T_{n+1}, \pi_{n+1}) = \overline{\mathbb{V}}(T_n, \pi_n)$, so that $T_{n+1} = \mathbb{F}\mathbb{U}T_n$ and $\pi_{n+1} : T_{n+1} \longrightarrow S$ is the morphism such that $\mathbb{U}\pi_{n+1} \circ \eta_{\mathbb{U}T_n} = \mathbb{U}\pi_n$.

An n-**cochain** is an element of

$$C^n(\overline{T}, \overline{G}) \;=\; \mathbb{A}\overline{\mathbb{V}}^n(\overline{T}) \;=\; \mathrm{Hom}_{\overline{\mathcal{C}}}(\overline{T}_n, \overline{G}),$$

that is, a morphism $u : T_n \longrightarrow G$ of $\mathcal{C}$ such that $\rho \circ u = \pi_n$. Next, $\overline{\epsilon}^{n,i}_{\overline{T}} = \epsilon^{n,i}_T : T_{n+1} \longrightarrow T_n$ and $\mathbb{A}\epsilon^{n,i}_T = \mathrm{Hom}_{\overline{\mathcal{C}}}(\epsilon^{n,i}_T, \overline{G}) : C^n(\overline{T}, \overline{G}) \longrightarrow C^{n+1}(\overline{T}, \overline{G})$ sends $u : T_n \longrightarrow G$ to $u \circ \epsilon^{n,i}_T$. Hence

$$\delta_n \;=\; \textstyle\sum_{0 \leqq i \leqq n} (-1)^i \mathbb{A}\epsilon^{n,i}_T : C^n(\overline{T}, \overline{G}) \longrightarrow C^{n+1}(\overline{T}, \overline{G})$$

sends $u : T_n \longrightarrow G$ to

$$\delta_n u \;=\; \textstyle\sum_{0 \leqq i \leqq n} (-1)^i (u \circ \epsilon^{n,i}_T).$$

By definition, $H^n(\overline{T}, \overline{G})$ is the n-th homology group of the cochain complex

$$0 \longrightarrow C^1(\overline{T}, \overline{G}) \longrightarrow \dots \longrightarrow C^n(\overline{T}, \overline{G}) \xrightarrow{\delta_n} C^{n+1}(\overline{T}, \overline{G}) \longrightarrow \dots$$

An n-**cocycle** is an element of $Z^n(\overline{T}, \overline{G}) = \mathrm{Ker}\, \delta_n \subseteq C^n(\overline{T}, \overline{G})$. An n-**coboundary** is an element of $B^n(\overline{T}, \overline{G}) = \mathrm{Im}\, \delta_{n-1} \subseteq Z^n(\overline{T}, \overline{G})$ if $n \geqq 2$,

$B^1(\overline{T},\overline{G}) = 0$ if $n = 1$. The Beck cohomology groups of $\overline{T}$ are the groups

$$H^n(\overline{T},\overline{G}) \;=\; Z^n(\overline{T},\overline{G})\,/\,B^n(\overline{T},\overline{G}).$$

The Beck cohomology groups of S are $H^n(S,\overline{G}) = H^n((S,1_S),\overline{G})$.

A morphism $\tau : \overline{T} \longrightarrow \overline{T}'$ in $\overline{\mathcal{C}}$ induces a homomorphism

$$\tau^n \;=\; \mathrm{Hom}_{\overline{\mathcal{C}}}(\overline{\mathbb{V}}^n\tau,\overline{G}) : C^n(\overline{T}',\overline{G}) \longrightarrow C^n(\overline{T},\overline{G})$$

which sends $u \in C^n(\overline{T},\overline{G})$ to $u \circ \overline{\mathbb{V}}^n\tau$. Since composition with $\overline{\mathbb{V}}^n\tau$ preserves sums and $\epsilon_T^{n,i}$ is natural in T, we have $\tau^{n+1}(\delta_n u) = \delta_n(\tau^n u)$ for all $u \in C^n(\overline{T}',\overline{G})$; that is, $(\tau^n)_{n \geqq 1}$ is a chain transformation. Hence τ^n takes $Z^n(\overline{T}',\overline{G})$ into $Z^n(\overline{T},\overline{G})$, takes $B^n(\overline{T}',\overline{G})$ into $B^n(\overline{T},\overline{G})$, and induces a homomorphism $H^n(\tau,\overline{G}) : H^n(\overline{T}',\overline{G}) \longrightarrow H^n(\overline{T},\overline{G})$. Thus $H^n(-,\overline{G})$ is a contravariant abelian group valued functor on $\overline{\mathcal{C}}$.

Similarly, a morphism $\varphi : \overline{G} \longrightarrow \overline{G}'$ of abelian group objects induces a homomorphism $\varphi^n = \mathrm{Hom}_{\overline{\mathcal{C}}}(\overline{T}_n,\varphi) : C^n(\overline{T},\overline{G}) \longrightarrow C^n(\overline{T},\overline{G}')$ which sends $u \in C^n(\overline{T},\overline{G})$ to $\varphi \circ u$. Again $\varphi^{n+1}(\delta_n u) = \delta_n(\varphi^n u)$ for all $u \in C^n(\overline{T},\overline{G})$, and φ induces a homomorphism $H^n(\overline{T},\varphi) : H^n(\overline{T},\overline{G}) \longrightarrow H^n(\overline{T},\overline{G}')$. Thus $H^n(\overline{T},-)$ is a functor. In fact $H^n(\overline{T},\varphi)$ is natural in $\overline{T}$, so $H^n(-,-)$ is a bifunctor.

5. The main properties of Beck cohomology are as follows.

Theorem 1.1. *When $\overline{T} = \overline{\mathbb{F}}\overline{X}$ for some object $\overline{X}$ of $\overline{\mathfrak{U}}$, then $H^n(\overline{T},\overline{G}) = 0$ for all $n \geqq 2$; also $H^1(\overline{\mathbb{V}}\overline{C},\overline{G}) \cong C^1(\overline{C},\overline{G})$ for every object $\overline{C}$ of $\overline{\mathcal{C}}$.*

A sequence $\overline{G} \longrightarrow \overline{G}' \longrightarrow \overline{G}''$ of abelian group objects and morphisms of $\overline{\mathcal{C}}$ is **short $\mathbb{V}$-exact** in case

$$0 \longrightarrow \mathrm{Hom}_{\overline{\mathcal{C}}}(\overline{\mathbb{V}}\overline{C},\overline{G}) \longrightarrow \mathrm{Hom}_{\overline{\mathcal{C}}}(\overline{\mathbb{V}}\overline{C},\overline{G}') \longrightarrow \mathrm{Hom}_{\overline{\mathcal{C}}}(\overline{\mathbb{V}}\overline{C},\overline{G}'') \longrightarrow 0$$

is a short exact sequence (in Ab) for every object $\overline{C}$ of $\overline{\mathcal{C}}$.

Theorem 1.2. *Every short $\mathbb{V}$-exact sequence $\mathcal{E} : \overline{G} \longrightarrow \overline{G}' \longrightarrow \overline{G}''$ of abelian group objects of $\mathcal{C}{\downarrow}S$ induces an exact sequence*

$$\cdots\; H^n(\overline{T},\overline{G}) \longrightarrow H^n(\overline{T},\overline{G}') \longrightarrow H^n(\overline{T},\overline{G}'') \longrightarrow H^{n+1}(\overline{T},\overline{G}) \;\cdots$$

which is natural in $\mathcal{E}$.

Theorems 1.1 and 1.2 constitute Theorem 2 of Beck [1967].

Up to natural isomorphisms, $H^n(\overline{T},-)$ is the only sequence for which Theorems 1.1 and 1.2 hold (Barr and Beck [1969], Theorem 3.3). Another useful characterization of $H^n(\overline{T},\overline{G})$ was given by Barr and Beck [1969] (Proposition 11.2). We give the contravariant version in Wells [1978]:

Theorem 1.3. *Let*

$$\mathbb{C} : 0 \longrightarrow \mathrm{Hom}_{\overline{\mathcal{C}}}(-,\overline{G}) \longrightarrow \mathbb{C}^1 \longrightarrow \cdots \longrightarrow \mathbb{C}^n \longrightarrow \mathbb{C}^{n+1} \longrightarrow \cdots$$

be a complex of abelian group valued contravariant functors on $\overline{\mathcal{C}}$ *and* $\mathbb{H}^n$ *be the* n*-th homology functor of* $\mathbb{C}$*. Assume that* $\mathbb{H}^n(\overline{\mathbb{V}}\overline{T}) = 0$ *for all* $\overline{T}$ *and* $n \geqq 2$*, that* $\mathbb{H}^1(\overline{\mathbb{V}}\overline{T})$ *is naturally isomorphic to* $C^1(\overline{T},\overline{G})$*, and that there is for each* $n \geqq 1$ *a natural transformation* $\tau^n : \mathbb{C}^n \circ \overline{\mathbb{V}} \longrightarrow \mathbb{C}^n$ *such that* $\tau^n \circ \mathbb{C}^n\overline{\epsilon} = 1_{\mathbb{C}^n}$*. Then* $\mathbb{H}^n$ *is naturally isomorphic to* $H^n(-,\overline{G})$*.*

6. The fourth property of Beck cohomology requires additional definitions. As above, let $(\mathbb{F},\mathbb{U},\eta,\epsilon) : \mathcal{U} \longrightarrow \mathcal{C}$ be an adjunction, S be an object of $\mathcal{C}$, and $\overline{G}$ be an abelian group object of $\mathcal{C}{\downarrow}S$. A **Beck extension** of $\overline{G}$ by S (called a $\overline{G}$-module in Beck [1967], Definition 6) is an object $\overline{E} = (E,\pi)$ of $\overline{\mathcal{C}} = \mathcal{C}{\downarrow}S$ together with an action $\centerdot$ of $\overline{G}$ on $\overline{E}$ such that

(BE1) $\mathbb{U}\pi \circ \sigma = 1_{\mathbb{U}S}$ for some $\sigma : \mathbb{U}S \longrightarrow \mathbb{U}E$;

(BE2) for every object $\overline{C}$ of $\overline{\mathcal{C}}$, the action of $\mathrm{Hom}_{\overline{\mathcal{C}}}(\overline{C},\overline{G})$ on $\mathrm{Hom}_{\overline{\mathcal{C}}}(\overline{C},\overline{E})$ preserves projection to S: $\pi \circ (g \centerdot e) = \pi \circ e$ whenever $g : \overline{C} \longrightarrow \overline{G}$ and $e : \overline{C} \longrightarrow \overline{E}$ in $\overline{\mathcal{C}}$;

(BE3) for every object $\overline{C}$ of $\overline{\mathcal{C}}$, $\mathrm{Hom}_{\overline{\mathcal{C}}}(\overline{C},\overline{G})$ acts simply and transitively on $\mathrm{Hom}_{\overline{\mathcal{C}}}(\overline{C},\overline{E})$: for every $e, f : \overline{C} \longrightarrow \overline{E}$, there exists a unique $g : \overline{C} \longrightarrow \overline{G}$ such that $g \centerdot e = f$.

A morphism $\varphi : \overline{E} \longrightarrow \overline{F}$ of Beck extensions of $\overline{G}$ by S is a morphism in $\overline{\mathcal{C}}$ which preserves the action of $\overline{G}$: $\varphi \circ (g \centerdot e) = g \centerdot (\varphi \circ e)$ for all $g : \overline{C} \longrightarrow \overline{G}$ and $e : \overline{C} \longrightarrow \overline{E}$.

Beck ([1967], Theorem 6) proved that $H^2(S,\overline{G})$ **classifies** Beck extensions of $\overline{G}$ by S in the following sense:

Theorem 1.4. *When* $\mathcal{C}$ *is tripleable over* $\mathcal{U}$*, there is a one-to-one correspondence between elements of* $H^2(S,\overline{G})$ *and isomorphy classes of Beck extensions of* $\overline{G}$ *by* S*, which is natural in* $\overline{G}$*.*

2. ABELIAN GROUP OBJECTS.

For a more concrete definition of triple cohomology for commutative semigroups we investigate abelian group objects in the category of commutative semigroups over a given commutative semigroup. The main result is from Grillet [1991C], [1995C]; a similar result for monoids was proved by Wells [1978].

1. Let S be a commutative semigroup.

Proposition 2.1. *Let $\mathcal{C}$ be the category of commutative semigroups. An abelian group object of $\mathcal{C}{\downarrow}S$ is a split commutative group coextension of S. The category $\mathcal{O}$ of abelian group objects of $\mathcal{C}{\downarrow}S$ is equivalent to the category $\mathcal{A}$ of abelian group valued functors on $H(S)$.*

Proof. First we show:

Lemma 2.2. *Let G be an abelian group object of a category $\mathcal{C}$ with finite products. Let p_1, $p_2 : G \times G \longrightarrow G$ be the projections and*

$$m = p_1 + p_2 : G \times G \longrightarrow G.$$

Then $g + h = m \circ (g,h)$ whenever $g,h : C \longrightarrow G$.

Proof. $(g,h) : C \longrightarrow G \times G$ is the morphism such that $p_1 \circ (g,h) = g$ and $p_2 \circ (g,h) = h$. Then $g + h = (p_1 \circ (g,h)) + (p_2 \circ (g,h)) = (p_1 + p_2) \circ (g,h) = m \circ (g,h)$. □

To probe c.s. over S we use the additive semigroup $\mathbb{N}^+$ and the following construction. For every $a \in S$ let $\pi_a : \mathbb{N}^+ \longrightarrow S$ be the unique homomorphism such that $\pi_a(1) = a$, namely $\pi_a(n) = a^n$; then $\overline{\mathbb{N}}_a^+ = (\mathbb{N}^+, \pi_a)$ is a c.s. over S.

Lemma 2.3. *Let $\overline{T} = (T, \tau)$ be a commutative semigroup over S. For every $a \in S$ let $T_a = \{t \in T \mid \tau t = a\}$. Evaluation at 1 is a bijection $\varphi \longmapsto \varphi(1)$ of $\mathrm{Hom}_{\overline{\mathcal{C}}}(\overline{\mathbb{N}}_a^+, \overline{T})$ onto T_a.*

Proof. If $\varphi : \overline{\mathbb{N}}_a^+ \longrightarrow \overline{T}$ is a morphism, then $\tau \circ \varphi = \pi_a$ and $\varphi(1) \in T_a$. Conversely there is for every $t \in T_a$ a unique semigroup homomorphism $\bar{t} : \mathbb{N}^+ \longrightarrow T$ such that $\bar{t}(1) = t$, namely, $t \longmapsto t^n$; then $\tau \circ \bar{t} = \pi_a$ and $\bar{t}$ is a morphism $\overline{\mathbb{N}}_a^+ \longrightarrow \overline{T}$ in $\overline{\mathcal{C}}$. This defines mutually inverse bijections. □

2. Now let $\mathcal{C}$ be the category of c.s. and $\overline{\mathcal{C}} = \mathcal{C}{\downarrow}S$. Let $\overline{G} = (G, \rho)$ be an abelian group object of $\overline{\mathcal{C}} = \mathcal{C}{\downarrow}S$; in particular, G is a c.s., which we write multiplicatively, and ρ is a multiplicative homomorphism.

We use Lemma 2.2 to construct a partial addition on G such that addition on every $\mathrm{Hom}_{\overline{\mathcal{C}}}(\overline{C},\overline{G})$ is pointwise. In $\overline{\mathcal{C}}$, the direct product $\overline{P} = (P,\pi) = \overline{G} \times \overline{G}$ and its projections p_1, $p_2 : \overline{P} \longrightarrow \overline{G}$ are given by the pullback

$$\begin{array}{ccc} P & \xrightarrow{p_1} & G \\ {\scriptstyle p_2}\downarrow & & \downarrow{\scriptstyle \rho} \\ G & \xrightarrow[\rho]{} & S \end{array}$$

where $P = \{(x,y) \in G \times G \mid \rho x = \rho y\}$, $p_1(x,y) = x$, and $p_2(x,y) = y$. Then $m = p_1 + p_2 : P \longrightarrow G$ is a partial addition on G; $x + y = m(x,y)$ is defined if and only if $\rho x = \rho y$, if and only if x and y belong to the same set $G_a = \rho^{-1}a$. Thus m provides an addition on every G_a.

Lemma 2.4. *When $\overline{G} = (G,\rho)$ is an abelian group object of $\overline{\mathcal{C}}$:*

$$(x+y)(z+w) \quad = \quad xz + yw \tag{1}$$

whenever $\rho x = \rho y$ and $\rho z = \rho w$; addition on $\mathrm{Hom}_{\overline{\mathcal{C}}}(\overline{C},\overline{G})$ is pointwise; every G_a is an abelian group under addition; for every $a \in S$, $\overline{g} \longmapsto \overline{g}(1)$ is an additive isomorphism of $\mathrm{Hom}_{\overline{\mathcal{C}}}(\overline{\mathbb{N}}_a^+,\overline{G})$ onto G_a; and ρ is surjective.

Proof. Since m is a multiplicative homomorphism, $(x+y)(z+w) = m(x,y)\,m(z,w) = m(xz,yw) = xz + yw$ whenever $\rho x = \rho y$ and $\rho z = \rho w$.

Let $\overline{C} = (C,\pi)$ and $g,h : \overline{C} \longrightarrow \overline{G}$. By Lemma 2.2, $g + h = m \circ (g,h)$. Now $(g,h)(c) = (g(c),h(c)) \in P$ for every $c \in C$, and $(g+h)(c) = m(g(c),h(c)) = g(c) + h(c)$. Thus, addition on $\mathrm{Hom}_{\overline{\mathcal{C}}}(\overline{C},\overline{G})$ is pointwise. Then the bijection $\mathrm{Hom}_{\overline{\mathcal{C}}}(\overline{\mathbb{N}}_a^+,\overline{G}) \longrightarrow G_a$, $\varphi \longmapsto \varphi(1)$ in Lemma 2.3 is an additive isomorphism. Hence G_a is an abelian group under addition. In particular, $G_a \neq \emptyset$ and ρ is surjective. □

Lemma 2.5. *Let p_a be the identity element of G_a under addition. An abelian group valued functor $\mathbb{A}\overline{G} = (G,\gamma)$ on $H(S)$ is defined by*

$$\gamma_{a,t}\, g \quad = \quad g p_t \tag{2}$$

for every $a \in S$, $t \in S^1$, $g \in G_a$, where $p_t = 1 \in G^1$ if $t = 1 \in S^1$. Moreover, $\overline{G}$ is a split group coextension of S by $\mathbb{A}\overline{G}$.

Proof. If $t = 1 \in S^1$, then $\gamma_{a,t}$ is the identity on G_a; otherwise $\gamma_{a,t} g \in G_{at}$, since ρ is a homomorphism, and $\gamma_{a,t}$ is an additive homomorphism, since

$$gp_t + hp_t = (g+h)(p_t + p_t) = (g+h)p_t$$

for all $g, h \in G_a$, by (1). For all $t, u \in S$ we have

$$p_t p_u = (p_t + p_t)(p_u + p_u) = p_t p_u + p_t p_u$$

by (1), which in the abelian group G_{tu} implies $p_t p_u = p_{tu}$. This also holds if $t = 1$ or $u = 1$. Hence $\gamma_{at,u} \circ \gamma_{a,t} = \gamma_{a,tu}$ for all $t, u \in S^1$. Thus $\mathbb{A}\overline{G} = (G, \gamma)$ is an abelian group valued functor on $H(S)$.

The abelian group G_a acts simply and transitively on itself by left addition: $g \centerdot x = g + x$. By (1),

$$(g \centerdot x)\, y = (g + x)(p_b + y) = gp_b + xy = (\gamma_{a,b} g) \centerdot xy$$

for all $g, x \in G_a$ and $y \in G_b$. Thus $(G, \rho, \centerdot)$ is a commutative group coextension of S by $\mathbb{A}\overline{G}$, which splits since $p_a p_b = p_{ab}$ for all $a, b \in S$. □

Lemma 2.6. *Let $\varphi : \overline{G} \longrightarrow \overline{H} = (H, \sigma)$ be a morphism of abelian group objects. Then $\varphi(G_a) \subseteq H_a$ for every $a \in S$ and $\mathbb{A}\varphi = (\varphi_{|G_a})_{a \in S}$ is a natural transformation from $\mathbb{A}\overline{G}$ to $\mathbb{A}\overline{H}$.*

Proof. By definition, φ is a multiplicative homomorphism, $\sigma \circ \varphi = \rho$, and $\varphi_* = \mathrm{Hom}_{\overline{\mathcal{C}}}(\overline{C}, \varphi)$ is a homomorphism for every $\overline{C}$. In particular $\varphi(G_a) \subseteq H_a$ for every $a \in S$. Let $\varphi_a = \varphi_{|G_a} : G_a \longrightarrow H_a$ be the restriction of φ to G_a. For every $a \in S$, Lemma 2.4 provides additive isomorphisms $\mathrm{Hom}_{\overline{\mathcal{C}}}(\overline{\mathbb{N}}_a^+, \overline{G}) \cong G_a$, $\mathrm{Hom}_{\overline{\mathcal{C}}}(\overline{\mathbb{N}}_a^+, \overline{H}) \cong H_a$. The square

$$\begin{array}{ccc}
\mathrm{Hom}_{\overline{\mathcal{C}}}(\overline{\mathbb{N}}_a^+, \overline{G}) & \xrightarrow{\cong} & G_a \\
{\scriptstyle \varphi_*}\big\downarrow & & \big\downarrow{\scriptstyle \varphi_a} \\
\mathrm{Hom}_{\overline{\mathcal{C}}}(\overline{\mathbb{N}}_a^+, \overline{H}) & \xrightarrow[\cong]{} & H_a
\end{array}$$

commutes, since $(\varphi_*(\overline{g}))(1) = (\varphi \circ \overline{g})(1) = \varphi(\overline{g}(1)) = \varphi_a(\overline{g}(1))$ for all $\overline{g} : \overline{\mathbb{N}}_a^+ \longrightarrow \overline{G}$. Since φ_* is an additive homomorphism, it follows that φ_a is an additive homomorphism. In particular $\varphi_a(p_a) = q_a$, the identity element of the abelian group H_a. Since φ is a multiplicative homomorphism, we have $\varphi_{at}(gp_t) = (\varphi_a g)\, q_t$ for all $a, t \in S$ and $g \in G_a$; thus $\mathbb{A}\varphi = (\varphi_a)_{a \in S}$ is a natural transformation from $\mathbb{A}\overline{G}$ to $\mathbb{A}\overline{H}$. □

We now have a functor $\mathbb{A} : \mathcal{O} \longrightarrow \mathcal{A}$.

3. Conversely let (G, γ) be an abelian group valued functor on $H(S)$. Let

g^t denote $\gamma_{a,t} g$ when $g \in G_a$ and $t \in S^1$. As in Theorem V.4.1, there is a split commutative group coextension $(E, \rho, .)$ of S by (G, γ), in which E is the disjoint union $\bigcup_{a \in S} (G_a \times \{a\})$ with multiplication

$$(g,a)(h,b) \;=\; (g^b + h^a,\, ab),$$

projection $\rho : (g,a) \longmapsto a$ to S, and action $g \mathbin{.} (h,a) = (g+h,\, a)$ of G_a on $E_a = G_a \times \{a\}$. Then $\overline{G} = (E, \rho)$ is a c.s. over S.

Let $\overline{C} = (C, \pi)$ be an object of $\overline{\mathcal{C}}$ and $\overline{g}, \overline{h} : \overline{C} \longrightarrow \overline{G}$ be morphisms in $\overline{\mathcal{C}}$. Then $\rho \circ \overline{g} = \rho \circ \overline{h} = \pi$. For every $c \in C$ there exist unique $g_c, h_c \in G_a$ such that $\overline{g}(c) = (g_c, a)$ and $\overline{h}(c) = (h_c, a)$, where $a = \pi c$. Define

$$(\overline{g} + \overline{h})(c) = (g_c + h_c,\, a).$$

Lemma 2.7. *With the addition defined above, $\overline{G}$ is an abelian group object over S. Moreover, $\mathbb{A}\overline{G} \cong (G, \gamma)$.*

Proof. Let $\overline{g}, \overline{h} : \overline{C} \longrightarrow \overline{G}$ be morphisms in $\overline{\mathcal{C}}$. Since $\overline{g}$ and $\overline{h}$ are multiplicative homomorphisms, we have

$$(g_{cd},\, ab) \;=\; \overline{g}(cd) \;=\; \overline{g}(c)\,\overline{g}(d) \;=\; (g_c, a)(g_d, b) \;=\; (g_c^b + g_d^a,\, ab),$$

where $a = \pi c$ and $b = \pi d$, and $g_{cd} = g_c^b + g_d^a$; similarly $h_{cd} = h_c^b + h_d^a$. Hence

$$\begin{aligned}(\overline{g} + \overline{h})(c)\,(\overline{g} + \overline{h})(d) &= (g_c + h_c,\, a)(g_d + h_d,\, b)\\ &= (g_c^b + h_c^b + g_d^a + h_d^a,\, ab)\\ &= (g_{cd} + h_{cd},\, ab) \;=\; (\overline{g} + \overline{h})(cd).\end{aligned}$$

and $\overline{g} + \overline{h}$ is a multiplicative homomorphism. We now have an addition on $\mathrm{Hom}_{\overline{\mathcal{C}}}(\overline{C}, \overline{A})$. It is immediate that $\mathrm{Hom}_{\overline{\mathcal{C}}}(\overline{C}, \overline{A})$ is an abelian group, in which the identity element is $c \longmapsto (0, \pi c)$ and the opposite of $\overline{g} : c \longmapsto (g_c, \pi c)$ is $c \longmapsto (-g_c, \pi c)$. Moreover $(\overline{g} + \overline{h}) \circ \delta = (\overline{g} \circ \delta) + (\overline{h} \circ \delta)$ for every morphism $\delta : \overline{D} \longrightarrow \overline{C}$. Thus $\overline{G}$ is an abelian group object of $\overline{\mathcal{C}}$.

We show that $\mathbb{A}\overline{G} \cong (G, \gamma)$. Since $\overline{G}$ is an abelian group object, it induces, as in the first part of the proof, an abelian group addition on every E_a. For every $a \in S$, Lemma 2.4 provides an additive isomorphism $\overline{g} \longmapsto \overline{g}(1)$ of $\mathrm{Hom}_{\overline{\mathcal{C}}}(\overline{\mathbb{N}}_a^+, \overline{G})$ onto E_a. For every $(g,a), (h,a) \in E_a$ we now have homomorphisms $\overline{g}, \overline{h} : \overline{\mathbb{N}}_a^+ \longrightarrow \overline{G}$ such that $\overline{g}(1) = (g,a)$ and $\overline{h}(1) = (h,a)$; since addition on

$\mathrm{Hom}_{\overline{\mathcal{C}}}(\overline{\mathbb{N}}_a^+, \overline{G})$ is pointwise,

$$(g,a) + (h,a) \;=\; \overline{g}(1) + \overline{h}(1) \;=\; (\overline{g} + \overline{h})(1) \;=\; (g+h,\, a).$$

Thus $g \longmapsto (g,a)$ is an isomorphism of G_a onto E_a.

In particular, the identity element of E_a is $p_a = (0,a)$. The homomorphism $\epsilon_{a,t}$ in the functor $\mathbb{A}\overline{G} = (E,\epsilon)$ is then given by (2): $\epsilon_{a,t}(g,a) = (g,a)(0,t) = (g^t,\, at)$ for all $t \in S$. Thus $\mathbb{A}\overline{G}$ is naturally isomorphic to (G,γ). □

Lemma 2.8. *Let $(\varphi_a)_{a\in S}$ be a natural transformation from (G,γ) to (H,δ). Then $\varphi : \overline{G} \longrightarrow \overline{H}$, defined by $\varphi(g,a) = (\varphi_a g,\, a)$, is a morphism of abelian group objects.*

Proof. Since $(\varphi_a)_{a\in S}$ is a natural transformation, we have $(\varphi_a g)^t = \varphi_{at}(g^t)$ whenever $g \in G_a$. Hence φ is a multiplicative homomorphism:

$$\begin{aligned} \varphi((g,a)(h,b)) &= \varphi(g^b + h^a,\, ab) \;=\; (\varphi_{ab} g^b + \varphi_{ab} h^a,\, ab) \\ &= ((\varphi_a g)^b + (\varphi_b h)^a,\, ab) \;=\; \varphi(g,a)\,\varphi(h,b) \end{aligned}$$

and a morphism in $\overline{\mathcal{C}}$. Moreover, for any $\overline{g}, \overline{h} : \overline{C} \longrightarrow \overline{G}$, we have, with the notation as above, $\varphi(\overline{g}(c)) = (\varphi_a g_c, a)$, $\varphi(\overline{h}(c)) = (\varphi_a h_c, a)$, and

$$(\varphi \circ \overline{g} + \varphi \circ \overline{h})(c) \;=\; (\varphi_a g_c + \varphi_a h_c,\, a) \;=\; \varphi(g_c + h_c,\, a) \;=\; \varphi\big((\overline{g} + \overline{h})(c)\big);$$

thus φ is a morphism of abelian group objects. □

We now have a functor $\mathbb{O} : \mathcal{A} \longrightarrow \mathcal{O}$; $\mathbb{O}(G,\gamma)$ is $\overline{G}$ in Lemma 2.7, and Lemma 2.8 constructs $\mathbb{O}(\varphi_a)_{a\in S} = \varphi$.

4. We saw that $\mathbb{A}\mathbb{O}G$ is isomorphic to G. If conversely $\overline{G} = (G,\pi)$ is any abelian group object of $\overline{\mathcal{C}}$, then G is a split group coextension of S by $\mathbb{A}\overline{G}$ and $\overline{G}$ is isomorphic to $\mathbb{O}\mathbb{A}\overline{G}$ as a semigroup over S and as an abelian group object. If not exhausted our reader will verify that these isomorphisms are natural, which completes the proof of Proposition 2.1. □

3. BECK EXTENSIONS.

Continuing Section 2 we now investigate Beck extensions in the category of commutative semigroups over a given commutative semigroup. The main result is from Grillet [1991C], [1995C]; a similar result for monoids was proved by Wells [1978].

1. Let S be a c.s. and $\mathcal{C}$ be the category of c.s. and homomorphisms.

Proposition 3.1. *Let $\overline{G} = (G, \rho)$ be an abelian group object of $\mathcal{C}{\downarrow}S$ and $\mathbb{G} = \mathbb{A}\overline{G} = (G, \gamma)$ be the corresponding abelian group valued functor on $H(S)$. A Beck extension of $\overline{G}$ by S is a commutative group coextension of S by $\mathbb{G}$. The category $\mathcal{B}$ of Beck extensions of $\overline{G}$ by S is isomorphic to the category $\mathcal{E}$ of commutative group coextensions of S by $\mathbb{G}$.*

A morphism in $\mathcal{E}$ (necessarily an isomorphism) is an equivalence of commutative group coextensions.

Proof. The proof is rather similar to that of Proposition 2.1. First we show:

Lemma 3.2. *Let G be an abelian group object of a category $\mathcal{C}$ with finite products and $\centerdot$ be an action of G on an object E of $\mathcal{C}$. Let $p_1 : G \times E \longrightarrow G$, $p_2 : G \times E \longrightarrow E$ be the projections and*

$$q = p_1 \centerdot p_2 : G \times E \longrightarrow E.$$

Then $g \centerdot e = n \circ (g, e)$ for all $g : C \longrightarrow G$ and $e : C \longrightarrow E$.

Proof. $(g, e) : C \longrightarrow G \times E$ is the morphism such that $p_1 \circ (g, e) = g$ and $p_2 \circ (g, e) = e$. Since $\centerdot$ is an action, $g \centerdot e = (p_1 \circ (g, e)) \centerdot (p_2 \circ (g, e)) = (p_1 \centerdot p_2) \circ (g, e) = n \circ (g, e)$. □

2. Now let $\overline{G}$ be an abelian group object over a c.s. S and $\overline{E} = (E, \pi)$ be a Beck extension of $\overline{G}$ by S, so that there is an action of $\overline{G}$ on $\overline{E}$ and (BE1), (BE2), (BE3) hold. In particular, E is a c.s., which we write multiplicatively, and π is a multiplicative homomorphism. By (BE1), $\pi \circ \sigma = 1_{\mathbb{U}S}$ for some mapping $\sigma : S \longrightarrow E$; hence π is surjective, and (E, π) is a coextension of S.

We use Lemma 3.2 to construct a partial action of G on E such that $\mathrm{Hom}_{\overline{\mathcal{C}}}(\overline{C}, \overline{G})$ acts pointwise on $\mathrm{Hom}_{\overline{\mathcal{C}}}(\overline{C}, \overline{E})$ for every $\overline{C} \in \overline{\mathcal{C}}$. In $\overline{\mathcal{C}}$, the direct product $\overline{P} = (P, r) = \overline{G} \times \overline{E}$ and its projections $p_1 : \overline{P} \longrightarrow \overline{G}$, $p_2 : \overline{P} \longrightarrow \overline{E}$ are given by the pullback

$$\begin{array}{ccc} P & \xrightarrow{p_1} & G \\ {\scriptstyle p_2}\big\downarrow & & \big\downarrow{\scriptstyle \rho} \\ E & \xrightarrow[\pi]{} & S \end{array}$$

where $P = \{(g, x) \in G \times E \mid \rho g = \pi x\}$, $p_1(g, x) = g$, and $p_2(g, x) = x$. Then $q = p_1 \centerdot p_2 : P \longrightarrow E$ is a partial action of G on E, for which $g \centerdot x = q(g, x)$ is defined if and only if $\rho g = \pi x$, if and only if $g \in G_a = \rho^{-1} a$ and $x \in E_a = \pi^{-1} a$ for some $a \in S$; thus q provides a set action of G_a on E_a, for every $a \in S$.

Lemma 3.3. *When $\overline{E} = (E,\pi)$ is a Beck extension of S by $\overline{G} = (G,\rho)$:*

$$(g\mathbin{.}x)(h\mathbin{.}y) \;=\; gh\mathbin{.}xy \tag{3}$$

whenever $\rho g = \pi x$ and $\rho h = \pi y$; $\mathrm{Hom}_{\overline{\mathcal{C}}}(\overline{C},\overline{G})$ acts on $\mathrm{Hom}_{\overline{\mathcal{C}}}(\overline{C},\overline{E})$ pointwise; the action of G_a on E_a is a simply transitive group action; and $\mathbb{E}\overline{E} = (E,\pi,.)$ is a commutative group coextension of S by $\mathbb{G}$.

Proof. Since q is a multiplicative homomorphism, we have $(g\mathbin{.}x)(h\mathbin{.}y) = q(g,x)\,q(h,y) = q(gh,xy) = gh\mathbin{.}xy$ whenever $\rho g = \pi x$ and $\rho h = \pi y$.

Let $\overline{C} = (C,\sigma)$ and $g : \overline{C} \longrightarrow \overline{G}$, $e : \overline{C} \longrightarrow \overline{E}$ be morphisms in $\overline{\mathcal{C}}$. Then $(g,e)(c) = (g(c),e(c)) \in P$ for every $c \in C$, and Lemma 3.2 yields $g\mathbin{.}e = n \circ (g,e)$ and

$$(g\mathbin{.}e)(c) \;=\; q(g(c),e(c)) \;=\; g(c)\mathbin{.}h(c).$$

Thus $\mathrm{Hom}_{\overline{\mathcal{C}}}(\overline{C},\overline{G})$ acts on $\mathrm{Hom}_{\overline{\mathcal{C}}}(\overline{C},\overline{E})$ pointwise.

For every $a \in S$, Lemmas 2.4 and 2.3 provide an additive isomorphism $\overline{g} \longmapsto \overline{g}(1)$ of $\mathrm{Hom}_{\overline{\mathcal{C}}}(\overline{\mathbb{N}}_a^+,\overline{G})$ onto G_a and a bijection $\overline{e} \longmapsto \overline{e}(1)$ of $\mathrm{Hom}_{\overline{\mathcal{C}}}(\overline{\mathbb{N}}_a^+,\overline{E})$ onto E_a. Moreover $(\overline{g}\mathbin{.}\overline{e})(1) = \overline{g}(1)\mathbin{.}\overline{e}(1)$, since $\mathrm{Hom}_{\overline{\mathcal{C}}}(\overline{C},\overline{G})$ acts on $\mathrm{Hom}_{\overline{\mathcal{C}}}(\overline{C},\overline{E})$ pointwise. Therefore the action of G_a on E_a is a group action, since the action of $\mathrm{Hom}_{\overline{\mathcal{C}}}(\overline{C},\overline{G})$ on $\mathrm{Hom}_{\overline{\mathcal{C}}}(\overline{C},\overline{E})$ is a group action; and G_a acts simply and transitively on E_a, by (BE3).

Let p_a be the identity element of G_a. Then $p_a\mathbin{.}x = x$ for all $x \in E_a$, and (3), (2) yield

$$(g\mathbin{.}x)y \;=\; (g\mathbin{.}x)(p_b\mathbin{.}y) \;=\; gp_b\mathbin{.}xy \;=\; \gamma_{a,b}g\mathbin{.}xy$$

whenever $g \in G_a$, $x \in E_a$, and $y \in E_b$. Thus $\mathbb{E}\overline{E} = (E,\pi,.)$ is a commutative group coextension of S by $\mathbb{G}$. □

Lemma 3.4. *Every morphism $\varphi : \overline{E} \longrightarrow \overline{E}' = (E',\pi')$ of Beck extensions of S by $\overline{G}$ is an equivalence of commutative group coextensions of $\mathbb{E}\overline{E}$ to $\mathbb{E}\overline{E}'$.*

Proof. By definition, φ is a multiplicative homomorphism of E into E'; $\pi' \circ \varphi = \pi$; and $\varphi \circ (\overline{g}\mathbin{.}\overline{e}) = \overline{g}\mathbin{.}(\varphi \circ \overline{e})$ for all $\overline{g} : \overline{C} \longrightarrow \overline{G}$ and $\overline{e} : \overline{C} \longrightarrow \overline{E}$. In particular, $\varphi(E_a) \subseteq E'_a$ for every $a \in S$. For every $g \in G_a$ and $x \in E_a$, Lemma 2.3 yields morphisms $\overline{g} : \overline{\mathbb{N}}_a^+ \longrightarrow \overline{G}$ and $\overline{x} : \overline{\mathbb{N}}_a^+ \longrightarrow \overline{E}$ such that $\overline{g}(1) = g$ and $\overline{x}(1) = x$. Then $\varphi \circ (\overline{g}\mathbin{.}\overline{x}) = \overline{g}\mathbin{.}(\varphi \circ \overline{x})$ and evaluation at 1 yields $\varphi(g\mathbin{.}x) = g\mathbin{.}\varphi(x)$. Thus φ preserves the action of G_a. Then φ is a bijection of E_a onto E'_a, since G_a acts simply and transitively on E_a and on E'_a, and

φ is an equivalence of commutative group coextensions. □

We now have a functor $\mathbb{E} : \mathcal{B} \longrightarrow \mathcal{E}$.

3. Conversely let $(E, \pi, .)$ be a commutative group coextension of S by G. Then $\overline{E} = (E, \pi)$ is a c.s. over S. By (1), (2),

$$gh \;=\; (g + p_a)(p_b + h) \;=\; gp_b + p_a h \;=\; \gamma_{a,b} g + \gamma_{b,a} h \;=\; g^b + h^a$$

for all $g \in G_a$ and $h \in G_b$; then

$$(g \,.\, x)(h \,.\, y) \;=\; g^b \,.\, (x\,(h \,.\, y)) \;=\; (g^b + h^a)\,.xy \;=\; gh \,.\, xy \tag{4}$$

for all $g \in G_a$, $h \in G_b$, $x \in E_a$, and $y \in G_b$, since E is a group coextension.

Let $\overline{C} = (C, \tau)$ be an object of $\overline{\mathcal{C}}$ and $\overline{g} : \overline{C} \longrightarrow \overline{G}$, $\overline{e} : \overline{C} \longrightarrow \overline{E}$ be morphisms in $\overline{\mathcal{C}}$. Then $\rho \circ g = \pi \circ e = \tau$ and $\overline{g}(c) \,.\, \overline{e}(c)$ is defined in E for every $c \in C$. Define

$$(\overline{g} \,.\, \overline{e})(c) \;=\; \overline{g}(c) \,.\, \overline{e}(c).$$

Lemma 3.5. *With the action defined above,* $\mathbb{B}(E, \pi, .) = \overline{E}$ *is a Beck extension of* S *by* $\overline{G}$*. Moreover,* $\mathbb{E}\overline{E}$ *is the given coextension* $(E, \pi, .)$.

Proof. Since $\overline{g}$ and $\overline{e}$ are multiplicative homomorphisms, we have, by (4),

$$\big(\overline{g}(c) \,.\, \overline{e}(c)\big)\big(\overline{g}(d) \,.\, \overline{e}(d)\big) \;=\; \overline{g}(c)\overline{g}(d) \,.\, \overline{e}(c)\overline{e}(d) \;=\; \overline{g}(cd) \,.\, \overline{e}(cd)$$

where $a = \tau c$, $b = \tau d$; hence $\overline{g} \,.\, \overline{e}$ is a multiplicative homomorphism. Now $\mathrm{Hom}_{\overline{\mathcal{C}}}(\overline{C}, \overline{G})$ acts on $\mathrm{Hom}_{\overline{\mathcal{C}}}(\overline{C}, \overline{E})$; this is a group action since addition on $\mathrm{Hom}_{\overline{\mathcal{C}}}(\overline{C}, \overline{G})$ is pointwise. (BE1) holds since π is surjective. Also

$$\pi\left((\overline{g} \,.\, \overline{e})(c)\right) \;=\; \pi\left((\overline{g}(c) \,.\, \overline{e}(c)\right) \;=\; \pi\left(\overline{e}(c)\right)$$

for all $c \in C$ and (BE2) holds.

We show that $\mathrm{Hom}_{\overline{\mathcal{C}}}(\overline{C}, \overline{G})$ acts simply and transitively on $\mathrm{Hom}_{\overline{\mathcal{C}}}(\overline{C}, \overline{E})$. Let $\overline{e}, \overline{f} : \overline{C} \longrightarrow \overline{E}$. For every $c \in C$ there exists a unique $\overline{g}(c) \in G_a$ such that $\overline{f}(c) = \overline{g}(c) \,.\, \overline{e}(c)$, where $a = \tau c$. Since $\overline{e}$ and $\overline{f}$ are multiplicative homomorphisms, we have

$$\begin{aligned} \overline{g}(cd) \,.\, \overline{e}(cd) \;&=\; \overline{f}(cd) \;=\; \overline{f}(c)\overline{f}(d) \\ &=\; \big(\overline{g}(c) \,.\, \overline{e}(c)\big)\big(\overline{g}(d) \,.\, \overline{e}(d)\big) \;=\; \overline{g}(c)\overline{g}(d) \,.\, \overline{e}(cd) \end{aligned}$$

for all $c, d \in C$, by (4). Hence $\overline{g}$ is a homomorphism. Then $\overline{g}$ is the unique morphism $\overline{C} \longrightarrow \overline{G}$ such that $\overline{g} \,.\, \overline{e} = \overline{f}$. Thus $\overline{E}$ (with the action $.$ of $\overline{G}$ on $\overline{E}$) is a Beck extension of S by $\overline{G}$.

Since $\overline{E}$ is a Beck extension, it induces, as in the first part of the proof, a simply transitive group action of G_a on E_a, which makes $\overline{E}$ a commutative group coextension $\mathbb{E}\overline{E}$ of S by G. Let $a \in S$. For every $g \in G_a$ and $e \in E_a$, Lemma 2.3 yields morphisms $\overline{g} : \overline{C}_a \longrightarrow \overline{G}$ and $\overline{e} : \overline{C}_a \longrightarrow \overline{G}$ such that $\overline{g}(1) = g$ and $\overline{e}(1) = e$. Then the action of G_a on E_a in $\mathbb{E}\overline{E}$ satisfies

$$g \centerdot e \;=\; \overline{g}(1) \centerdot \overline{e}(1) \;=\; (\overline{g} \centerdot \overline{e})(1);$$

so does the action of G_a on E_a in the given coextension $(E, \pi, .)$. Hence the two actions coincide and $\mathbb{E}\overline{E} = (E, \pi, .)$. □

Lemma 3.6. *Every equivalence of commutative group coextensions from* $(E, \pi, .)$ *to* $(E', \pi', .)$ *is a morphism of Beck extensions from* $\mathbb{B}(E, \pi, .)$ *to* $\mathbb{B}(E', \pi', .)$.

Proof. Let φ be an equivalence of commutative group coextensions from $(E, \pi, .)$ to $(E', \pi', .)$. Then φ is a multiplicative homomorphism and preserves projection to S and action of G. For every $\overline{g} : \overline{C} \longrightarrow \overline{G}$ and $\overline{e} : \overline{C} \longrightarrow \overline{E}$ we then have $\varphi\big(\overline{g}(c) \centerdot \overline{e}(c)\big) = \overline{g}(c) \centerdot \varphi(\overline{e}(c))$ and $\varphi \circ (\overline{g} \centerdot \overline{e}) = \overline{g} \centerdot (\varphi \circ \overline{e})$. Thus φ is a morphism of Beck extensions. □

4. We now have a functor $\mathbb{B} : \mathcal{E} \longrightarrow \mathcal{B}$. We saw (Lemma 3.5) that $\mathbb{E}\mathbb{B} = 1_{\mathcal{E}}$. If conversely $\overline{E}$ is a Beck extension of $\overline{G}$ by S, then the action of $\mathrm{Hom}_{\overline{\mathcal{C}}}(\overline{C}, \overline{G})$ on $\mathrm{Hom}_{\overline{\mathcal{C}}}(\overline{C}, \overline{E})$ is the pointwise action induced by the action of G on $\mathbb{E}\overline{E}$ and coincides with the action of $\mathrm{Hom}_{\overline{\mathcal{C}}}(\overline{C}, \overline{G})$ on $\mathrm{Hom}_{\overline{\mathcal{C}}}(\overline{C}, \mathbb{B}\mathbb{E}\overline{E})$. Hence $\mathbb{B}\mathbb{E}\overline{E} = \overline{E}$. Thus $\mathbb{B}$ and $\mathbb{E}$ are mutually inverse isomorphisms. □

4. COMMUTATIVE SEMIGROUP COHOMOLOGY.

Triple cohomology in the category $\mathcal{C}$ of commutative semigroups provides a good definition of commutative semigroup cohomology. This section brings a more concrete definition, based on the results in Sections 2 and 3. The results are from Grillet [1991C], [1995C].

Other cohomology theories have been considered for commutative semigroups. Inasaridze extended the construction of n-extensions and Ext^n from abelian groups to commutative cancellative monoids [1964], [1965] and to commutative Clifford semigroups [1964], [1967]. Kruming [1982] characterized finite commutative semigroups whose Eilenberg-MacLane cohomology vanishes; Novikov [1990] showed that cancellative c.s. with this property are subsemigroups of $\mathbb{N}$.

See also Carbonne [1983]. For a survey of semigroup cohomology in general, see Grillet & Novikov [2002].

1. Let F_X be the free c.s. on a set X, which we write multiplicatively. For what follows it is best to regard the elements of F_X as **commutative words** in X, which are nonempty unordered sequences $[x_1, \dots, x_m]$ of elements of X; unordered means

$$[x_{\sigma 1}, \dots, x_{\sigma m}] \;=\; [x_1, \dots, x_m]$$

for every permutation σ. It is customary to write $[x_1, \dots, x_m]$ as a product $x_1 \cdots x_m$, but this would quickly become very confusing in what follows. Multiplication in F_X is by concatenation:

$$[x_1, \dots, x_m]\,[y_1, \dots, y_n] \;=\; [x_1, \dots, x_m,\, y_1, \dots, y_n].$$

The injection $\eta_X : X \longrightarrow F_X$ takes $x \in X$ to $[x] \in F_X$. Every mapping of a set X into a c.s. factors uniquely through η_X.

The free c.s. functor $\mathbb{F} : \mathit{Sets} \longrightarrow \mathcal{C}$ assigns to a set X the free c.s. $\mathbb{F}X = F_X$ on X, and assigns to a mapping $f : X \longrightarrow Y$ the unique homomorphism $\mathbb{F}f : \mathbb{F}X \longrightarrow \mathbb{F}Y$ such that $\mathbb{F}f \circ \eta_X = \eta_Y \circ f$; $\mathbb{F}f$ sends $[x_1, x_2, \dots, x_m] = [x_1]\,[x_2] \cdots [x_m] \in \mathbb{F}X$ to $[ft_1, ft_2, \cdots, ft_m] \in \mathbb{F}Y$. $\mathbb{F}$ is a left adjoint of the forgetful functor $\mathbb{U} : \mathcal{C} \longrightarrow \mathit{Sets}$.

Lemma 4.1. *In the adjunction* $(\mathbb{F}, \mathbb{U}, \eta, \epsilon) : \mathit{Sets} \longrightarrow \mathcal{C}$, $\epsilon_T : \mathbb{FU}T \longrightarrow T$ *sends* $[t_1, t_2, \dots, t_m] \in \mathbb{FU}T$ *to* $t_1\, t_2 \cdots t_m \in T$, *for every c.s.* T.

Proof. Since $(\mathbb{F}, \mathbb{U}, \eta, \epsilon) : \mathit{Sets} \longrightarrow \mathcal{C}$ is an adjunction, we have $\mathbb{U}\epsilon_T \circ \eta_{\mathbb{U}T} = 1_{\mathbb{U}T}$ (see e.g. Theorem IV.1 in MacLane [1971]). Thus $\epsilon_T : \mathbb{FU}T \longrightarrow T$ is the homomorphism such that $\epsilon_T\,[t] = \epsilon_T\,(\eta_{\mathbb{U}T}\, t) = t$ for all $t \in T$. Hence ϵ_T sends $[t_1, t_2, \dots, t_m] = [t_1]\,[t_2] \cdots [t_m] \in \mathbb{FU}T$ to $t_1\, t_2 \cdots t_m \in T$. $\square$

In the cotriple $(\mathbb{V}, \epsilon, \nu)$ induced by the adjunction $(\mathbb{F}, \mathbb{U}, \eta, \epsilon) : \mathit{Sets} \longrightarrow \mathcal{C}$, $\mathbb{V} = \mathbb{FU}$ sends a c.s. T to the free c.s. $\mathbb{V}T = \mathbb{FU}T$ on the set T. If $f : T \longrightarrow T'$ is a homomorphism, then $\mathbb{V}f : [x_1, \dots, x_m] \longmapsto [fx_1, \dots, fx_m]$. Lemma 4.1 describes $\epsilon_T : \mathbb{V}T \longrightarrow T$; ν will not be used.

For every c.s. S the cotriple $(\mathbb{V}, \epsilon, \nu)$ lifts to a cotriple $(\overline{\mathbb{V}}, \overline{\epsilon}, \overline{\nu})$ on $\overline{\mathcal{C}} = \mathcal{C}{\downarrow}S$; if $\overline{T} = (T, \tau)$ is a c.s. over S, then $\overline{\mathbb{V}}\,\overline{T} = (\mathbb{V}T, \overline{\tau})$, where $\mathbb{U}\overline{\tau} \circ \eta_{\mathbb{U}T} = \mathbb{U}\tau$; that is, $\overline{\tau}[t] = \tau t$ for every $t \in T$ and

$$\overline{\tau}[x_1, \dots, x_m] \;=\; \tau x_1 \cdots \tau x_m \;=\; \tau\,(x_1 \cdots x_m)$$

for all $m > 0$ and $x_1, \dots, x_m \in T$.

2. The next Lemma describes $\mathrm{Hom}_{\overline{\mathcal{C}}}(\overline{\mathbb{V}}\,\overline{T}, \overline{G})$ when $\overline{G}$ is an abelian group

object of $\overline{\mathfrak{C}}$. When $\mathbb{G} = (G, \gamma)$ is an abelian group valued functor on $H(S)$, and $\overline{T} = (T, \tau)$ is a c.s. over S, let

$$C(\overline{T}, \mathbb{G}) = \prod_{t \in T} G_{\tau t};$$

$C(\overline{T}, \mathbb{G})$ consists of all families $u = (u_t)_{t \in T}$ such that $u_t \in G_{\tau t}$ for all $t \in T$, under pointwise addition (that is, 1-cochains on T with values in $\mathbb{G}$).

Lemma 4.2. *Let $\overline{G} = (G, \rho)$ be an abelian group object of $\overline{\mathfrak{C}}$ and $\mathbb{G}$ be the corresponding abelian group valued functor on $H(S)$. For every object $\overline{T}$ of $\overline{\mathfrak{C}}$ there is a natural isomorphism* $\mathrm{Hom}_{\overline{\mathfrak{C}}}(\overline{\mathbb{V}T}, \overline{G}) \cong C(\overline{T}, \mathbb{G})$.

Proof. By Proposition 2.1 we may assume that $\overline{G}$ is the split commutative group coextension $\bigcup_{a \in S} G_a \times \{a\}$ of S by the corresponding abelian group valued functor $\mathbb{G}$. Then addition on $\mathrm{Hom}_{\overline{\mathfrak{C}}}(\overline{C}, \overline{G})$ is as follows: if $f, g : \overline{C} \longrightarrow \overline{G}$ and $f(c) = (f_c, a)$, $g(c) = (g_c, a)$, then $(f + g)(c) = (f_c + g_c,\ a)$.

We have $\overline{\mathbb{V}T} = (\mathbb{V}T, \overline{\tau})$, where $\mathbb{V}T$ is the free c.s. on the set T and $\overline{\tau}[x_1, \dots, x_m] = \tau(x_1 \cdots x_m)$ for all $m > 0$ and $x_1, \dots, x_m \in T$, in particular $\overline{\tau}[t] = \tau t$ for all $t \in T$. If $f : \overline{\mathbb{V}T} \longrightarrow \overline{G}$ is a morphism in $\overline{\mathfrak{C}}$, then $\rho \circ f = \overline{\tau}$ and there is for every $t \in T$ a unique $u_t \in G_{\overline{\tau}[t]} = G_{\tau t}$ such that $f[t] = (u_t, \tau t)$.

Since $\overline{\mathbb{V}T}$ is free on T there is for every $u = (u_t)_{t \in T} \in C(\overline{T}, \mathbb{G})$ a unique semigroup homomorphism $f : \overline{\mathbb{V}T} \longrightarrow \overline{G}$ such that $f[t] = (u_t, \tau t)$. Then

$$f\,[x_1, \dots, x_m] = (u_{x_1}, \tau x_1) \cdots (u_{x_m}, \tau x_m) = \left(\textstyle\sum_{1 \leqq i \leqq m} u_{x_j}^{\tau \widehat{x}_j},\ \tau x \right),$$

where $\widehat{x}_j = x_1 \cdots x_{j-1}\, x_{j+1} \cdots x_m$ and $x = x_1 \cdots x_m$, for all $m > 0$ and $x_1, \dots, x_m \in T$. (If $m = 1$, then $\widehat{x}_j$ is an empty product and $\widehat{x}_j = 1 \in T^1$.) This provides a bijection $\Theta : \mathrm{Hom}_{\overline{\mathfrak{C}}}(\overline{\mathbb{V}T}, \overline{G}) \longrightarrow C(\overline{T}, \mathbb{G})$. If $f[t] = (u_t, \tau t)$ and $g[t] = (v_t, \tau t)$, then $(f + g)[t] = (u_t + v_t,\ \tau t)$; hence Θ is an isomorphism. It is immediate that Θ is natural in $\overline{T}$ and $\mathbb{G}$. □

3. The triple cohomology of $\overline{T} = (T, \tau) \in \overline{\mathfrak{C}}$, and the triple cohomology of S, which is that of $(S, 1_S)$, can now be obtained as follows. Let $\overline{G} = (G, \rho)$ be an abelian group object of $\overline{\mathfrak{C}}$.

Let $\overline{T}_n = (T_n, \pi_n) = \overline{\mathbb{V}}^n \overline{T}$. Then $T_0 = T$, $\pi_0 = \tau$ ($T_0 = S$, $\pi_0 = 1_S$ for the cohomology of S); $T_{n+1} = \mathbb{V}T_n$ is the semigroup of all commutative words $[x_1, \dots, x_m]$ with $m > 0$ and $x_1, \dots, x_m \in T_n$, and

$$\pi_{n+1}\,[x_1, \dots, x_m] = \pi_n(x_1 \cdots x_m) \in S. \qquad (\pi)$$

For all $0 \leqq i \leqq n$ the morphisms $\overline{\epsilon}_{\overline{T}}^{n,i} = \epsilon_T^{n,i} : \overline{T}_{n+1} \longrightarrow \overline{T}_n$ in $\overline{\mathcal{C}}$, $\epsilon_T^{n,i} = \mathbb{V}^{n-i}\, \epsilon_{\mathbb{V}^i T}$ may be defined by induction:

$$\epsilon^{n,n}\,[x_1,\dots,x_m] \;=\; \epsilon_{\mathbb{V}^n T}\,[x_1,\dots,x_m] \;=\; x_1 \cdots x_m \in T_n \qquad (\epsilon')$$

for all $[x_1,\dots,x_m] \in T_{n+1}$, by Lemma 4.1; when $i < n$, $\epsilon_T^{n,i} = \mathbb{V}\,\epsilon_T^{n-1,\,i}$ and

$$\epsilon^{n,i}\,[x_1,\dots,x_m] \;=\; [\epsilon^{n-1,\,i}x_1,\,\dots,\,\epsilon^{n-1,\,i}x_m] \in T_n \qquad (\epsilon'')$$

for all $[x_1,\dots,x_m] \in T_{n+1}$.

The equality

$$\pi_n \circ \epsilon^{n,i} \;=\; \pi_{n+1} \qquad (\pi\epsilon)$$

is proved by induction on n: for every $[x_1,\dots,x_m] \in T_{n+1}$,

$$\pi_n\,\epsilon^{n,n}\,[x_1,\dots,x_m] \;=\; \pi_n\,(x_1 \cdots x_m) \;=\; \pi_{n+1}\,[x_1,\dots,x_m]$$

and

$$\begin{aligned} \pi_n\,\epsilon^{n,i}\,[x_1,\dots,x_m] \;&=\; \pi_n\,[\epsilon^{n-1,i}x_1,\,\dots,\,\epsilon^{n-1,i}x_m] \\ &=\; \pi_{n-1}\,(\epsilon^{n-1,i}x_1 \cdots \epsilon^{n-1,i}x_m) \;=\; \pi_{n-1}\,\epsilon^{n-1,i}\,(x_1 \cdots x_m) \\ &=\; \pi_n\,(x_1 \cdots x_m) \;=\; \pi_{n+1}\,[x_1,\dots,x_m] \end{aligned}$$

for all $i < n$, since $\epsilon^{n,i}$ is a homomorphism.

An n-cochain is an element of $C^n(\overline{T},\overline{G}) = \mathrm{Hom}_{\overline{\mathcal{C}}}(\overline{T}_n,\overline{G})$. The coboundary homomorphism

$$\delta_n \;=\; \textstyle\sum_{0 \leqq i \leqq n} (-1)^i \mathrm{Hom}_{\overline{\mathcal{C}}}(\epsilon_T^{n,i},\overline{G}) : C^n(\overline{T},\overline{G}) \longrightarrow C^{n+1}(\overline{T},\overline{G})$$

sends $v : \overline{T}_n \longrightarrow \overline{G}$ to

$$\delta_n v \;=\; \textstyle\sum_{0 \leqq i \leqq n} (-1)^i (v \circ \epsilon_T^{n,i}).$$

An n-cocycle is an element of $Z^n(\overline{T},\overline{G}) = \mathrm{Ker}\,\delta_n \subseteq C^n(\overline{T},\overline{G})$; if $n \geqq 2$, an n-coboundary is an element of $B^n(\overline{T},\overline{G}) = \mathrm{Im}\,\delta_{n-1} \subseteq Z^n(\overline{T},\overline{G})$. The Beck cohomology groups of $\overline{T}$ are the groups

$$H^n(\overline{T},\overline{G}) \;=\; Z^n(\overline{T},\overline{G}) \,/\, B^n(\overline{T},\overline{G})$$

where $n \geqq 2$, and $H^1(\overline{T},\overline{G}) = Z^1(\overline{T},\overline{G})$.

By Proposition 2.1, $\overline{G}$ is a split commutative group coextension of S by the corresponding abelian group valued functor $\mathbb{G}$. By Lemma 4.2, there is a natural isomorphism Θ of $C^n(\overline{T},\overline{G}) = \mathrm{Hom}_{\overline{\mathfrak{C}}}(\overline{T}_n, \overline{G})$ onto

$$C^n(S,\mathbb{G}) = C(\overline{T}_{n-1},\mathbb{G}) = \textstyle\prod_{t\in T_{n-1}} G_{\pi_{n-1}t}\,;$$

when $v : \overline{T}_n \longrightarrow \overline{G}$, then $\Theta v = u = (u_t)_{t\in T_{n-1}} \in C^n(S,\mathbb{G})$ is given by

$$v\,[t] = (u_t, \pi_{n-1}t).$$

When $u = (u_t)_{t\in T_{n-1}} \in C^n(S,\mathbb{G})$, then $\Theta^{-1}u = v : \overline{T}_n \longrightarrow \overline{G}$ is given by

$$v\,[x_1,\dots,x_m] = \left(\textstyle\sum_{1\leqq j\leqq m} u_{x_j}^{\pi_{n-1}\widehat{x}_j}\,,\ \pi_{n-1}x\right),$$

where $\widehat{x}_j = x_1\cdots x_{j-1}\,x_{j+1}\cdots x_m$ and $x = x_1\cdots x_m$, for all $m > 0$ and $x_1,\dots,x_m \in T_{n-1}$; then $\delta'_n u = \Theta(\delta_n v) \in C^{n+1}(S,\mathbb{G})$ is given for all $t = [x_1,\dots,x_m] \in T_n$ by

$$\begin{aligned}
\left((\delta'_n u)_t,\ \pi_n t\right) &= (\delta_n v)\,[t] = \textstyle\sum_{0\leqq i\leqq n} (-1)^i \left(v\,\epsilon^{n,i}\,[t]\right)\\
&= \textstyle\sum_{0\leqq i\leqq n-1} (-1)^i \left(v\,[\epsilon^{n-1,i}\,t]\right) + (-1)^n\, vt\\
&= \textstyle\sum_{0\leqq i\leqq n-1} (-1)^i \left(u_{\epsilon^{n-1,i}\,t},\ \pi_{n-1}\,\epsilon^{n-1,i}t\right)\\
&\quad + (-1)^n \left(\textstyle\sum_{1\leqq j\leqq m} u_{x_j}^{\pi_{n-1}\widehat{x}_j}\,,\ \pi_{n-1}x\right),\\
&= \textstyle\sum_{0\leqq i\leqq n-1} (-1)^i \left(u_{\epsilon^{n-1,i}\,t},\ \pi_n t\right)\\
&\quad + (-1)^n \left(\textstyle\sum_{1\leqq j\leqq m} u_{x_j}^{\pi_{n-1}\widehat{x}_j}\,,\ \pi_{n-1}x\right),
\end{aligned}$$

since $\pi_{n-1}\circ\epsilon^{n-1,i} = \pi_n$, where $\widehat{x}_j = x_1\cdots x_{j-1}\,x_{j+1}\cdots x_m$ and $x = x_1\cdots x_m$, so that $\pi_{n-1}x = \pi_n t$. Thus

$$(\delta'_n u)_t = \textstyle\sum_{0\leqq i\leqq n-1} (-1)^i\, u_{\epsilon^{n-1,i}\,t} + (-1)^n \sum_{1\leqq j\leqq m} u_{x_j}^{\pi_{n-1}\widehat{x}_j}$$

for all $t = [x_1,\dots,x_m] \in T_n$ and we have proved:

Theorem 4.3. *Let S be a commutative semigroup. Up to natural isomorphisms, the Beck cohomology of S has coefficients in an abelian group valued functor $\mathbb{G}$ on $H(S)$, and $H^n(S,\mathbb{G})$ is the n-th homology group $\operatorname{Ker}\delta_n / \operatorname{Im}\delta_{n-1}$ of the complex*

$$0 \longrightarrow C^1(S,\mathbb{G}) \longrightarrow \cdots \xrightarrow{\delta_{n-1}} C^n(S,\mathbb{G}) \xrightarrow{\delta_n} C^{n+1}(S,\mathbb{G}) \longrightarrow \cdots$$

where $C^n(S,\mathbb{G}) = \prod_{t\in T_{n-1}} G_{\pi_{n-1}t}$ and

$$(\delta_n u)_t \;=\; \textstyle\sum_{0 \leqq i \leqq n-1} (-1)^i \, u_{\epsilon^{n-1,i}\, t} \;+\; (-1)^n \sum_{1 \leqq j \leqq m} u_{x_j}^{\pi_{n-1} \widehat{x}_j} \qquad (\delta)$$

for all $t = [x_1, \ldots, x_m] \in T_n$, *where* $\widehat{x}_j = x_1 \cdots x_{j-1}\, x_{j+1} \cdots x_m$.

If $m = 1$, then $\widehat{x}_j$ is an empty product and $\widehat{x}_j = 1 \in T^1_{n-1}$.

4. Theorem 4.3 describes $H^1(S, \mathbb{G})$ and $H^2(S, \mathbb{G})$ as follows.

A 1-cochain $u \in C^1(S, \mathbb{G})$ is a family $u = (u_a)_{a \in S}$ such that $u_a \in G_a$ for all $a \in S$ ($u \in \prod_{s \in S} G_a$). When $t = [a_1, \ldots, a_m] \in T_1$, so that $m > 0$ and $a_1, \ldots, a_m \in S$, and $a = a_1 \cdots a_m$, then $\pi_1 t = \pi_0(a_1 \cdots a_m) = a$ by (π) and $\epsilon^{0,0} t = a_1 \cdots a_m = a = \pi_1 t$ by (ϵ'). Thus (δ) reads

$$(\delta_1 u)_t \;=\; u_a \;-\; \textstyle\sum_{1 \leqq j \leqq m} u_{a_j}^{\widehat{a}_j}$$

for all $t = [a_1, \ldots, a_m] \in T_1$, where $a = a_1 \cdots a_m$ and $\widehat{a}_j = a_1 \cdots a_{j-1}\, a_{j+1} \cdots a_m$. Hence u is a 1-cocycle if and only if

$$u_{a_1 \cdots a_m} \;=\; \textstyle\sum_{1 \leqq j \leqq m} u_{a_j}^{\widehat{a}_j}$$

for all $a_1, \ldots, a_m \in S$; equivalently, if $u_{ab} = u_a^b + u_b^a$ for all $a, b \in S$.

A 2-cochain $u \in C^2(S, \mathbb{G}) = \prod_{x \in T_1} G_{\pi_1 x}$ is a family $u = (u_x)_{x \in T_1}$ such that $u_x \in G_{\pi_1 x}$ for all $x \in T_1$. When $A = [x_1, \ldots, x_m] \in T_2$, so that $m > 0$ and $x_1, \ldots, x_m \in T_1$, and $x = x_1 \cdots c_m$, then $\pi_2 t = \pi_1(x_1 \cdots x_m) = \pi_1 x$, $\epsilon^{1,1} t = x_1 \cdots x_m = x \in T_1$ by (ϵ'), and

$$\epsilon^{1,0} t \;=\; [\epsilon^{0,0} x_1, \;\ldots,\; \epsilon^{0,0} x_m] \;=\; [\pi_1 x_1, \;\ldots,\; \pi_1 x_m]$$

by (ϵ''). Thus (δ) reads

$$(\delta_2 u)_{[x_1, \ldots, x_m]} \;=\; u_{[\pi x_1, \ldots, \pi x_m]} \;-\; u_{x_1 \cdots x_m} \;+\; \textstyle\sum_{1 \leqq j \leqq m} u_{x_j}^{\pi \widehat{x}_j}$$

for all $x_1, \ldots, x_m \in T_1$, where $\pi = \pi_1$ and $\widehat{x}_j = x_1 \cdots x_{j-1}\, x_{j+1} \cdots x_m$. Hence u is a 2-cocycle if and only if

$$u_{x_1 \cdots x_m} \;=\; u_{[\pi x_1, \ldots, \pi x_m]} \;+\; \textstyle\sum_{1 \leqq j \leqq m} u_{x_j}^{\pi \widehat{x}_j}$$

for all $m > 0$ and $x_1, \ldots, x_m \in T_1$; u is a 2-coboundary if and only if $u = \delta v$ for some 1-cochain v:

$$u_{[a_1, \ldots, a_m]} \;=\; v_{a_1 \cdots a_m} \;-\; \textstyle\sum_{1 \leqq j \leqq m} v_{a_j}^{\pi \widehat{a}_j}\,.$$

In the above, m is unbounded; computing $H^2(S,\mathbb{G})$ by Theorem 4.3 is therefore an infinite task, even when S is finite. Fortunately, more effective techniques are just around the corner.

5. Commutative semigroup cohomology inherits a number of properties from triple cohomology in general. Theorems 1.1, 1.2, and 4.3 yield:

Theorem 4.4. *When S is a free commutative semigroup, then $H^n(S,\mathbb{G})=0$ for all $n \geqq 2$.*

Theorem 4.5. *Every short exact sequence $\mathcal{G} : \mathbb{G} \longrightarrow \mathbb{G}' \longrightarrow \mathbb{G}''$ of abelian group valued functors on $H(S)$ induces an exact sequence*

$$\cdots\ H^n(S,\mathbb{G}) \longrightarrow H^n(S,\mathbb{G}') \longrightarrow H^n(S,\mathbb{G}'') \longrightarrow H^{n+1}(S,\mathbb{G})\ \cdots$$

which is natural in $\mathcal{G}$.

Proposition 3.1 and Theorem 1.4 yield

Theorem 4.6. *There is a one-to-one correspondence between elements of $H^2(S,\mathbb{G})$ and equivalence classes of commutative group coextensions of S by $\mathbb{G}$, which is natural in $\mathbb{G}$.*

Theorem 4.6 provides a bijection between $H^2(S,\mathbb{G})$ and the abelian group $\mathrm{Ext}\,(S,\mathbb{G})$ from Chapter V. In the next section we shall prove that the two groups are in fact isomorphic. Hence $H^2(S,\mathbb{G})$ can be calculated from factor sets and split factor sets, which one may assume are normalized.

6. We conclude this section by constructing a projective complex of which $H^n(S,\mathbb{G})$ is the cohomology; this takes place in the abelian category $\mathcal{A}$ of abelian group valued functors on $H(S)$.

Let $\overline{T} = (T,\pi)$ be a c.s. over S. For every $a \in S$ let $K_a = K_a(\overline{T})$ be the free abelian group generated by the set

$$X_a \;=\; X_a(\overline{T}) \;=\; \{(t,z) \in T \times S^1 \mid (\pi t)\, z = a\}.$$

When $v \in S^1$, $(t,z) \in X_a$ implies $(t,zv) \in X_{av}$ and there is a unique homomorphism $\kappa_{a,v} : K_a \longrightarrow K_{av}$ such that

$$\kappa_{a,v}(t,z) \;=\; (t,zv)$$

whenever $(t,z) \in X_a$. We see that $\kappa_{a,1}$ is the identity on K_a and that $\kappa_{av,w} \circ \kappa_{a,v} = \kappa_{a,vw}$. Thus $\mathbb{K}\overline{T} = (K,\kappa)$ is an abelian group valued functor on $H(S)$.

If $f : \overline{T} \longrightarrow \overline{T}'$ is a morphism in $\overline{\mathcal{C}}$, then $(t,z) \in X_a(\overline{T})$ implies $(ft,z) \in X_a(\overline{T}')$ and f induces a homomorphism $(\mathbb{K}f)_a : K_a(\overline{T}) \longrightarrow K_a(\overline{T}')$. It is immediate that $\mathbb{K}f : \mathbb{K}\overline{T} \longrightarrow \mathbb{K}\overline{T}'$ is a natural transformation. This constructs a functor $\mathbb{K} : \overline{\mathcal{C}} \longrightarrow \mathcal{A}$.

Lemma 4.7. *For every object $\overline{T} = (T,\pi)$ of $\overline{\mathcal{C}}$ and abelian group valued functor $\mathbb{G}$ on $H(S)$ there is an isomorphism $\mathrm{Hom}_{\mathcal{A}}(\mathbb{K}\overline{T},\mathbb{G}) \cong C(\overline{T},\mathbb{G})$ which is natural in $\overline{T}$ and $\mathbb{G}$. Hence $\mathbb{K}\overline{T}$ is projective.*

Lemma 4.7 shows that the functor $C(\overline{T},-)$ is representable. Hence the isomorphism $\mathrm{Hom}_{\mathcal{A}}(\mathbb{K}\overline{T},\mathbb{G}) \cong C(\overline{T},\mathbb{G})$ determines $\mathbb{K}$ uniquely up to natural isomorphism.

Proof. Let $\varphi : K \longrightarrow \mathbb{G}$ be a natural transformation, where $K = \mathbb{K}\overline{T}$. For every $t \in T$, $(t,1) \in X_{\pi t}$ and

$$\overline{\varphi}_t \;=\; \varphi_{\pi t}(t,1) \;\in\; G_{\pi t}\,.$$

This constructs $\overline{\varphi} \in C(\overline{T},\mathbb{G}) = \prod_{t\in T} G_{\pi t}$. Since φ is natural we have

$$\varphi_a(t,z) \;=\; \varphi_a\,\kappa_{\pi t,z}(t,1) \;=\; \gamma_{\pi t,z}\,\varphi_{\pi t}(t,1) \;=\; \gamma_{\pi t,z}\,\overline{\varphi}_t$$

for all $(t,z) \in X_a$. Thus φ is uniquely determined by $\overline{\varphi}$.

Conversely let $u \in C(\overline{T},\mathbb{G})$. Define $\varphi_a : K_a \longrightarrow G_a$ by:

$$\varphi_a(t,z) \;=\; \gamma_{\pi t,z}\,u_t$$

for all $(t,z) \in X_a$. Then

$$\gamma_{a,v}\,\varphi_a(t,z) = \gamma_{a,v}\,\gamma_{\pi t,z}\,u_t = \gamma_{\pi t,vz}\,u_t = \varphi_{av}(t,vz) = \varphi_{av}\,\kappa_{a,v}(t,z)$$

for all $(t,z) \in X_a$. Hence $\varphi : K \longrightarrow \mathbb{G}$ is a natural transformation. We see that $\overline{\varphi} = u$. Thus $\mathrm{Hom}_{\mathcal{A}}(K,\mathbb{G}) \cong C(\overline{T},\mathbb{G})$. It is immediate that this isomorphism is natural in $\overline{T}$ and $\mathbb{G}$.

If $\sigma : \mathbb{G} \longrightarrow \mathbb{G}'$ is an epimorphism in $\mathcal{A}$, then every $\sigma_a : G_a \longrightarrow G'_a$ is surjective, $C(\overline{T},\sigma) = \prod_{t\in T} \sigma_{\pi t} : \prod_{t\in T} G_{\pi t} \longrightarrow \prod_{t\in T} G'_{\pi t}$ is surjective, and $\mathrm{Hom}_{\mathcal{A}}(K,\sigma)$ is an epimorphism. Thus K is projective in $\mathcal{A}$. □

Proposition 4.8. *$H^n(S,\mathbb{G})$ is the n-th cohomology group of the projective complex*

$$0 \longleftarrow \mathbb{C}_1(S) \longleftarrow \cdots \longleftarrow \mathbb{C}_n(S) \longleftarrow \mathbb{C}_{n+1}(S) \longleftarrow \;\cdots$$

where $\mathbb{C}_n(S) = \mathbb{K}\overline{T}_{n-1}$ and $\partial : \mathbb{C}_{n+1}(S) \longrightarrow \mathbb{C}_n(S)$ is given by

$$\partial_a(t,z) \;=\; \textstyle\sum_{0\leqq i\leqq n-1} (-1)^i\,(\epsilon^{n-1,i}\,t,\, z) \;+\; (-1)^n \sum_{1\leqq j\leqq m} (x_j,\, (\pi_{n-1}\widehat{x}_j)\,z)$$

for all $(t,z) \in X_a$, *where* $t = [x_1,\dots,x_m] \in T_n$ *and* $\widehat{x}_j = x_1 \cdots x_{j-1}\, x_{j+1} \cdots x_m$.

Proof. By Lemma 4.7 there are natural isomorphisms

$$\mathrm{Hom}_{\mathcal{A}}(\mathbb{C}_n(S),\mathbb{G}) = \mathrm{Hom}_{\mathcal{A}}(\mathbb{K}\overline{T}_{n+1},\mathbb{G}) \xrightarrow{\cong} C(\overline{T}_{n+1},\mathbb{G}) = C^n(S,\mathbb{G}).$$

We show that these isomorphisms take the coboundary homomorphism

$$\mathrm{Hom}_{\mathcal{A}}(\partial,\mathbb{G}) : \mathrm{Hom}_{\mathcal{A}}(\mathbb{C}_n,\mathbb{G}) \longrightarrow \mathrm{Hom}_{\mathcal{A}}(\mathbb{C}_{n+1},\mathbb{G}),$$

$\varphi \longmapsto \varphi \circ \partial$, to the coboundary homomorphism δ_n in Theorem 4.3.

Let $u \in C^n(S,\mathbb{G})$. To u corresponds the natural transformation φ from $\mathbb{K}\overline{T}_{n-1}$ to $\mathbb{G}$ defined by

$$\varphi_a(t,z) \;=\; \gamma_{\pi t,z}\, u_t \;=\; u_t^z$$

for all $(t,z) \in X_a(\overline{T}_{n-1})$. Let $v \in C^{n+1}(S,\mathbb{G})$ correspond to $\varphi \circ \partial$. For every $t = [x_1,\dots,x_m] \in T_n$,

$$\begin{aligned} v_t &= \varphi_{\pi t}\big(\partial_{\pi t}(t,1)\big) \\ &= \varphi_{\pi t}\big(\textstyle\sum_{0 \leqq i \leqq n-1} (-1)^i\, (\epsilon^{n-1,i}\, t,\, 1) \\ &\qquad + (-1)^n \textstyle\sum_{1 \leqq j \leqq m} (x_j,\, (\pi_{n-1}\widehat{x}_j)\, 1)\big) \\ &= \textstyle\sum_{0 \leqq i \leqq n-1} (-1)^i\, u_{\epsilon^{n-1,i}\, t} \;+\; (-1)^n \textstyle\sum_{1 \leqq j \leqq m} u_{x_j}^{\pi_{n-1}\widehat{x}_j} \\ &= (\delta_n u)_t. \end{aligned}$$

Thus $v = \delta_n u$. □

The complex in Proposition 4.8 is not very barlike, since the generators (t,v) of $\mathbb{C}_n(S)$ include sequences $t \in T_{n-1}$ of unbounded length when $n \geqq 2$. It is not known in general whether there is a commutative "bar" complex in which t is replaced by a sequence of length n. Results in the next section indicate how the first groups of such a complex might be constructed.

7. Simpler chains can be used when coefficient functors are constant, or nearly constant. We call an abelian group valued functor $\mathbb{G} = (G,\gamma)$ on S **constant** when there is an abelian group A such that $G_a = A$ and $\gamma_{a,t} = 1_A$ for all $a \in S$ and $t \in S^1$. Then $\mathbb{G}$ and A may be identified, and we denote $C^n(S,\mathbb{G})$ by $C^n(S,A)$, and similarly for B^n, Z^n, and H^n. Constant functors are thin and surjecting. Cohomology with constant coefficients is the commutative analogue of the Eilenberg-MacLane cohomology for monoids.

When $\overline{T} = (T,\pi)$ is a c.s. over S let $K(\overline{T})$ be the free abelian group generated by the set T. For every abelian group G we have $\mathrm{Hom}\,(K(\overline{T}),G) \cong \prod_{t\in T} G = C(\overline{T},G)$. Hence $H^n(S,G)$ is, as in Proposition 4.8, the n-th cohomology group of the complex

$$C(S):\ 0 \longleftarrow C_1(S) \longleftarrow \cdots \longleftarrow C_n(S) \longleftarrow C_{n+1}(S) \longleftarrow \cdots$$

where $C_n(S) = K(\overline{T}_{n-1})$ and $\partial : C_{n+1}(S) \longrightarrow C_n(S)$ is given by

$$\partial t \;=\; \textstyle\sum_{0\leqq i\leqq n-1} (-1)^i\, \epsilon^{n-1,i}\, t \;+\; (-1)^n \sum_{1\leqq j\leqq m} x_j$$

for all $t = [x_1,\dots,x_m] \in T_n$. Since $C(S)$ is a complex of free abelian groups, there is for every abelian group G a Universal Coefficient Theorem

$$H^n(S,G) \;\cong\; \mathrm{Ext}\,(H_{n-1}(S),G) \oplus \mathrm{Hom}\,(H_n(S),G)$$

where $H_n(S)$ is the n-th homology group of $C(S)$, with $H_1(S) = C^1(S)/\mathrm{Im}\,\partial^1$ and $H_0(S) = 0$ (MacLane [1963]). We leave the details to our tireless reader.

When S has a zero element, then $H^n(S,G) = 0$ for every abelian group G, at least when $n \leqq 3$. This can be remedied by using functors that are not quite constant. When S has a zero element, an abelian group valued functor $\mathbb{G}$ on S is **almost constant** when there exists an abelian group A such that $G_a = A$ for all $a \neq 0$, $G_0 = 0$, and $\gamma_{a,t} = 1_A$ whenever $at \neq 0$. Then $\mathbb{G}$ may be identified with the abelian group A, and we denote the cohomology groups $H^n(S,\mathbb{G})$ by $H^n_0(S,A)$, and similarly for cochains, cocycles, and coboundaries. Almost constant functors are thin and surjecting.

As noted in Grillet [1974], almost constant functors arise naturally in the construction of homogeneous elementary semigroups. The Universal Coefficient Theorem can be saved if in the above we replace $K(\overline{T})$ by the almost constant functor $\mathbb{K}^0(\overline{T})$ in which the abelian group is the free abelian group generated by $\{t \in T \mid \pi t \neq 0\}$. Again we leave the details to our reader.

5. SYMMETRIC COCHAINS.

Commutative cohomology cries out for an equivalent description in which n-cochains are functions of n variables. This has been found only for $n \leqq 4$.

1-cochains already are functions $u = (u_a)_{a\in S}$ of one variable $a \in S$.

1. In dimension 2, Theorem 4.6 provides a one-to-one correspondence be-

tween the elements of $H^2(S, \mathbb{G})$ and the elements of the abelian group $\mathrm{Ext}\,(S, \mathbb{G})$ of all equivalence classes of commutative group coextensions of S by $\mathbb{G}$, which is also the abelian group of all equivalence classes of commutative factor sets on S with values in $\mathbb{G}$.

We now construct a more direct connection between factor sets and 2-cocycles, which induces an isomorphism $H^2(S, \mathbb{G}) \cong \mathrm{Ext}\,(S, \mathbb{G})$.

Let T_1 be the free commutative semigroup on the set S and $\pi = \pi_1 : T_1 \longrightarrow S$, so that $\pi\,[a_1, a_2, \ldots, a_m] = a_1\, a_2 \cdots a_m$ for all $a_1, \ldots, a_m \in S$. As we saw in Section 4, a 2-cochain $u \in C^2(S, \mathbb{G}) = \prod_{x \in T_1} G_{\pi x}$ is a family $u = (u_x)_{x \in T_1}$ such that $u_x \in G_{\pi x}$ for all $x \in T_1$; a 2-cocycle is a 2-cochain u such that

$$u_{x_1 \cdots x_m} \;=\; u_{[\pi x_1, \ldots, \pi x_m]} \;+\; \textstyle\sum_{1 \leqq j \leqq m} u_{x_j}^{\pi \widehat{x}_j} \tag{Z}$$

for all $m > 0$ and $x_1, \ldots, x_m \in T_1$, where $\widehat{x}_j = x_1 \cdots x_{j-1}\, x_{j+1} \cdots x_m$; a 2-coboundary is a 2-cochain u (necessarily a 2-cocycle) of the form $u = \delta v$,

$$(\delta v)_{[a_1, \ldots, a_l]} \;=\; v_{a_1 \cdots a_l} \;-\; \textstyle\sum_{1 \leqq i \leqq l} v_{a_i}^{\pi \widehat{a}_i} \tag{B}$$

for some 1-cochain v, where $\widehat{a}_i = a_1 \cdots a_{i-1}\, a_{i+1} \cdots a_l$.

A **symmetric** 2-**cochain** on S with values in $\mathbb{G}$ is a family $s = (s_{a,b})_{a,b \in S}$ such that $s_{a,b} \in G_{ab}$ and

$$s_{b,a} \;=\; s_{a,b}$$

for all $a, b \in S$. Under pointwise addition, symmetric 2-cochains constitute an abelian group $SC^2(S, \mathbb{G}) = \prod_{a,b \in S} G_{ab}$. A **symmetric** 2-**cocycle** on S with values in $\mathbb{G}$ is a symmetric 2-cochain s such that

$$s_{a,b}^{c} \;+\; s_{ab,c} \;=\; s_{a,bc} \;+\; s_{b,c}^{a} \tag{A}$$

for all $a, b, c \in S$; equivalently, a commutative factor set on S with values in $\mathbb{G}$ as defined in Section V.4. A **symmetric** 2-**coboundary** on S with values in $\mathbb{G}$ is a symmetric 2-cochain (necessarily, a symmetric 2-cocycle) s of the form

$$s_{a,b} \;=\; v_a^b \;+\; v_b^a \;-\; v_{ab}$$

for some 1-cochain v; equivalently, a split factor set. Under pointwise addition, symmetric 2-coboundaries and 2-cocycles form abelian groups $SB^2(S, \mathbb{G}) \subseteq SZ^2(S, \mathbb{G}) \subseteq SC^2(S, \mathbb{G})$.

We saw in Section V.4 that

$$\mathrm{Ext}\,(S, \mathbb{G}) \;\cong\; SZ^2(S, \mathbb{G}) \,/\, SB^2(S, \mathbb{G}).$$

2. Restriction to $S \times S$ yields a **trimming homomorphism** Σ of $C^2(S,\mathbb{G})$ into $SC^2(S,\mathbb{G})$; when u is a 2-cochain, Σu is given by

$$(\Sigma u)_{a,b} = u_{[a,b]}$$

for all $a,b \in S$; Σu is symmetric since $[b,a] = [a,b]$ in T_1.

Lemma 5.1. *When u is a 2-cocycle, then $u_{[a]} = 0$ for all $a \in S$ and*

$$u_{[a_1,\dots,a_l]} = \sum_{1 \leqq i \leqq l-1} u^{a''_{i+1}}_{[a'_i,\, a_{i+1}]} \tag{Z'}$$

for all $l > 0$ and $a_1,\dots,a_l \in S$, where $a'_i = a_1 \cdots a_i$ and $a''_{i+1} = a_{i+2} \cdots a_l$.

Proof. When $m = 1$ and $x_1 = [a]$, then $\pi x_1 = [a]$, (Z) reads: $u_{x_1} = u_{x_1} + u^1_{x_1}$, and $u_{x_1} = 0$.

Now let $i \geqq 1$ and $a_1,\dots,a_{i+1} \in S$. With $m = 2$, $x_1 = [a_1,\dots,a_i]$, and $x_2 = [a_{i+1}]$, then $\pi x_1 = a'_i$, $\pi x_2 = a_{i+1}$, $u_{x_2} = 0$, and (Z) reads

$$u_{[a_1,\dots,a_{i+1}]} = u_{[a'_i,\, a_{i+1}]} + u^{a_{i+1}}_{[a_1,\dots,a_i]}\,. \tag{Z''}$$

(Z′) is proved by induction on l, using (Z″). First, (Z′) is trivial when $l = 1$ (then the right hand side is the empty sum 0) and when $l = 2$ (then a''_2 is the empty product and $a''_2 = 1 \in S^1$). If $l = 3$, then (Z″) yields

$$u_{[a_1,a_2,a_3]} = u^{a_3}_{[a_1,a_2]} + u_{[a'_2,a_3]}$$

and (Z′) holds. If $l > 3$, then, with $b''_{i+1} = a_{i+2} \cdots a_{l-1}$, (Z″) and the induction hypothesis yield

$$\begin{aligned} u_{[a_1,\dots,a_l]} &= u^{a_l}_{[a_1,\dots,a_{l-1}]} + u_{[a'_{l-1},\, a_l]} \\ &= \Big(\sum_{1 \leqq i \leqq l-2} u^{b''_{i+1}}_{[a'_i,\, a_{i+1}]}\Big)^{a_l} + u_{[a'_{l-1},\, a_l]} \\ &= \sum_{1 \leqq i \leqq l-1} u^{a''_{i+1}}_{[a'_i,\, a_{i+1}]}. \ \square \end{aligned}$$

Lemma 5.1 shows that $\Sigma : Z^2(S,\mathbb{G}) \longrightarrow SC^2(S,\mathbb{G})$ is injective.

Lemma 5.2. $\operatorname{Im} \Sigma = SZ^2(S,\mathbb{G})$.

Proof. Let u be a 2-cocycle. With $m = 2$, $x_1 = [a]$, and $x_2 = [b,c]$, (Z) reads

$$u_{[a,b,c]} = u_{[a,\, bc]} + u^a_{[b,c]}$$

(since $u_{[a]} = 0$). With $m = 2$, $x_1 = [a,b]$, and $x_2 = [c]$, (Z) reads

$$u_{[a,b,c]} = u_{[ab,c]} + u^c_{[a,b]} \, .$$

Therefore $\Sigma u \in SZ^2(S,G)$.

Conversely let $s \in SZ^2(S,\mathbb{G})$. Since Σ is injective, there is at most one $u \in Z^2(S,\mathbb{G})$ such that $\Sigma u = s$, and it is given by (Z′). Accordingly, define $t = (t_{a_1,\dots,a_l})_{l>0,\, a_1,\dots,a_l \in S}$ by

$$t_{a_1,\dots,a_l} = \textstyle\sum_{1 \leqq i \leqq l-1} s^{a''_{i+1}}_{a'_i,\, a_{i+1}}$$

for all $l > 0$ and $a_1,\dots,a_l \in S$, where $a'_i = a_1 \cdots a_i$ and $a''_{i+1} = a_{i+2} \cdots a_l$. In particular, $t_a = 0$ (if $l = 1$, then the right hand side is the empty sum 0) and $t_{a,b} = s_{a,b}$, for all $a,b \in S$. Also

$$\begin{aligned} t_{a_1,\dots,a_l,b} &= \textstyle\sum_{1 \leqq i \leqq l-1} s^{a''_{i+1}b}_{a'_i,\, a_{i+1}} + s_{a'_l,b} \\ &= \left(\textstyle\sum_{1 \leqq i \leqq l-1} s^{a''_{i+1}}_{a'_i,\, a_{i+1}}\right)^b + s_{a'_l,b} \\ &= t^b_{a_1,\dots,a_l} + t_{a'_l,b} \end{aligned}$$

for all $l > 0$ and $a_1,\dots,a_l, b \in S$; thus t satisfies (Z″).

We show by induction on $l > 0$ that

$$t_{a_{\sigma 1},\dots,a_{\sigma l}} = t_{a_1,\dots,a_l} \tag{P}$$

for every $a_1,\dots,a_l \in S$ and permutation σ. This is trivial if $l = 1$ and follows from $t_{a,b} = s_{a,b}$ if $l = 2$. When $l > 2$, σ is a product of transpositions of the form $(j \;\; j+1)$ and it suffices to prove (P) when $\sigma = (j \;\; j+1)$. If $j < l-1$, then (P) follows from the induction hypothesis and (Z″). If $j = l-1$, then, with $a'_{l-2} = b$, $a_{l-1} = c$, $a_l = d$, we have $a'_{l-1} = bc$, $a''_{l-1} = d$, and

$$\begin{aligned} t_{a_1,\dots,a_l} &= \textstyle\sum_{1 \leqq i \leqq l-3} s^{a''_{i+1}}_{a'_i,\, a_{i+1}} + s^d_{b,c} + s_{bc,d} \, , \\ t_{a_{\sigma 1},\dots,a_{\sigma l}} &= \textstyle\sum_{1 \leqq i \leqq l-3} s^{a''_{i+1}}_{a'_i,\, a_{i+1}} + s^c_{b,d} + s_{bd,c} \, , \end{aligned}$$

and (P) holds, since $s \in SZ^2(S,\mathbb{G})$ yields

$$s^d_{b,c} + s_{bc,d} = s^d_{c,b} + s_{cb,d} = s_{c,bd} + s^c_{b,d} = s^c_{b,d} + s_{bd,c} \, .$$

By (P), a 2-cochain $u \in C^2(S,\mathbb{G})$ is well defined by

$$u_{[a_1,\dots,a_l]} \;=\; t_{a_1,\dots,a_l}$$

for all $l > 0$ and $a_1,\dots,a_l \in S$. We show that $u \in Z^2(S,\mathbb{G})$; that is,

$$u_{x_1\cdots x_m} \;=\; u_{[\pi x_1,\dots,\pi x_m]} \;+\; \textstyle\sum_{1\leqq j\leqq m} u_{x_j}^{\pi\widehat{x}_j} \tag{Z}$$

holds for all $m > 0$ and $x_1,\dots,x_m \in T_1$, with $\widehat{x}_j = x_1\cdots x_{j-1}\,x_{j+1}\cdots x_m$.

(Z) is trivial if $m = 1$; for $m > 1$ we proceed by induction on m. Assume that (Z) holds for m and let $y_1,\dots,y_m, z \in T_1$. Let $x = y_1\cdots y_m = [a_1,\dots,a_l]$, $z = [c_1,\dots,c_n]$, $\pi y_j = b_j$, $\pi x = a = a_1\cdots a_n = b_1\cdots b_m$, and $\pi z = c = c_1\cdots c_n$. By the induction hypothesis,

$$u_x \;=\; u_{[b_1,\dots,b_l]} \;+\; \textstyle\sum_{1\leqq j\leqq m} u_{y_j}^{\widehat{b}_j}$$

where $\widehat{b}_j = b_1\cdots b_{j-1}\,b_{j+1}\cdots b_m$. Hence

$$\begin{aligned}
u_{xz} \;&=\; u_{[a_1,\dots,a_l,\,c_1,\dots,c_n]}\\
&=\; \textstyle\sum_{1\leqq i\leqq l-1} s_{a'_i,\,a_{i+1}}^{a''_{i+1}c} \;+\; \sum_{0\leqq k\leqq n-1} s_{ac'_k,\,c_{k+1}}^{c''_{k+1}}\\
&=\; u_x^c \;+\; s_{a,c_1}^{c''_1} \;+\; \textstyle\sum_{1\leqq k\leqq n-1} s_{ac'_k,\,c_{k+1}}^{c''_{k+1}}\\
&=\; u_{[b_1,\dots,b_l]}^c \;+\; \textstyle\sum_{1\leqq j\leqq m} u_{y_j}^{\widehat{b}_j c} \;+\; s_{a,c_1}^{c''_1}\\
&\qquad +\; \textstyle\sum_{1\leqq k\leqq n-1} \big(-\, s_{a,c'_k}^{c_{k+1}} \;+\; s_{a,\,c'_k c_{k+1}} \;+\; s_{c'_k\,,c_{k+1}}^{a}\big)^{c''_{k+1}}
\end{aligned}$$

by the induction hypothesis and (A). Hence

$$\begin{aligned}
u_{xz} \;&=\; u_{[b_1,\dots,b_l]}^c \;+\; \textstyle\sum_{1\leqq j\leqq m} u_{y_j}^{\widehat{b}_j c} \;+\; s_{a,c_1}^{c''_1} \;-\; \sum_{1\leqq k\leqq n-1} s_{a,c'_k}^{c''_k}\\
&\qquad +\; \textstyle\sum_{2\leqq k\leqq n} s_{a,c'_k}^{c''_k} \;+\; \big(\sum_{1\leqq k\leqq n-1} s_{c'_k\,,c_{k+1}}^{c''_{k+1}}\big)^a\\
&=\; u_{[b_1,\dots,b_l]}^c \;+\; \textstyle\sum_{1\leqq j\leqq m} u_{y_j}^{\widehat{b}_j c} \;+\; s_{a,c} \;+\; u_z^a\\
&=\; u_{[b_1,\dots,b_l,c]} \;+\; \textstyle\sum_{1\leqq j\leqq m} u_{y_j}^{\widehat{b}_j c} \;+\; u_z^a
\end{aligned}$$

by (Z"), and (Z) holds for $m+1$. Thus $u \in Z^2(S,\mathbb{G})$. Then $\Sigma u = s$. □

Lemma 5.3. $\Sigma\left(B^2(S,\mathbb{G})\right) = SB^2(S,\mathbb{G})$.

Proof. When $v \in C^1(S,\mathbb{G})$, then (B) yields $(\delta v)_{[a,b]} = v_{ab} - v_a^b - v_b^a$. □

3. Lemmas 5.1, 5.2, and 5.3 provide an isomorphism $Z^2(S,\mathbb{G}) \cong SZ^2(S,\mathbb{G})$ which is natural in G, sends $B^2(S,\mathbb{G})$ to $SB^2(S,\mathbb{G})$, and induces an isomorphism $Z^2(S,\mathbb{G}) \,/\, B^2(S,\mathbb{G}) \cong SZ^2(S,\mathbb{G}) \,/\, SB^2(S,\mathbb{G})$. This proves:

Theorem 5.4. *For every commutative semigroup S there is an isomorphism $H^2(S,\mathbb{G}) \cong \operatorname{Ext}(S,\mathbb{G})$ which is natural in $\mathbb{G}$.*

The isomorphism $H^2(S,\mathbb{G}) \cong SZ^2(S,\mathbb{G}) \,/\, SB^2(S,\mathbb{G})$ can be refined, using Proposition V.4.5, into an isomorphism

$$H^2(S,\mathbb{G}) \;\cong\; NSZ^2(S,\mathbb{G}) \,/\, NSB^2(S,\mathbb{G}),$$

where $NSZ^2(S,\mathbb{G})$ and $NSB^2(S,\mathbb{G})$ are the groups of symmetric 2-cocycles and 2-coboundaries s that are normalized ($s_{e,a} = 0$ whenever $e^2 = e \geqq_{\mathcal{H}} a$). Normalization can be confined to a single idempotent; when applied to the identity element, it yields:

Corollary 5.5. *When the commutative semigroup S has no identity element, then $H^2(S,\mathbb{G}) \cong H^2(S^1,\mathbb{G}')$, where $\mathbb{G}'$ extends $\mathbb{G}$ to S^1 so that $G_1' = 0$.*

The study of $H^2(S,\mathbb{G})$ may therefore be limited to monoids. Theorems 5.6 and 5.7 below have similar corollaries for dimensions 3 and 4. It is probable that Corollary 5.5 extends to all dimensions.

When S is finite, the computation of $H^2(S,\mathbb{G})$ using Theorem 5.4 is a finite task, since a symmetric 2-cocycle consists of finitely many group elements subject to finitely many conditions, and a symmetric 2-coboundary is determined by finitely many group elements. Thus Theorem 5.4 is a marked improvement upon Theorem 4.3. But further improvement would not hurt.

For example, let S be the Volkov nilsemigroup

$$S \;=\; \langle\, a,b \;\big|\; a^3 = a^2b = ab^2 = b^4 = 0,\; a^2 = ab = b^3 \,\rangle.$$

(Example III.3.6); the elements of S are a, b, $c = b^2$, $d = a^2 = ab = b^3$, and 0. A normalized symmetric 2-cochain s on S consists of $s_{a,a}$, $s_{a,b}$, $s_{a,c}$, $s_{a,d}$, $s_{a,0}$, $s_{b,b}$, $s_{b,c}$, $s_{b,d}$, $s_{b,0}$, $s_{c,c}$, $s_{c,d}$, $s_{c,0}$, $s_{d,d}$, $s_{d,0}$, and $s_{0,0} = 0$ (since s is normalized).

Then s is a symmetric 2-cocycle if and only if

$$s_{x,y}^{z} \;+\; s_{xy,z} \;=\; s_{x,yz} \;+\; s_{y,z}^{x} \qquad A(x,y,z)$$

holds in G_{xyz} for all $x,y,z \in S$. We note that $A(x,y,z)$ is trivial if $x = z$ and remains unchanged if x and z are interchanged; moreover, $A(x,y,z)$ follows from

$$s^z_{y,x} + s_{xy,z} = s_{y,xz} + s^y_{x,z} \qquad A(y,x,z)$$

and

$$s^x_{y,z} + s_{yz,x} = s_{y,zx} + s^y_{z,x} \qquad A(y,z,x).$$

Hence it suffices to state $A(x,y,z)$ when $x \leqq y$ and $x < z$ in the lexicographic order $a < b < c < d < 0$ on S. If $x = a$, there are 5 choices for y and 4 choices for z; if $x = b$, there are 4 choices for y and 3 choices for z; if $x = c$, there are 3 choices for y and 2 choices for z; if $x = d$, there are 2 choices for y and 1 choice for z; this yields $20 + 12 + 6 + 2 = 40$ conditions.

We leave the actual conditions and their solutions to northern readers, who are blessed with long winter evenings. $H^2(S, \mathbb{G})$ will be computed in the next chapter when $\mathbb{G}$ is thin, using more efficient methods.

4. A **symmetric** 3-**cochain** on S with values in $\mathbb{G}$ is a family $t = (t_{a,b,c})_{a,b,c \in S}$ such that $t_{a,b,c} \in G_{abc}$ and

$$t_{c,b,a} = -t_{a,b,c} \quad \text{and} \quad t_{a,b,c} + t_{b,c,a} + t_{c,a,b} = 0$$

for all $a, b, c \in S$. These conditions are satisfied by the coboundary

$$(\delta s)_{a,b,c} = s^c_{a,b} - s_{a,bc} + s_{ab,c} - s^a_{b,c}$$

of every symmetric 2-cochain $s \in SC^2(S, \mathbb{G})$. Under pointwise addition, symmetric 3-cochains constitute an abelian group $SC^3(S, \mathbb{G}) \subseteq \prod_{a,b,c \in S} G_{abc}$.

A **symmetric** 3-**cocycle** on S with values in $\mathbb{G}$ is a symmetric 3-cochain t such that

$$t^d_{a,b,c} - t_{a,b,cd} + t_{a,bc,d} - t_{ab,c,d} + t^a_{b,c,d} = 0$$

for all $a, b, c, d \in S$. This condition is satisfied by the coboundary $t = \delta s$ of every symmetric 2-cochain s; such a coboundary is a **symmetric** 3-**coboundary** on S with values in G. Under pointwise addition, symmetric 3-coboundaries and 3-cocycles form abelian groups $SB^3(S, \mathbb{G}) \subseteq SZ^3(S, \mathbb{G}) \subseteq SC^3(S, \mathbb{G})$. The following result is due to Grillet [1991C], [1997C]:

Theorem 5.6. *For every commutative semigroup S there is an isomorphism $H^3(S, \mathbb{G}) \cong SZ^3(S, \mathbb{G}) \,/\, SB^3(S, \mathbb{G})$ which is natural in $\mathbb{G}$.*

A **symmetric** 4-**cochain** on S with values in $\mathbb{G}$ is a family

$u = (t_{a,b,c,d})_{a,b,c,d \in S}$ such that $u_{a,b,c,d} \in G_{abcd}$ and

$$u_{a,b,b,a} = 0, \quad u_{d,c,b,a} = -u_{a,b,c,d} ,$$
$$u_{a,b,c,d} - u_{b,c,d,a} + u_{c,d,a,b} - u_{d,a,b,c} = 0 ,$$
$$u_{a,b,c,d} - u_{b,a,c,d} + u_{b,c,a,d} - u_{b,c,d,a} = 0$$

for all $a, b, c, d \in S$. These conditions are satisfied by the coboundary

$$(\delta t)_{a,b,c,d} = t^d_{a,b,c} - t_{a,b,cd} + t_{a,bc,d} - t_{ab,c,d} + t^a_{b,c,d}$$

of every symmetric 3-cochain $s \in SC^3(S, \mathbb{G})$ (this is shown in the next section). Under pointwise addition, symmetric 4-cochains constitute an abelian group $SC^4(S, \mathbb{G}) \subseteq \prod_{a,b,c,d \in S} G_{abcd}$.

A **symmetric** 4-**cocycle** on S with values in $\mathbb{G}$ is a symmetric 4-cochain u such that

$$u^e_{a,b,c,d} - u_{a,b,c,de} + u_{a,b,cd,e} - u_{a,bc,d,e} + u_{ab,c,d,e} - u^a_{b,c,d,e} = 0$$

for all $a, b, c, d, e \in S$. A **symmetric** 4-**coboundary** on S with values in $\mathbb{G}$ is the coboundary $u = \delta t$ of a symmetric 3-cochain t. Under pointwise addition, symmetric 4-coboundaries and 4-cocycles form abelian groups $SB^4(S, \mathbb{G}) \subseteq SZ^4(S, \mathbb{G}) \subseteq SC^4(S, \mathbb{G})$. The following result is due to Grillet [2001H]:

Theorem 5.7. *For every commutative semigroup S there is an isomorphism $H^4(S, \mathbb{G}) \cong SZ^4(S, \mathbb{G}) \,/\, SB^4(S, \mathbb{G})$ which is natural in $\mathbb{G}$.*

The author's proofs of Theorem 5.6 and 5.7 are computational like the proof of Theorem 5.4, but much longer, and very likely to tax the reader's patience (even during long winter evenings). Better proofs would rely on Theorem 1.3, or on the underlying spectral sequence argument, to show that symmetric cochains define the same cohomology. This requires a general definition of symmetric cochains and a proof that $SZ^n(S, \mathbb{G}) = SB^n(S, \mathbb{G})$ when S is free. The author has proofs of this last fact when $n \leqq 3$, which are, unfortunately, entirely comparable in length and spirit to the proofs of Theorems 5.4 and 5.6.

Theorems 5.6 and 5.7 strongly suggest that symmetric cochains, cocycles, and coboundaries can be defined in every dimension n so that $H^n(S, \mathbb{G}) \cong SZ^n(S, \mathbb{G}) \,/\, SB^n(S, \mathbb{G})$. How to do this is still an open problem. In the above, the coboundary homomorphisms for symmetric cochains are essentially the same as in Leech cohomology and can be defined in all dimensions. The symmetry conditions in dimension $n + 1$ can be defined by induction as all the symmetry conditions inherited by coboundaries of symmetric n-cochains. The difficulty lies in proving that $H^n(S,\mathbb{G}) \cong SZ^n(S,\mathbb{G}) \,/\, SB^n(S,\mathbb{G})$ (or that $SZ^n(S,\mathbb{G}) = SB^n(S,\mathbb{G})$ when S is free). There the matter rests, for now.

Chapter XIII.

THE OVERPATH METHOD.

We saw in Chapter XII that the cohomology group $H^2(S,\mathbb{G})$ can be computed by means of functions of two variables in S, namely, symmetric 2-cocycles. The overpath method computes $H^2(S,\mathbb{G})$ by means of functions of one variable, one for every defining relation of S in any suitable presentation of S. This makes the computation of $H^2(S,\mathbb{G})$ a finite task whenever S is finitely generated.

Applications compute $H^2(S,\mathbb{G})$ when S is cyclic or, more generally, has only one defining relation, and when S is partially free. This last application depends rather heavily on the construction of group-free congruences in Chapter X. We also show that strand bases give rise to minimal cocycles. The main results are from Grillet [1992], [1995F], [1996C], [1995P], [2000Z].

1. OVERPATHS.

The overpath method depends on certain properties of free commutative semigroups and congruences on these semigroups.

1. In what follows $F = F_X$ is the free commutative monoid on a set X. We return to the additive notation for F and write the elements of F as finite linear combinations $a = \sum_{x \in X} a_x x$ of elements of X, with the usual order:

$$a \leqq b \text{ if and only if } a_x \leqq b_x \text{ for all } x \in X;$$

the length of $a = \sum_{x \in X} a_x x$ is $|a| = \sum_{x \in X} a_x$.

Proposition 1.1. *On every free commutative monoid F there exists an order relation $\sqsubseteq$ such that:*

(1) *$(F, \sqsubseteq)$ is well-ordered;*

(2) *if $a \leqq b$ in F, then $a \sqsubseteq b$;*

(3) *if $a \sqsubseteq b$ in F, then $a + c \sqsubseteq b + c$ for all $c \in F$.*

Proof. Well-order X, then define $\sqsubseteq$ as follows: $a \sqsubset b$ if and only if either $|a| < |b|$, or $|a| = |b|$, $a \neq b$ and the least $x \in X$ such that $a_x \neq b_x$ satisfies $a_x > b_x$. (This is the **degree lexicographic order** on F.)

An element of F of length l is the sum $x_1 + \ldots + x_l$ of l elements $x_1 \sqsubseteq x_2 \sqsubseteq \ldots \sqsubseteq x_l$ of X. Each set $F_l = \{a \in F \mid |a| = l\}$ is a subset of the lexicographic product $X^l = X \times \ldots \times X$ and is well-ordered by $\sqsubseteq$. Then F is the ordinal sum of $F_0, F_1, \ldots, F_l, \ldots$ and is well-ordered by $\sqsubseteq$.

If $a \leqq b$ in F, then either $a = b$ or $|a| < |b|$; in either case $a \sqsubseteq b$.

Finally let $a \sqsubset b$ and $c \in F$; then either $|a| < |b|$, or $|a| = |b|$ and the least x such that $a_x \neq b_x$ satisfies $a_x > b_x$. In the first case, $|a+c| < |b+c|$. In the second case, the least x such that $a_x \neq b_x$ is also the least x such that $a_x + c_x \neq b_x + c_x$, and satisfies $a_x + c_x > b_x + c_x$. In either case $a + c \sqsubseteq b + c$. □

We call an order relation $\sqsubseteq$ on F a **compatible well order** when it has properties (1), (2), and (3) in Proposition 1.1. Explicit compatible well orders can be constructed in various ways, besides the degree lexicographic order, particularly if $X = \{x_1, x_2, \ldots, x_n\}$ is finite (see e.g. Adams & Loustaunau [1994]); if for instance $p_1, \ldots, p_n$ are the first n prime numbers, then

$$a_1 x_1 + \cdots + a_n x_n \sqsubseteq b_1 x_1 + \cdots + b_n x_n \iff {p_1}^{a_1} \cdots {p_n}^{a_n} \leqq {p_1}^{b_1} \cdots {p_n}^{b_n}$$

is a compatible well order on F, the **prime order** on F of Rosales [1995].

A **lexicographic order** $\sqsubseteq$ on F is defined from a well order $\preccurlyeq$ on X by: $a \sqsubset b$ if and only if $a \neq b$ and the least $x \in X$ such that $a_x \neq b_x$ has $a_x < b_x$. Then $a < b$ implies $a \sqsubset b$. Also, $x \prec y$ in X implies $x \sqsupset y$ in F. (The usual definition requires $a_x > b_x$, but then $a < b$ does not imply $a \sqsupset b$.)

Proposition 1.2. *If F is finitely generated, then every lexicographic order on F is a compatible well order.*

Proof. If $a \sqsubset b$ and $c \in F$, then, in the above, the least x such that $a_x \neq b_x$ is also the least x such that $a_x + c_x \neq b_x + c_x$, so that $a_x + c_x > b_x + c_x$ and $a + c \sqsubset b + c$. If X is finite, then $(F, \sqsubseteq)$ is a finite lexicographic product of copies of $\mathbb{N}^+$ and is well ordered. □

When F is finitely generated, every total order $\sqsubseteq$ on F with property (2) is a well order: if indeed there is an infinite sequence $a_1 \sqsupset a_2 \sqsupset \cdots \sqsupset a_n \sqsupset a_{n+1} \sqsupset \cdots$, then $A_n = \{t \in F \mid t \sqsupseteq a_n\}$ is an ideal of F for every n, by (2), and $A_1 \subsetneqq A_2 \subsetneqq \cdots \subsetneqq A_n \subsetneqq A_{n+1} \subsetneqq \cdots$, a flagrant contradiction of Corollary VI.1.3; therefore the totally ordered set $(F, \sqsubseteq)$ satisfies the descending chain condition and is well ordered. Compatible well orders are then also known as a **linear admissible orders**.

If on the other hand X is infinite, then the first generators $x_1 \prec x_2 \prec \cdots \prec x_n \prec \cdots$ of X yield a nonempty subset $x_1 \sqsupset x_2 \sqsupset \cdots \sqsupset x_n \sqsupset \cdots$ of F with no least element and F is not well ordered by its lexicographic orders, even though they satisfy (2) and (3).

2. In what follows, $\sqsubseteq$ is any compatible well order on F.

Let $\mathcal{C}$ be a congruence on F. Under $\sqsubseteq$ the $\mathcal{C}$-class C_a of $a \in F$ has a least element $q(a)$ (the **function minimum** of Rosales [1995]). By definition,

$$a \mathrel{\mathcal{C}} q(a);\ c \mathrel{\mathcal{C}} a \text{ implies } c \sqsupseteq q(a); \text{ and } a \mathrel{\mathcal{C}} b \iff q(a) = q(b).$$

Then F is the disjoint union $F = P \cup Q$, where $Q = Q(\mathcal{C}) = \{q(a) \mid a \in F\} = \{q \in F \mid a \mathrel{\mathcal{C}} q \implies a \sqsupseteq q\}$ is the set of all least elements of all $\mathcal{C}$-classes, and

$$P = P(\mathcal{C}) = F \backslash Q = \{a \in F \mid a \sqsupset q(a)\}.$$

If $\mathcal{C}$ is the equality, then $P = \emptyset$.

Lemma 1.3. *P is an ideal of F.*

Proof. When $a \in P$ and $c \in F$, then $a \sqsupset q(a)$, $a + c \mathrel{\mathcal{C}} q(a) + c$, $a + c \sqsupset q(a) + c \sqsupseteq q(a + c)$, and $a + c \in P$. □

In what follows, $M = M(\mathcal{C})$ is the set of all minimal elements of P, under the usual order $\leqq$; since $(F, \leqq)$ satisfies the descending chain condition, P is generated as an ideal of F by M.

Proposition 1.4. *The congruence $\mathcal{C}$ is generated by all pairs $(m, q(m))$ with $m \in M(\mathcal{C})$.*

Proof. Let $\mathcal{M}$ be the congruence on F generated by all pairs $(m, q(m))$ with $m \in M$. Then $\mathcal{M} \subseteq \mathcal{C}$, since $m \mathrel{\mathcal{C}} q(m)$ for all $m \in M$. We show by artinian induction that $a \mathrel{\mathcal{M}} q(a)$ for every $a \in F$ (this also follows from Proposition 1.5 below); then $a \mathrel{\mathcal{C}} b$ implies $a \mathrel{\mathcal{M}} q(a) = q(b) \mathrel{\mathcal{M}} b$, so that $\mathcal{C} = \mathcal{M}$.

We have $a \mathrel{\mathcal{M}} q(a)$ for all $a \in Q$ (then $a = q(a)$) and for all $a \in M$ (by definition of $\mathcal{M}$). Let $a \in P$. Then $a \geqq m$ for some $m \in M$ and $a = m + t$ for some $t \in F$. Let $b = q(m) + t$. Then $a \mathrel{\mathcal{M}} b$, since $m \mathrel{\mathcal{M}} q(m)$, $a \mathrel{\mathcal{C}} b$ since $m \mathrel{\mathcal{C}} q(m)$, and $a \sqsupset b$, since $m \sqsupset q(m)$. Then $b \mathrel{\mathcal{M}} q(b)$ by the induction hypothesis, and $a \mathrel{\mathcal{M}} b \mathrel{\mathcal{M}} q(b) = q(a)$. □

Proposition 1.4 implies Rédei's Theorem. If indeed F is finitely generated, then M is finite by Dickson's Theorem (Corollary VI.1.3), and Proposition 1.4 shows that $\mathcal{C}$ is finitely generated (Grillet [1993R]).

Conversely Rosales [1995] devised an algorithm which constructs q from any finite set of generators of $\mathcal{C}$; this provides an explicit algorithm for the solution of the word problem in any finite commutative presentation.

3. Given the congruence $\mathcal{C}$, we now regard the free c.m. F as a directed graph with labeled edges, in which the vertices are the elements of F and an edge $a \xrightarrow{m} b$ from a to b, labeled by m, is an ordered pair (a, m) such that $m \in M(\mathcal{C})$, $m \leqq a$, and $a - b = m - q(m)$. Then $a = m + t$ and $b = q(m) + t$, where $t = a - m = b - q(m) \in F$; hence $q(m) \leqq b$, $a \,\mathcal{C}\, b$ (since $m \,\mathcal{C}\, q(m)$), and $a \sqsupset b$ (since $m \sqsupset q(m)$).

A descending **path** from $a \in F$ to $b \in F$ is a sequence $a = p^0, \ldots, p^k = b$ of elements of F and edges

$$a = p^0 \xrightarrow{m^1} p^1 \xrightarrow{m^2} \cdots \xrightarrow{m^k} p^k = b,$$

where $k \geqq 0$. (We index sequences of elements of F by superscripts, to keep subscripts for coordinates.) Equivalently, a path from a to b consists of a sequence $a = p^0, \ldots, p^k = b$ of elements of F and a sequence $m^1, \ldots, m^k$ of elements of $M(\mathcal{C})$ such that $m^i \leqq p^{i-1}$ and $p^{i-1} - p^i = m^i - q(m^i)$ for all $1 \leqq i \leqq k$. Then $q(m^i) \leqq p^i$, $p^0, \ldots, p^k \in C_a$ and $p^0 \sqsupset \ldots \sqsupset p^k$; in particular, $a \,\mathcal{C}\, b$ and $a \sqsupseteq b$ (with $a = b$ if $k = 0$). Also $a - b = \sum_{1 \leqq i \leqq k} (p^{i-1} - p^i) = \sum_{1 \leqq i \leqq k} (m^i - q(m^i))$.

An **overpath** from $a \in F$ to $b \in F$ is the sequence $p : m^1, \ldots, m^k \in M(\mathcal{C})$ of labels in a path

$$a = p^0 \xrightarrow{m^1} p^1 \xrightarrow{m^2} \cdots \xrightarrow{m^k} p^k = b,$$

from a to b. A path from a to b is determined by $p^0 = a$ and its overpath, since in the above the relation $p^{i-1} - p^i = m^i - q(m^i)$ determines p^i from p^{i-1} and m^i. In particular,

$$a - b = \sum_{1 \leqq i \leqq k} (m^i - q(m^i)).$$

The empty sequence is an overpath from any $a \in F$ to itself. If

$$a = p^0 \xrightarrow{m^1} p^1 \xrightarrow{m^2} \cdots \xrightarrow{m^k} p^k = b$$

is a path from a to b, and

$$b = q^0 \xrightarrow{n^1} q^1 \xrightarrow{n^2} \cdots \xrightarrow{n^k} q^k = c$$

is a path from b to c, then

$$a = p^0 \xrightarrow{m^1} \dots \xrightarrow{m^k} p^k = b = q^0 \xrightarrow{n^1} \dots \xrightarrow{n^k} q^k = c$$

is a path from a to c. Hence if $p : m^1, \dots, m^k$ is an overpath from a to b, and $q : n^1, \dots, n^l$ is an overpath from b to c, then $p + q : m^1, \dots, m^k, n^1, \dots, n^l$ is an overpath from a to c.

Let $c \in F$. If (a, m) is an edge from a to b, then $(a + c, m)$ is an edge from $a + c$ to $b + c$. Hence if

$$a = p^0 \xrightarrow{m^1} p^1 \xrightarrow{m^2} \dots \xrightarrow{m^k} p^k = b$$

is a path from a to b, then there is a path with the same labels from $a + c$ to $b + c$. Thus if $m^1, \dots, m^k$ is an overpath from a to b, then $m^1, \dots, m^k$ is an overpath from $a + c$ to $b + c$.

The following result is a well-ordered version of Proposition I.2.9, and shows how $\mathcal{C}$ is generated by all pairs $(m, q(m))$ with $m \in M$.

Proposition 1.5. *For every $a \in F$, there exist a path from a to $q(a)$ and an overpath from a to $q(a)$.*

Proof. This is proved by artinian induction on a. If $a = q(a) \in Q$, then there is an empty path from a to $q(a)$. Now let $a \in P$. Then $a \geqq m$ for some $m \in M$. Let $b = q(m) + t$, where $a = m + t$. Then (a, m) is an edge from a to b. Hence $a \,\mathcal{C}\, b$, $a \sqsupset b$, and the induction hypothesis yields a path from b to $q(b)$. Adding $a \xrightarrow{m} b$ yields a path from a to $q(a) = q(b)$. □

4. The process of well ordering F to select "minimal" generators of $\mathcal{C}$ (as in Proposition 1.4) is reminiscent of Gröbner bases. Indeed let K be a field and $K[X]$ be the polynomial ring with the set X of commuting indeterminates. Ordering F also orders the monomials $X^a = \prod_{x \in X} x^{a_x} \in K[X]$ (where $a = \sum_{x \in X} a_x x \in F$).

Proposition 1.6. *Let $\mathcal{C}$ be a congruence on F and $I(\mathcal{C})$ be the ideal of $K[X]$ generated by all $X^a - X^b$ with $a \,\mathcal{C}\, b$. The set*

$$G(M) = \{ X^m - X^{q(m)} \mid m \in M \}$$

is a Gröbner basis of $I(\mathcal{C})$.

Proof. First we note that $I = I(\mathcal{C})$ is generated by all $X^a - X^{q(a)}$, since $a \,\mathcal{C}\, b$ implies $q(a) = q(b)$ and $X^a - X^b = (X^a - X^{q(a)}) - (X^b - X^{q(b)})$.

We show that the ideal $L(I)$ generated by the leading terms of polynomials in I coincides with the ideal $L(G)$ generated by the leading terms of polynomials in $G(M)$; this is one of the criteria for Gröbner bases (see e.g. Adams & Loustaunau [1994], Theorem 1.6.2).

When $a \in P$, then $a \sqsupset q(a)$ and the leading term of $X^a - X^{q(a)}$ is X^a. Since P is an ideal of F by Lemma 1.3, $L(I)$ is generated by all X^a with $a \in P$. Now $a \in P$ implies, as above, $a \geqq m$ for some $m \in M$, $a = m + t$ for some $t \in F$, and $X^a = X^m X^t \in L(G)$. Therefore $L(I) \subseteq L(G)$; conversely $L(G) \subseteq L(I)$ since $G(M) \subseteq I$. Thus $L(G) = L(I)$. □

We give a direct proof that $I(\mathcal{C})$ is generated by $G(M)$. Let J be the ideal of $K[X]$ generated by $G(M)$. We show by induction on a that $X^a - X^{q(a)} \in J$ for all $a \in F$. When $a \in Q$, then $a = q(a)$ and $X^a - X^{q(a)} \in J$. Let $a \in P$. As in the proof of Proposition 1.4, $a \geqq m$ for some $m \in M$ and $a = m + t$ for some $t \in F$. Let $b = q(m) + t$. Then $a \sqsupset b$ since $m \sqsupset q(m)$, $X^b - X^{q(b)} \in J$ by the induction hypothesis, $X^a - X^b = X^t \left(X^m - X^{q(m)}\right) \in J$, and $X^a - X^{q(a)} = \left(X^a - X^b\right) - \left(X^b - X^{q(b)}\right) \in J$. Thus $X^a - X^{q(a)} \in J$ for all $a \in F$; therefore $I = J$.

2. MAIN RESULT.

The main result in this chapter is the computation of $H^2(S, \mathbb{G})$ by the overpath method. As a first application we find $H^2(S, \mathbb{G})$ when S has a presentation with only one defining relation; for instance, when S is cyclic. We also relate $H^2(S, \mathbb{G})$ to the strand bases in Chapter XI.

1. When S is a commutative semigroup which does not have an identity element, we saw that $H^2(S, \mathbb{G}) \cong H^2(S^1, \mathbb{G}')$, where $\mathbb{G}'$ extends $\mathbb{G}$ to $H(S^1)$ so that $G'_1 = 0$ (Corollary XII.4.5). Hence we may as well start with a monoid S.

In what follows S is a commutative monoid and $\mathbb{G} = (G, \gamma)$ is an abelian group valued functor on $H(S)$; $\pi : F \longrightarrow S$ is a surjective homomorphism, where F is the free c.m. on some set X, and $\mathcal{C} = \ker \pi$; $\sqsubseteq$ is any compatible well order on F; M and q are as in Section 1. By Proposition 1.4, $\mathcal{C}$ is generated by all $(m, q(m))$ with $m \in M(\mathcal{C})$; this provides a presentation of S as the c.m. generated by X subject to all relations $m = q(m)$ with $m \in M$.

A **minimal cochain** on S with values in $\mathbb{G}$ (short for minimal 1-cochain) is a family $u = (u_m)_{m \in M}$ such that $u_m \in G_{\pi m}$ for all $m \in M$.

Let u be a minimal cochain. Let $a \in F$, $p : m^1, \dots, m^k$ be an overpath from a to b, and

$$a = p^0 \xrightarrow{m^1} p^1 \xrightarrow{m^2} \dots \xrightarrow{m^k} p^k = b$$

be the corresponding path from a to b, in which $p^0, \dots, p^k \in C_a$, so that $\pi p^i = \pi a$ for all i. Define

$$u_{a;p} = \textstyle\sum_{1 \leqq i \leqq k} u_{m^i}^{\pi t^i} \in G_{\pi a}$$

where $t^i = p^i - q(m^i) = p^{i-1} - m^i$. A **minimal cocycle** on S with values in $\mathbb{G}$ is a minimal cochain u such that $u_{a;p} = u_{a;q}$ whenever p and q are overpaths from a to $q(a)$ (so that $u_{a;p}$ does not depend on p).

Let $g = (g_x)_{x \in X} \in \prod_{x \in X} G_{\pi x}$ be a family such that $g_x \in G_{\pi x}$ for every generator $x \in X$ of F. A minimal cochain δg is defined by

$$(\delta g)_m = \textstyle\sum_{x \in X,\, x \leqq m} m_x\, g_x^{\pi(m-x)} - \sum_{x \in X,\, x \leqq q(m)} q(m)_x\, g_x^{\pi(q(m)-x)}$$

for every $m = \sum_{x \in X} m_x x \in M$. A minimal cochain constructed in this fashion is a **minimal coboundary**. Under pointwise addition, minimal coboundaries, minimal cocycles, and minimal cochains constitute abelian groups

$$MB^1(S, \mathbb{G}) \subseteq MZ^1(S, \mathbb{G}) \subseteq MC^1(S, \mathbb{G}) = \textstyle\prod_{m \in M} G_{\pi m} \,.$$

The main result in this chapter is:

Theorem 2.1. *For every commutative monoid S there is an isomorphism*

$$H^2(S, \mathbb{G}) \cong MZ^1(S, \mathbb{G}) \,/\, MB^1(S, \mathbb{G})$$

which is natural in $\mathbb{G}$.

2. The proof of Theorem 2.1 occupies the next section. First we consider an example: when S has a commutative presentation (as a semigroup or as a monoid)

$$S \cong \langle\, a_1, \dots, a_n, \dots \mid a_1^{r_1}\, a_2^{r_2} \cdots a_n^{r_n} = a_1^{s_1}\, a_2^{s_2} \cdots a_n^{s_n} \,\rangle$$

with a single defining relation, in which we assume, not unreasonably, that $r_i + s_i > 0$ for all $i \leqq n$ and that $r_i \neq s_i$ for some i. Other examples are given in Grillet [2000T] and in Sections 4 and 5.

We can set up the surjective homomorphism $\pi : F = F_X \longrightarrow S$ so that X

contains distinct elements $x_1, \dots, x_n$ such that $\pi x_1 = a_1, \dots, \pi x_n = a_n$. Then $\mathcal{C} = \ker \pi$ is the congruence on F generated by the single pair (r,s), where

$$r = \sum_{1 \leqq i \leqq n} r_i x_i \quad \text{and} \quad s = \sum_{1 \leqq i \leqq n} s_i x_i .$$

The congruence $\mathcal{C}$ is readily described:

Lemma 2.2. *$a \,\mathcal{C}\, b$ if and only if there exists a sequence $p^0, \dots, p^k$ of elements of F such that $k \geqq 0$, $a = p^0$, $p^k = b$, and either*

$$p^{i-1} - r = p^i - s \geqq 0 \quad \textit{for all } i \geqq 1 \tag{A}$$

or

$$p^{i-1} - s = p^i - r \geqq 0 \quad \textit{for all } i \geqq 1 \tag{B}$$

Proof. By Proposition I.2.9, $a \,\mathcal{C}\, b$ if and only if there exists a sequence $p^0, \dots, p^k$ of elements of F such that $k \geqq 0$, $a = p^0$, $p^k = b$, and, for every $i \geqq 1$, either

$$p^{i-1} - r = p^i - s \geqq 0 \tag{a}$$

or

$$p^{i-1} - s = p^i - r \geqq 0. \tag{b}$$

If (a) holds for $i < k$ and (b) holds for $i+1$, then $p^{i-1} - r = p^i - s = p^{i+1} - r$ and p^i, p^{i+1} may be deleted from the sequence. Similarly if (b) holds for $i < k$ and (a) holds for $i+1$, then $p^{i-1} - s = p^i - r = p^{i+1} - s$ and again p^i, p^{i+1} may be deleted from the sequence. After all such deletions, either (a) holds for all i, or (b) holds for all i. □

Now let $\sqsubseteq$ be any compatible well order on F. We may assume that $r \sqsupset s$. When $a \,\mathcal{C}\, b$ and (A) holds, then in Lemma 2.2 $p^{i-1} - r = p^i - s \geqq 0$ implies $p^{i-1} = r + t$ and $p^i = s + t$, where $t = p^{i-1} - r = p^i - s \geqq 0$, so that $p^{i-1} \sqsupset p^i$ for all $i \geqq 1$ and $a \sqsupset b$; if (B) holds, then similarly $a \sqsubset b$. If therefore $a \in P$, so that $a \,\mathcal{C}\, q(a)$ and $a \sqsupset q(a)$, then (A) holds, $a - r = p^0 - r \geqq 0$, and $a \geqq r$. On the other hand, $r \,\mathcal{C}\, s$ and $r \sqsupset s$, so that $r \in P$. This proves:

Lemma 2.3. *$M(\mathcal{C})$ has just one element, namely r; and $q(r) = s$.*

Let

$$d = a_1^{r_1} a_2^{r_2} \cdots a_n^{r_n} = a_1^{s_1} a_2^{s_2} \cdots a_n^{s_n} = \pi r = \pi s \in S.$$

By Lemma 2.3, a minimal cochain consists of $u \in G_d$, and $MC^1(S, \mathbb{G}) = G_d$. Moreover there is only one overpath from any $c \in F$ to $q(c)$, which is a sequence of r's. Hence every minimal cochain is a minimal cocycle and $MZ^1(S, \mathbb{G}) = G_d$.

A minimal cochain $u \in G_d$ is a minimal coboundary if and only if there exists a family $g = (g_x)_{x \in X}$ such that $g_x \in G_{\pi x}$ for every $x \in X$ and

$$\begin{aligned} u &= \textstyle\sum_{x \in X,\, x \leqq r} r_x\, g_x^{\pi(r-x)} - \sum_{x \in X,\, x \leqq q(r)} q(r)_x\, g_x^{\pi(q(r)-x)} \\ &= \textstyle\sum_{i \leqq n,\, r_i > 0} r_i\, g_{x_i}^{\pi(r-x_i)} - \sum_{i \leqq n,\, s_i > 0} s_i\, g_{x_i}^{\pi(s-x_i)}. \end{aligned}$$

Let $\gamma_i : G_{a_i} \longrightarrow G_d$ be defined by:

$$\gamma_i = \begin{cases} r_i\, \gamma_{a_i, d_i'} - s_i\, \gamma_{a_i, d_i''} & \text{if } r_i, s_i > 0, \\ r_i\, \gamma_{a_i, d_i'} & \text{if } r_i > 0,\ s_i = 0, \\ -s_i\, \gamma_{a_i, d_i''} & \text{if } r_i = 0,\ s_i > 0, \end{cases}$$

where

$$\begin{aligned} d_i' &= a_1^{r_1} \cdots a_{i-1}^{r_{i-1}}\, a_i^{r_i - 1}\, a_{i+1}^{r_{i+1}} \cdots a_n^{r_n} \quad \text{(when } r_i > 0\text{), and} \\ d_i'' &= a_1^{s_1} \cdots a_{i-1}^{s_{i-1}}\, a_i^{s_i - 1}\, a_{i+1}^{s_{i+1}} \cdots a_n^{s_n} \quad \text{(when } s_i > 0\text{).} \end{aligned}$$

Then u is a minimal coboundary if and only if there exist $g_i = g_{x_i} \in G_{a_i}$ such that $u = \sum_{1 \leqq i \leqq n} \gamma_i g_i$. Hence $MB^1(S, \mathbb{G}) = \sum_{1 \leqq i \leqq n} \operatorname{Im} \gamma_i$ and:

Proposition 2.4. *When S has a commutative semigroup or monoid presentation with a single nontrivial defining relation, then, with the notation as above,*

$$H^2(S, \mathbb{G}) \cong G_d / \big(\textstyle\sum_{1 \leqq i \leqq n} \operatorname{Im} \gamma_i\big).$$

Corollary 2.5. *When $S = \langle\, a \mid a^r = a^{r+p} \,\rangle$ is cyclic with index r and period p, then $H^2(S, \mathbb{G}) \cong G_{a^r} / p \operatorname{Im} \gamma_{a, a^{r-1}}$.*

Proposition 2.4 becomes simpler when $\mathbb{G}$ is thin. Then $d \leqq_{\mathcal{H}} a_i$ for all i (since $r_i + s_i > 0$) and $\gamma_{a_i, d_i'} = \gamma_d^{a_i}$ when $r_i > 0$, $\gamma_{a_i, d_i''} = \gamma_d^{a_i}$ when $s_i > 0$, and $\gamma_i = (r_i - s_i)\, \gamma_d^{a_i}$ for all i. Hence

Corollary 2.6. *When S has a commutative semigroup or monoid presentation with a single nontrivial defining relation and $\mathbb{G}$ is thin, then, with the notation as above,*

$$H^2(S, \mathbb{G}) \cong G_d / \big(\textstyle\sum_{1 \leqq i \leqq n} (r_i - s_i) \operatorname{Im} \gamma_d^{a_i}\big).$$

If for instance $S = \langle\, a \mid a^r = a^{r+p} \,\rangle$ is cyclic with index r and period p and $\mathbb{G} = A$ is constant ($G_s = A$ and $\gamma_{s,t} = 1_A$ for all $s \in S$ and $t \in S^1$), then

$H^2(S,\mathbb{G}) \cong A/pA$; thus $H^2(S,\mathbb{G}) \cong \operatorname{Ext}(H,A)$, where H is the subgroup $\{a^k \mid k \geqq r\} \cong \mathbb{Z}_p$ of S.

3. Finally we show that strand bases in Chapter XI give rise to minimal cocycles. This result is from Grillet [1996C], [2001C].

In what follows, $\mathcal{C}$ is a subcomplete congruence on a free commutative monoid F and $\mathcal{C}^*$ is its group-free hull; $\pi : F \longrightarrow S$ and $\pi^* : F \longrightarrow S^*$ are surjective homomorphisms which induce $\mathcal{C}$ and $\mathcal{C}^*$ respectively. If $\mathcal{C}$ is complete, then $S^* \cong S/\mathcal{H}$ and one expects the cohomology of S^* to show up somewhere in the construction of S and $\mathcal{C}$. Minimal cocycles provide this connection.

The direction set, extent cells, strand groups, strand bases, and notation are as in Chapter XI. Also $\sqsubseteq$ is a compatible well order on F; the mapping q and sets M and Q are those of $\mathcal{C}^*$, not of $\mathcal{C}$.

Lemma 2.7. *Let s be a strand base of $\mathcal{C}$. For every $m \in M(\mathcal{C}^*)$ let*

$$\overline{s}_m \;=\; s_m - m - s_{q(m)} + q(m) \;\in\; G_m .$$

If $m^1, m^2, \dots, m^k$ is an overpath from a to b, then

$$s_a - s_b - a + b - \overline{s}_{m^1} - \cdots - \overline{s}_{m^k} \;\in\; R_a = R_b .$$

Proof. Let

$$a = p^0 \xrightarrow{m^1} p^1 \xrightarrow{m^2} \cdots \xrightarrow{m^k} p^k = b$$

be a path from a to b, so that $p^{i-1} = m^i + t^i$, $p^i = q(m^i) + t^i$ for some $t^i \in F$ and $p^{i-1} - p^i = m^i - q(m^i)$, for every $1 \leqq i \leqq k$. Then $R_a = R_{p^0} = R_{p^1} = \cdots = R_{p^k} = R_b$ by (R2), since $a = p^0, p^1, \dots, p^k = b$ are all in the same $\mathcal{C}^*$-class, and

$$\begin{aligned} s_{p^{i-1}} - s_{p^i} - p^{i-1} + p^i - \overline{s}_{m^i} &= s_{p^{i-1}} - s_{p^i} - m^i + q(m^i) - \overline{s}_{m^i} \\ &= s_{p^{i-1}} - s_{p^i} - s_{m^i} + s_{q(m^i)} \;\in\; R_{p^i} = R_a \end{aligned}$$

by (S+) in Lemma XI.6.1. Adding these equalities yields $s_a - s_b - a + b - \overline{s}_{m^1} - \cdots - \overline{s}_{m^k} \in R_a$. □

With $b = q(a)$, Lemma 2.7 implies that a strand base of $\mathcal{C}$ is completely determined modulo strand groups by its values on $M \cup Q$.

2. The strand group functor $\mathbb{K} = (K, \psi)$ of $\mathcal{C}$ is the thin abelian group valued functor on $F/\mathcal{C}^*$ defined as follows (Section XI.4). To every $\mathcal{C}^*$-class

C^*, K assigns the group $K_a = G_a/R_a$, which does not depend on the choice of $a \in C^*$. When $C^* \geqq_{\mathcal{H}} D^*$ in $F/\mathcal{C}^*$, then $a \leqq b$ for some $a \in C^*$, $b \in D^*$, $G_a \subseteq G_b$, $R_a \subseteq R_b$ by (R3), and $\psi_b^a : K_a \longrightarrow K_b$ sends $g + R_a$ to $g + R_b$ and does not depend on the choice of $a \in C^*$ and $b \in D^*$ (as long as $a \leqq b$). Since $S^* \cong F/\mathcal{C}^*$ we may regard $\mathbb{K}$ as a thin abelian group valued functor on S^*; then $\mathbb{K}$ is isomorphic to the extended Schützenberger functor of S, which is the usual Schützenberger functor if S is complete (Proposition XI.4.8).

Proposition 2.8. *Let $\mathcal{C}$ be a subcomplete congruence on F and s be a strand base of $\mathcal{C}$. For every $m \in M(\mathcal{C}^*)$ let*

$$s_m^* \;=\; \overline{s}_m + R_m \;=\; s_m - m - s_{q(m)} + q(m) + R_m \;\in\; G_m/R_m.$$

Then s^* is a minimal 1-cocycle on $F/\mathcal{C}^*$ with values in the strand group functor $\mathbb{K}$ of $\mathcal{C}$. Moreover, two strand bases s and t define the same congruence if and only if $s^* = t^*$.

Proof. s^* is a minimal cochain. When $p : m^1, \dots, m^k$ is an overpath from a to b and

$$a = p^0 \xrightarrow{m^1} p^1 \xrightarrow{m^2} \cdots \xrightarrow{m^k} p^k = b$$

is the corresponding path, then

$$s_{a;p;b}^* \;=\; \textstyle\sum_i \psi_{p^i}^{m_i} s_{m_i}^* \;=\; \sum_i \overline{s}_{m^i} + R_a \;=\; s_a - s_b - a + b + R_a$$

by Lemma 2.7. Hence $s_{a;p;b}^*$ is independent of path and s^* is a minimal cocycle.

By Proposition XI.5.2, two strand bases s and t define the same congruence if and only if $a \;\mathcal{C}^*\; b$ implies $s_a - s_b - t_a + t_b \in R_a (= R_b)$. Since $m \;\mathcal{C}^*\; q(m)$ this implies

$$\big(s_m - m - s_{q(m)} + q(m)\big) - \big(t_m - m - t_{q(m)} + q(m)\big) \;\in\; R_m$$

for all $m \in M$ and $s^* = t^*$. Conversely assume that $s^* = t^*$. Then $s_{a;p;b}^* = t_{a;p;b}^*$ and

$$s_a - s_b - a + b + R_a \;=\; t_a - t_b - a + b + R_a$$

whenever p is an overpath from a to b. Hence $s_a - s_{q(a)} - t_a + t_{q(a)} \in R_a$ for all a. If $a \;\mathcal{C}^*\; b$, then $q(a) = q(b)$, $s_b - s_{q(a)} - t_b + t_{q(a)} \in R_b = R_a$, and $s_a - s_b - t_a + t_b \in R_a$. Thus s and t are equivalent. □

Proposition 2.8 embeds the set of equivalence classes of strand bases (and the set of all subcomplete congruence with the given strand groups) into the abelian group $MZ^1(F/\mathcal{C}^*, \mathbb{K})$.

3. PROOF OF MAIN RESULT.

1. In what follows, S is a commutative monoid, $\mathbb{G} = (G, \gamma)$ is an abelian group valued functor on $H(S)$, $F = F_X$ is the free c.s. on a set X, and $\pi : F \longrightarrow S$ is a surjective homomorphism; we prove Theorem 2.1.

We begin the proof by lifting 1- and 2-cochains from S to F.

The homomorphism $\pi : F \longrightarrow S$ extends to a functor $\pi : H(F) \longrightarrow H(S)$. Hence $\mathbb{G} = (G, \gamma)$ lifts to an abelian group valued functor $\mathbb{G}\pi = (\mathbb{G}\pi, \gamma\pi) = \mathbb{G} \circ \pi$ on $H(F)$; $\mathbb{G}\pi$ assigns $G_{\pi a}$ to $a \in F$ and $\gamma_{\pi a, \pi t}$ to $(a,t) : a \longrightarrow at$. Thus $g^t = g^{\pi t}$, where $g \in G_{\pi a}$, g^t is provided by $\mathbb{G}\pi$, and $g^{\pi t}$ is provided by $\mathbb{G}$. Note that $\mathbb{G}\pi$ is thin, since F is cancellative.

Every 1-cochain $u = (u_a)_{a \in S} \in C^1(S, \mathbb{G})$ lifts to a 1-cochain $\pi^* u = u \circ \pi \in C^1(F, \mathbb{G}\pi)$ defined by

$$(\pi^* u)_a = u_{\pi a} \in G_{\pi a}$$

for all $a \in F$. If u is a 1-cocycle, so that $u_{ab} = u_a^b + u_b^a$ for all $a, b \in S$, then

$$(\pi^* u)_{ab} = u_{(\pi a)(\pi b)} = u_{\pi a}^{\pi b} + u_{\pi b}^{\pi a} = (\pi^* u)_a^b + (\pi^* u)_b^a$$

and $\pi^* u$ is a 1-cocycle; thus $\pi^* Z^1(S, \mathbb{G}) \subseteq Z^1(F, \mathbb{G}\pi)$.

Similarly every symmetric 2-cochain $u = (u_a)_{a \in S} \in SC^2(S, \mathbb{G})$ lifts to a symmetric 2-cochain $\pi^* u = u \circ \pi \in C^2(F, \mathbb{G}\pi)$ defined by

$$(\pi^* u)_{a,b} = u_{\pi a, \pi b} \in G_{\pi(ab)}$$

for all $a, b \in F$. If u is a symmetric 2-cocycle, so that $u_{a,b}^c + u_{ab,c} = u_{a,bc} + u_{b,c}^a$ for all $a, b, c \in S$, then

$$\begin{aligned} (\pi^* u)_{a,b}^c + (\pi^* u)_{ab,c} &= u_{\pi a, \pi b}^{\pi c} + u_{(\pi a)(\pi b), \pi c} \\ &= u_{\pi a, (\pi b)(\pi c)} + u_{\pi b, \pi c}^{\pi a} = (\pi^* u)_{a,bc} + (\pi^* u)_{b,c}^a \end{aligned}$$

for all $a, b, c \in F$, and $\pi^* u$ is a symmetric 2-cocycle. If $u = \delta v$ is a symmetric 2-coboundary, so that $u_{a,b} = v_a^b - v_{ab} + v_b^a$ for all $a, b \in S$, then

$$(\pi^* u)_{a,b} = v_{\pi a}^{\pi b} - v_{(\pi a)(\pi b)} + v_{\pi b}^{\pi a} = (\delta \pi^* v)_{a,b}$$

for all $a, b \in S$, and $\pi^* u$ is a symmetric 2-coboundary. Thus

$$\pi^* SZ^2(S, \mathbb{G}) \subseteq SZ^2(F, \mathbb{G}\pi) \quad \text{and} \quad \pi^* SB^2(S, \mathbb{G}) \subseteq SB^2(F, \mathbb{G}\pi).$$

Since $H^2(F,\mathbb{G}\pi) = 0$ (Theorem XII.3.4), symmetric 2-cocycles on S lift to symmetric 2-coboundaries on F and can therefore be constructed by projecting the coboundaries of 1-cochains on F. This marks the birthplace of Theorem 2.1.

2. A 1-cochain $u \in C^1(F,\mathbb{G}\pi)$ is **consistent** (relative to π) when

$$\pi a = \pi b \quad \text{implies} \quad u_{a+c} - u_{b+c} = u_a^c - u_b^c \quad \text{for all} \quad c \in F.$$

Under pointwise addition consistent 1-cochains form a subgroup $K^1(F,\mathbb{G}\pi)$ of $C^1(F,\mathbb{G}\pi)$. We shall see that consistent 1-cochains on F are precisely those whose coboundaries project to symmetric 2-cocycles on S. First we show:

Lemma 3.1. *$K^1(F,\mathbb{G}\pi)$ contains $Z^1(F,\mathbb{G}\pi)$ and $\pi^* C^1(S,\mathbb{G})$.*

Proof. If $u \in Z^1(F,\mathbb{G}\pi)$, then $\pi a = \pi b$ implies $\gamma_{\pi a,\pi c} = \gamma_{\pi b,\pi c}$ and $u_a^c - u_{a+c} = u_c^a = u_c^b = u_b^c - u_{b+c}$.

If $u = \pi^* v$, then $\pi a = \pi b$ implies $u_a = v_{\pi a} = v_{\pi b} = u_b$, $\gamma_{\pi a,\pi c} = \gamma_{\pi b,\pi c}$, $\pi(a+c) = \pi(b+c)$, $u_{ac} = u_{bc}$, and $u_a^c - u_{a+c} = u_b^c - u_{b+c}$. □

When $u \in K^1(F,\mathbb{G}\pi)$, then $\pi a = \pi b$, $\pi c = \pi d$ imply $\gamma_{\pi c,\pi a} = \gamma_{\pi c,\pi b}$, $\gamma_{\pi b,\pi c} = \gamma_{\pi b,\pi d}$, and

$$u_a^c - u_{a+c} + u_c^a \;=\; u_b^c - u_{b+c} + u_c^b \;=\; u_b^d - u_{b+d} + u_d^b .$$

Hence a homomorphism $\Delta : K^1(F,\mathbb{G}\pi) \longrightarrow SC^2(S,\mathbb{G})$ is well defined by

$$(\Delta u)_{\pi a,\pi c} \;=\; u_a^c - u_{a+c} + u_c^a \;=\; (\delta u)_{a,c}$$

for all $a,c \in F$. It is immediate that Δ is natural in $\mathbb{G}$.

Lemma 3.2. $\operatorname{Im}\Delta = SZ^2(S,\mathbb{G})$.

Proof. Δu is a symmetric 2-cocycle, since δu is a cocycle and

$$\begin{aligned}(\Delta u)^{\pi c}_{\pi a,\pi b} + (\Delta u)_{(\pi a)(\pi b),\,\pi c} &= (\delta u)^c_{a,b} + (\delta u)_{a+b,\,c}\\ = (\delta u)_{a,\,b+c} + (\delta u)^a_{b,c} &= (\Delta u)_{\pi a,\,(\pi b)(\pi c)} + (\Delta u)^{\pi a}_{\pi b,\pi c}.\end{aligned}$$

Conversely let $s \in SZ^2(S,\mathbb{G})$. Then $\pi^* s \in SZ^2(F,\mathbb{G}\pi)$. Since $H^2(F,\mathbb{G}\pi) = 0$ we have $\pi^* s = \delta u$ for some $u \in C^1(F,\mathbb{G}\pi)$, so that

$$s_{\pi a,\pi c} \;=\; u_a^c - u_{a+c} + u_c^a$$

for all $a,c \in F$. If $\pi a = \pi b$, then

$$u_a^c - u_{a+c} + u_c^a \;=\; s_{\pi a,\pi c} \;=\; s_{\pi b,\pi c} \;=\; u_b^c - u_{b+c} + u_c^b,$$

$\gamma_{\pi c,\pi b} = \gamma_{\pi c,\pi a}$, $u^a_c = u^b_c$, and $u^c_a - u_{a+c} = u^c_b - u_{b+c}$. Thus $u \in K^1(F,\mathbb{G}\pi)$. We see that $\Delta u = s$. □

3. Lemma 3.2 shows that $H^2(S,\mathbb{G})$ is determined by consistent cochains.

Lemma 3.3. *When* $u \in K^1(F,\mathbb{G}\pi)$, *then* $\Delta u \in SB^2(S,\mathbb{G})$ *if and only if* $u \in Z^1(F,\mathbb{G}\pi) + \pi^* C^1(S,\mathbb{G})$.

Proof. If $u = v + \pi^* w$, where $v \in Z^1(F,\mathbb{G}\pi)$ and $w \in C^1(S,\mathbb{G})$, then $u \in K^1(F,\mathbb{G}\pi)$ by Lemma 3.1 and

$$(\Delta u)_{\pi a,\pi b} = v^b_a - v_{a+b} + v^a_b + w^{\pi b}_{\pi a} - w_{(\pi a)(\pi b)} + w^{\pi a}_{\pi b} = (\delta w)_{\pi a,\pi b}$$

for all $a,b \in F$, since $v \in Z^1(F,\mathbb{G}\pi)$, so that $\Delta u = \delta w \in SB^2(S,\mathbb{G})$.

Conversely assume $\Delta u = \delta w \in SB^2(S,\mathbb{G})$, where $w \in C^1(S,\mathbb{G})$. Then

$$u^b_a - u_{a+b} + u^a_b = w^{\pi b}_{\pi a} - w_{(\pi a)(\pi b)} + w^{\pi a}_{\pi b} = (\pi^* w)^b_a - (\pi^* w)_{a+b} + (\pi^* w)^a_b$$

for all $a,b \in F$, and $v = u - \pi^* w \in Z^1(F,\mathbb{G}\pi)$. □

Corollary 3.4. *There is an isomorphism*

$$H^2(S,\mathbb{G}) \cong K^1(F,\mathbb{G}\pi) / \big(Z^1(F,\mathbb{G}\pi) + \pi^* C^1(S,\mathbb{G})\big)$$

which is natural in $\mathbb{G}$.

Proof. The isomorphism follows from Lemmas 3.2 and 3.3, since $H^2(S,\mathbb{G}) \cong SZ^2(S,\mathbb{G}) / SB^2(S,\mathbb{G})$; it is natural in $\mathbb{G}$ since Δ is natural in $\mathbb{G}$. □

4. Now let $\sqsubseteq$ be any compatible well order on F and $\mathcal{C} = \ker \pi$. We use P, Q, M, q from Section 1 to trim consistent cochains.

A **partial 1-cochain** on F with values in $\mathbb{G}$ is a 1-cochain $u = (u_a)_{a\in F} \in C^1(F,\mathbb{G}\pi)$ such that $u_c = 0$ for all $c \in Q$. (Thus u is, in effect, a cochain on P only.) Under pointwise addition, partial 1-cochains constitute a subgroup $P^1(F,\mathbb{G}\pi) \cong \prod_{a\in P} G_{\pi a}$ of $C^1(F,\mathbb{G}\pi)$. Consistent partial 1-cochains constitute an abelian group $KP^1(F,\mathbb{G}\pi) = K^1(F,\mathbb{G}\pi) \cap P^1(F,\mathbb{G}\pi)$.

When $u \in C^1(F,\mathbb{G}\pi)$, define Πu by

$$(\Pi u)_a = u_a - u_{q(a)} \in G_{\pi a}$$

for all $a \in F$. If $a \in Q$, then $a = q(a)$ and $(\Pi u)_a = 0$. Thus Π is a homomorphism of $C^1(F,\mathbb{G}\pi)$ into $P^1(F,\mathbb{G}\pi)$. In fact $\operatorname{Im}\Pi = P^1(F,\mathbb{G}\pi)$, since every partial cochain v satisfies $v_{q(a)} = 0$ and $\Pi v = v$. We see that Π is natural in $\mathbb{G}$.

Lemma 3.5. *When* $u \in C^1(F, \mathbb{G}\pi)$, *then* $u \in K^1(F, \mathbb{G}\pi)$ *if and only if* $\Pi u \in KP^1(F, \mathbb{G}\pi)$.

Proof. Let $u \in C^1(F, \mathbb{G}\pi)$ and $v = \Pi u$. Let $a, b, c \in F$ satisfy $\pi a = \pi b$. Let $r = q(a) = q(b)$ and $s = q(a+c) = q(b+c)$. Then

$$v_{a+c} - v_a^c = (u_{a+c} - u_a^c) - (u_s - u_r^c), \quad \text{and}$$
$$v_{b+c} - v_b^c = (u_{b+c} - u_b^c) - (u_s - u_r^c).$$

If u is consistent, then $u_{a+c} - u_a^c = u_{b+c} - u_b^c$ and $v_{a+c} - v_a^c = v_{b+c} - v_b^c$; hence v is consistent. If conversely v is consistent, then $v_{a+c} - v_a^c = v_{b+c} - v_b^c$ and $u_{a+c} - u_a^c = u_{b+c} - u_b^c$; hence u is consistent. □

5. Given a family $g = (g_x)_{x \in X} \in \prod_{x \in X} G_{\pi x}$ (with $g_x \in G_{\pi x}$ for all $x \in X$), define $Dg \in C^1(F, \mathbb{G}\pi)$ by

$$(Dg)_a = \textstyle\sum_{x \in X,\, x \leqq a} a_x\, g_x^{a-x} - \sum_{x \in X,\, x \leqq q(a)} q(a)_x\, g_x^{q(a)-x}$$

for all $a = \sum_{x \in X} a_x x \in F$. We see that $(Dg)_a \in G_{\pi a} = G_{\pi q(a)}$, and that D is a homomorphism of $\prod_{x \in X} G_{\pi x}$ into $C^1(F, \mathbb{G}\pi)$ and is natural in $\mathbb{G}$.

Lemma 3.6. $\operatorname{Im} D \subseteq KP^1(F, \mathbb{G}\pi)$. *When* $u \in C^1(F, \mathbb{G}\pi)$, *then* $u \in Z^1(F, \mathbb{G}\pi) + \pi^* C^1(S, \mathbb{G})$ *if and only if* $\Pi u \in \operatorname{Im} D$.

Proof. If $a \in Q$, then $a = q(a)$ and $(Dg)_a = 0$; thus $Dg \in P^1(F, \mathbb{G}\pi)$. That $Dg \in K^1(F, \mathbb{G}\pi)$ can be proved directly but follows from Lemma 3.1 and the rest of the statement, as $u \in \operatorname{Im} D$ implies $\Pi u = u$ and $u \in Z^1(F, \mathbb{G}\pi) + \pi^* C^1(S, \mathbb{G}) \subseteq K^1(F, \mathbb{G}\pi)$.

Let $u = z + \pi^* w \in Z^1(F, \mathbb{G}\pi) + \pi^* C^1(S, \mathbb{G})$, where $z \in Z^1(F, \mathbb{G}\pi)$ and $w \in C^1(S, \mathbb{G})$, so that $u_a = z_a + w_{\pi a}$ for all $a \in F$. Since z is a 1-cocycle we have $z_{a+b} = z_a^b + z_b^a$ for all $a, b \in F$; hence

$$z_a = \textstyle\sum_{x \in X,\, x \leqq a} a_x\, z_x^{a-x},$$

for every $a = \sum_{x \in X} a_x x \in F$. Since $\pi\, q(a) = \pi a$,

$$\begin{aligned} (\Pi u)_a &= z_a + w_{\pi a} - z_{q(a)} - w_{\pi q(a)} \\ &= \textstyle\sum_{x \in X,\, x \leqq a} a_x\, z_x^{a-x} - \sum_{x \in X,\, x \leqq q(a)} q(a)_x\, z_x^{q(a)-x} = (Dg)_a, \end{aligned}$$

where $g = (z_x)_{x \in X}$. Thus $\Pi u \in \operatorname{Im} D$.

Conversely assume $\Pi u \in \operatorname{Im} D$, so that there exists $g = (g_x)_{x\in X} \in \prod_{x\in X} G_{\pi x}$ such that

$$u_a - u_{q(a)} = \textstyle\sum_{x\in X,\, x\leqq a} a_x\, g_x^{a-x} - \sum_{x\in X,\, x\leqq q(a)} q(a)_x\, g_x^{q(a)-x}$$

for all $a \in F$. For every $a \in F$ let

$$z_a = \textstyle\sum_{x\in X,\, x\leqq a} a_x\, g_x^{a-x} \in G_{\pi a}\,.$$

If $a,b \in F$, and $x \leqq a+b$ (equivalently, $(a+b)_x > 0$), then $x \leqq a$ or $x \leqq b$ (or both); hence

$$\begin{aligned} z_{a+b} &= \textstyle\sum_{x\in X,\, x\leqq a,\, x\nleqq b} a_x\, g_x^{a+b-x} \\ &\quad + \textstyle\sum_{x\in X,\, x\nleqq a,\, x\leqq b} b_x\, g_x^{a+b-x} \\ &\quad + \textstyle\sum_{x\in X,\, x\leqq a,\, x\leqq b} (a_x + b_x)\, g_x^{a+b-x} = z_a^b + z_b^a\,; \end{aligned}$$

thus $z \in Z^1(F,\mathbb{G}\pi)$. Also $u_a - u_{q(a)} = z_a - z_{q(a)}$ for all a. Let $v_a = u_a - z_a$. Then $v \in C^1(F,\mathbb{G}\pi)$ and $v_a = v_{q(a)}$ for all a. Hence $\pi a = \pi b$ implies $v_a = v_b$ and $v = \pi^* w$, where $w \in C^1(S,\mathbb{G})$ is well defined by $w_{\pi a} = v_a$. Thus $u = z + \pi^* w \in Z^1(F,\mathbb{G}\pi) + \pi^* C^1(S,\mathbb{G})$. □

Corollary 3.7. *There is an isomorphism*

$$H^2(S,\mathbb{G}) \cong KP^1(F,\mathbb{G}\pi)\,/\operatorname{Im} D$$

which is natural in $\mathbb{G}$.

Proof. We saw that $\Pi : C^1(F,\mathbb{G}\pi) \longrightarrow P^1(F,\mathbb{G}\pi)$ is a surjective homomorphism. Now $K^1(F,\mathbb{G}\pi) = \Pi^{-1}\, KP^1(F,\mathbb{G}\pi)$ by Lemma 3.5 and $Z^1(F,\mathbb{G}\pi) + \pi^* C^1(S,\mathbb{G}) = \Pi^{-1} \operatorname{Im} D$ by Lemma 3.6; therefore

$$K^1(F,\mathbb{G}\pi)\,/\,\big(Z^1(F,\mathbb{G}\pi) + \pi^* C^1(S,\mathbb{G})\big) \cong KP^1(F,\mathbb{G}\pi)\,/\operatorname{Im} D.$$

This isomorphism is natural in $\mathbb{G}$ since Π and D are natural in $\mathbb{G}$. The natural isomorphism $H^2(S,\mathbb{G}) \cong KP^1(F,\mathbb{G}\pi)\,/\operatorname{Im} D$ then follows from Corollary 3.4. □

6. Recall that a minimal cochain on S with values in $\mathbb{G}$ is a family $u = (u_m)_{m\in M}$ such that $u_m \in G_{\pi m}$ for all $m \in M$. Under pointwise addition, minimal cochains constitute a subgroup $M^1(F,\mathbb{G}\pi) \cong \prod_{m\in M} G_{\pi m}$ of $C^1(F,\mathbb{G}\pi)$.

Every partial 1-cochain $u = (u_a)_{a\in F}$ has a restriction $Ru = (u_m)_{m\in M}$ to M, which is a minimal cochain. This defines a restriction homomorphism

$R : P^1(F,\mathbb{G}\pi) \longrightarrow M^1(F,\mathbb{G}\pi)$ which is natural in $\mathbb{G}$.

Lemma 3.8. *R is injective on $KP^1(F,\mathbb{G}\pi)$.*

Proof. Let $u = (u_a)_{a\in F}$ be a consistent partial cochain such that $Ru = 0$ (such that $u_m = 0$ for all $m \in M$). We use artinian induction on a to prove that $u_a = 0$ for all $a \in P$ (so that $u = 0$). Already $u_a = 0$ for all $a \in Q$ and for all $a \in M$. Let $a \in P\backslash M$. Then $a > m$ for some $m \in M$, $a = m + c$ for some $c \in F$, and $b = q(m) + c$ satisfies $\pi b = \pi a$ and $b \sqsubset a$. Since u is consistent we have $u_{m+c} - u_{q(m)+c} = u^c_m - u^c_{q(m)}$, with $u_m = 0$, $u_{q(m)} = 0$, and $u_{q(m)+c} = u_b = 0$ by the induction hypothesis; hence $u_a = u_{m+c} = 0$. □

7. Lemma 3.8 shows that a consistent partial cochain is determined by its restriction to M. Therefore Corollary 3.7 can be restated in terms of minimal cochains; this will yield the main result. First we reconstruct consistent partial cochains from their restrictions.

Let u be a minimal cochain. When $a \sqsupseteq b$ in F and $p : m^1,\dots,m^k$ is an overpath from a to b, let

$$u_{a;p;b} \;=\; \textstyle\sum_{1\leqq i\leqq k} u^{t^i}_{m^i} \;\in\; G_{\pi a},$$

where t^i is obtained from the corresponding path

$$a = p^0 \xrightarrow{m^1} p^1 \xrightarrow{m^2} \dots \xrightarrow{m^k} p^k = b$$

by $t^i = p^i - q(m^i) = p^{i-1} - m^i$. Recall that $p^0,\dots,p^k \in C_a$, so that $\pi p^i = \pi a$ and $u^{t^i}_{m^i} \in G_{\pi a}$ for all i.

We denote $u_{a;p;q(a)}$ by $u_{a;p}$. If $a \in Q$, then p is empty and $u_{a;p} = 0$. If $a = m \in M$, then $p = \{m\}$ and $u_{m;p} = u_m$.

When p is an overpath from a to b, then p is an overpath from $a + c$ to $b + c$ for any $c \in F$; the corresponding path is

$$a + c = p^0 + c \xrightarrow{m^1} p^1 + c \xrightarrow{m^2} \dots \xrightarrow{m^k} p^k + c = b + c,$$

with $p^i + c - q(m^i) = p^{i-1} + c - m^i = t^i + c$, and

$$u_{a+c;\,p;\,b+c} \;=\; u^c_{a;p;b}\,.$$

If $a = b$, then p is empty and $u_{a;p;b} = 0$. If $p : m^1,\dots,m^k$ is an overpath from a to b, and $q : n^1,\dots,n^l$ is an overpath from b to c, then $p + q$:

$m^1, \dots, m^k$, $n^1, \dots, n^l$ is an overpath from a to c, and

$$u_{a;\,p+q;\,c} \;=\; u_{a;p;b} \;+\; u_{b;q;c}\,.$$

In particular, $u_{a;p;b} = u_{a;p+q} - u_{b;q}$ when q is an overpath from b to $q(b) = q(a)$.

8. **Independence of path** for $u_{a;p;b}$ means that $u_{a;p;b} = u_{a;q;b}$ whenever p and q are overpaths from a to b; and similarly for $u_{a;p}$. These properties are equivalent and characterize the restrictions of consistent partial cochains:

Lemma 3.9. *When $u \in M^1(F, \mathbb{G}\pi)$, then $u \in R\big(KP^1(F, \mathbb{G}\pi)\big)$ if and only if $u_{a;p}$ is independent of path; and then $u_{a;p;b}$ is independent of path.*

Proof. First let $v \in KP^1(F, \mathbb{G}\pi)$ and $u = Rv$. Let $p : m^1, \dots, m^k$ be an overpath from a to b; let

$$a = p^0 \xrightarrow{\;m^1\;} p^1 \xrightarrow{\;m^2\;} \cdots \xrightarrow{\;m^k\;} p^k = b$$

be the corresponding path and $t^i = p^i - q(m^i) = p^{i-1} - m^i$. Since $v \in KP^1(F, \mathbb{G}\pi)$ and $\pi m^i = \pi q(m^i)$ we have

$$v^{t^i}_{m^i} \;-\; v_{p^{i-1}} \;=\; v^{t^i}_{q(m^i)} \;-\; v_{p^i}$$

and $u^{t^i}_{m^i} = v_{p^{i-1}} - v_{p^i}$. Therefore

$$u_{a;p;b} \;=\; \textstyle\sum_{1 \leqq i \leqq k}\, u^{t^i}_{m^i} \;=\; v_{p^0} - v_{p^k} \;=\; v_a - v_b\,.$$

Hence $u_{a;p;b}$ is independent of path. In particular $u_{a;p}$ is independent of path.

Conversely let $u \in M^1(F, \mathbb{G}\pi)$. Assume that $u_{a;p}$ is independent of path. Then $v \in C^1(F, \mathbb{G}\pi)$ is well defined by

$$v_a \;=\; u_{a;p}$$

whenever p is an overpath from a to $q(a)$. If $a \in Q$, then p is empty and $v_a = u_{a;p} = 0$; thus $v \in P^1(F, \mathbb{G}\pi)$. If $a = m \in M$, then $p = \{m\}$ and $v_m = u_{m;p} = u_m$; thus $u = Rv$. It remains to show that v is consistent: $v_{a+c} - v_{b+c} = v^c_a - v^c_b$ whenever $\pi a = \pi b$ and $c \in F$.

First let $b = q(a)$. Let p be an overpath from a to b and q be an overpath from $b + c$ to $q(b + c) = q(a + c)$. Then p is an overpath from $a + c$ to $b + c$ and $p + q$ is an overpath from $a + c$ to $q(a + c)$. Hence

$$\begin{aligned} v_{a+c} &= u_{a+c;\,p+q} \\ &= u_{a+c;\,p;\,b+c} + u_{b+c;\,q} = u^c_{a;p} + u_{b+c;\,q} = v^c_a + v_{b+c} \end{aligned}$$

and $v^c_a - v_{a+c} = -v_{b+c} = -v_{q(a)+c}$. If now we assume only $\pi a = \pi b$, then $q(a) = q(b)$ and

$$v^c_a - v_{a+c} = -v_{q(a)+c} = -v_{q(b)+c} = v^c_b - v_{b+c}\,.$$

Thus v is consistent. □

9. Finally, recall that a minimal cocycle is a minimal cochain u such that $u_{a;p}$ is independent of path. Under pointwise addition minimal cocycles constitute a subgroup $MZ^1(F,\mathbb{G}\pi)$ of $MC^1(F,\mathbb{G}\pi)$.

A minimal coboundary is a minimal cochain u for which there exists $g = (g_x)_{x\in X} \in \prod_{x\in X} G_{\pi x}$ such that $u = RDg$; equivalently,

$$u_m = \textstyle\sum_{x\in X,\, x\leqq m} m_x\, g_x^{m-x} - \sum_{x\in X,\, x\leqq q(m)} q(m)_x\, g_x^{q(m)-x}$$

for all $m \in M$. Under pointwise addition minimal coboundaries constitute a subgroup $MB^1(F,\mathbb{G}\pi)$ of $MC^1(F,\mathbb{G}\pi)$.

Lemma 3.9 shows that R induces an isomorphism of $KP^1(F,\mathbb{G}\pi)$ onto $MZ^1(F,\mathbb{G}\pi)$. Since $\operatorname{Im} D \subseteq KP^1(F,\mathbb{G}\pi)$ it follows that $MB^1(F,\mathbb{G}\pi) = \operatorname{Im} RD \subseteq MZ^1(F,\mathbb{G}\pi)$. Then Corollary 3.7 yields

$$H^2(S,\mathbb{G}) \cong KP^1(F,\mathbb{G}\pi)/\operatorname{Im} D \cong MZ^1(F,\mathbb{G}\pi)\,/\,MB^1(F,\mathbb{G}\pi)$$

which is natural in $\mathbb{G}$ since R is natural in $\mathbb{G}$. This proves Theorem 2.1. □

4. DEFINING VECTORS.

In this section we show that minimal cocycles are determined by relations between certain integer vectors, and that the computation of $H^2(S,\mathbb{G})$ is a finite task when S is finitely generated and $\mathbb{G}$ is thin.

1. As before, S is a commutative monoid, $F = F_X$ is the free commutative semigroup on a set X, and $\pi : F \longrightarrow S$ is a surjective homomorphism; $G = G_X$ is the free abelian group on X, whose elements are finite linear combinations $a = \sum_{x\in X} a_x\, x$ with integer coefficients and can be regarded as integer vectors.

By Proposition 1.4, $\mathcal{C}$ is generated by all pairs $(m,\, q(m))$ with $m \in M$, which may be regarded as defining relations of S. The **defining vectors** of $\mathcal{C}$

(or of S) are the integer vectors

$$v(m) = m - q(m) \in G$$

with $m \in M$.

Proposition 4.1. *The subgroup of G generated by the defining vectors is the Rédei group R of $\mathcal{C}$; the universal group of S is isomorphic to G/R.*

Proof. Recall that the Rédei group of $\mathcal{C}$ is

$$R = \{a - b \in G \mid a \;\mathcal{C}\; b\}.$$

Since $\mathcal{C}$ is generated by all pairs $(m, q(m))$ with $m \in M$, it follows from Proposition I.2.9 that $a - b$ is a sum of differences $m - q(m)$ and $q(m) - m$ when $a \;\mathcal{C}\; b$, so that $a - b$ belongs to the subgroup K of G generated by the defining vectors. Hence $R \subseteq K$. Conversely every defining vector $v(m) = m - q(m)$ is in R, since $m \;\mathcal{C}\; q(m)$; hence $K \subseteq R$.

Since $\mathcal{C}$ is generated by all pairs $(m, q(m))$ with $m \in M$, $S \cup \{0\}$ is generated, as a commutative monoid with zero, by the set X subject to all relations $m = q(m)$ $(m \in M)$, with $m, q(m) \neq 0$ in $S \cup \{0\}$. By Proposition III.3.4, $G(S)$ is the abelian group generated by X subject to all relations $m = q(m)$; that is, $G(S) \cong G/K$. □

Proposition 4.2. *Let S have a zero element and $Z = \pi^{-1}0 \subseteq F$ be the zero class. Let K be the subgroup of G generated by all defining vectors $v(m)$ with $m \notin Z$. Then G/K is the universal abelian group $G(S\backslash 0)$ of the partial semigroup $S\backslash 0$.*

Proof. Since Z is a $\mathcal{C}$-class, $m \in Z$ implies $q(m) \in Z$. Since $\mathcal{C}$ is generated by all pairs $(m, q(m))$ with $m \in M$, S is generated, as a commutative monoid with zero, by the set X subject to all relations $m = q(m)$ $(m \in M\backslash Z)$ and $m = 0$ $(m \in M \cap Z)$. By Proposition III.3.4, $G(S\backslash 0)$ is the abelian group generated by X subject to all relations $m = q(m)$ $(m \in M\backslash Z)$; that is, $G(S\backslash 0) \cong G/K$. □

2. We now consider relations between defining vectors. We distinguish **vector relations**

$$R(r): \qquad \textstyle\sum_{m\in M} r_m\, v(m) = 0,$$

in which every r_m is an integer and $r_m = 0$ for almost all m, and **positive relations**

$$R(r,s): \qquad \textstyle\sum_{m\in M} r_m\, v(m) = \sum_{m\in M} s_m\, v(m),$$

in which every r_m and every s_m is a nonnegative integer and $r_m = s_m = 0$ for almost all m. The two types are essentially equivalent.

Lemma 4.3. *Every positive relation is a trivial consequence of a finite sum of minimal positive relations; if* X *is finite, then there are only finitely many minimal positive relations.*

Proof. When $R = R(r,s) : \sum_{m\in M} r_m\, v(m) = \sum_{m\in M} s_m\, v(m)$ is a positive relation, let the *weight* of R be $\sum_{m\in M} (r_m + s_m)$; call R *nontrivial* when $r_m \neq s_m$ for some m, and *essential* when it is nontrivial but there is no m such that $r_m > 0$ and $s_m > 0$ (so that no cancellation is possible in $R(r,s)$). Every nontrivial positive relation can be simplified by cancellation in G into an essential positive relation; hence every nontrivial positive relation $R(r,s)$ is a *trivial consequence* of an essential positive relation $R(e,f)$ ($r_m - e_m = s_m - f_m \geqq 0$ for all m).

Positive relations are ordered coefficientwise: $R(p,q) \leqq R(r,s)$ if and only if $p_m \leqq r_m$ and $q_m \leqq s_m$ for all m. A *minimal* positive relation is a minimal nontrivial positive relation. Minimal positive relations are essential.

When $R(p,q) < R(r,s)$, then $\sum_{m\in M} p_m\, v(m) = \sum_{m\in M} q_m\, v(m)$ and $\sum_{m\in M} r_m\, v(m) = \sum_{m\in M} s_m\, v(m)$ imply

$$\textstyle\sum_{m\in M} (r_m - p_m)\, v(m) \;=\; \sum_{m\in M} (s_m - q_m)\, v(m),$$

so that $R(r-p,\ s-q)$ is a positive relation; then $R(r,s)$ is the sum of $R(p,q)$ and $R(r-p,\ s-q)$. If $R(r,s)$ is essential, then so are $R(p,q)$ and $R(r-p,\ s-q)$. Thus an essential positive relation which is not minimal is a sum of essential positive relations of lesser weight. Therefore every essential positive relation is a finite sum of minimal positive relations, and every positive relation is a consequence of a finite sum of minimal positive relations.

Under pointwise addition, pairs (r,s) of families $r = (r_m)_{m\in M}$, $s = (s_m)_{m\in M}$ of nonnegative integers constitute a finitely generated free commutative monoid F'. Minimal positive relations constitute an antichain of F'. If X is finite, then, by Dickson's Theorem, M is finite, all antichains of F' are finite, and there are only finitely many minimal positive relations. □

3. Relations between defining vectors arise when there is more than one path from an element of F to another. When $p : m^1, \ldots, m^k$ is an overpath from a to b, we saw that

$$a - b \;=\; \textstyle\sum_{1\leqq i\leqq k} (m^i - q(m^i)) \;=\; \sum_{1\leqq i\leqq k} v(m^i)$$

We write this equality as

$$a - b \;=\; \textstyle\sum_{m\in M} p_m\, v(m),$$

where p_m is the number of appearances of m in the sequence $p : m^1, \ldots, m^k$.

When p and q are two overpaths from a to b, then

$$R(p,q) : \sum_{m \in M} p_m \, v(m) = a - b = \sum_{m \in M} q_m \, v(m)$$

is a positive relation and there is a vector relation

$$R(p-q) : \sum_{m \in M} (p_m - q_m) \, v(m) \;=\; 0.$$

A relation between defining vectors is **realized at** $a \in F$ when it arises in this fashion from a pair of overpaths from a to some b. Thus a vector relation $R(r)$: $\sum_{m \in M} r_m \, v(m) = 0$ is realized at a when there exist $b \in F$ and overpaths p and q from a to b such that $r_m = p_m - q_m$ for all m (so that $R(r) = R(p-q)$); a positive relation $R(r,s) : \sum_{m \in M} r_m \, v(m) = \sum_{m \in M} s_m \, v(m)$ is realized at $a \in F$ when there exist $b \in F$ and overpaths p and q from a to b such that $r_m - p_m = s_m - q_m \geqq 0$ for all m (so that $R(r,s)$ is a trivial consequence of $R(p,q)$).

Lemma 4.4. *When $R(r,s)$ is realized at a, then $r_m = s_m$ whenever $\pi m \ngeqq_{\mathcal{H}} \pi a$.*

Proof. Let $b \in F$ and p, q be overpaths from a to b such that $r_m - p_m = s_m - q_m \geqq 0$ for all $m \in M$. If $p_m > 0$ or $q_m > 0$ (if m appears in p or in q), then $m \geqq c$ for some $c \in C_a$ and $\pi m \geqq_{\mathcal{H}} \pi a$. Therefore $\pi m \ngeqq_{\mathcal{H}} \pi a$ implies $p_m = q_m = 0$ and $r_m = s_m$. □

A relation of either kind is **realized in** a $\mathcal{C}$-class C when it is realized at some $a \in C$ (then $b \in C$ in the above). In the case of a vector relation it may be assumed that $b = q(a)$, since an overpath from b to $q(a) = q(b)$ can be added to p and q if necessary, without changing $p_m - q_m$.

The trivial relation $0 = 0$ is realized at every $a \in F$. A relation which is realized at a is realized at $a + c$ for every $c \in F$, since overpaths from a to b are also overpaths from $a + c$ to $b + c$. More surprisingly:

Proposition 4.5. *Every relation between defining vectors is realized in some $\mathcal{C}$-class (in the zero class, if S has a zero element).*

Proof. First we prove the following: for every family $r = (r_m)_{m \in M}$ of nonnegative integers (with $r_m = 0$ for almost all $m \in M$) there exists an overpath p such that $p_m = r_m$ for all m. This is shown by induction on $|r| = \sum_{m \in M} r_m$. If $|r| = 0$, then the empty ovepath from any a to a serves. If $|r| > 0$, then $r_n > 0$ for some $n \in M$, and the induction hypothesis yields an overpath $p : m^1, \ldots, m^k$ from some $a \in F$ to some $b \in F$ such that $p_m = r_m$

for all $m \neq n$ and $p_n = r_n - 1$. In F there is an element c such that $c \geqq b$ and $c \geqq n$ (for instance, $b \vee n$). Then $p : m^1, \dots, m^k$ is an overpath from $a + (c - b)$ to $b + (c - b) = c$, $\{n\}$ is an overpath from $n + (c - n) = c$ to $q(n) + (c - n)$, and $q = p + \{n\} : m^1, \dots, m^k, n$ is an overpath from $a + (c - b)$ to $q(n) + (c - n)$. We see that $q_m = r_m$ for all m.

If now $R(r, s)$ is a positive relation, then the above provides an overpath p from some a to some b such that $p_m = r_m$ for all m and an overpath q from some c to some d such that $q_m = s_m$ for all m. Then

$$\begin{aligned} a - b &= \textstyle\sum_{m \in M} p_m\, v(m) = \sum_{m \in M} r_m\, v(m) \\ &= \textstyle\sum_{m \in M} s_m\, v(m) = \sum_{m \in M} q_m\, v(m) = c - d. \end{aligned}$$

In F there is an element e such that $e \geqq a$ and $e \geqq c$; if S has a zero element, then the zero class Z is an ideal of F and we can arrange that $e \in Z$. Let $f = (e - a) + b = (e - c) + d$. Then p and q are overpaths from e to f, and $p_m = r_m$, $q_m = s_m$ for all m. In particular $R(r, s)$ is realized in C_e. Thus every positive relation is realized in some $\mathcal{C}$-class (in the zero class, if S has a zero element). Then so is every vector relation. □

4. We now show that minimal cocycles are determined by relations between the defining vectors, when the coefficient functor is thin.

Let $\mathbb{G} = (G, \gamma)$ be an abelian group valued functor on $H(S)$. When $u = (u_m)_{m \in M}$ is a minimal cochain and $p : m^1, \dots, m^k$ is as overpath from a to b, then

$$u_{a;p;b} = \textstyle\sum_{1 \leqq i \leqq k} u_{m^i}^{t^i}$$

where $t^i = p^i - q(m^i) = p^{i-1} - m^i$ is provided by the corresponding path

$$a = p^0 \xrightarrow{m^1} p^1 \xrightarrow{m^2} \cdots \xrightarrow{m^k} p^k = b.$$

If $\mathbb{G}$ is thin, then $\gamma_{\pi m^i,\, \pi t^i}$ depends only on m^i and $\pi(m^i + t^i) = \pi p^{i-1} = \pi a$ and is denoted by $\gamma_{\pi a}^{\pi m^i}$; hence

$$\begin{aligned} u_{a;p;b} &= \textstyle\sum_{1 \leqq i \leqq k} u_{m^i}^{t^i} = \sum_{m \in M,\, \pi m \geqq_{\mathcal{H}} \pi a} p_m\, \gamma_{\pi a}^{\pi m} u_m \\ &= \textstyle\sum_{m \in M_a} p_m\, \gamma_{\pi a}^{\pi m} u_m\,, \end{aligned}$$

where

$$M_a = \{ m \in M \mid \pi m \geqq_{\mathcal{H}} \pi a \}.$$

Recall that $\pi m \geqq_{\mathcal{H}} \pi a$ when m appears in p (when $p_m > 0$). Thus u is a minimal cocycle if and only if

$$Z(p,q,a): \qquad \sum_{m\in M_a} p_m\, \gamma_{\pi a}^{\pi m} u_m \;=\; \sum_{m\in M_a} q_m\, \gamma_{\pi a}^{\pi m} u_m$$

whenever p and q are overpaths from $a \in F$ to $q(a)$.

Proposition 4.6. *When $\mathbb{G}$ is thin, a minimal cochain u is a minimal cocycle if and only if*

$$Z(r,s,a): \qquad \sum_{m\in M_a} r_m\, \gamma_{\pi a}^{\pi m} u_m \;=\; \sum_{m\in M_a} s_m\, \gamma_{\pi a}^{\pi m} u_m$$

holds whenever $a \in F$ and the positive relation $R(r,s)$ is realized at a.

Proof. If $R(r,s)$ is realized at a, then $R(r,s)$ is a trivial consequence of $R(p,q)$ for some overpaths p and q from a to some $b \in C_a$; then $Z(r,s,a)$ is a trivial consequence of $Z(p,q,a)$, since $p_m = q_m = 0$ and $r_m = s_m$ when $m \notin M_a$ by Lemma 4.5. By cancellation in $G_{\pi a}$, $Z(r,s,a)$ holds if and only if $Z(p,q,a)$ holds. Similarly, when an overpath from b to $q(b) = q(a)$ is added to p and q (to obtain overpaths from a to $q(a)$), then $Z(p,q,a)$ is replaced by an equivalent condition. Hence u is a minimal cocycle if and only if every $Z(r,s,a)$ holds. □

On the other hand, minimal coboundaries satisfy relations between defining vectors regardless of whether they are realized. Call a positive relation $R(r,s)$ **verifiable at** $a \in F$ when $r_m = s_m$ whenever $m \notin M_a$ (whenever $\pi m \not\geqq_{\mathcal{H}} \pi a$). By Lemma 4.4, every positive relation which is realized at a is verifiable at a.

Proposition 4.7. *When $\mathbb{G}$ is thin and u is a minimal coboundary, then*

$$Z(r,s,a): \qquad \sum_{m\in M_a} r_m\, \gamma_{\pi a}^{\pi m} u_m \;=\; \sum_{m\in M_a} s_m\, \gamma_{\pi a}^{\pi m} u_m$$

holds whenever $a \in F$ and the positive relation $R(r,s)$ is verifiable at a.

Proof. Let u be a minimal coboundary, so that

$$u_m \;=\; \sum_{x\in X,\, x\leqq m} m_x\, g_x^{\pi(m-x)} \;-\; \sum_{x\in X,\, x\leqq q(m)} q(m)_x\, g_x^{\pi(q(m)-x)}$$

for all $m = \sum_{x\in X} m_x x \in M$, where $g_x \in G_{\pi x}$ for all $x \in X$. Since $\mathbb{G}$ is thin and $\pi m = \pi\, q(m)$,

$$u_m \;=\; \sum_{x\in X,\, x\leqq m} m_x\, \gamma_{\pi m}^{\pi x} g_x \;-\; \sum_{x\in X,\, x\leqq q(m)} q(m)_x\, \gamma_{\pi m}^{\pi x} g_x$$

for all $m \in M$.

Let $R(r,s): \sum_{m\in M} r_m\, v(m) = \sum_{m\in M} s_m\, v(m)$ be verifiable at a. Then $\sum_{m\notin M_a} r_m\, v(m) = \sum_{m\notin M_a} s_m\, v(m)$, since $r_m = s_m$ when $m \notin M_a$;

hence $\sum_{m\in M_a} r_m\, v(m) = \sum_{m\in M_a} s_m\, v(m)$ and

$$\textstyle\sum_{m\in M_a} r_m\, v(m)_x \;=\; \sum_{m\in M_a} s_m\, v(m)_x$$

for every $x \in X$. Let $X_a = \{x \in X \mid \pi x \geqq_{\mathcal{H}} \pi a\}$. Then $x \leqq m \in M_a$ implies $\pi x \geqq_{\mathcal{H}} \pi m \geqq_{\mathcal{H}} \pi a$ and $x \in X_a$; $x \leqq q(m) \in M_a$ implies $x \in X_a$; and

$$\begin{aligned}
&\textstyle\sum_{m\in M_a} r_m\, \gamma_{\pi a}^{\pi m} u_m \\
&= \textstyle\sum_{m\in M_a} \sum_{x\in X,\, x\leqq m} r_m\, m_x\, \gamma_{\pi a}^{\pi x} g_x \\
&\quad - \textstyle\sum_{m\in M_a} \sum_{x\in X,\, x\leqq q(m)} r_m\, q(m)_x\, \gamma_{\pi a}^{\pi x} g_x \\
&= \textstyle\sum_{m\in M_a} \sum_{x\in X_a} r_m\, m_x\, \gamma_{\pi a}^{\pi x} g_x \\
&\quad - \textstyle\sum_{m\in M_a} \sum_{x\in X_a} r_m\, q(m)_x\, \gamma_{\pi a}^{\pi x} g_x\,,
\end{aligned}$$

since $m_x = 0$ if $x \nleqq m$ and $q(m)_x = 0$ if $x \nleqq q(m)$,

$$\begin{aligned}
&= \textstyle\sum_{m\in M_a} \sum_{x\in X_a} r_m\, v(m)_x\, \gamma_{\pi a}^{\pi x} g_x \\
&= \textstyle\sum_{m\in M_a} \sum_{x\in X_a} s_m\, v(m)_x\, \gamma_{\pi a}^{\pi x} g_x \\
&= \textstyle\sum_{m\in M_a} \sum_{x\in X,\, x\leqq m} s_m\, m_x\, \gamma_{\pi a}^{\pi x} g_x \\
&\quad - \textstyle\sum_{m\in M_a} \sum_{x\in X,\, x\leqq q(m)} s_m\, q(m)_x\, \gamma_{\pi a}^{\pi x} g_x \\
&= \textstyle\sum_{m\in M_a} r_m\, \gamma_{\pi a}^{\pi m} u_m\,,
\end{aligned}$$

since $m_x = 0$ if $x \nleqq m$ and $q(m)_x = 0$ if $x \nleqq q(m)$. Thus u satisfies $Z(r,s,a)$. □

5. Computing $H^2(S,\mathbb{G})$ with Theorem 2.1 still looks like an infinite task even when S is finite, since independence of path must be established at every $a \in F$. When F is finitely generated and $\mathbb{G}$ is thin, we show that minimal cocycles are characterized by finitely many conditions $u_{a;p} = u_{a;q}$; hence the computation of $H^2(S,\mathbb{G})$ a finite task. It seems likely that this result holds even if $\mathbb{G}$ is not thin.

Proposition 4.8. *When $\mathbb{G}$ is thin and X is finite, a minimal cochain u is a minimal cocycle if and only if if satisfies finitely many conditions $Z(r,s,a)$, in which $a \in F$ and the positive relation (r,s) is realized at a.*

Proof. Let F'' be the set of all ordered pairs (r,a) where $r = (r_m)_{m\in M}$ is a family of nonnegative integers and $a \in F$. Under pointwise addition, F'' is a free commutative monoid $F'' \cong F_M \times F$.

Realizability yields a binary relation $\mathcal{R}$ on F'':

$$(r,a)\,\mathcal{R}\,(s,b) \iff a = b \text{ and } R(r,s) \text{ is realized at } a.$$

We see that $\mathcal{R}$ is reflexive and symmetric. If moreover $(r,a)\,\mathcal{R}\,(s,b)$, so that $R(r,s)$ is realized at a, then, for any t, $R(r+t,\ s+t)$ is a trivial consequence of $R(r,s)$ and is realized at a, $R(r+t,\ s+t)$ is realized at $a+c$ for any $c \in F$, and $(r+t,\ a+c)\,\mathcal{R}\,(s+t,\ b+c)$; thus $\mathcal{R}$ admits addition.

By Proposition I.2.9, the congruence $\overline{\mathcal{R}}$ on F'' generated by $\mathcal{R}$ is given by: $(r,a)\,\overline{\mathcal{R}}\,(s,b)$ if and only if there exist $k \geqq 0$ and $(r^0,a^0),\dots,(r^k,a^k) \in F''$ such that $(r,a) = (r^0,a^0)$, $(r^{i-1},a^{i-1})\,\mathcal{R}\,(r^i,a^i)$ for all $1 \leqq i \leqq k$, and $(r^k,a^k) = (s,b)$. Then $a = a^0 = \dots = a^k = b$ and the equalities

$$\textstyle\sum_{m\in M} r^0_m\, v(m) \;=\; \sum_{m\in M} r^1_m\, v(m) \;=\; \dots \;=\; \sum_{m\in M} r^k_m\, v(m)$$

show that $R(r,s)$ is a consequence of $R(r^0,r^1)$, $R(r^1,r^2)$, ..., $R(r^{k-1},r^k)$.

By Proposition 4.6, a minimal cochain u is a minimal cocycle if and only if it satisfies $Z(r,s,a)$ whenever $(r,a)\,\mathcal{R}\,(s,a)$. If $(r,a)\,\overline{\mathcal{R}}\,(s,a)$, then in the above $Z(r^0,r^1,a)$, $Z(r^1,r^2,a)$, ..., $Z(r^{k-1},r^k,a)$ hold in $G_{\pi a}$; by the equalities

$$\begin{aligned}\textstyle\sum_{m\in M_a} r^0_m\, \gamma^{\pi m}_{\pi a} u_m \;&=\; \textstyle\sum_{m\in M_a} r^1_m\, \gamma^{\pi m}_{\pi a} u_m \\ &=\; \dots \;=\; \textstyle\sum_{m\in M_a} r^k_m\, \gamma^{\pi m}_{\pi a} u_m\,,\end{aligned}$$

$Z(r,s,a)$ is a consequence of $Z(r^0,r^1,a)$, $Z(r^1,r^2,a)$, ..., $Z(r^{k-1},r^k,a)$ and holds in $G_{\pi a}$ if u is a minimal cocycle. Hence a minimal cochain u is a minimal cocycle if and only if it satisfies $Z(r,s,a)$ whenever $(r,a)\,\overline{\mathcal{R}}\,(s,a)$.

Since M is finite it follows from Rédei's Theorem that $\overline{\mathcal{R}}$ is finitely generated. Therefore a minimal cochain u is a minimal cocycle if and only if it satisfies finitely many conditions $Z(r,s,a)$, with $a \in F$ and $(r,a)\,\overline{\mathcal{R}}\,(s,a)$, each of which is a consequence of finitely many conditions $Z(r,s,a)$, with $a \in F$ and (r,s) realized at a. □

A more explicit choice of conditions to verify is given in Grillet [1995F] but no longer seems particularly helpful.

6. We conclude this section with an example. More general examples are given in Grillet [2000T].

Example 4.9. Let S be the commutative nilmonoid

$$S \;=\; \langle\, c,d \mid c^3 = c^2d = cd^2 = d^4 = 0,\ c^2 = cd = d^3 \,\rangle;$$

S is the Volkov semigroup (Example III.3.6) with an identity adjoined.

Let $X = \{x,y\}$ and $\pi x = c$, $\pi y = d$. Then $\mathcal{C} = \ker \pi$ has four one element classes, one three element class $C = \{2x, x+y, 3y\} = \pi^{-1}(cd)$, and one infinite class $J = \pi^{-1}0$ which is the ideal of F generated by $\{3x, 2x+y, x+2y, 4y\}$.

The lexicographic order $\sqsubseteq$ on F

$$ix + jy \sqsubset kx + ly \iff i < k, \quad \text{or} \quad i = k,\ j < l$$

is a compatible well order on F. Under $\sqsubseteq$ the least element of C is $3y$; the least element of J is $4y$.

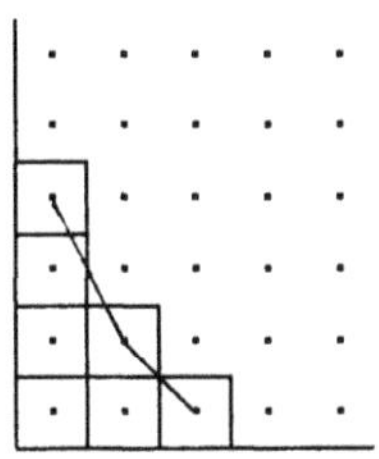

Example 4.9

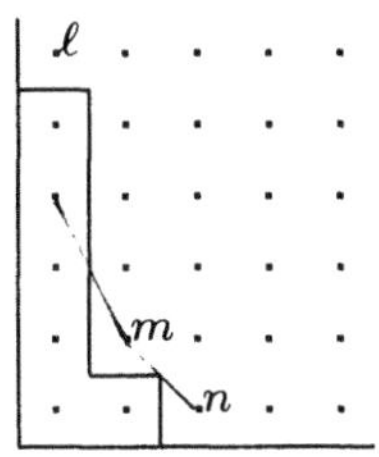

Q and M

Thus Q is the coideal generated by $4y$ and x; M and the defining vectors are given by the table

m	$q(m)$	$v(m)$
$l = 5y$	$4y$	y
$m = x+y$	$3y$	$x - 2y$
$n = 2x$	$3y$	$2x - 3y$

The defining vectors $v(m)$ with $m \notin J$ are $v(m)$ and $v(n)$. They constitute a basis of G since $\begin{vmatrix} 1 & -2 \\ 2 & -3 \end{vmatrix} = 1$. Hence the universal abelian group of $S\backslash 0$ is trivial, by Proposition 4.2.

We see that $v(n) = 2v(m) + v(l)$. The only $\mathcal{C}$-class in which nontrivial positive relations are realized is J; by Proposition 4.5, every positive relation is realized in J. (More sophisticated examples are given in the next chapter.)

Let $\mathbb{G}$ be a thin abelian group valued functor on $H(S)$. A minimal cochain u consists of $u_l \in G_0$, $u_m \in G_{cd}$, and $u_n \in G_{cd}$.

In J, $2x + y \xrightarrow{n} 4y$ and $2x + y \xrightarrow{m} x + 3y \xrightarrow{m} 5y \xrightarrow{l} 4y$ are paths from $a = 2x + y$ to $q(a) = 4y$. Hence $p : n$ and $q : m, m, l$ are overpaths from a to $q(a)$ (and the relation $v(n) = 2v(m) + v(l)$ is realized at $2x + y$).

Therefore minimal cocycles satisfy

$$u_n^0 = u_{a;p} = u_{a;q} = 2u_m^0 + u_l\,;$$

this is according to Proposition 4.6.

Conversely let u be a minimal cochain such that $u_n^0 = 2u_m^0 + u_l$. Let $a \in F$ and let p and q be overpaths from a to $q(a)$. We may assume that $a \in J$ (otherwise $p = q$). We have

$$\begin{aligned} a - q(a) &= p_l\, y + p_m\,(x - 2y) + p_n\,(2x - 3y) \\ &= q_l\, y + q_m\,(x - 2y) + q_n\,(2x - 3y); \end{aligned}$$

hence $p_m + 2p_n = q_m + 2q_n$ and $p_l - 2p_m - 3p_n = q_l - 2q_m - 3q_n$. Adding twice the first equality to the second yields $p_l + p_n = q_l + q_n$. Hence

$$\begin{aligned} u_{a;p} &= p_l\, u_l \;+\; p_m\, u_m^0 \;+\; p_n\, u_n^0 \\ &= (p_l + p_n)\, u_l \;+\; (p_m + 2p_n)\, u_m^0 \\ &= (q_l + q_n)\, u_l \;+\; (q_m + 2q_n)\, u_m^0 \\ &= q_l\, u_l \;+\; q_m\, u_m^0 \;+\; q_n\, u_n^0 \;=\; u_{a;q}\,. \end{aligned}$$

Thus minimal cocycles are characterized by the single condition $u_n^0 = 2u_m^0 + u_l$; this is according to Proposition 4.8. Hence there is an isomorphism $u \longmapsto (u_m, u_n)$ of $MZ^1(F, \mathbb{G}\pi)$ onto $G_{cd} \oplus G_{cd}$. A peek at minimal coboundaries suggests that u_m and u_n are uniquely determined by $g = u_n - 2u_m$ and $h = 2u_n - 3u_m$ (namely, $u_m = h - 2g$ and $u_n = 2h - 3g$), and provides a more useful isomorphism $\theta : u \longmapsto (u_n - 2u_m,\; 2u_n - 3u_m)$ of $MZ^1(F, \mathbb{G}\pi)$ onto $G_{cd} \oplus G_{cd}$.

Next, u is a minimal coboundary if and only if there exist $g \in G_c$ and $h \in G_d$ such that $u_l = 5g^0 - 4g^0 = g^0$, $u_m = (g^d + h^c) - 3g^d = h^c - 2g^d$, and $u_n = 2h^c - 3g^d$. Then $\theta(u) = (g^d, h^c)$ and θ sends $MB^1(F, \mathbb{G}\pi)$ onto $\operatorname{Im}\gamma_{c,d} \oplus \operatorname{Im}\gamma_{d,c} \subseteq G_{cd} \oplus G_{cd}$. Hence

$$H^2(S, \mathbb{G}) \;\cong\; (G_{cd} / \operatorname{Im}\gamma_{c,d}) \oplus (G_{cd} / \operatorname{Im}\gamma_{d,c}).$$

If $\mathbb{G}$ is surjecting (as well as thin), then $\gamma_{1,c}$, $\gamma_{1,d}$, and $\gamma_{1,cd}$ are surjective; hence $\gamma_{c,d}$, and $\gamma_{d,c}$ are surjective and $H^2(S, \mathbb{G}) = 0$.

5. PARTIALLY FREE SEMIGROUPS.

Partially free semigroups were defined in Section X.6. At this time they constitute the only large class of finite commutative semigroups with a formula for $H^2(S,\mathbb{G})$: namely,

$$H^2(S,\mathbb{G}) \;\cong\; \bigoplus_{c\in \mathrm{Irr}\,(S)} G_{e(c)}/\mathrm{Im}\,\gamma^c_{e(c)}\,,$$

where $\mathbb{G}$ is thin, Irr (S) is the set of all irreducible elements of S, and $e(c)$ is the idempotent in the archimedean component of c. This result is from Grillet [1995P].

Other formulas yield $H^2(S,\mathbb{G})$ when S has one defining relation (Proposition 2.4) or is cyclic (Corollary 2.5). But no such formula seems to exist for semigroups with two generators (Grillet [2000T]).

1. Let S be finite and partially free. By Corollary XII.5.5 we may assume that S is a monoid. Since S is group-free, $\leqq_{\mathcal{H}}$ is a partial order relation on S, which we denote by $\leqq$.

By Proposition X.2.2, S is generated by Irr (S) and has a standard presentation $\pi : F = F_X \longrightarrow S$, where X is finite, π is injective on X, and $\pi(X) = \mathrm{Irr}\,(S)$. The direction set $\mathcal{D}$, extent cells E_A, coideals H_A, and trace congruences $\mathcal{C}_A$ of the congruence $\mathcal{C}$ induced by π are as in Chapter X. The idempotents of S are all $e_A = \pi(I_A)$ with $A \in \mathcal{D}$, where $I_A = E_A \cap A$. Since S is partially free, all trace congruences are Rees congruences and

$$a \;\mathcal{C}\; b \text{ if and only if } a, b \in E_A \text{ and } p'_A a = p'_A b \text{ for some } A \in \mathcal{D}.$$

Put any total order $\preccurlyeq$ on X and order F lexicographically: $\sum_{x\in X} a_x\, x \sqsubset \sum_{x\in X} b_x\, x$ if and only if there exists $t \in X$ such that $a_x = b_x$ for all $x \prec t$ and $a_t < b_t$. (Then $x \sqsubset y$ in F if and only if $x \succ y$ in X.) Since X is finite, $\sqsubseteq$ is a compatible well order on F.

We show that the defining vectors contain a basis of G.

Lemma 5.1. *Let $x \in X$, $c = \pi x$, and D be the smallest element of $\mathcal{D}$ that contains x. There exists $m(x) \in M$ such that: $m(x) \in E_D \cap D$; $\pi\, m(x) = e(c)$; if $x \in A \in \mathcal{D}$, then $\pi\, m(x) \geqq \pi a$ for all $a \in E_A$; and*

$$v(m(x)) \;=\; x \;+\; \textstyle\sum_{y\in X} t_{x,y}\, y$$

with $t_{x,y} = 0$ unless $y \prec x$ and $y \in D$.

Proof. First take any $m \in I_D$, so that $\pi m = e_D$.

If $A \in \mathcal{D}$, $A \subset D$, then $x \notin A$ by the choice of D. By (E2), $kx \in E_D$ for some $k > 0$. Then $c^k = \pi(kx) = e_D$ and $\pi m = e(c)$.

If $x \in A \in \mathcal{D}$, then $D \subseteq A$, $\epsilon(\pi a) = e_A$ for all $a \in E_A$ by Proposition X.3.4, and $\pi m = e_D \geqq e_A \geqq \pi a$ by Proposition X.3.3.

We now choose $m(x) = m \in I_D$ as follows. Let w be the least element of the $\mathcal{C}$-class I_D under $\sqsubseteq$. Since $x \in D = D(w)$ we have $w + x \in I_D$. Let $m(x) = m$ be least under $\sqsubseteq$ such that $m \in I_D$ and $m_x = w_x + 1$. In particular, $w \sqsubset m \sqsubseteq w + x$.

We show that $m \in M$. First $m \notin Q = \{a \in F \mid q(a) = a\}$, since $w \,\mathcal{C}\, m$ and $w \sqsubset m$. To prove that m is minimal with this property (under $\leqq$) it suffices to show that $m - y \in Q$ for every $y \in X$, $y \leqq m$. Note that $y \leqq m$ implies $y \in D$.

Assume $m - y \in E_D$. If $y \neq x$, then $(m - y)_x = m_x = w_x + 1$; since $m - y \in I_D$ and $m - y \sqsubset m$ this contradicts the choice of m. Therefore $y = x$. Hence $m - x \in I_D$, $m - x \sqsupseteq w$, $m \sqsupseteq w + x$, $m = w + x$, and $m - y = m - x = w \in Q$.

Now let $m - y \notin E_D$. Then $y \neq x$. Also $m - y \in E_B$ for some $B \in \mathcal{D}$, $B \subset D$. Then $D \nsubseteq B$ and $x \notin B$. Let $q = q(m - y)$. Then $q \sqsubseteq m - y$ and $q \,\mathcal{C}\, m - y$; hence $q \in E_B$, and $q \in D$ since D is a union of $\mathcal{C}$-classes. Since S is partially free we also have $p'_B q = p'_B(m - y)$. In particular $q_x = (m - y)_x = m_x$. Hence $q + y$ has the following properties: $q + y \in D$, $q + y \in E_D$ (since $q + y \;\mathcal{C}\; m$), $q + y \sqsubseteq m$, and $(q + y)_x = q_x = m_x = w_x + 1$. By the choice of m, $q + y = m$, and $m - y = q \in Q$.

This proves $m \in M$. We have $w = q(m)$. Hence $v(m)_x = m_x - w_x = 1$. Also $v(m)_y = m_y - w_y = 0$ if $y \notin D$, since $m, w \in D$. Since $w \sqsubset m$ there exists $t \in X$ such that $w_y = m_y$ for all $y \prec t$ and $w_t < m_t$; in particular $t \preccurlyeq x$. If $t \prec x$, then $w_y = m_y$ for all $y \prec t$ and $w_t < m_t$ implies $w + x \sqsubset m$, whereas $m \sqsubseteq w + x$; therefore $t = x$, and $v(m)_y = m_y - w_y = 0$ for all $y \prec x$. □

Lemma 5.1 implies that $\big(v(m(x))\big)_{x \in X}$ is a basis of G. In fact:

Corollary 5.2. *For every $A \in \mathcal{D}$, $\big(v(m(x))\big)_{x \in X \cap A}$ is a basis of G_A.*

Proof. $G_A = G(A) \subseteq G$ is the free abelian group on $X \cap A$. When $m \in M \cap E_A$, then $q(m) \in E_A$ and $m \,\mathcal{C}\, q(m)$ implies $p'_A m = p'_A\, q(m)$ since S is partially free; hence $v(m) \in G_A$. In particular $v(m(x)) \in G_A$ for all

$x \in X \cap A$. When $x \in X \cap A$, then $D \subseteq A$ in Lemma 5.1 and

$$v(m(x)) = x + \textstyle\sum_{y \in X \cap A,\, y \prec x} t_{x,y}\, y.$$

Hence the defining vectors $v(m(x))$ with $x \in X \cap A$ constitute a basis of G_A. □

2. By Corollary 5.2 there is for every $m \in M \cap E_A$ an equality

$$v(m) = \textstyle\sum_{x \in A} k_{m,x}\, v(m(x)) \in G_A$$

with integer coefficients $k_{m,x}$.

Lemma 5.3. *For every $m \in M \cap E_A$ the vector relation*

$$v(m) = \textstyle\sum_{x \in A} k_{m,x}\, v(m(x)) \qquad (*)$$

is realized in the $\mathcal{C}$-class C_m of m.

Proof. If $m = m(x)$ for some $x \in A$, then $(*)$ is trivial, and is realized in C_m. Hence we may assume that $m \neq m(x)$ for all $x \in A$.

We have $m + A \subseteq C_m$, since $m \in E_A$. Let $p : m^1, \dots, m^k$ consist of m and of $-k_{m,x}$ copies of $m(x)$ for every $x \in X \cap A$ with $k_{m,x} < 0$, arranged in any order. Let $q : n^1, \dots, n^l$ similarly consist of $k_{m,x}$ copies of $m(x)$ for every $x \in X \cap A$ with $k_{m,x} > 0$, arranged in any order. Then $\sum_i v(m^i) = \sum_j v(n^j)$, by $(*)$. Also $p_m - q_m = 1$ and $q_{m(x)} - p_{m(x)} = k_{m,x}$ for every $x \in X \cap A$. If we can show that p and q are overpaths from some $a \in C_m$ to some $b \in C_m$, then they will realize $(*)$.

We have $m(x) \in A$ for all $x \in X \cap A$, and $m^i \in A$ for all $i > 1$. Also $v(m^i) \in G_A$ by Corollary 5.2. If $a \in m + A$ has sufficiently large coordinates, we can arrange by induction on i that $p^{i-1} \geqq m^i$ and $p^i - m = p^{i-1} - m + v(m^i) \in A$ for all i. Then p is an overpath from a to $b = a + \sum_i v(m^i) \in m + A$. Similarly, if $a \in m + A$ has sufficiently large coordinates, then q is an overpath from a to $a + \sum_j v(n^j)$, and $\sum_j v(n^j) = b$ since $\sum_i v(m^i) = \sum_j v(n^j)$). □

3. Now let $\mathbb{G}$ be a thin abelian group valued functor on $H(S)$.

Proposition 5.4. *When S is partially free and $\mathbb{G}$ is thin, a minimal cochain $u = (u_m)_{m \in M}$ is a minimal cocycle if and only if*

$$u_m = \textstyle\sum_{x \in X \cap A} k_{m,x}\, \gamma_{\pi m}^{\pi m(x)} u_{m(x)} \qquad (**)$$

whenever $A \in \mathcal{D}$ and $m \in M \cap E_A$.

Proof. By Proposition 4.6, a minimal cocycle $u = (u_m)_{m \in M}$ must satisfy

(**) for all $A \in \mathcal{D}$ and $m \in M \cap E_A$, since (*) is realized at C_m by Lemma 5.3. (Note that $\pi m(x) \geqq \pi m$ for all $x \in X \cap A$, by Lemma 5.1, since $m \in E_A$.)

Conversely assume that (**) holds for all $A \in \mathcal{D}$ and $m \in M \cap E_A$. Let p and q be overpaths from a to b. Then $\sum_{m \in M} p_m\, v(m) = b - a = \sum_{m \in M} q_m\, v(m)$, with $m \in M_a = \{m \in M \mid \pi m \geqq \pi a\}$ whenever $p_m > 0$ or $q_m > 0$. Also $a \,\mathcal{C}\, b$ and $a, b \in E_A$ for some $A \in \mathcal{D}$.

If m appears in either p or q, then $m \leqq c$ for some $c \in C_a \subseteq E_A$ and $m \in E_B$ for some $B \subseteq A$ by (E2). Then

$$v(m) = \textstyle\sum_{x \in X \cap B} k_{m,x}\, v(m(x)) \quad \text{and} \quad u_m = \textstyle\sum_{x \in X \cap B} k_{m,x}\, \gamma_{\pi m}^{\pi m(x)}\, u_{m(x)}\,.$$

(By Lemma 5.1, $\pi m(x) \geqq \pi m$ when $x \in X \cap B$ and $m \in E_B$). Let $k_{m,x} = 0$ when $m \in E_B$ and $x \in X \backslash B$. Then $v(m) = \sum_{x \in X \cap A} k_{m,x}\, v(m(x))$ and

$$\begin{aligned} \textstyle\sum_{m \in M} p_m\, v(m) &= \textstyle\sum_{m \in M,\, x \in X \cap A} p_m\, k_{m,x}\, v(m(x)), \\ \textstyle\sum_{m \in M} q_m\, v(m) &= \textstyle\sum_{m \in M,\, x \in X \cap A} q_m\, k_{m,x}\, v(m(x)). \end{aligned}$$

Since the defining vectors $v(m(x))$ with $x \in X \cap A$ constitute a basis of G_A (Corollary 5.2), this implies

$$\textstyle\sum_{m \in M} p_m\, k_{m,x} = \textstyle\sum_{m \in M} q_m\, k_{m,x}$$

for all $x \in X \cap A$ and

$$\textstyle\sum_{m \in M_a} p_m\, k_{m,x} = \textstyle\sum_{m \in M_a} q_m\, k_{m,x}$$

for all $x \in X \cap A$, since $m \in M_a$ whenever $p_m > 0$ or $q_m > 0$. Since $\pi m(x) \geqq \pi a$ when $x \in A$ by Lemma 5.1, we now have

$$\begin{aligned} u_{a;p;b} &= \textstyle\sum_{m \in M_a}\ p_m\, \gamma_{\pi a}^{\pi m} u_m \\ &= \textstyle\sum_{B \subseteq A}\ \sum_{m \in E_B \cap M_a}\ p_m\, \gamma_{\pi a}^{\pi m} u_m \\ &= \textstyle\sum_{B \subseteq A}\ \sum_{m \in E_B \cap M_a}\ \sum_{x \in X \cap B}\ p_m\, k_{m,x}\, \gamma_{\pi a}^{\pi m(x)}\, u_{m(x)} \\ &= \textstyle\sum_{B \subseteq A}\ \sum_{m \in E_B \cap M_a}\ \sum_{x \in X \cap A}\ p_m\, k_{m,x}\, \gamma_{\pi a}^{\pi m(x)}\, u_{m(x)} \\ &= \textstyle\sum_{m \in M_a}\ \sum_{x \in X \cap A}\ p_m\, k_{m,x}\, \gamma_{\pi a}^{\pi m(x)}\, u_{m(x)} \\ &= \textstyle\sum_{m \in M_a}\ \sum_{x \in X \cap A}\ q_m\, k_{m,x}\, \gamma_{\pi a}^{\pi m(x)}\, u_{m(x)} \\ &= u_{a;q;b}\,. \end{aligned}$$

Thus $u_{a;p;b}$ is independent of path. □

4. Proposition 5.4 shows that a minimal cocycle is determined by its values on all $m(x)$. Hence

$$MZ^1(S,\mathbb{G}) \;\cong\; \bigoplus_{x\in X} G_{\pi m(x)}\,;$$

the isomorphism $\Psi : MZ^1(S,\mathbb{G}) \longrightarrow \bigoplus_{x\in X} G_{\pi m(x)}$ sends $u = (u_m)_{m\in M} \in MZ^1(S,\mathbb{G})$ to $(u_{m(x)})_{x\in X}$. We now compute $MB^1(S,\mathbb{G})$.

Lemma 5.5. $\Psi\big(MB^1(S,\mathbb{G})\big) = \Theta\big(\bigoplus_{x\in X} \operatorname{Im} \gamma^{\pi x}_{\pi m(x)}\big)$, *where* Θ *is the automorphism of* $\bigoplus_{x\in X} G_{\pi m(x)}$ *defined for all* $v = (v_x)_{x\in X} \in \bigoplus_{x\in X} G_{\pi m(x)}$ *by*

$$(\Theta v)_x \;=\; v_x \;+\; \textstyle\sum_{y\in X,\, t_{x,y}\neq 0} t_{x,y}\; \gamma^{\pi m(y)}_{\pi m(x)}\, v_y\,.$$

Proof. Let D be the smallest element of $\mathcal{D}$ containing x. By Lemma 5.1, $\pi x \geqq \pi m(x)$ and $v(m(x)) = x + \sum_{y\in X} t_{x,y}\, y$, where $t_{x,y} \neq 0$ implies $y \prec x$, $y \in D$, and (by Lemma 5.1 applied to y) $\pi m(y) \geqq \pi m(x)$, since $m(x) \in E_D$. Hence Θ is well defined. Since $t_{x,y} \neq 0$ implies $y \prec x$, the matrix of Θ is triangular with 1's on the diagonal and Θ is an isomorphism.

Let $u = \delta g \in MB^1(S,\mathbb{G})$, where $g = (g_x)_{x\in X} \in \bigoplus_{x\in X} G_{\pi x}$. Then

$$u_m \;=\; \textstyle\sum_{x\in X,\, x\leqq m}\; m_x\; \gamma^{\pi x}_{\pi m} g_x \;-\; \sum_{x\in X,\, x\leqq q(m)}\; q(m)_x\; \gamma^{\pi x}_{\pi q(m)}\,.$$

Let $x\in X$, D be the smallest element of $\mathcal{D}$ containing x, $y\in X$, and $m = m(x)$. By Lemma 5.1, $m \in I_D = E_D \cap D$, $y \leqq m$ implies $y \in D$, $q(m) \in I_D$, and $y \leqq q(m)$ implies $y \in D$. Since $\pi y \geqq \pi m(y) \geqq \pi m(x)$ for all $y \in D$ and $m - q(m) = v(m) = x + \sum_{y\in X} t_{x,y}\, y$, we have

$$\begin{aligned}
u_{m(x)} &= \textstyle\sum_{y\in X,\, y\leqq m}\; m_y\; \gamma^{\pi y}_{\pi m} g_y \;-\; \sum_{y\in X,\, y\leqq q(m)}\; q(m)_y\; \gamma^{\pi y}_{\pi q(m)} \\
&= \textstyle\sum_{y\in X\cap D}\; m_y\; \gamma^{\pi y}_{\pi m} g_y \;-\; \sum_{y\in X\cap D}\; q(m)_y\; \gamma^{\pi y}_{\pi q(m)} \\
&= \textstyle\sum_{y\in X\cap D}\; v(m)_y\; \gamma^{\pi y}_{\pi m} g_y \;=\; \gamma^{\pi x}_{\pi m} g_x \;+\; \sum_{y\in D}\; t_{x,y}\; \gamma^{\pi y}_{\pi m} g_y \\
&= \gamma^{\pi x}_{\pi m} g_x \;+\; \textstyle\sum_{y\in X,\, t_{x,y}\neq 0}\; t_{x,y}\; \gamma^{\pi y}_{\pi m} g_y \;=\; (\Theta v)_x\,,
\end{aligned}$$

where $v = \big(\gamma^{\pi x}_{\pi m(x)}\, g_x\big)_{x\in X}$. Thus $\Psi u = \Theta v$. Hence $\Psi\big(MB^1(S,\mathbb{G})\big) = \Theta\big(\bigoplus_{x\in X} \operatorname{Im} \gamma^{\pi x}_{\pi m(x)}\big)$. □

We can now prove

Theorem 5.6. *When* S *is a finite partialy free commutative semigroup and*

$\mathbb{G}$ is thin, there is an isomorphism

$$H^2(S,\mathbb{G}) \cong \bigoplus_{c \in \mathrm{Irr}\,(S)} G_{e(c)}/\mathrm{Im}\,\gamma^c_{e(c)}\,,$$

which is natural in $\mathbb{G}$.

Proof. We may assume that S is a monoid. Since Ψ and Θ are isomorphisms,

$$\begin{aligned} H^2(S,\mathbb{G}) &\cong MZ/MB \cong \Theta^{-1}\Psi MZ\,/\,\Theta^{-1}\Psi MB \\ &= \big(\textstyle\bigoplus_{x\in X} G_{\pi m(x)}\big)/\big(\bigoplus_{x\in X} \mathrm{Im}\,\gamma^{\pi x}_{\pi m(x)}\big) \\ &= \textstyle\bigoplus_{x\in X} \big(G_{\pi m(x)}/\ \mathrm{Im}\,\gamma^{\pi x}_{\pi m(x)}\big) \\ &= \textstyle\bigoplus_{c\in \mathrm{Irr}\,(S)} G_{e(c)}/\mathrm{Im}\,\gamma^c_{e(c)}\,, \end{aligned}$$

since $\pi : F \longrightarrow S$ induces a bijection of X onto $\mathrm{Irr}\,(S)$ and $x \in X$ implies $c = \pi x \in \mathrm{Irr}\,(S)$ and $\pi m(x) = e(c)$, by Lemma 5.1. The isomorphism is natural in $\mathbb{G}$, since Ψ and Θ are natural in $\mathbb{G}$. □

Chapter XIV.

SEMIGROUPS WITH ZERO COHOMOLOGY.

Like other cohomology theories, commutative semigroup cohomology gives rise to the following problem:

(1) For which commutative semigroups S does $H^n(S,\mathbb{G}) = 0$ for all $n \geqq 2$ and all $\mathbb{G}$?

By Theorem XII.4.4, free commutative semigroups have this property.

The special role of H^2 suggests two additional problems:

(2) For which commutative semigroups S does $H^2(S,\mathbb{G}) = 0$ for all $\mathbb{G}$?

(3) For which finite (more generally, complete) group-free semigroups S does $H^2(S,\mathbb{G}) = 0$ whenever $\mathbb{G}$ is thin and surjecting (or thin, finite, and surjecting)?

Since $H^2(S,\mathbb{G}) \cong \mathrm{Ext}\,(S,\mathbb{G})$, $H^2(S,\mathbb{G})$ vanishes for all $\mathbb{G}$ if and only if every commutative group coextension of S splits. Free commutative semigroups and free commutative monoids have this property; so do semilattices, by Proposition V.4.4, and free abelian groups, by Proposition V.4.6. Problem (2) asks if there are any other semigroups with this property.

When S is complete group-free, then $H^2(S,\mathbb{G})$ vanishes whenever $\mathbb{G}$ is thin and surjecting if and only if every exact $\mathcal{H}$-coextension of S splits; equivalently, if every complete semigroup T with $T/\mathcal{H} \cong S$ splits as a coextension of S. By Proposition V.4.4, semilattices have this property. Problem (3) asks what other complete group-free semigroups have this property.

From the point of view of the structure and construction of commutative semigroups, problem (2) is at this time more interesting than problem (1), and problem (3) is most interesting of all.

Problem (1) is still unsolved, but problems (2) and (3) have been solved in some major particular cases. In this chapter we solve problem (2) when S is finite group-free and, after some preliminary results, we solve problem (3) when S is a finite nilmonoid. The results are due to the author [1997Z], [2001Z]. Problem

(3) was also solved by the author [2000T] for semigroups with two generators, in which case the solutions are nilmonoids or semilattices.

1. GROUP-FREE MONOIDS.

In this section, S is a finite commutative group-free monoid. We show that $H^2(S,\mathbb{G}) = 0$ for all $\mathbb{G}$ if and only if S is a semilattice. This was proved by the author [1997Z].

1. If S is a semilattice, then $H^2(S,\mathbb{G}) = 0$ for all $\mathbb{G}$ by Proposition V.4.4. We now let S be a finite commutative group-free semigroup, but not a semilattice, and cook up a functor $\mathbb{G}$ such that $H^2(S,\mathbb{G}) \neq 0$. By Corollary XII.5.5 we may assume that S is a monoid.

Call an abelian group valued functor $\mathbb{G}$ on $H(S)$ **selective** if there exists an element c of S such that $G_c \neq 0$ and $G_s = 0$ for all $s \neq c$. Then $\gamma_{s,t} = 0$ unless $s = st = c$, in which case $\gamma_{s,t}$ is an endomorphism of G_c; and $\mathbb{G}$ is thin if and only if $\gamma_{s,t}$ is the identity on G_c when $s = st = c$.

Lemma 1.1. *If S is partially free, but not a semilattice, then $H^2(S,\mathbb{G}) \neq 0$ for some thin finite surjecting and selective functor $\mathbb{G}$.*

Proof. By Theorem XIII.5.6 there is for every thin functor $\mathbb{G} = (G,\gamma)$ an isomorphism

$$H^2(S,\mathbb{G}) \cong \bigoplus\nolimits_{c\in \mathrm{Irr}\,(S)} \left(G_{e(c)} / \operatorname{Im} \gamma^c_{e(c)}\right),$$

where $\mathrm{Irr}\,(S)$ is the set of all irreducible elements of S and $e(c)$ is the idempotent in the archimedean component of c. If S is not a semilattice, then $\mathrm{Irr}\,(S)$, which generates S, contains an element c which is not idempotent. Then $e = e(c) \neq c$. Let $\mathbb{G}$ be the thin selective functor in which $G_e \neq 0$ is any nontrivial finite abelian group and $G_s = 0$ for all $s \neq e$. $\mathbb{G}$ is also finite and surjecting. Moreover $H^2(S,\mathbb{G}) \neq 0$, since its direct summand $G_{e(c)} / \operatorname{Im} \gamma^c_{e(c)} \cong G_e \neq 0$. □

2. We now let the finite monoid S be group-free but not partially free. Green's preorder $\leqq_{\mathcal{H}}$ is a partial order relation on S, which we denote as before by just $\leqq$. Let $\pi : F = F_X \longrightarrow S$ be the standard presentation of S and $\mathcal{C} = \ker \pi$. Then π induces a bijection of X onto $\mathrm{Irr}\,(S)$, and F is finitely generated. The direction set $\mathcal{D}$, extent cells E_A, and trace congruences $\mathcal{C}_A$ of $\mathcal{C}$ are as in Chapter X.

Since S is not partially free, one of the trace congruences $\mathcal{C}_B$ is not a Rees congruence and has a nontrivial class other than the ideal $B\backslash H_B$; then there is a $\mathcal{C}$-class $C \subseteq E_B$ whose projection $p'_B C \subseteq H_B$ is not trivial. Let $c \in S$ be maximal (under $\leqq$) such that the $\mathcal{C}$-class $C = \pi^{-1} c$ has a nontrivial projection $p'_B C \subseteq H_B$ (where $B \in \mathcal{D}$ is determined by $C \subseteq E_B$).

Lemma 1.2. *C does not contain elements a, b such that $p'_B a < p'_B b$; hence $0 \notin p'_B C$.*

Proof. If $a, b \in C$ and $p'_B a < p'_B b$, then $p'_B a \ \mathcal{C}_B \ p'_B b = p'_B a + t$ for some $t \in B'$, $t > 0$; hence

$$p'_B a \ \ \mathcal{C}_B \ \ p'_B a + t \ \ \mathcal{C}_B \ \ p'_B a + 2t \ \ \mathcal{C}_B \ \ \cdots \ \ \mathcal{C}_B \ \ p'_B a + kt$$

for all $k > 0$ and $p'_B C \subseteq H_B$ contains $p'_B a + kt$ for all $k > 0$. This contradicts the finiteness of H_B (Lemma X.4.1).

Assume $0 \in p'_B C$. Since $p'_B C$ is nontrivial, C contains elements a, b such that $p'_B a = 0 \neq p'_B b$; then $p'_B a < p'_B b$, which we just saw is impossible. □

Lemma 1.3. *The element c is not idempotent and is not irreducible; hence $C \cap X = \varnothing$.*

Proof. If c is idempotent, then c, $2c \in C$ with either $p'_B c < p'_B(2c)$ or $p'_B c = 0$, which is impossible by Lemma 1.2.

Since $\pi : F \longrightarrow S$ is the standard presentation, c is irreducible if and only if $c = \pi y$ for some $y \in X$, if and only if $C \cap X \neq \varnothing$. Assume that C contains some $y \in X$. Then $y \notin B$, otherwise $0 = p'_B y \in p'_B C$. Since $p'_B C$ is nontrivial there is some $a = \sum_{x \in X} a_x x \in C$ with $p'_B a \neq y$. By Lemma 1.2 we cannot have $y = p'_B y < p'_B a$; therefore $y \nleqq p'_B a$, $y \nleqq a$, $a_y = 0$, and $a = \sum_{x \in X,\, x \neq y} a_x x$. Also $a \neq 0$ by Lemma 1.2. Since π is injective on X this makes $c = \pi a$ a product of irreducible elements $\pi x \neq \pi y$. If $|a| = \sum_{x \in X} a_x > 1$, then $c = \pi y$ is not irreducible; otherwise $|a| = 1$, $a = x \neq y$, and π is not injective on X; this is the required contradiction. □

3. We now call upon the overpath method. Let $\preccurlyeq$ be any total order on X in which $X\backslash B$ precedes $X \cap B$ ($x \prec y$ for all $x \in X\backslash B$ and $y \in X \cap B$). Order $G = G_X$ lexicographically: let $\sum_{x \in X} a_x x \sqsubset \sum_{x \in X} b_x x$ if and only if there exists $t \in X$ such that $a_x = b_x$ for all $x \prec t$ but $a_t < b_t$. (Then $x \sqsubset y$ in G if and only if $x \succ y$ in X.) Then $\sqsubseteq$ is a compatible total order on G, and induces a compatible well order on F. Since $X\backslash B$ precedes $X \cap B$, $p'_B a \sqsubset p'_B b$ implies $a \sqsubset b$.

Let $\epsilon(c)$ be the least idempotent $e \geqq_{\mathcal{H}} c$ of S.

Lemma 1.4. *If $m \in M$ and $m \leqq a \in C$, then either $p'_B m = p'_B q(m)$, or $m \in C$ and $p'_B m \sqsupset p'_B q(m)$. If $m \in M$ and $m \leqq a \in F$, where $c < \pi a \leqq \epsilon(c)$ in S, then $p'_B m = p'_B q(m)$.*

Proof. In either case $\epsilon(c) = e(\pi a)$ and $a \in E_B$ (Proposition X.3.4). Assume $p'_B m \neq p'_B q(m)$. Then $p'_B m \sqsupset p'_B q(m)$ (otherwise $m \sqsubset q(m)$). Since $m \leqq a \in E_B$, we have $m \in E_D$ for some $D \in \mathfrak{D}$, $D \subseteq B$ by (E2). Then $X \backslash B \subseteq X \backslash D$; since $X \backslash B$ precedes $X \cap B$, $p'_B m \sqsupset p'_B q(m)$ implies $p'_D m \sqsupset p'_D q(m)$. Thus the $\mathcal{C}$-class $C_m \subseteq E_D$ has a nontrivial projection $p'_D C$. Since $\pi m \geqq c$ the choice of c implies $\pi m = c$, and $m \in C$. Then $c \not< \pi a$, since $\pi a \leqq \pi m$. □

In what follows

$$\begin{aligned} M_B &= \{m \in M \mid p'_B m = p'_B q(m)\} \quad \text{and} \\ M_C &= \{m \in M \mid m \in C \text{ and } p'_B m \sqsupset p'_B q(m)\}. \end{aligned}$$

By Lemma 1.4, when $m \in M$ and $m \leqq a$, then $m \in M_B \cup M_C$ if $a \in C$, and $m \in M_B$ if $c < \pi a \leqq \epsilon(c)$.

Lemma 1.5. *Let p be an overpath from $a \in C$ to b. If $p'_B a = p'_B b$, then p consists solely of elements of M_B. If $p'_B a \neq p'_B b$, then p consists of elements of M_B and one element m of M_C such that $p'_B v(m) = p'_B(a - b)$.*

Proof. Let $p : m^1, \dots, m^k$ be an overpath from $a \in C$ to b and

$$a = p^0 \xrightarrow{m^1} p^1 \xrightarrow{m^2} \cdots \xrightarrow{m^k} p^k = b$$

be the corresponding path, so that $p^{i-1} \geqq m^i$ and $p^i - p^{i-1} = v(m^i)$ for all $i > 0$. Since $a \in C$, then $m^i \leqq p^{i-1} \in C$ and $m^i \in M_B \cup M_C$ for all i, by Lemma 1.4. Also

$$a - b = \textstyle\sum_i v(m^i);$$

since $p'_B v(m^i) = 0$ when $m^i \in M_B$,

$$p'_B a - p'_B b = \textstyle\sum_{m^i \in M_C} p'_B v(m^i),$$

with $p'_B v(m^i) \sqsupset 0$ in G since $p'_B m^i \sqsupset p'_B q(m^i)$ for all $m^i \in M_C$.

If $p'_B a = p'_B b$, then $0 = \sum_{m^i \in M_C} p'_B v(m^i)$ is a sum of positive elements of G (ordered by $\sqsubseteq$), which is not possible unless the sum is empty; hence p consists solely of elements of M_B.

If $p'_B a \neq p'_B b$, then $\sum_{m^i \in M_C} p'_B v(m^i) \neq 0$ and there is some $m^i \in M_C$. Let j be the least i such that $m^i \in M_C$. For all $i < j$ we have $m^i \in M_B$ for all $i < j$ and $p'_B p^i - p'_B p^{i-1} = p'_B v(m^i) = 0$. Therefore $p'_B a = p'_B p^0 = p'_B p^{j-1} \leqq p'_B m^j$. Since $m^j \in C$ this implies $p'_B a = p'_B m^j$, by Lemma 1.2. Also $q(m^j) = q(a) \sqsubseteq b$, so that $p'_B q(m^j) \sqsubseteq p'_B b$ (otherwise $q(m^j) \sqsupset b$) and $p'_B q(m^j) = p'_B b$ by Lemma 1.2. (In fact, $p'_B q(m^j) = p'_B q(a)$.) Hence $p'_B a - p'_B b = p'_B v(m^j)$ and

$$\textstyle\sum_{m^i \in M_C,\, m^i \neq m^j} p'_B v(m^i) \;=\; 0.$$

As before, this sum must be empty. Hence m^j is the only element of M_C which appears in p. □

Corollary 1.6. $M_C \neq \emptyset$.

Proof. $p'_B a \neq p'_B q(a)$ for some $a \in C$, since $p'_B C$ is nontrivial; there exists an overpath from a to $q(a)$, which by Lemma 1.5 includes some $m \in M_C$. □

4. Now let $\mathbb{G}$ be the thin selective functor in which G_c is any finite abelian group and $G_s = 0$ for all $s \neq c$. Then $\mathbb{G}$ is finite, but not surjecting since $\operatorname{Im} \gamma_c^{\epsilon(c)} = 0$. By Lemma 1.3, $G_s = 0$ when s is idempotent and when s is irreducible. Hence $\prod_{x \in X} G_{\pi x} = 0$ and $MB^1(S, \mathbb{G}) = 0$. We use Lemma 1.5 to construct nontrivial minimal cocycles.

Lemma 1.7. *Given $g_x \in G_c$ for every $x \in X \backslash B$ let*

$$u_m \;=\; \begin{cases} 0 \in G_{\pi m} & \text{if } m \notin M_C, \\ \sum_{x \in X \backslash B} v(m)_x g_x \in G_c & \text{if } m \in M_C. \end{cases}$$

Then u is a minimal cocycle.

Proof. Let $a \in F$ and p be an overpath from a to b. We show that

$$u_{a;p;b} \;=\; \textstyle\sum_{m \in M,\, \pi m \geqq \pi a} p_m \, \gamma_{\pi a}^{\pi m} \, u_m \;\in\; G_{\pi a}$$

is independent of path (where p_m is the number of appearances of m in p).

If $a \notin C$, then $G_{\pi a} = 0$ and $u_{a;p;b} = 0$. Now let $a \in C$. Since $u_m = 0$ when $m \notin M_C$ and $\pi m = \pi a$ when $m \in M_C$ we have

$$u_{a;p;b} \;=\; \textstyle\sum_{m \in M_C} p_m \, u_m \,.$$

We now invoke Lemma 1.5. If $p'_B a = p'_B b$, then p consists solely of elements of M_B and $u_{a;p;b} = 0$. If $p'_B a \neq p'_B b$, then p consists of elements of M_B and one element n of M_C such that $p'_B v(n) = p'_B (a - b)$. Then $v(n)_x = a_x - b_x$

for all $x \in X\backslash B$ and

$$u_{a;p;b} \;=\; u_n \;=\; \sum_{x \in X\backslash B} (a_x - b_x)\, g_x \,.$$

In either case $u_{a;p;b}$ depends only on a and b. □

5. Since $M_C \neq \varnothing$ (Corollary 1.6) it is possible to choose the finite abelian group G_c and $g_x \in G_c$ so that $u_m \neq 0$ for some $m \in M_C$. (For instance take any $n \in M_C$; then $v(n)_x \neq 0$ for some $x \in X\backslash B$; let G_c be cyclic of order p, where p does not divide $v(n)_x$, and let $g_x \neq 0$, $g_y = 0$ for all $y \neq x$; then $u_n = v(n)_x\, g_x \neq 0$.) Then $MZ^1(S,\mathbb{G}) \neq 0$; since we saw that $MB^1(S,\mathbb{G}) = 0$ it follows that $H^2(S,\mathbb{G}) \neq 0$, and we have proved that $H^2(S,\mathbb{G}) \neq 0$ for some $\mathbb{G}$ if S is not partially free. Since $\mathbb{G}$ is thin finite and selective we have in fact proved:

Theorem 1.8. *For a finite group-free commutative semigroup S the following conditions are equivalent:*

(1) $H^2(S,\mathbb{G}) = 0$ *for all* $\mathbb{G}$*;*

(2) $H^2(S,\mathbb{G}) = 0$ *whenever* $\mathbb{G}$ *is thin, finite, and selective;*

(3) S *is a semilattice.*

2. THE ZERO GROUP.

In this section we assume that S has a zero element; for instance, that S is finite group-free. We study how $H^2(S,\mathbb{G})$ depends on the zero group G_0. This yields necessary conditions that $H^2(S,\mathbb{G})$ vanish when $\mathbb{G}$ is Schützenberger.

1. When S has a zero element, an abelian group valued functor $\mathbb{G}$ on $H(S)$ is **almost null** if $G_a = 0$ for all $a \neq 0$ and **reduced** if $G_0 = 0$.

When $\mathbb{G}$ is thin and almost null, Proposition V.4.7 provides isomorphisms $H^2(S,\mathbb{G}) \cong \mathrm{PHom}(S\backslash 0, G_0) \cong \mathrm{Hom}(G(S\backslash 0), G_0)$; the partial homomorphism φ which corresponds to the cohomology class cls s of $s \in SZ^2(S,\mathbb{G})$ sends $a \in S\backslash 0$ to $\varphi(a) = s_{a,0}$.

Proposition 2.1. *Let S have a zero element. For every abelian group valued functor $\mathbb{G}$ on $H(S)$ there is a short exact sequence*

$$0 \longrightarrow \mathbb{G}' \longrightarrow \mathbb{G} \longrightarrow \mathbb{G}'' \longrightarrow 0$$

which is natural in $\mathbb{G}$, in which $\mathbb{G}'$ is almost null and $\mathbb{G}''$ is reduced. If $\mathbb{G}$ is thin, then $\mathbb{G}'$ and $\mathbb{G}''$ are thin. If $\mathbb{G}$ is finite, then $\mathbb{G}'$ and $\mathbb{G}''$ are finite. If S is complete group-free and $\mathbb{G}$ is thin and surjecting, then $\mathbb{G}'$ and $\mathbb{G}''$ are thin and surjecting.

Proof. $\mathbb{G}' = (G', \gamma')$ and $\mathbb{G}'' = (G'', \gamma'')$ are defined as follows. Let $G'_a = 0$ for all $a \neq 0$ and $G'_0 = G_0$; let $\gamma'_{a,t} = 0$ if $a \neq 0$ and $\gamma'_{0,t} = \gamma_{0,t}$. Then $\mathbb{G}'$ is almost null, and $\mathbb{G}'$ is thin if $\mathbb{G}$ is thin (so that $\gamma_{0,t}$ is the identity on G_0 for all t), and is finite if $\mathbb{G}$ is finite. If S is complete group-free and $\mathbb{G}$ is thin and surjecting, then $\mathbb{G}'$ is thin and surjecting.

Let $G''_a = G_a$ for all $a \neq 0$ and $G''_0 = 0$; let $\gamma''_{a,t} = 0$ if $at = 0$, $\gamma''_{a,t} = \gamma_{a,t}$ if $at \neq 0$. Then $\mathbb{G}''$ is reduced, and is thin (finite, surjecting) if $\mathbb{G}$ is thin (finite, surjecting).

The exact sequence $\mathbb{G}' \xrightarrow{\alpha} \mathbb{G} \xrightarrow{\beta} \mathbb{G}''$ is defined as follows: if $a \neq 0$, then $\alpha_a = 0$ and $\beta_a = 1_{G_a}$; $\alpha_0 = 1_{G_0}$ and $\beta_0 = 0$. The following diagrams commute:

$$\begin{array}{ccccc} G'_a = 0 & \longrightarrow & G_a & = & G''_a \\ \downarrow{\scriptstyle \gamma'_{a,t}} & & \downarrow{\scriptstyle \gamma_{a,t}} & & \downarrow{\scriptstyle \gamma''_{a,t}} \\ G'_{at} = 0 & \longrightarrow & G_{at} & = & G''_{at} \end{array}$$

whenever $a, at \neq 0$;

$$\begin{array}{ccccc} G'_a = 0 & \longrightarrow & G_a & = & G''_a \\ \downarrow{\scriptstyle \gamma'_{a,t}} & & \downarrow{\scriptstyle \gamma_{a,t}} & & \downarrow{\scriptstyle \gamma''_{a,t}} \\ G'_{at} & = & G_0 & \longrightarrow & 0 = G''_0 \end{array}$$

whenever $a \neq 0$ and $at = 0$; and

$$\begin{array}{ccccc} G'_0 & = & G_0 & \longrightarrow & 0 = G''_0 \\ \downarrow{\scriptstyle \gamma'_{0,t}} & & \downarrow{\scriptstyle \gamma_{0,t}} & & \downarrow{\scriptstyle \gamma''_{0,t}} \\ G'_0 & = & G_0 & \longrightarrow & 0 = G''_0. \end{array}$$

Thus α and β are natural transformations. Naturality in $\mathbb{G}$ is similar. □

2. By Theorem XII.4.5 there is an exact sequence

$$H^2(S, \mathbb{G}') \longrightarrow H^2(S, \mathbb{G}) \longrightarrow H^2(S, \mathbb{G}'')$$

which is natural in $\mathbb{G}$, in which the homomorphisms are induced by α and β.

Proposition 2.2. *If S has a zero element, then $H^2(S,\mathbb{G}) = 0$ whenever $\mathbb{G}$ is thin and surjecting if and only if*

(N) $H^2(S,\mathbb{G}) = 0$ *whenever $\mathbb{G}$ is thin and almost null, and*

(R) $H^2(S,\mathbb{G}) = 0$ *whenever $\mathbb{G}$ is thin, surjecting, and reduced;*

also, $H^2(S,\mathbb{G}) = 0$ whenever $\mathbb{G}$ is thin finite and surjecting if and only if

(Nf) $H^2(S,\mathbb{G}) = 0$ *whenever $\mathbb{G}$ is thin finite and almost null, and*

(Rf) $H^2(S,\mathbb{G}) = 0$ *whenever $\mathbb{G}$ is thin finite surjecting and reduced.*

Proof. In the exact sequence

$$0 \longrightarrow \mathbb{G}' \longrightarrow \mathbb{G} \longrightarrow \mathbb{G}'' \longrightarrow 0$$

in Proposition 2.1, if $\mathbb{G}$ is thin (finite, surjecting), then $\mathbb{G}'$ and $\mathbb{G}''$ are thin (finite, surjecting). If therefore $H^2(S,\mathbb{G}) = 0$ whenever $\mathbb{G}$ is thin and surjecting, then (N) and (R) hold; if $H^2(S,\mathbb{G}) = 0$ whenever $\mathbb{G}$ is thin finite and surjecting, then (Nf) and (Rf) hold.

If conversely (N) and (R) hold, and $\mathbb{G}$ is thin and surjecting, then $H^2(S,\mathbb{G}') = 0$, $H^2(S,\mathbb{G}'') = 0$, and the exact sequence

$$H^2(S,\mathbb{G}') \longrightarrow H^2(S,\mathbb{G}) \longrightarrow H^2(S,\mathbb{G}'')$$

shows that $H^2(S,\mathbb{G}) = 0$. Conditions (Nf) and (Rf) similarly imply $H^2(S,\mathbb{G}) = 0$ whenever $\mathbb{G}$ is thin finite and surjecting. □

Conditions (N) and (Nf) are easily settled when S is finite. Let $\pi : F = F_X \longrightarrow S$ be a presentation of S, $\mathcal{C}$ be the congruence induced by π, and $Z = \pi^{-1}0$ be the zero class. In the following result, K is the subgroup of G generated by all differences $a - b$ with $a \mathcal{C} b$ and $a, b \notin Z$; relative to any compatible well order on F, K is also generated by all defining vectors $v(m)$ with $m \notin Z$.

Proposition 2.3. *Let S be a finite commutative semigroup with a zero element. The following conditions on S are equivalent:*

(N) $H^2(S,\mathbb{G}) = 0$ *whenever $\mathbb{G}$ is thin and almost null;*

(Nf) $H^2(S,\mathbb{G}) = 0$ *whenever $\mathbb{G}$ is thin finite and almost null;*

(K) *there is a presentation of S in which $K = G$;*

(K+) $K = G$ *in every presentation of S.*

Proof. By Proposition V.4.7, $H^2(S,\mathbb{G}) \cong \mathrm{Hom}\,(G(S\backslash 0), G_0)$ whenever $\mathbb{G}$

is thin and almost null. Now $G(S\backslash 0) \cong G/K$ by Proposition XIII.4.2, in any presentation of S. Hence (K) implies (N), which in turn implies (Nf). If on the other hand S is finite and $G/K \neq 0$, then $G/K \cong G(S\backslash 0)$ is finitely generated, there exists a finite abelian group A such that $\mathrm{Hom}(G/K,\ A) \neq 0$, and (Nf) does not hold; therefore (Nf) implies (K+). □

3. NILMONOIDS.

In this section, we characterize finite commutative nilmonoids S such that $H^2(S,\mathbb{G}) = 0$ whenever $\mathbb{G}$ is thin finite and surjecting; equivalently, every elementary semigroup T such that $T/\mathcal{H} \cong S$ splits as a coextension of S. This result is due to the author [2001Z].

1. To explore nilmonoids we use certain simple coefficient functors. When J is an ideal of a commutative semigroup S, an abelian group valued functor $\mathbb{G} = (G,\gamma)$ on $H(S)$ is **semiconstant over** J when there is an abelian group A such that

$$G_s = A \text{ for all } s \in S\backslash J,\ G_s = 0 \text{ for all } s \in J,$$

$\gamma_{s,t} = 1_A$ if $st \notin J$, and $\gamma_{s,t} = 0$ if $st \in J$. If $J = \varnothing$, then $\mathbb{G}$ is constant. If S has a zero element, then almost constant functors are semiconstant over 0; conversely one can view a semiconstant functor $\mathbb{G}$ on S over an ideal $J \neq \varnothing$ as an almost constant functor on S/J. In general, a semiconstant functor is thin, surjecting, and (if S has a zero element and $J \neq \varnothing$) reduced.

Let $\pi : F \longrightarrow S$ be any presentation of S, $\mathcal{C} = \ker \pi$ be the congruence induced by π, and $\sqsubseteq$ be a compatible well order on F. Let J be an ideal of S. Let M_J and X_J be the sets

$$M_J = \{m \in M \mid \pi m \notin J\},\quad X_J = \{x \in X \mid \pi x \notin J\}.$$

Lemma 3.1. *When $\mathbb{G}$ is semiconstant over J, then:*

(1) *a minimal cochain u is determined by its values $(u_m)_{m\in M_J}$ on M_J;*

(2) *u is a minimal cocycle if and only if*

$$\textstyle\sum_{m\in M_J} r_m\, u_m = \sum_{m\in M_J} s_m\, u_m$$

for every positive relation $\sum_{m\in M_J} r_m\, v(m) = \sum_{m\in M_J} s_m\, v(m)$ which is realized in a $\mathcal{C}$-class $C \subsetneqq \pi^{-1}J$;

(3) *u is a minimal coboundary if and only if there exists $g = (g_x)_{x \in X_J}$ such that $g_x \in A$ for all $x \in X_J$ and*

$$u_m = \textstyle\sum_{x \in X_J} v(m)_x \, g_x$$

for all $m \in M_J$;

(4) *when u is a minimal coboundary, then*

$$\textstyle\sum_{m \in M_J} r_m \, u_m = 0$$

for every vector relation $\sum_{m \in M_J} r_m \, v(m) = 0$ which holds in G.

Proof. (1) is clear since $u_m = 0$ whenever $\pi m \notin J$.

(2) follows from Proposition XIII.4.6: u is a minimal cocycle if and only if

$$Z(a,r,s) : \textstyle\sum_{m \in M_a} r_m \, \gamma_{\pi a}^{\pi m} u_m = \sum_{m \in M_a} s_m \, \gamma_{\pi a}^{\pi m} u_m$$

whenever $a \in S$ and the positive relation $\sum_{m \in M} r_m \, v(m) = \sum_{m \in M} s_m \, v(m)$ is realized in C_a, where $M_a = \{ m \in M \mid \pi m \geqq_{\mathcal{H}} \pi a \}$. If $\pi a \in J$, then $Z(a,r,s)$ is trivial. If $\pi a \notin J$, then $M_a \subseteq M_J$ and $Z(a,r,s)$ is equivalent to

$$\textstyle\sum_{m \in M_J} r_m \, u_m = \sum_{m \in M_J} s_m \, u_m$$

since $r_m > 0$ implies $\pi m \geqq_{\mathcal{H}} \pi a$ and similarly for $s_m > 0$.

(3) follows from the definition of δg: when $g = (g_x)_{x \in X}$, then $u = \delta g$ is given by

$$u_m = \textstyle\sum_{x \in X,\, x \leqq m} m_x \, \gamma_{\pi m}^{\pi x} g_x - \sum'_{x \in X,\, x \leqq q(m)} q(m)_x \, \gamma_{\pi m}^{\pi x} g_x$$

for all $m \in M$. Again $u_m = 0$ if $m \notin M_J$. If $m \in M_J$, then $x \leqq m$ implies $x \in X_J$, and so does $x \leqq q(m)$, and

$$\begin{aligned} u_m &= \textstyle\sum_{x \in X_J,\, x \leqq m} m_x \, g_x - \sum_{x \in X_J,\, x \leqq q(m)} q(m)_x \, g_x \\ &= \textstyle\sum_{x \in X_J} (m_x - q(m)_x) \, g_x = \sum_{x \in X_J} v(m)_x \, g_x , \end{aligned}$$

since $m_x > 0$ implies $x \leqq m$ and similarly for $q(m)_x > 0$.

(4) follows from (3). Assume that $\sum_{m \in M_J} r_m \, v(m) = 0$ holds in G. Then $\sum_{m \in M_J} r_m \, v(m)_x = 0$ for every $x \in X$. If u is a minimal coboundary, then

$$\textstyle\sum_{m \in M_J} r_m \, u_m = \sum_{m \in M_J} \sum_{x \in X_J} r_m \, v(m)_x \, g_x = 0. \ \square$$

2. As a consequence of Lemma 3.1, we prove:

Lemma 3.2. *Let J be a nonempty ideal of S which contains every element $s \in S$ such that a nontrivial positive relation is realized in $C_s = \pi^{-1} s$. If* (Rf) *holds, then the defining vectors $v(m)$ with $m \in M_J$ are distinct and linearly independent.*

Proof. Let A be any finite abelian group and $\mathbb{G}$ be the corresponding semi-constant functor over J, which is thin finite surjecting and reduced.

Every minimal cochain $u = (u_m)_{m \in M}$ is a minimal cocycle: if p and q are overpaths from a to b, then either p and q consist of the same elements of M, in which case $u_{a;p;b} = u_{a;q;b}$, or $\sum_{m \in M} p_m\, v(m) = \sum_{m \in M} q_m\, v(m)$ is a nontrivial relation which is realized in $C_a = C_b$, in which case $\pi a = \pi b \in J$, $G_{\pi a} = 0$, and $u_{a;p;b} = u_{a;q;b} = 0$. Hence $MZ^1(S, \mathbb{G}) = MC^1(S, \mathbb{G}) \cong \prod_{m \in M_J} A$.

If there is a nontrivial vector relation $r : \sum_{m \in M_J} r_m\, v(m) = 0$ between the vectors $v(m)$ with $m \in M_J$, then

$$\textstyle\sum_{m \in M_J} r_m\, u_m = 0$$

for every minimal coboundary u, by Lemma 3.1. If A is a cyclic group of suitable prime order p, then p does not divide every nonzero r_m and there is a minimal cochain u such that $\sum_{m \in M_J} r_m\, u_m \neq 0$; for instance, let $u_m \neq 0$, where p does not divide r_m, and $u_n = 0$ for all $n \neq m$). Then u is a minimal cocycle but not a minimal coboundary, $H^2(S, \mathbb{G}) \cong MZ^1(S, \mathbb{G}) \,/\, MB^1(S, \mathbb{G}) \neq 0$, and (Rf) does not hold. □

3. We now let S be a finite nilmonoid and assume that F is finitely generated. A thin abelian group valued functor $\mathbb{G}$ on S is surjecting if and only every homomorphism γ_s^1 is surjective.

A vector relation $\sum_{m \in M} r_m\, v(m) = 0$ is **reachable** in a $\mathcal{C}$-class C when it follows from relations that are realized in C (when it is a linear combination with integer coefficients of relations that are realized in C). By Proposition XIII.4.5, every vector relation is reachable in some $\mathcal{C}$-class and is reachable in the zero class $Z = \pi^{-1}0$.

Let J be a nonempty ideal of S. Let K_J be the subgroup of G generated by all defining vectors $v(m)$ with $m \in M_J$. G is a finitely generated free abelian group and so is $K_J \subseteq G$. A **defining basis** of K_J (relative to J) is a subset B of M_J such that

(1) the defining vectors $v(m)$ with $m \in B$ are distinct and constitute a basis of K_J, so that for every $m \in M_J \backslash B$ there is a unique vector relation $v(m) = \sum_{n \in B} r_n\, v(n)$ (with integer coefficients); and

(2) for every $m \in M_J \backslash B$ the relation $v(m) = \sum_{n \in B} r_n\, v(n)$ is reachable in C_m.

If $m \notin B$ and the relation $v(m) = \sum_{n \in B} r_n\, v(n)$ is reachable in a $\mathcal{C}$-class $C_s = \pi^{-1} s$, then some vector relation containing m is realized in C_s and $\pi m \geqq_{\mathcal{H}} s$; thus (2) states that the relation $v(m) = \sum_{n \in B} r_n\, v(n)$ is reachable in the highest possible $\mathcal{C}$-class.

Lemma 3.2 shows that in some cases M_J itself is a defining basis of K_J. Our main lemma is:

Lemma 3.3. *Let S be a finite nilmonoid and J be a nonempty ideal of S. If* (Rf) *holds, then K_J has a defining basis.*

Proof. We assume (Rf) and proceed by induction on $|S \backslash J|$. Let J_0 be the set of all $s \in S$ such that a nontrivial positive relation is realized in $C_s = \pi^{-1} s$. Then J_0 is an ideal of S, since a relation which is realized in C_s is realized in every C_{st}. By Lemma 3.2, K_J has a defining basis whenever J contains J_0. This kickstarts the induction.

For the general case we expand $S \backslash J$ from the bottom, which matches what writing this book is doing to the author. Let J be a nonempty ideal of S. Assume that K_J has a defining basis B (relative to J) and that $J \neq 0$. Let s be a maximal element of J (under $\leqq_{\mathcal{H}}$), so that $J' = J \backslash \{s\}$ is an ideal of S and s is a minimal element of $S \backslash J' = (S \backslash J) \cup \{s\}$. We construct a defining basis of $K_{J'}$.

We have $M_{J'} = M_J \cup M_s$, where

$$M_s = M \cap C_s = \{m \in M \mid \pi m = s\}.$$

Since S is a nilmonoid, the $\mathcal{C}$-class C_s, which is not the zero class, cannot contain comparable elements $a < b$. Therefore an overpath $p : m^1, \dots, m^k$ from $a \in C_s$ to $b \in C_s$ contains at most one element of M_s which must be its last element m^k: if $a = p^0, \dots, p^k = b$ is the corresponding path and $m^j \in M_s$, then $p^{j-1} \leqq m^j$ implies $p^{j-1} = m^j$ in C_s, and then $p^j = q(m^j)$, so that the path $p^0, \dots, p^j$ has reached the least element of C_s and continues no further.

Therefore the positive relations which are realized in C_s contain at most two elements of M_s and are of three types:

(a) relations $\sum_{n \in M_J} r_n\, v(n) = \sum_{m \in M_J} s_m\, v(m)$ containing no element of M_s;

(b) relations $v(m_1) + \sum_{n \in M_J} r_n\, v(n) = \sum_{m \in M_J} s_m\, v(m)$ containing one element m_1 of M_s, with coefficient 1;

(c) relations $v(m_1) + \sum_{n \in M_J} r_n\, v(n) = v(m_2) + \sum_{m \in M_J} s_m\, v(m)$ containing two elements $m_1 \neq m_2$ of M_s, with coefficients 1.

From $B \cup M_s$ we extract a defining basis $B \cup D$ of $K_{J'}$. Starting from $D = M_s$ we trim D, one element at a time, as follows. If $m_1 \in D$ appears in a relation $v(m_1) = \sum_{n \in M_J} r_n\, v(n)$ of type (b), then remove m_1 from D and replace $v(m_1)$ by $\sum_{n \in M_J} r_n\, v(n)$ in every other relation (of type (b) or (c)) in which $v(m_1)$ appears; this yields relations of type (a) or (b) which are reachable in C_s. If $m_1 \in D$ appears in a relation $v(m_1) = v(m_2) + \sum_{n \in M_J} r_n\, v(n)$ of type (c) (with $m_1 \neq m_2$), then remove m_1 from D and replace $v(m_1)$ by $v(m_2) + \sum_{n \in M_J} r_n\, v(n)$ in every other relation (of type (b) or (c)) in which $v(m_1)$ appears; this yields relations of type (b) or (c) which are reachable in C_s. This process terminates since M_s is finite. Then all relations of type (b) or (c) have been used and D has the following properties:

(A) no relation $v(m_1) + \sum_{n \in M_J} r_n\, v(n) = \sum_{m \in M_J} s_m\, v(m)$ with $m_1 \in D$, or $v(m_1) + \sum_{n \in M_J} r_n\, v(n) = v(m_2) + \sum_{n \in M_J} s_n\, v(n)$ with $m_1, m_2 \in D$ and $m_1 \neq m_2$, can be realized in C_s;

(B) for every $m \in M_s \backslash D$ there is a relation $v(m) = \sum_{n \in M_J \cup D} r_n\, v(n)$ which is reachable in C_s; in particular,

(C) the defining vectors $v(m)$ with $m \in M_J \cup D$ generate $K_{J'}$.

These properties imply:

(A*) If $\pi a \notin J'$, then no nontrivial relation

$$R : \sum_{m \in B \cup D} r_m\, v(m) = \sum_{m \in B \cup D} s_m\, v(m)$$

can be realized in C_a. Indeed assume that R can be realized in C_a. If $r_m \neq s_m$ for some $m \in D$, then $\pi m \geqq_{\mathcal{H}} \pi a$, $\pi a = s$ since s is minimal in $S \backslash J'$, the given relation R is realized in C_s, and R is of the form $v(m_1) + \sum_{n \in M_J} r_n\, v(n) = \sum_{n \in M_J} s_n\, v(n)$ with $m_1 \in D$, or of the form $v(m_1) + \sum_{n \in M_J} r_n\, v(n) = v(m_2) + \sum_{n \in M_J} s_n\, v(n)$ with $m_1, m_2 \in D$ and $m_1 \neq m_2$, which contradicts (A). Therefore $r_m = s_m$ for all $m \in D$; then $r_m = s_m$ for all $n \in B \cup D$, since the vectors $v(n)$ with $n \in B$ are distinct and linearly independent, and R is trivial.

(B*) for every $m \in M_{J'} \backslash (B \cup D)$ there is a relation $v(m) = \sum_{n \in B \cup D}$

$r_n\, v(n)$ which is reachable in C_m. This follows from (B) if $m \in M_s$; if $m \in M_J$ there is a relation $v(m) = \sum_{n \in B} r_n\, v(n)$ which is reachable in C_m since B is a defining basis of K_J.

(C*) the defining vectors $v(m)$ with $m \in B \cup D$ generate $K_{J'}$.

We show that the defining vectors $v(m)$ with $m \in B \cup D$ are distinct and linearly independent; then (B*) and (C*) show that $B \cup D$ is a defining basis of $K_{J'}$ (relative to J').

As in the proof of Lemma 3.2, let A be any finite abelian group and $\mathbb{G}$ be the corresponding semiconstant functor over J', which is thin finite surjecting and reduced.

By Lemma 3.1, a minimal cochain $u = (u_m)_{m \in M}$ is determined by its values $(u_m)_{m \in M_{J'}}$ on $M_{J'}$, and $u = (u_m)_{m \in M}$ is a minimal cocycle if and only if $\sum_{m \in M_{J'}} r_m\, u_m = \sum_{m \in M_{J'}} s_m\, u_m$ whenever the positive relation $\sum_{m \in M_{J'}} r_m\, v(m) = \sum_{m \in M_{J'}} s_m\, v(m)$ is realized in a $\mathcal{C}$-class $C \nsubseteq \pi^{-1} J'$. Hence u is a minimal cocycle if and only if $\sum_{m \in M_{J'}} r_m\, u_m = 0$ whenever the vector relation $\sum_{m \in M_{J'}} r_m\, v(m) = 0$ is reachable in a $\mathcal{C}$-class $C \nsubseteq \pi^{-1} J'$.

For every $m \in M_{J'}$ there is by (B*) a relation $v(m) = \sum_{n \in B \cup D} r_n\, v(n)$ which is reachable in C_m. Hence every minimal cocycle u satisfies $u_m = \sum_{n \in B \cup D} r_n\, u_n$. Therefore a minimal cocycle u is uniquely determined by its values $(u_n)_{n \in B \cup D}$ on $B \cup D$, which can be chosen arbitrarily since no nontrivial vector relation $\sum_{n \in B \cup D} r_n\, v(n) = 0$ can be realized in any $\mathcal{C}$-class $C \nsubseteq \pi^{-1} J'$, by (A*). Thus $MZ^1(S, \mathbb{G}) \cong \prod_{n \in B \cup D} A$ has $|A|^{|B \cup D|}$ elements.

If there is a nontrivial vector relation $r : \sum_{n \in B \cup D} r_n\, v(n) = 0$ between the vectors $v(n)$ with $n \in B \cup D$, then

$$\sum_{n \in B \cup D} r_n\, u_n \;=\; 0$$

holds for every minimal coboundary u, by Lemma 2.3. If A is a cyclic group of suitable prime order p, then p does not divide every r_n and there is a minimal cocycle u such that $\sum_{n \in B \cup D} r_n\, u_n \neq 0$; for instance, let $u_m \neq 0$, where p does not divide r_m, and $u_n = 0$ for all $n \neq m$, $n \in B \cup D$. Then u is not a minimal coboundary and $H^2(S, \mathbb{G}) = MZ^1(S, \mathbb{G}) \,/\, MB^1(S, \mathbb{G}) \neq 0$. If therefore (Rf) holds, then there can be no nontrivial relation $\sum_{n \in B \cup D} r_n\, v(n) = 0$. Hence $B \cup D$ is a defining basis of $K_{J'}$. □

4. We can now prove:

Theorem 3.4. *For a finite commutative nilmonoid S the following conditions are equivalent:*

(1) $H^2(S,\mathbb{G}) = 0$ *whenever* $\mathbb{G}$ *is thin and surjecting;*

(2) $H^2(S,\mathbb{G}) = 0$ *whenever* $\mathbb{G}$ *is thin finite and surjecting;*

(3) *in some presentation* $\pi : F \longrightarrow S$ *(with* F *finitely generated),* G *has a defining basis (relative to* 0*);*

(4) *in every presentation* $\pi : F \longrightarrow S$ *(with* F *finitely generated),* G *has a defining basis (relative to* 0*).*

If for example S is the Volkov nilmonoid (Example XII.4.9), we saw in Section XIII.4 that G has a defining basis; we also saw that $H^2(S,\mathbb{G}) = 0$ whenever $\mathbb{G}$ is thin finite and surjecting.

Proof. If (2) holds, then $G = K = K_0$ by Lemma 3.1 and $G = K_0$ has a defining basis by Lemma 3.2, applied to the ideal $J = \{0\}$. Thus (2) implies (4). It remains to show that (3) implies (1).

Assume that G has a defining basis B relative to 0, in some presentation $\pi : F \longrightarrow S$ where F is finitely generated. Then $K = G$ and $H^2(S,\mathbb{G}) = 0$ whenever $\mathbb{G}$ is thin and almost null, by Proposition 2.2. Now let $\mathbb{G}$ be thin, surjecting, and reduced ($G_0 = 0$).

Since B is a defining basis of G there is for every $m \in M_0 \backslash B$ a relation

$$v(m) \;=\; \textstyle\sum_{n\in B} r_n \, v(n) \;=\; \sum_{n\in B,\, r_n\neq 0} r_n \, v(n)$$

which is reachable in C_m; in particular $\pi n \geqq_{\mathcal{H}} \pi m$ when $r_n \neq 0$. Then

$$u_m \;=\; \textstyle\sum_{n\in B,\, r_n\neq 0} r_n \, \gamma^{\pi n}_{\pi m} u_n$$

for every minimal cocycle u. Also $u_m = 0$ whenever $\pi m = 0$ since $\mathbb{G}$ is reduced. Thus a minimal cocycle is determined by its values $(u_n)_{n\in B}$ on B. (The latter can be chosen arbitrarily, as readily shown, so that $MZ^1(S,\mathbb{G}) \cong \prod_{n\in B} G_{\pi n}$.)

Since the vectors $v(n)$ with $n \in B$ constitute a basis of G, their coordinate matrix $\big(v(n)_x\big)_{n\in B,\, x\in X}$ has an inverse, which is an integer matrix $\big(t_{n,x}\big)_{n\in B,\, x\in X}$ such that

$$\textstyle\sum_{x\in X} v(m)_x \, t_{n,x} \;=\; 1 \text{ if } m = n, \;\; 0 \text{ if } m \neq n$$

for all $m,n \in B$. Let u be any minimal cocycle. Since $\mathbb{G}$ is surjecting there is for every $n \in B$ some $h_n \in G_1$ such that $u_n = \gamma^1_{\pi n} h_m$. For every $x \in X$ let

$$g_x = \sum_{n\in B} t_{n,x}\, \gamma^1_{\pi x} h_n \in G_{\pi x}\,.$$

Since $v(m)_x \neq 0$ implies $x \leqq m$ or $x \leqq q(m)$, and $\pi x \geqq_{\mathcal{H}} \pi m$, we have

$$\begin{aligned}(\delta g)_m &= \textstyle\sum_{x\in X,\, x\leqq m} m_x\, \gamma^{\pi x}_{\pi m} g_x - \sum_{x\in X,\, x\leqq q(m)} q(m)_x\, \gamma^{\pi x}_{\pi m} g_x \\ &= \textstyle\sum_{x\in X,\, \pi x\geqq \pi m} v(m)_x\, \gamma^{\pi x}_{\pi m} g_x \\ &= \textstyle\sum_{x\in X,\, \pi x\geqq \pi m} \sum_{n\in B} v(m)_x\, t_{n,x}\, \gamma^1_{\pi m} h_n \\ &= \textstyle\sum_{x\in X} \sum_{n\in B} v(m)_x\, t_{n,x}\, \gamma^1_{\pi m} h_n = \gamma^1_{\pi m} h_m = u_m\end{aligned}$$

for every $m \in B$. Since u and δg are minimal cocycles, this implies $u = \delta g$. Thus $H^2(S,\mathbb{G}) = 0$ whenever $\mathbb{G}$ is thin, surjecting, and reduced. By Proposition 2.1, $H^2(S,\mathbb{G}) = 0$ whenever $\mathbb{G}$ is thin and surjecting. □

Theorem 3.4 does not extend immediately to every finite group-free semigroup S. If indeed S has two generators and $H^2(S,\mathbb{G}) = 0$ whenever $\mathbb{G}$ is thin finite and surjecting, then S is a either a semilattice or a nilmonoid (Grillet [2000T]). Theorem XIII.5.6 also shows that a partially free semigroup S does not in general satisfy $H^2(S,\mathbb{G}) = 0$ whenever $\mathbb{G}$ is thin finite and surjecting, even though G always has a defining basis, by Lemmas XIII.5.1 and XIII.5.3.

5. We conclude this section with some examples. All examples have two generators c and d; in their standard presentation we let $X = \{x, y\}$, with $\pi x = c$, $\pi y = d$, and order F lexicographically, with $ix + jy \sqsubset kx + ly$ if and only if either $i < k$, or $i = k$ and $j < l$.

Example 3.5. Let S be the nilmonoid with the presentation

$$S \cong \langle\, c,d \mid c^3d = d^5,\ c^4 = c^2d^3,\ c^5 = c^3d^2 = c^2d^5 = d^6 = 0\,\rangle.$$

The nontrivial $\mathcal{C}$-classes (other than the zero class) are $\{5y,\, 3x+y\}$, $\{2x + 3y,\, 4x\}$, and $\{x+5y,\, 2x+4y,\, 4x+y\}$. Hence M_0 consists of $m_1 = 3x+y$, $m_2 = 4x$, $m_3 = 2x+4y$. We have $q_1 = q(m_1) = 5y$, $q_2 = q(m_2) = 2x+3y$, $q_3 = q(m_3) = x+5y$, and $v(m_1) = 3x-4y$, $v(m_2) = 2x-3y$, $v(m_3) = x-y$. Thus $v(m_1)$ and $v(m_2)$ constitute a basis of G (since $\begin{vmatrix} 3 & -4 \\ 2 & -3 \end{vmatrix} = -1$).

There are two paths from $4x+y$ to $x+5y$:

$$4x+y \xrightarrow{m_1} x+5y \quad\text{and}\quad 4x+y \xrightarrow{m_2} 2x+4y \xrightarrow{m_3} x+5y;$$

the corresponding overpaths are m_1 and m_2, m_3. Thus the relation $v(m_3) = v(m_1) - v(m_2)$ is realized in C_{m_3}, and $\{m_1, m_2\}$ is a defining basis of G. Therefore $H^2(S,\mathbb{G}) = 0$ whenever $\mathbb{G}$ is thin finite and surjecting. □

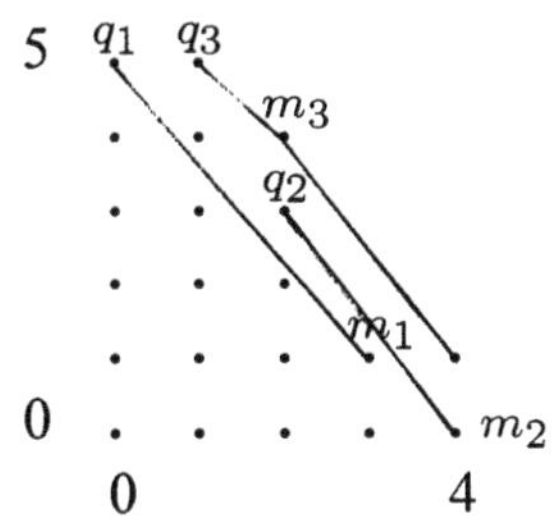

Example 3.5

Example 3.6. Let S be the nilmonoid with the presentation

$$S \cong \langle\, c, d \mid c^3 = c^2d^2 = d^3,\ c^4 = c^3d = cd^3 = d^4 = 0 \,\rangle.$$

There is one nontrivial $\mathcal{C}$-class (other than the zero class): $\{\,3y,\ 2x+2y,\ 3x\,\}$. Hence M_0 consists of $m_1 = 2x+2y$ and $m_2 = 3x$, with $q = q(m_1) = q(m_2) = 3y$ and $v(m_1) = 2x - y$ and $v(m_2) = 3x - 3y$.

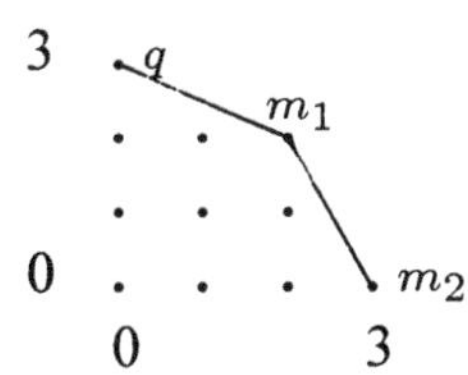

Example 3.6

The defining vectors $v(m_1)$ and $v(m_2)$ are linearly independent but do not constitute a basis of G (since $\begin{vmatrix} 2 & -1 \\ 3 & -3 \end{vmatrix} = -3$). Hence $K \neq G$ and (Nf) does not hold: there exists an almost null functor $\mathbb{G}$ such that $H^2(S, \mathbb{G}) \neq 0$. Since $G/K \cong \mathbb{Z}_3$ the almost null functor with $G_0 = \mathbb{Z}_3$ has $H^2(S, \mathbb{G}) \cong \operatorname{Hom}(\mathbb{Z}_3, \mathbb{Z}_3) \cong \mathbb{Z}_3$. However, $\{m_1, m_2\}$ is a defining basis of K. □

Example 3.7. Let S be the nilmonoid with the presentation

$$S \cong \langle\, c, d \mid c^6 = cd^7,\ c^5d = c^3d^4 = d^8,\ c^7 = c^5d^2 = c^2d^5 = d^9 = 0 \,\rangle.$$

The nontrivial $\mathcal{C}$-classes (other than the zero class) are $\{\,8y,\ 3x+4y,\ 5x+y\,\}$, $\{\,x+7y,\ 6x\,\}$, and $\{\,x+8y,\ 4x+4y,\ 6x+y\,\}$. Hence M_0 consists of $m_1 = 3x+4y$, $m_2 = 5x+y$, and $m_3 = 6x$. We have $q = q(m_1) = q(m_2) = 8y$, $q_3 = q(m_3) = x+7y$, and $v(m_1) = 3x-4y$, $v(m_2) = 5x-7y$, $v(m_3) = 5x-7y$; $v(m_1)$ and $v(m_2)$ constitute a basis of G, since $\begin{vmatrix} 3 & -4 \\ 5 & -7 \end{vmatrix} = -1$.

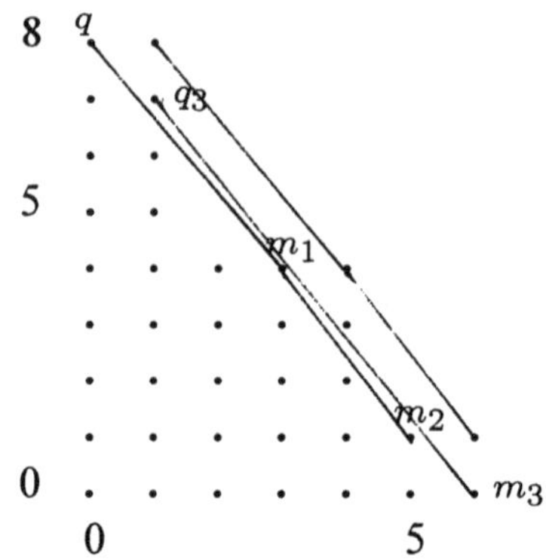

Example 3.7

The only $\mathcal{C}$-class with two overpaths (other than the zero class) is $C = \{x + 8y,\ 4x + 4y,\ 6x + y\}$, which does not contain m_2 or m_3. The relation $v(m_3) = v(m_2)$ is realized in C but it is not reachable in C_{m_3}. Therefore $\{m_1, m_2\}$ is not a defining basis of G. Similarly $\{m_1, m_3\}$ is not a defining basis of G. Therefore G does not have a defining basis and (Rf) does not hold; $H^2(S, \mathbb{G}) \neq 0$ for some functor $\mathbb{G}$ which is thin finite surjecting and reduced.

The proof of Lemma 3.2 provides such a functor. Let $t = c^5 d = c^3 d^4 = d^8$, so that $J_0 = \{0, t\}$ and πm_1, πm_2, $\pi m_3 \notin J_0$. Since $v(m_1)$, $v(m_2)$, $v(m_3)$ are not linearly independent, there is a finite abelian group A such that $H^2(S, \mathbb{G}) \neq 0$ when G is semiconstant with $G_0 = G_t = 0$ and $G_s = A$ for all other $s \in S$. The proof of Lemma 3.2 shows that every minimal cochain is a minimal cocycle, whereas a minimal coboundary u must satisfy $u_{m_2} = u_{m_3}$; accordingly $H^2(S, \mathbb{G}) \cong A$ and any finite abelian group $A \neq 0$ serves. □

Example 3.8. If we delete the relation $c^6 = cd^7$ from the presentation of S in Example 3.7, we obtain a nilmonoid

$$S \cong \langle\, c, d \mid c^5 d = c^3 d^4 = d^8,\ c^7 = c^5 d^2 = c^2 d^5 = d^9 = 0\,\rangle$$

for which M_0 consists only of $m_1 = 3x + 4y$ and $m_2 = 5x + y$, and is a defining basis of G. Then $H^2(S, \mathbb{G}) = 0$ whenever $\mathbb{G}$ is thin finite and surjecting.

REFERENCES.

Abdali, S.K. [1993] *Parallel computations in $*$-semirings,* Computational Algebra (Fairfax VA, 1993), 1-16. Marcel Dekker, New York 1994.

Abel, N.H. [1826] *Untersuchung der Funktionen zweier unabhanging veranderlicher Grossen x und y, wie $f(x,y)$, welche die Eigenschaft haben, dass $f(z,f(x,y))$ eine symmetrische Funktion von z,x und y ist,* J. Reine Angew. Math. 1 (1826) 11-15.

Abrgan, I. [1984] *Prime ideals in groupoids and multiplicative semigroups of residue classes (mod m)* [Russian], Math. Slovaca 34 (1984) 121-133.

Abellanas, P. [1965] *Structure of commutative semi-groups* [Spanish], Rev. Mat. Hisp.-Amer. (4) 25 (1965) 3-44.

Adámek, J. & Koubek, V. [1977] *On a representation of semigroups by products of algebras and relations,* Colloq. Math. 38 (1977) 7-25.

Adams, W.W. & Loustaunau, P. [1994] *An introduction to Gröbner bases,* Amer. Math. Soc., Providence RI, 1994.

Ahsan, J., Khan, M.F., Shabir, M., & Takahashi, M. [1991] *Characterizations of monoids by p-injective and normal S-systems,* Kobe J. Math. 8 (1991) 173-190.

Almqvist, G. [1968] *Fractional categories,* Arkiv för Math. 7 (1968) 449-475.

Altmann, K. [1996] *Torsion of differentials on toric varieties,* Semigroup Forum 53 (1996) 89-97.

Anderson, D.D. & Johnson, E.W. [1984] *Ideal theory in commutative semigroups,* Semigroup Forum 30 (1984) 127-158.

Anderson, D.D. & Stickles, J. [2000] *Commutative rings with finitely generated multiplicative semigroups,* Semigroup Forum 60 (2000) 436-443.

Anderson, D.F. & Scherpenisse, C. [1995] *Factorization in $K[S]$,* Commutative ring theory (Fès 1995), 45-56. Marcel Dekker, New York 1997.

Anderson, J.A. [1974] *The tensor product of a semigroup with a union of groups,* Semigroup Forum 8 (1974) 65-68.

[1978] *Tensor product of commutative semigroups,* Semigroup Forum 16 (1978) 153-164.

[1979A] *Characters of commutative semigroups. I,* Math. Sem. Notes Kobe Univ. 7 (1979) 301-308.

[1997] *The structure of the tensor product of commutative regular semigroups generated by posets of groups,* Math. Japon. 46 (1997) 57-62.

[1998] *Tensor product of commutative regular semigroups generated by posets of groups,* Math. Japon. 48 (1998) 181-186.

Anderson, J.A. & Kimura, N. [1978] *The tensor product of semilattices,* Semigroup Forum 16 (1978) 83-88.

[1978] *Tensor product of commutative unions of groups,* Proc. Amer. Math. Soc. 73 (1979) 149-153.

[1979B] *Characters of commutative semigroups, II. Topological semigroups,* Math. Sem. Notes Kobe Univ. 7 (1979) 491-498.

[1979C] *Characters of commutative semigroups, III. Bicharacters,* Math. Sem. Notes Kobe Univ. 7 (1979) 499-508.

Anderson, M. & Feil, T. [1988] *Lattice-ordered groups,* D. Reidel, Dordrecht-Boston 1988.

Angermüller, G. [1977] *Die Wertehalbgruppe einer ebenen irreduziblen algebroiden Kurve,* Math. Z. 153 (1977) 267-282.

Apéry, R. [1946] *Sur les branches superlinéaires des courbes algébriques,* C. R. Acad. Sci. Paris 222 (1946) 1198-1200.

Arendt, B.D. [1975] *On semisimple commutative semigroups,* Trans. Amer. Math. Soc. 208 (1975) 341-351.

Arendt, B.D. & Stuth, C.J. [1970A] *On the structure of commutative periodic semigroups,* Pacific J. Math. 35 (1970) 1-6.

[1970B] *On partial homomorphisms of semigroups,* Pacific J. Math. 35 (1970) 7-9.

[1972] *Translations of commutative unique factorization semigroups,* Semigroup Forum 4 (1972) 87-92.

Aubert, K.E. [1953] *On the ideal theory of commutative semigroups,* Math. Scand. 1 (1953) 39-54.

[1962] *Theory of x-ideals,* Acta Math. 107 (1962) 1-52.

Aucoin, K.D. [1995] *The structure of commutative ideal semigroups,* Semigroup Forum 50 (1995) 295-300.

[1999] *The structure of commutative semigroups with the ideal extension property,* Semigroup Forum 58 (1999) 175-189.

Aull, C.E. [1966] *Ideals and filters,* Compositio Math. 18 (1966) 79-86.

Backelin, J. [1990] *On the number of semigroups of natural numbers,* Math. Scand. 66 (1990) 197-215.

Baransky, V.A. [1975] *Structural isomorphism of commutative Archimedean semigroups without idempotents* [Russian], Ural. Gos. Univ. Mat. Zap. 9, no.3, 8-13, 133 (1975).

Barr, M. & Beck, J. [1969] *Homology and standard constructions,* Seminar on Triples and Categorical Homology Theory, 245-335. Lecture Notes in Math. no.80, Springer, Berlin-NewYork 1969.

Barucci, V., Dobbs, D.E., & Fontana, M., [1997] *Maximality properties in numerical semigroups and applications to one-dimensional analytically irreducible local domains,* Mem. Amer. Math. Soc. 125 (1997) no. 598.

Barucci, V. & Fröberg, R. [1997] *One-dimensional almost Gorenstein rings,* J. Algebra 188 (1997) 418-442.

Batbedat, A. [1978A] *Le language des schémas pour les demi-groupes commutatifs,* Semigroup Forum 16 (1978) 463-471.

[1978B] *Des schémas en demi-groupes commutatifs,* Semigroup Forum 16 (1978) 473-481.

Bazzoni, S. & Salce, L. [1996] *Groups in the class semigroups of valuation domains,* Israel J. Math. 95 (1996) 135-155.

Beck, J. [1967] *Triples, algebras and cohomology,* Doct. Diss., Columbia Univ. 1967.

Berg, L. [1985] *Distributive Verbände als relativ invertierbare Halbgruppen,* Rostock. Math. Kolloq. No.28 (1985) 66-74.

Berthiaume, P. [1964] *Rational completions of monoids,* Doct. Diss., McGill Univ. 1964.

[1971] *Generalized semigroups of quotients,* Glasgow Math. J. 12 (1971) 150-161.

[1973] *Une remarque sur les monoïdes de quotients généralisés,* Canad. Math. Bull. 16 (1973) 23-25.

Bertin, J. & Carbonne, P. [1975] *Sur la structure des semi-groupes d'entiers et application aux branches,* C. R. Acad. Sci. Paris Sér. A-B 280 (1975) A1745-A1748.

[1977] *Semi-groupes d'entiers et application aux branches,* J. Algebra 49 (1977) 81-95.

Biggs, R., Sasaki M., & Tamura, T. [1965] *Non-negative integer valued functions on commutative groups, I.,* Proc. Japan Acad. 41 (1965) 564-569.

Birjukov, A.P. [1966] *Solution of certain algorithmic problems for finitely determined commutative semigroups* [Russian], Sibirsk. Math. Zh. 7 (1966) 523-530.

[1967] *Certain algorithmic problems for finitely determined commutative semigroups* [Russian], Sibirsk. Math. Zh. 8 (1967) 525-534.

Birkhoff, G. [1944] *Subdirect unions in universal algebra,* Bull. Amer. Math. Soc. 50 (1944) 764-768.

Bisgaard, T.M. [1998] *On perfect semigroups,* Acta Math. Hungar. 79 (1998) 269-294.

Blagojević, D. [1982] *More on anti-inverse semigroups,* Publ. Inst. Math. (Beograd) (N.S.) 31 (45) (1982) 9-13.

Bogacheva, E.V. [1988] *Duality of commutative semigroups* [Russian], Homomorphisms [Russian], 3-9. Leningrad. Gos. Ped. Inst., Leningrad 1988.

[1989] *Absolute approximation of semigroups by bicharacters* [Russian], Semigroup studies of mappings [Russian], 3-7. Leningrad. Gos. Ped. Inst., Leningrad 1989.

[1990A] *On the absolute approximability of commutative semigroups by bicharacters* [Russian], Algorithmic problems in the theory of groups and semigroups [Russian], 43-46. Tul'sk. Gos. Ped. Inst., Tula 1990.

[1990D] *Weak duality of commutative semigroups* [Russian], Studies of semigroups [Russian], 3-9. Leningrad. Gos. Ped. Inst., Leningrad 1990.

Bogomolov, V.S. & Mustafin, T.G. [1988] *Description of commutative monoids over which all polygons are ω-stable* [Russian], Algebra and Logic 28 (1989) 239-247.

Boguta, B.K. [1976] *The determination of commutative semigroups by their endomorphism semigroups* [Russian], Modern algebra no. 4 [Russian] 26-30, Leningrad. Gos. Ped. Inst., Leningrad 1976.

Borisov, A.A. [1982] *Lattice determinacy of a class of commutative semigroups without idempotents* [Russian], Izv. Vysš. Učebn. Zaved. Mat. 1982 no.12, 11-16.

[1984] *Commutative semigroups that are lattice-isomorphic to commutative Archimedean semigroups without idempotents* [Russian], Izv. Vysš. Učebn. Zaved. Mat. 1984 no.9, 14-20.

[1986] *Lattice isomorphisms of commutative semigroups without idempotents* [Russian], Izv. Vysš. Učebn. Zaved. Mat. 1986 no.2, 54-56, 84.

[1992] *Lattice isomorphisms of commutative holoidal semigroups without idempotents* [Russian], Partitions and holomorphic mappings of semigroups [Russian], 3-14. "Obrazovanie", St. Petersburg 1992.

Bouvier, A. [1969] *Idéaux d'un demi-groupe de fractions,* Séminaire P. Lefebvre, 1968/69, no.10. Dép. Math., Univ. Claude Bernard, Villeurbanne 1969.

[1970] *Sur la factorisation dans les demi-groupes commutatifs,* Publ. Dép. Math. (Lyon) 7 (1970) 187-216.

Bouvier, A. & Faisant, A. [1970] *Propriétés des demi-groupes de fractions. III,* Séminaire P. Lefebvre, 1969/70, no.11. Dép. Math., Univ. Lyon I, Villeurbanne 1970.

Bouvier, C. & Gonzales-Springberg, G. [1992] *G-désingularisations de variétés toriques,* C. R. Acad. Sci. Paris Sér. I Math. 315 (1992) 817-820.

[1995] *Système générateur minimal, diviseurs essentiels et G-désingularisations de variétés toriques,* Tôkoku Math. J. (2) 47 (1995) 125-149.

Brameret, M.P. [1962] *Sur les demi-groupes d'homomorphismes,* C. R. Acad. Sci. Paris 254 (1962) 2115-2116.

Brauer, A. [1942] *On a problem of partitions,* Amer. J. Math. 64 (1942) 299-312.

Bredihin, B.M. [1953] *On characters of numerical semigroups with a sufficiently sparse base* [Russian], Dokl. Akad. Nauk. SSSR 90 (1953) 707-710.

[1954] *On sum functions of characters of numerical semigroups* [Russian], Dokl. Akad. Nauk. SSSR 94 (1954) 609-612.

[1958A] *Free numerical semigroups with power densities* [Russian], Dokl. Akad. Nauk. SSSR 118 (1958) 855-857.

[1958B] *Free numerical semigroups with power densities* [Russian], Mat. Sb. (N.S.) 46 (88) (1958) 143-158.

[1960] *Elementary solutions of inverse problems on bases of free semigroups* [Russian], Mat. Sb. (N.S.) 50 (92) (1960) 221-232.

Bredihin, D.A. [1976] *Bands of congruences of semigroups* [Russian], Modern algebra no. 4 [Russian] 31-47, Leningrad. Gos. Ped. Inst., Leningrad 1976.

Bresinsky, H. [1972] *Semigroups corresponding to algebroid branches in the plane,* Proc. Amer. Math. Soc. 32 (1972) 381-384.

[1975] *Symmetric semigroups of integers generated by 4 elements,* Manuscripta Math. 17 (1975) 205-219.

[1979] *Monomial Gorenstein curves in A^4 as set-theoretic complete intersections,* Manuscripta Math. 27 (1979) 353-358.

[1984] *Monomial Buchsbaum ideals in P^r,* Manuscripta Math. 47 (1984) 105-132.

[1988] *Binomial generating sets for monomial curves, with applications in A^4,* Rend. Sem. Mat. Univ. Politec. Torino 46 (1988) 353-370.

Briales, E., Campillo, A., Marijuán, C., & Pisón, P. [1998] *Minimal systems of generators for ideals of semigroups,* J. Pure Appl. Algebra 124 (1998) 7-30.

Brillouet-Belluot, N. [1996] *Multiplicative symmetry and reated functional equations,* Aequationes Math. 33 (1996) 187-191.

Brøndsted, A. [1983] *An introduction to convex polytopes,* Springer-Verlag 1983.

Brown, D.R. & LaTorre, J.G. [1966] *A characterization of uniquely divisible semigroups,* Pacific J. Math. 18 (1966) 57-60.

Brown, W.C. & Curtis, F. [1991] *Numerical semigroups of maximal and almost maximal length,* Semigroup Forum 42 (1991) 219-235.

Bruns, W. & Gubeladze, J. [1996] *Rectangular simplicial semigroups,* Commutative algebra, algebraic geometry, and computational methods (Hanoi 1996), 201-213. Springer, Singapore 1999.

[1999] *A regularity criterion for semigroup rings,* Georgian Math. J. 6 (1999) 259-262.

Buchweitz, R.O. [1980] *On Zariski's criterion for equisingularity and nonsmoothable monomial curves,* Preprint 113, Univ. of Hannover, 1980.

Budach, L. [1964] *Struktur Noethersche kommutativer Halbgruppen,* Monatsb. Deutsch. Akad. Wiss. Berlin 6 (1964) 85-88.

Bulman-Fleming, S. & McDowell, K. [1980] *Flatness and projectivity in commutative regular semigroups,* Semigroup Forum 19 (1980) 247-259.

Burgess, D.A. [1968] *On a problem of Baayen and Kruyswijk,* Proc. Edinburgh Math. Soc. (2) 15 (1968/69) 145-149.

[1969] *A problem on semi-groups,* Studia Sci. Math. Hungar. 4 (1969) 9-11.

Butzbach, P. [1972] *Les 1-convexes du monoïde commutatif* $\mathbb{N}^n$, J. Comput. System Sci. 6 (1972) 283-289.

Buzashi, K. & Vishnyakova, N.I. [1986] *Characterization of locally cyclic nilsemigroups* [Russian], Publ. Math. Debrecen 33 (1986) 99-106.

Byeveld, S. & Easdown, D. [1992] *The minimal faithful degree of a finite commutative inverse semigroup,* Computational algebra and number theory (Sidney 1992), 19-27. Kluwer Acad. Publ., Dordrecht 1995.

Byrd, R.D., Lloyd, J.T., Pedersen, F.D., & Stepp, J.W. [1977] *Automorphisms of the semigroup of finite complexes of a periodic locally cyclic group,* Pacific J. Math. 72 (1977) 27-39.

Caillot, J.F. & Wehrung, F. [2000] *Finitely presented, coherent, and ultrasimplicial ordered abelian groups,* Semigroup Forum 61 (2000) 116-137.

Campillo, A. & Marijuan, C. [1991] *Higher order relations for a numerical semigroup,* Sém. Théor. Nombres Bordeaux (2) 3 (1991) 249-260.

Campillo, A. & Pisón, P. [1993] *Generators of a monomial curve and graphs for the associated semigroup,* Bull. Soc. Math. Belg. Sér. A 45 (1993) 45-58.

Carbonne, P. [1983] *Cohomologie des semi-groupes d'entiers,* J. Algebra 84 (1983) 1-13.

Carlisle, W.H. [1971] *Residual finiteness of finitely generated commutative semigroups,* Pacific J. Math. 36 (1971) 99-101.

Cavaliere, M.P. & Niesi, G. [1983] *On monomial curves and Cohen-Macaulay type,* Manuscripta Math. 42 (1983) 147-159.

[1984] *Complete intersection projective monomial curves* [Italian], Boll. Un. Mat. Ital. D (6) 3 (1984) 189-200.

Chacron, J. [1981] *Toutes les congruences du demigroupe additif des entiers naturels,* Proc. 106th National Congress of Learned Societies (Perpignan 1981), 63-70, Bib. Nat. Paris 1982.

Chandran, V.R. [1987] *On a problem of Zelinka, I, II,* Czechoslovak Math. J. 37 (112) (1987) 124, 125-127.

Chapman, S., Krause, U., & Oeljeklaus, E. [2001] *On Diophantine monoids and their class groups,* to appear in Pacific J. Math.

Chen, Q.H. [1988] *On well-semimodules,* Fujian Shifan Daxue Xiebao Ziran Kexue Ban 4 (1988) no.3, 12-17.

Cho, J.R. [1990] *Representations of certain medial algebras,* J. Korean Math. Soc. 27 (1990) 69-76.

Chrislock, J.L. [1969] *On medial semigroups,* J. Algebra 12 (1969) 1-9.

Ch'üan, J.C. & Lee, W.S. [1982] *Permuation of* $\mathbb{N}$ *that preserves semigroups,* Math. Japon. 27 (1982) 597-598.

Cibul'skite, D. [1970] *The distribution of generating elements in free numerical semigroups. I, II, III* [Russian], Litovsk. Mat. Sb. 10 (1970) 397-415, 593-610, 835-844.

Cirić, M. [2002] *Decompositions of semigroups,* Kluwer Acad. Publ., Dordrecht 2002.

Čistov, A.L. [1978] *The number of generators of the semigroup of stable equivalence classes of algebraic tori* [Russian], Dokl. Akad. Nauk. SSSR 242 (1978) 1027-1029.

Clark, W.E., Holland, W.C., & Székely, G.J. [1998] *Decompositions in discrete semigroups,* Studia Sci. Math. Hungar. 34 (1998) 15-23.

Clarke, J.L. [1980] *A study of commutative cancellative idempotent-free semigroups,* Doct. Diss., Univ. California Davis, 1980.

[1984] *Rees quotients of $\mathcal{N}$-semigroups,* Semigroup Forum 29 (1984) 259-270.

Clifford, A.H. [1938] *Arithmetic and ideal theory of commutative semigroups,* Annals of Math. 39 (1938) 594-610.

[1941] *Semigroups admitting relative inverses,* Annals of Math. 42 (1941) 1037-1049.

[1950] *Extensions of semigroups,* Trans. Amer. Math. Soc. 68 (1950) 165-173.

[1954] *Bands of semigroups,* Proc. Amer. Math. Soc. 5 (1954) 499-504.

Clifford, A.H., & Preston, G.B. [1961] *The algebraic theory of semigroups, vol.I,* Amer. Math. Soc., Providence RI, 1961.

[1967] *The algebraic theory of semigroups, vol.II,* Amer. Math. Soc., Providence RI, 1967.

Comfort, W.W. [1959] *The isolated points in the dual of a commutative semigroup,* Proc. Amer. Math. Soc. 11 (1960) 227-233.

Comfort, W.W. & Hill, P. [1966] *On extending nonvanishing semicharacters,* Proc. Amer. Math. Soc. 17 (1966) 936-941.

Coppens, M. [1988] *Weierstrass points with two prescribed nongaps,* Pacific J. Math. 131 (1988) 71-104.

[1995] *The existence of base point free linear systems on smooth plane curves,* J. Algebraic Geom. 4 (1995) 1-15.

Croisot, R. [1953] *Demi-groupes inversifs et demi-groupes réunions de demi-groupes simples,* Ann. Sci. Ecole Norm. Sup. (3) 70 (1953) 361-379.

[1957] *Equivalences principales bilatères dans les demi-groupes,* J. Math. Pures Appl. (9) 36 (1957) 373-417.

Čudakov, N.G. & Bredihin, B.M. [1956] *Application of Parseval's equality for the estimation of sum functions of characters of numerical semigroups* [Russian], Ukrain. Mat. Zh. 8 (1956) 347-360.

Čudakov, N.G. & Pavlyučuk, A.K. [1951] *On summation functions of characters of numerical semigroups with a finite basis* [Russian], Trudy Mat. Inst. Steklov., v.38, 366-381. Izdat. Akad. Nauk. SSSR, Moscow 1951.

Curtis, F. [1990] *On formulas for the Frobenius number of a numerical semigroup,* Math. Scand. 67 (1990) 190-192.

D'Anna, M. [1997] *Type sequences of numerical semigroups,* Semigroup Forum 56 (1998) 1-31.

D'Anna, M. & Delfino, D. [1997] *Integrally closed ideals and type sequences in one-dimensional local rings,* Rocky Mountain J. Math. 27 (1997) no.4, 1065-1073.

Darnel, M.R. [1995] *Theory of lattice-ordred groups,* Marcel Dekker, New York 1995.

de Bruijn, N.G. [1956] *On number systems,* Nieuw Arch. Wisk. (3) 4 (1956) 15-17.

Deddens, J.A. [1979] *A combinatorial identity involving relatively prime integers,* J. Combin. Theory Ser. A 26 (1979) 189-192.

de Luca, A. & Restivo, A. [1984] *A finiteness condition for finitely generated semigroups,* Semigroup Forum 28 (1984) 123-134.

Delgado, F. [1988] *Gorenstein curves and symmetry of the semigroup of values,* Manuscripta Math. 61 (1988) 285-296.

[1990] *The symmetry of the Weierstrass generalized semigroups and affine embeddings,* Proc. Amer. Math. Soc. 108 (1990) no.3, 627-631.

Delorme, C. [1976] *Sous-monoïdes d'intersection complète de N,* Ann. Sci. Ecole Norm. Sup. (4) 9 (1976) 145-154.

Dickinson, R.P., Jr. [1972] *Noncancellative congruences on N-semigroups,* Proc. Amer. Math. Soc. 36 (1972) 317-325.

Dickson, L.E. [1913] *Finiteness of the odd perfect and primitive abundant numbers with n distinct prime factors,* Amer. J. Math. 35 (1913) 413-422.

Dobbertin, H. [1982] *On Vaught's criterion for isomorphisms of countable Boolean algebras,* Algebra Universalis 15 (1982) 95-114.

[1983] *Refinement monoids, Vaught monoids, and Boolean algebras,* Math. Ann. 265 (1983) 473-487.

[1984] *Primely generated regular refinement monoids,* J. Algebra 91 (1984) 166-175.

Dorofeeva, M.P. [1972] *Herediary and semi-hereditary monoids,* Semigroup Forum 4 (1972) 301-311.

Drbohlav, K. [1963] *Concerning congruence relations on commutative semigroups,* Comment. Math. Univ. Carolinae 4 (1963) 11-13.

[1964] *Zur Theorie der Kongruenzrelationen auf kommutativen Halbgruppen,* Math. Nachr. 26 (1963/64) 233-245.

Dress, A. & Grabmeier, J. [1991] *The interpolation problem for k-sparse polynomials and character sums,* Adv. in Appl. Math. 12 (1991) 57-75.

Dubreil, P. [1941] *Contribution à la théorie des demi-groupes,* Mém. Acad. Sci. Inst. France (2) 63, no.3 (1941).

Duske, J. [1976] *Zur Theorie kommutativer Automaten,* Acta Cybernet. 3 (1976) 15-23.

Dyadchenko, G.G. & Shokuev, V.N. [1985] *Idempotents of a multiplicative semigroup of the ring $\mathbb{Z}_n$* [Russian], Algebraic systems with one action and relation [Russian], 18-20. Leningrad. Gos. Ped. Inst., Leningrad 1985.

Easdown, D. & Gould, V. [1996] *Commutative orders,* Proc. Roy. Soc. Edinburgh Sect. A, 126 (1996) no.6, 1201-1216.

Ebanks, B.R. [1979] *Branching measures of information on strings,* Houston J. Math. 8 (1982) 323-332.

Eberhart, C. [1982] *On abelian semigroups whose congruences form a simple lattice,* Houston J. Math. 8 (1982) 323-332.

Eigenthaler, G., Nöbauer, W. & Wiesenbauer, J. [1987] *Über Halbgruppen vertauschbarer Polynome,* Österreich. Akad. Wiss. Math.-Natur. Kl. Sitzungsber. II, 196 (1987) 227-247.

Eigenthaler, G. & Woracek, H. [1997] *A remark on permutable polynomials,* Contributions to general algebra, 10 (Klagenfurt 1997), 139-142. Heyn, Klagenfurt 1998.

Eilenberg, S. & Schützenberger, M.P. [1969] *Rational sets in commutative monoids,* J. Algebra 13 (1969) 173-191.

Ego, M. [1961] *Structure des demi-groupes dont le treillis des sous-demi-groupes est distributif,* C. R. Acad. Sci. Paris 252 (1961) 2490-2492.

[1962A] *Structure des demi-groupes dont le treillis des sous-demi-groupes est modulaire ou semi-modulaire,* C. R. Acad. Sci. Paris 254 (1962) 1723-1725.

[1962B] *Structure des demi-groupes dont le treillis des sous-demi-groupes satisfait aux conditions $\overline{C}_2$, C_2, $\overline{C}_1$, C_1, à la semi-modularité affaiblie, ou à la modularité affaiblie,* C. R. Acad. Sci. Paris 255 (1962) 1840-1842.

[1962C] *Structure des demi-groupes dont le treillis des sous-demi-groupes satisfait aux conditions $\overline{C}_2$, C_2, $\overline{C}_1$, C_1, à la semi-modularité affaiblie, ou à la modularité affaiblie,* C. R. Acad. Sci. Paris 255 (1962) 2699-2701.

[1963] *Structure des demi-groupes dont le treillis des sous-demi-groupes satisfait à certaines conditions,* Bull. Soc. Math. France 91 (1963) 137-201.

Emeličev, V.A. [1962] *The solution of certain algorithmic problems for commutative semigroups* [Russian], Dokl. Akad. Nauk. SSSR 144 (1962) 261-263.

[1966] *Regularity of a finitely determined commutative semigroup* [Russian], Vesci Akad. Navuk. BSSR Ser. Fiz.-Mat. Navuk 1966, no.2, 23-30.

Ermolina, N.G. [1988] *Approximation of commutative semigroups by real and finite bicharacters* [Russian], Homomorphisms [Russian], 27-43. Leningrad. Gos. Ped. Inst., Leningrad 1988.

Esterle, J. [1997] *Embedding semigroups into commutative radical Banach algebras,* Banach algebras '97 (Blaubeuren 1997), 153-168. de Gruyter, Berlin 1998.

Etayo, J.J. [1965] *A category of semigroups,* Fourth Annual Meeting of Spanish Mathematicians [Spanish], 79-88. Univ. Salamanca, Salamanca 1965.

Evseev, A.E. [1965] *Semigroups with ordinally decomposable semi-lattices of subsemigroups* [Russian], Izv. Vysš. Učebn. Zaved. Mat. 1965 no.6, 74-84.

[1991] *Commutative semigroups, every partition of which contains a subsemigroup* [Russian], Semigroups and their homomorphisms [Russian], 8-11. Ross. Gos. Ped. Univ., Leningrad 1991.

Ewald, G. & Wessels, U. [1991] *On the ampleness of invertible sheaves in complete projective toric varieties,* Results in Math. 15 (1991) 275-278.

Farrell, E.J. [1992] *An introduction to F-graphs, a graph-theoretic representation of natural numbers,* Internat. J. Math. Math. Sci. 15 (1992) 313-317.

Findlay, G.D. & Lambek, J. [1958] *A generalized ring of quotients. I, II,* Canad. Math. Bull. 1 (1958) 77-85.

Fischer, K.G., Morris, W., & Shapiro, J. [1997] *Affine semigroup rings that are complete intersections,* Proc. Amer. Math. Soc. 125 (1997) no.11, 3137-3145.

Fischer, K.G. & Shapiro, J. [1996] *Mixed matrices and binomial ideals,* J. Pure Appl. Algebra 113 (1996) 39-54.

Forsythe, G.E. [1955] *SWAC computes 126 distinct semigroups of order 4,* Proc. Amer. Math. Soc. 6 (1955) 443-447.

Fountain, J.B. & Petrich, M. [1986] *Completely 0-simple semigroups of quotients,* J. Algebra 101 (1986) 365-402.

Fraenkel, A.S., Porta, H., & Stolarsky, K.B. [1989] *Some arithmetical semigroups,* Analytic number theory (Allerton park 1989), 255-264.

Progr. Math. 85, Birkhäuser, Boston 1990.

Fraser, G.A. & Albert, J.P. [1984] *Associativity of the tensor product of semilattices,* Proc. Edinburgh Math. Soc. (2) 27 (1984) 337-340.

Freyd, P. [1968] *Rédei's finiteness theorem for commutative semigroups,* Proc. Amer. Math. Soc. 19 (1968) 1003.

Frobenius, F.G. [1895] *Über endliche Gruppen,* Sitzungsber. Preuss. Akad. Wiss. Berlin (1895) 163-194.

Fröberg, R. [1994] *The Frobenius number of some semigroups,* Comm. Algebra 22 (1994) 6021-6024.

Fröberg, R., Gottlieb, C., & Häggkvist, R. [1987] *On numerical semigroups,* Semigroup Forum 35 (1987) 63-83.

Fuchs, L. [1970] *Infinite abelian groups,* Academic Press, New York, 1970.

Fulp, R.O. [1966] *On homomorphisms of commutative inverse semigroups,* Czechoslovak Math. J. 16 (91) (1966) 72-75.

[1967] *On extending semigroup characters,* Proc. Edinburgh Math. Soc. (2) 15 (1967) 199-202.

[1970] *Tensor and torsion products of semigroups,* Pacific J. Math. 32 (1970) 685-696. correction, ibid. 39 (1971) 827.

Fulp, R.O. & Hill, P. [1965] *Reflexive semigroups,* Proc. Amer. Math. Soc. 16 (1965) 1096-1098.

Galanová, J. [1982] *The tensor product of a semigroup and semigroups of idempotents* [Russian], Math. Slovaca 32 (1982) 55-60.

[1987] *Some properties of tensor products of semigroups,* Demonstratio Math. 20 (1987) 57-65.

Galbiati, J.L. & Veronesi, M.L. [1983] *Permutable commutative semigroups* [Italian], Istit. Lombardo Accad. Sci. Lett. Rend. A 117 (1983) 43-56.

Gale, D. & Tamura, T. [1976] *Maximal cancellative subsemigroups and sub-archimedeanness,* Algebraic theory of semigroups (Szeged, 1976), 147-154. Colloq. Math. Soc. János Bolyai, North-Holland, Amsterdam, 1979.

[1980] *A theorem on commutative cancellative Archimedean idempotent-free semigroups,* Simon Stevin 54 (1980) 233-240.

García-Sánchez, P.A. & Rosales, J.C. [1999] *Numerical semigroups generated by intervals,* Pacific J. Math. 191 (1999) 75-83.

Gastinger, W., Kunz, E., & Waldi, R. [1991] *Relation matrices of monomial curves,* An. Univ. Bucuresti Mat. 40 (1991) 41-53.

Geroldinger, A. [1996] *On the structure and arithmetic of finitely primary monoids,* Czechoslovak Math. J. 46 (121) (1996) 677-695.

[1997] *Chains of factorizations in orders of global fields,* Colloq. Math. 72 (1997) 83-102.

Gilmer, R. [1984] *Commutative semigroup rings* Univ. of Chicago Press, 1984.

Gluskin, L.M. [1971D] *Dense extensions* [Russian], Dokl. Akad. Nauk SSSR 196 (1971) 270-273.

[1971S] *Separative semigroups* [Russian], Izv. Vysš. Učebn. Zaved. Mat. 1971 no.9, 30-39.

Gluskin, L.M. (cont'd) [1973] *Dense extensions of semigroups* [Russian], Mini-Conference on Semigroup Theory (Szeged, 1972) 1-6. Jószef Attila Univ., Szeged 1973.

[1976D] *Dense extensions of commutative semigroups* [Russian], Algebraic theory of semigroups (Szeged, 1976), 155-171. Colloq. Math. Soc. János Bolyai, North-Holland, Amsterdam, 1979. Correction, Semigroups (Szeged, 1981), 489-490. Colloq. Math. Soc. János Bolyai, North-Holland, Amsterdam, 1985.

[1976E] *Extensions of commutative semigroups* [Russian], Dokl. Akad. Nauk SSSR 230 (1976) 757-760.

[1979] *Extensions of commutative semigroups* [Russian], Izv. Vysš. Učebn. Zaved. Mat. 1979 no.11, 3-12. correction, ibid. 1982 no.7, 88.

[1983I] *Ideals of commutative semigroups* [Russian], Izv. Vysš. Učebn. Zaved. Mat. 1983 no.3, 31-39.

[1983S] *Subreductive commutative semigroups* [Russian], Associative actions [Russian] 18-25, Leningrad. Gos. Ped. Inst., Leningrad 1983.

[1985] *Normal extensions of commutative semigroups* [Russian], Izv. Vysš. Učebn. Zaved. Mat. 1985 no.9, 14-22.

Golubov, È.G. [1969] *Semigroups with certain finitely separable subsemigroups* [Russian], Ural. Gos. Univ. Mat. Zap. 7 (1969/70) tetrad' 1, 35-51.

[1970A] *Conditions for the finite approximability of naturally totally ordered commutative semigroups* [Russian], Izv. Vysš. Učebn. Zaved. Mat. 1970 no.2, 10-18.

[1970S] *Finite separability in semigroups* [Russian], Sibirsk. Mat. Ž. 11 (1970) 1247-1263.

Gonchigdorzh, R. & Mal'tsev, Yu.N. [1982] *On a variety of universal algebras for which the groupoid of subvarieties is a commutative semigroup* [Russian], Sibirsk. Mat. Ž. 23 (1982) no.2, 31-38, 215.

Goodearl, K.R. [1986] *Partially ordered abelian groups with interpolation,* Amer. Math. Soc., Providence RI, 1986.

Gould, M. & Iskra, J.A. [1984G] *Globally determined classes of semigroups,* Semigroup Forum 28 (1984) 1-11.

[1984R] *Embedding in globals of groups and semilattices,* Semigroup Forum 28 (1984) 61-71.

Gould, M., Iskra, J.A. & Pálfy, P.P. [1986] *Embedding in globals of finite semilattices,* Czechoslovak Math. J. 36 (111) (1986) 87-92.

Greco, S. [1990] *Orders of rational functions on algebraic curves,* Proc. 21st. Annual Iranian Math. Conf. (Isfahan 1990), 159-165. Univ. Isfahan, Isfahan.

Greco, S. & Raciti, G. [1991] *The Lüroth semigroup of plane algebraic curves,* Pacific J. Math. 151 (1991) 43-56.

Green, J.A. [1951] *On the structure of semigroups* Annals of Math. 54 (1951) 163-172.

Grillet, P.A. [1969F] *On free commutative semigroups,* J. Natur. Sci. and Math. 9 (1969) 71-78.

[1969T] *The tensor product of commutative semigroups,* Trans. Amer. Math. Soc. 138 (1969) 281-293.

[1970C] *Cancellative commutative semigroups,* Semigroup Forum 1 (1970) 249-253.

[1970E] *The free envelope of a finitely generated commutative semigroup,* Trans. Amer. Math. Soc. 149 (1970) 665-682.

[1970I] *Interpolation properties and tensor product of semigroups,* Semigroup Forum 1 (1970) 162-168.

[1970T] *The tensor product of cancellative semigroups,* J. Natur. Sci. and Math. 10 (1970) 199-208.

[1972] *Subdirect decompositions of finite commutative semigroups,* Semigroup Forum 4 (1972) 321-326.

[1974] *Left coset extensions,* Semigroup Forum 7 (1974) 200-263.

[1975C] *A completion theorem for finitely generated commutative semigroups,* J. Algebra 34 (1975) 25-53.

[1975P *Primary semigroups,* Michigan Math. J. 22 (1975) 321-326.

[1976] *Directed colimits of free commutative semigroups,* J. Pure Appl. Algebra 9 (1976/77) 73-87.

[1977] *On subdirectly irreducible commutative semigroups,* Pacific J. Math. 69 (1977) 55-71.

[1991C] *Commutative semigroup cohomology,* Semigroup Forum 43 (1991) 247-252.

[1991N] *A construction of finite commutative nilsemigroups,* Comm. Algebra 19 (1991) no.11, 3145-3172. errata, ibid. 20 (1992) no.9, 2789.

[1991T] *Nilsemigroups on trees,* Semigroup Forum 43 (1991) 187-201.

[1992] *The commutative cohomology of nilsemigroups,* J. Pure Appl. Algebra 82 (1992) 233-251.

[1993] *Some congruences on free commutative semigroups,* Acta Sci. Math. (Szeged) 58 (1993) 3-23.

[1993R] *A short proof of Rédei's Theorem,* Semigroup Forum 46 (1993) 126-127.

[1995] *Semigroups: an introduction to the structure theory,* Marcel Dekker, New York 1995.

[1995C] *Commutative semigroup cohomology,* Comm. Algebra 23 (1995) no.10, 3573-3587.

[1995F] *The commutative cohomology of finite semigroups,* J. Pure Appl. Algebra 102 (1995) 25-47.

[1995N] *The number of commutative semigroups of order n,* Semigroup Forum 50 (1995) 317-326.

[1995P] *Partially free commutative semigroups,* Acta Sci. Math. (Szeged) 61 (1995) 155-169.

[1995S] *Stratified semigroups,* Semigroup Forum 50 (1995) 25-36.

[1996C] *Finite congruences on free commutative semigroups,* Acta Sci. Math. (Szeged) 62 (1996) 391-405.

[1996N] *Computing finite commutative semigroups,* Semigroup Forum 53 (1996) 140-154.

[1997C] *Cocycles in commutative semigroup cohomology,* Comm. Algebra 25 (1997) no.11, 3427-3462.

[1997Z] *Commutative semigroups with zero cohomology,* Acta Sci. Math. (Szeged) 63 (1997) 463-469.

[1999] *Algebra,* Wiley, New York 1999.

Grillet, P.A. (cont'd) [2000T] *The commutative cohomology of two-generator semigroups,* Semigroup Forum 60 (2000) 4-15.

[2001C] *Congruences on free commutative semigroups,* submitted to Acta Sci. Math. (Szeged)

[2001F] *Fully invariant congruences on free commutative semigroups,* Acta Sci. Math. (Szeged) 67 (2001) 173-201.

[2001H] *4-Cocycles in commutative semigroup cohomology,* submitted to Comm. Algebra

[2001N] *A construction of commutative nilsemigroups,* to appear in Period. Math. Hungar.

[2001S] *Subcomplete commutative nilsemigroups,* submitted to Acta Sci. Math. (Szeged)

[2001Z] *Commutative nilsemigroups with zero cohomology,* Semigroup Forum 62 (2001) 66-78.

Grillet, P.A. & Novikov, B.V. [2002] *Cohomology of semigroups,* to appear in Semigroup Forum.

Grossman, P.A. [1982] *A note on local polynomial functions on commutative semigroups,* J. Austral. Math. Soc. Ser. A 33 (1982) 50-53.

Groza, Gh. [1995] *Semigroups of integral functions in valued fields,* Algebra, logic & discrete mathematics (Niš 1995). Filomat No.9, part 3 (1995) 675-687.

Gubeladze, J. [1998] *The isomorphism problem for commutative monoid rings,* J. Pure Appl. Algebra 129 (1998) 35-65.

Guričan, J. [1984] *The semigroup of general circulant matrices,* Acta Math. Univ. Comenian. 44/45 (1984) 13-21.

Halezov, V.A. [1966] *Algorithmic solution of certain problems in the theory of finitely defined commutative semigroups with cancellation* [Russian], Sibirsk. Math. Zh. 7 (1966) 419-436.

Hall, R.E. [1971] *Commutative cancellative semigroups with two generators,* Czechoslovak Math. J. 21 (1971) 449-452.

[1972] *The translational hull of an N-semigroup,* Pacific J. Math. 41 (1972) 379-389.

Halter-Koch, F. [1993] *The integral closure of a finitely generated monoid and the Frobenius problem in higher dimensions,* Semigroups. Algebraic Theory and applications to formal languages and codes, 86-93, World Scientific, 1993.

Hamilton, H. [1975] *Permutability of congruences on commutative semigroups,* Semigroup Forum 10 (1975/76) 55-66.

[1978L] *Modularity of the congruence lattice of a commutative cancellative semigroup,* Pacific J. Math. 79 (1978) 469-485.

[1978N] *Congruences on N-semigroups,* Pacific J. Math. 75 (1978) 423-448.

[1982] *Modularity and distributivity of the congruence lattice of a commutative separative semigroup,* Math. Japon. 27 (1982) 581-589.

Hamilton, H. & Nordahl, T. [1978] *Semigroups whose congruence lattice is Boolean,* Pacific J. Math. 77 (1978) 131-143.

Hamilton, H., Nordahl, T. & Tamura, T. [1975] *Commutative cancellative semigroups without idempotents,* Pacific J. Math. 61 (1975) 441-456.

Hancock, R. [1960] *On complete semimodules,* Proc. Amer. Math. Soc. 11 (1960) 71-76.

Hansen, R.T. [1969] *Complementing pairs of subsets of the plane,* Duke Math. J. 36 (1969) 441-449.

Hartshorne, R. [1992] *Generalized divisors on Gorenstein rings,* K-Theory 8 (1994) 287-339.

Head, T. [1967A] *Homomorphisms of commutative semigroups as tensor maps,* J. Natur. Sci. and Math. 7 (1967) 39-49.

[1967B] *The tensor product of a group with a semigroup,* J. Natur. Sci. and Math. 7 (1967) 155-159.

[1967C] *The tensor product of semigroups with minimal ideals,* J. Natur. Sci. and Math. 7 (1967) 161-165.

[1967D] *Functor properties of semigroup tensor products,* J. Natur. Sci. and Math. 7 (1967) 167-171.

[1968] *Problems and conjectures concerning tensor products of commutative semigroups,* J. Natur. Sci. and Math. 8 (1968) 211-215.

[1969] *Tensor products and maximal subgroups of commutative semigroups,* Publ. Math. Debrecen 16 (1969) 145-147.

[1971G] *A semigroup-theoretic view of projective class groups,* Proc. Japan Acad. 47 (1971) suppl. I, 831-836.

[1971H] *Commutative semigroups having greatest regular images,* Semigroup Forum 2 (1971) 130-137.

[1979] *Quotient monoids and greatest commutative monoid images of several types,* Semigroup Forum 17 (1979) 351-363.

Head, T. & Anderson, J.A. [1980] *On injective and flat commutative regular semigroups,* Semigroup Forum 21 (1980) 283-284.

Head, T. & Kuroki, N. [1974] *Greatest regular images of tensor products of commutative semigroups,* Kōdai Math. Sem. Rep. 26 (1974/75) 132-136.

Herzinger, K. [1999] *The number of generators for an ideal and its dual in a numerical semigroup,* Comm. Algebra 27 (1999) no.10, 4673-4690.

Herzog, J. [1970] *Generators and relations of abelian semigroups and semigroup rings,* Manuscripta Math. 3 (1970) 175-193.

Heuer, C.V. [1966] *On the equivalence of cancellative extensions of commutative cancellative semigroups by groups,* J. Austral. Math. Soc. 12 (1971) 187-192.

Heuer, C.V. & Miller, D.W. [1966] *An extension problem for cancellative semigroups,* Trans. Amer. Math. Soc. 122 (1966) 499-515.

Hewitt, E. & Zuckerman, H.S. [1955] *Finite dimensional convolution algebras,* Acta Math. 93 (1955) 67-119.

[1956] *The l_1-algebra of a commutative semigroup,* Trans. Amer. Math. Soc. 83 (1956) 70-97.

[1960] *The multiplicative semigroup of integers modulo m,* Pacific J. Math. 10 (1960) 1291-1308.

Higgins, J.C. [1966] *Finitely generated commutative Archimedean semigroups generated N-semigroups,* Doct. Diss., Univ. California Davis, 1966.

[1969I] *A faithful canonical representation for finitely generated N-semigroups,* Czechoslovak Math. J. 19 (1969) 375-379.

[1969S] *Representing N-semigroups,* Bull. Austral. Math. Soc. 1 (1969) 115-125.

Higgins, J.C. (cont'd) [1972] *Subsemigroups of the additive positive integers,* Fibonacci Quart. 10 (1972) 225-230.

[1983] *The varieties of commutative semigroups for which epis are onto,* Proc. Roy. Soc. Edinburgh Sect. A, 94 (1983) 1-7.

[1984] *Semigroup epimorphisms,* Proc. 1984 Marquette conf. on semigroups, 51-64. Marquette Univ., Milwaukee Wis. 1985.

[1985] *Epimorphisms, dominions, and semigroups,* Algebra Universalis 21 (1985) 225-233.

[1990] *A short proof of Isbell's zigzag theorem,* Pacific J. Math. 144 (1990) 47-50.

Higgins, J.C. & Tamura, T. [1973] *Finitely generated N-semigroup and quotient group,* Proc. Japan Acad. 49 (1973) 323-327.

Hildebrant, J.A. [1984] *The translational degree of a semigroup,* Semigroup Forum 30 (1984) 331-349.

Hill, P. [1966] *A solution to the nonvanishing character extension problem,* Proc. Amer. Math. Soc. 17 (1966) 1178-1182.

[1967] *Characters of commutative semigroups,* J. Algebra 5 (1967) 16-24.

Hindman, N. & Woan, W.J. [1992] *Central sets in commutative semigroups and partition regularity of systems of linear equations,* Mathematika 40 (1993) 169-186.

Hinkle, C.V., Jr. [1972] *Semigroups of right quotients of a semigroup which is a semilattice of groups,* Semigroup Forum 5 (1972/73) 167-173.

[1974] *Semigroups of right quotients of a semigroup which is a semilattice of groups,* J. Algebra 31(1974) 276-286.

Hippisley-Cox, S.D. [1992] *Finite point compactifications of* $(\mathbb{N},+)$, Semigroup Forum 44 (1992) 388-392.

Hirabuki, S. & Matsuda, R. [1999] *Note on the Krull-Akizuki theorem for semigroups,* Bull. Yamagata Univ. Natur. Sci. 14 (1999) 155-159.

Hmelnitsky, I.L. [1985] *On semigroups with the idealizer condition,* Semigroup Forum 32 (1985) 135-144.

Hoa, L.T. [1991] *A note on projective monomial surfaces,* Math. Nachr. 154 (1991) 183-188.

Hochster, M. [1972] *Rings of invariants of tori, Cohen-Macaulay rings generated by monomials, and polytopes,* Ann. of Math. 96 (1972) 318-337.

Hoehnke, H.-J. [1965] *Eine Charakterisierung des Radikals kommutativer Halbgruppen,* Arch. Math. 16 (1965) 412-413.

Holme, A. [1966] *A general theory of structure spaces,* Fund. Math. 68 (1966) 335-347.

Homma, M. [1996] *The Weierstrass semigroup of a pair of points on a curve,* Arch. Math. (Basel) 67 (1996) 337-348.

Hora, R.B. & Kimura, N. [1972] *Commutative semigroups of type* (m,n,p,q). *I. Monovariable cases,* Proc. Japan Acad. 48 (1972) 555-560.

Hora, R.B. [1973] *On factorizable simple and commutative semigroups,* J. Natur. Sci. and Math. 13 (1973) 51-61.

Howie, J.M. [1968] *Commutative semigroup amalgams,* J. Austral. Math. Soc. 8 (1968) 609-630.

[1976] *An introduction to semigroup theory,* Academic Press, London 1976.

[1995] *Fundamentals of semigroup theory,*
Clarendon Press, Oxford Univ. Press, New York 1995.

Howie, J.M. & Isbell, J.R. [1967] *Epimorphisms and dominions, II,*
J. Algebra 6 (1967) 7-21.

Hsieh, S.C. [1982] *A note on a saturated commutative semigroup,*
Soochow J. Math. 8 (1982) 55-57.

Huang, I.C. [1995] *A residue map and its applications to some one-dimensional rings,*
Proc. Amer. Math. Soc. 123 (1995) no.8, 2369-2372.

Huang, W.C. [1996] *Matrices which belong to an idempotent in a sandwich semigroup of circulant Boolean matrices,* Linear Algebra Appl. 249 (1996) 157-167.

Hule, H. [1976] *Über die Eindeutigkeit der Lösungen algebraischer Gleichungssysteme,*
J. Reine Angew. Math. 282 (1976) 157-161.

Hurwitz, A. [1893] *Über algebraischer Gebilde mit eindeutigen Transformationen in sich,* Math. Ann. 41 (1893) 403-442.

Huỳnh, D.T. [1986] *A superexponential lower bound for Gröbner bases and Church-Rosser commutative Thue systems,* Inform. and Control 68 (1986) 196-206.

Ignat'eva, I.V. [1996] *SH-approximatoin of semigroups by finite characters* [Russian], Modern algebra, No.I [Russian], 25-30. Rostov. Gos. Ped. Univ., Rostov-on-Don 1996.

Imaoka, T. [1976] *Free products with amalgamation of commutative inverse semigroups,*
J. Austral. Math. Soc. Ser. A 22 (1976) 246-251.

Inasaridze, H.N. [1964] *Extensions of semigroups* [Russian],
Soobšč. Akad. Nauk. Gruzin. SSR 33 (1964) 263-269.

[1965] *Extensions of regular semigroups* [Russian],
Soobšč. Akad. Nauk. Gruzin. SSR 39 (1965) 3-10.

[1967] *Extensions of commutative inverse semigroups* [Russian],
Sakharth, SSR Mecn. Akad. Moambe 46 (1967) 11-18.

Isbell, J.R. [1959] *On the multiplicative semigroup of a commutative ring,*
Proc. Amer. Math. Soc. 10 (1959) 908-909.

[1965] *Epimorphisms and dominions,*
Proc. Conf. Categorical Algebra (La Jolla 1965), 232-246, Springer, New York 1966.

[2000] *On multiplicative generators of commutative rings,*
Semigroup Forum 61 (2000) 315.

Iseki, K. [1957] *Contributions to the theory of semigroups, VI,*
Proc. Japan Acad. 33 (1957) 29-30.

Iwanik, A. & Plonka, J. [1975] *Linear independence in commutative semigroups,*
Mat. Časopis Sloven. Akad. Vied 25 (1975) 333-338.

Jain, R.K. [1982] *On Dedekind monoids,* Semigroup Forum 24 (1982) 91-92.

[1983] *Semigroups with primary ideals of prime power,*
Acta Math. Hungar. 42 (1983) 165-169.

[1988] *On Prüfer monoids,* Ranchi Univ. Math. J. 19 (1988) 49-52.

[1988] *Prüfer monoids and Dedekind monoids,* Ranchi Univ. Math. J. 20 (1989) 15-18.

Jarek, P. [1964] *Commutative regular semigroups,* Colloq. Math. 12 (1964) 195-208.

Jensen, B.A. & Miller, D.W. [1968] *Commutative semigroups which are almost finite,* Pacific J. Math. 27 (1968) 533-538.

Ježek, J. & Kepka, T. [1975] *Semigroup representations of commutative idempotent abelian groupoids,* Comment. Math. Univ. Carolinae 16 (1975) 487-500.

John, C.C., Jr. [1973] *The construction of finite commutative semigroups,* Doct. Diss., Kansas State Univ. 1973.

[1974] *A theorem on free envelopes,* Semigroup Forum 9 (1974/75) 74-75.

Johnson, C.S., Jr. & McMorris, F.R. [1977] *Commutative non-singular semigroups,* Canad. Math. Bull. 20 (1977) 41-67.

Johnson, E.W. [1986] *Primary factorization in semigroups,* Czechoslovak Math. J. 36 (111) (1986) 180-184.

Johnson, R.E. [1951] *The extended centralizer of a ring over a module,* Proc. Amer. Math. Soc. 2 (1951) 891-895.

Jones, P. & Ligh, S. [1980] *The multiplicative semigroup of integers modulo n,* Math. Japon. 25 (1980) 15-17.

Juan, L. [1993] *Some results about symmetric semigroups,* Comm. Algebra 21 (1993) no.10, 3637-3645.

Jürgensen, H. [1990] *Annotated tables of semigroups of orders 2 to 6,* Report 231, Dept. of Computer Sci., Univ. Western Ontario 1990.

Jürgensen, H., Migliorini, F., & Szép, J. [1991] *Semigroups,* Akad. Kiadó, Budapest 1991.

Jürgensen, H., Pallas, I., & Wick, P. [1976] *Halbgruppenprogramme,* Bericht 1/76, Inst. Informatik Prakt. Math., Univ. Kiel 1976.

Jürgensen, H. & Wick, P. [1973] *Bestimmung der Unterhabgruppenverbände für zwei klassen endlicher Halbgruppen,* Computing (Arch. Elektron. Rechnen) 11 (1973) 337-351.

[1977] *Die Halbgruppen der Ordnungen $\leqq 7$,* Semigroup Forum 14 (1977) 69-79.

Kaarli, K. & Márki, L. [1992] *Affine complete commutative inverse semigroups,* Semigroups (Luino 1992), 144-151. World Sci. Publishing, River Edge NJ, 1993.

[1995] *Affine complete commutative inverse semigroups,* Algebra Colloq. 2 (1995) 51-78.

Kaarli, K., Márki, L., & Schmidt, E.T. [1985] *Affine complete semilattices,* Monatsh. Math. 99 (1985) 297-309.

Kacman, S.I. [also spelled **Katsman**] [1979] *Commutative semigroups with self-dual lattice of subsemigroups,* Semigroup Forum 18 (1979) 119-161.

Kainrath, F. & Lettl, G. [2000] *Geometric notes on monoids,* Semigroup Forum 61 (2000) 298-302.

Kanemitsu, M. & Bansho, S. [1999] *On primary ideals of valuation semigroups,* Far East J. Math. Sci. 1 (1999) 27-32.

Kanemitsu, M., Okabe, A. & Matsuda, R. [1999] *On polynomial extensions of g-monoids,* Algebras Groups Geom. 16 (1999) 269-276.

Kapur, D. & Narendran, P. [1983] *The Knuth-Bendix completion procedure and Thue systems,* Fundations of software technology and theoretical computer science (Bangalore, 1983) 363-385, Tata Inst. Fund. Res., Bombay 1983.

[1985] *The Knuth-Bendix completion procedure and Thue systems,* SIAM J. Comput. 14 (1985) no.4, 1052-1072.

Karásek, J. [1985] *A remark concerning x-systems,* Arch. Math. (Brno) 21 (1985) 177-180.

Karhumäki, J. & Plandowski, W. [1996] *On the size of independent systems of equations in semigroups,* Theoret. Comput. Sci. 168 (1996) 105-119.

Kato, T. [1979] *On criteria of g̃-hyperellipticity,* Kodai Math. J. 2 (1979) 275-285.

Kaufman, R.P. [1966A] *Maximal semicharacters,* Proc. Amer. Math. Soc. 17 (1966) 1314-1316.

[1966B] *Interpolation of additive functionals,* Studia Math. 27 (1966) 269-272.

[1967] *Positive semicharacters,* J. London Math. Soc. 42 (1967) 264-266.

Keimel, K. [1971] *Baer extensions of rings and Stone extensions of semigroups,* Semigroup Forum 2 (1971) 55-63.

Khan, N.M. [1982] *Epimorphisms, dominions, and varieties of semigroups,* Semigroup Forum 25 (1982) 331-337.

Khovansky, A.G. [1992] *The Newton polytope, the Hilbert polynomial and sums of finite sets,* Functional Anal. Appl. 26 (1992) 276-281.

[1995] *Sums of finite sets, orbits of commutative semigroups and Hilbert functions,* Functional Anal. Appl. 29 (1995) 102-112.

Kil'p, M. [1973] *Commutative monoids all of whose principal ideals are projective,* Semigroup Forum 6 (1973) 334-339.

Kim, A.C. & Neumann, B.H. [1991] *Laws in a class of groupoids,* Discrete Math. 92 (1991) 145-158.

Kim, S.J. [1994] *On the index of the Weierstrass semigroup of a pair of points on a curve,* Arch. Math. (Basel) 62 (1994) 73-82.

[1996] *Semigroups which are not Weierstrass semigroups,* Bull. Korean Math. Soc. 33 (1996) 187-191.

Kimura, N. [1954] *Maximal subgroups of a semigroup,* Kōdai Math. Sem. Rep. 6 (1954) 85-88.

Kimura, N. & Tsai, Y.S. [1972] *On power cancellative archimedean semigroups,* Proc. Japan Acad. 48 (1972) 553-554.

Kisielewicz, A. [1991] *All pseudovarieties of commutative semigroups,* Semigroups with applications (Oberwolfach, 1991), 78-89; World Scientific, 1992.

[1994] *Varieties of commutative semigroups,* Trans. Amer. Math. Soc. 342 (1994) 275-306.

Kisielewicz, A. & Newrly, N. [1993] *Polynomial density of commutative semigroups,* Bull. Austral. Math. Soc. 48 (1993) 151-162.

Kist, J. [1963] *Minimal prime ideals in commutative semigroups,* Proc. London Math. Soc. (3) 13 (1963) 31-50.

Kitsuneda, Y., Matsuda, R., & Sugatani, T. [1999] *Note on 1-dimensional integrally closed Mori semigroups,* Tamkang. J. Math. 30 (1999) 295-297.

Knebl, H. [1984] *Ebene algebraische Kurven vom Typ* p,q, Manuscripta Math. 49 (1984) 165-175.

Knebusch, M. [1969] *Grothendieck- und Wittringe von nichtausgearteten symmetrischen Bilinearformen,* S.-B. Heidelberger Akad. Wiss. Math.-Natur. Kl. 1969/70, 93-157.

Kleitman, D.J., Rothschild, B.R., & Spencer, J.H. [1976] *The number of semigroups of order n,* Proc. Amer. Math. Soc. 55 (1976) 227-232.

Kobayashi, Y. [1974] *Irreducible bases of N-semigroups,* J. Math. Tokushima Univ. 8 (1974) 9-12.

Kobayashi, Y. (cont'd) [1977C] *Conditions for commutative semigroups to have non-trivial homomorphisms into nonnegative (positive) reals,* Proc. Amer. Math. Soc. 65 (1977) 199-203.

[1977E] *On extending functionals on commutative semigroups,* Sûrikaisekikenkyûsho Kôkyûroku No.292 (1977) 90-97.

[1978] *On a problem of commutative nil semigroups,* Proceedings of the 2nd Seminar on Semigroups, Tokyo Gakugei Univ., Tokyo, 1978, 25-28.

[1980] *An extension theorem of functionals on commutative semigroups,* Studia Math. 69 (1980) 87-90.

[1982] *Cancellative and Archimedean ideal-extensions by commutative nil semigroups,* Semigroup Forum 24 (1982) 361-371.

Kobayashi, Y. & Tamura, T. [1974] *Quasi-order preserving homomorphisms of commutative semigroups into the non-negative reals,* Math. Japon. 22 (1977) 267-285.

Kogalovsky, S.R. [1961] *The multiplicative semigroups of rings* [Russian], Dokl. Akad. Nauk. SSSR 140 (1961) 1005-1007.

Kolibiarová, B. [1958] *Über kommutative periodische Halbgruppen,* Mat.-Fyz. Časopis Sloven. Akad. Vied 8 (1958) 127-135.

Komeda, J. [1982] *On the existence of Weierstrass points with a certain semigroup generated by 4 elements,* Tsukuba J. Math. 6 (1982) 237-270.

[1983] *On Weierstrass points whose first nongaps are four,* J. Reine Angew. Math. 341 (1983) 68-86.

[1991] *On primitive Schubert indices of genus g and weight $g-1$,* J. Math. Soc. Japan 43 (1991) 437-445.

[1992] *On the existence of Weierstrass points whose first nongaps are five,* Manuscripta Math. 76 (1992) 193-211.

[1994] *On the existence of Weierstrass gap sequences on curves of genus $\leqq 8$,* J. Pure Appl. Algebra 97 (1994) 51-71.

[1997] *Remarks on non-Weierstrass numerical semigroups,* Algebras and combinatorics (Hong Kong 1997), 313-319, Springer, Singapore 1999.

[1998] *Non-Weierstrass numerical semigroups,* Semigroup Forum 57 (1998) 157-185.

Kominek, Z. [1990] *On approximately subadditive functions,* Demonstratio Math. 23 (1990) 155-160.

Kopytov, V.M. & Medvedev, N.Ya. [1994] *The theory of lattice-ordered groups,* Kluwer Acad. Publ., Dordrecht 1994.

[1996] *Right ordered groups,* Consultants Bureau, New York 1996.

Kowol, G. & Mitsch, H. [1976] *Polynomial functions over commutative semi-groups,* Semigroup Forum 7 (1976) 109-118.

Kozlov, K.P. [1976] *Semigroups with a minimal number of rigid cells in a variety of commutative semigroups* [Russian], Modern algebra no. 4 [Russian] 91-96, Leningrad. Gos. Ped. Inst., Leningrad 1976.

Kranz, P. [1974] *Sandwich and extension theorems on semigroups and lattices,* Comment. Math. Prace Mat. 18 (1974/75) 193-200.

[1979] *Extension of additive functionals and semicharacters on commutative semigroups,* Semigroup Forum 18 (1979) 293-305.

Krause, U. [1989] *On monoids of finite real character,* Proc. Amer. Math. Soc. 105 (1989) 546-554.

Krivenko, V.M. [1985] *Semigroups with normal monogenic subsemigroups* [Russian], Algebraic systems with one action and relation [Russian], 54-57. Leningrad. Gos. Ped. Inst., Leningrad 1985.

Krob, D. [1987] *Monoïdes et semi-anneaux complets,* Semigroup Forum 36 (1987) 323-339.

Kruming, P.D. [1982] *Homological and cohomological characteristics of generalized periodic semigroup* [Russian], Izv. Vysš. Učebn. Zaved. Mat. 1982 no.5, 30-34.

Kublanovsky, S.I. [1983] *Endomorphisms of finite semigroups* [Russian], Associative actions [Russian] 65-75, Leningrad. Gos. Ped. Inst., Leningrad 1983.

Kublanovsky, S.I. & Lesohin, M.M. [1976] *Separability of semigroups with respect to predicates* [Russian], Algebraic Theory of Semigroups (Szeged, 1976) 309-333. Colloq. Math. Soc. János Bolyai, North-Holland, Amsterdam, 1979.

Kulkarni, D.M. [1990] *On the semigroup of an ordinary multiple point,* Bull. Austral. Math. Soc. 42 (1990) 1-6.

Kunz, E. [1966] *Gruppenringe und Differentiale,* Math. Ann. 163 (1966) 346-350.

[1973] *The value-semigroup of a one-dimensional Gorenstein ring,* Proc. Amer. Math. Soc. 25 (1973) 748-751.

[1987] *Über die Klassifikation numerischer Halbgrupper,* Univ. Regensburg, Fachbereit Math., Regensburg 1987.

Kunze, M. [1978] *Die Halbgruppe der Grössenklassen eines geordneten lokalen Rings,* Semigroups (Oberwolfach, 1978), 19-25, Lecture Notes in Math. no.855, Springer, Berlin 1981.

Kupcov, A.I. [also spelled **Kuptsov**] [1979E] *A property of the semigroup of endomorphisms of commutative regular semigroups* [Russian], Semigroup varieties and semigroups of endomorphisms [Russian] 80, Leningrad. Gos. Ped. Inst., Leningrad 1979.

[1979H] *Some properties of the semigroup of homomorphisms of commutative regular semigroups* [Russian], Semigroup varieties and semigroups of endomorphisms [Russian] 67-79, Leningrad. Gos. Ped. Inst., Leningrad 1979.

[1980] *Periodicity of a semigroup of homomorphisms of commutative regular semigroups* [Russian], Modern algebra [Russian] 89-98, Leningrad. Gos. Ped. Inst., Leningrad 1980.

[1981B] *The non-triviality of the semigroup of bilinear mappings of semigroups* [Russian], Materials of VII Region conf. math. mech. [Russian], Tomsk 1981, 35-36.

[1981H] *Some properties of the functors h^A and h_C in the category of commutative regular semigroups* [Russian], Modern algebra [Russian] 17-22, Leningrad. Gos. Ped. Inst., Leningrad 1981.

Kupcov, A.I. & Lesohin, M.M. [1986] *Quasiduality of semigroups* [Russian], Semigroups with additional structures [Russian], 38-49. Leningrad. Univ., Leningrad 1986.

Kuroki, N. [1971] *Note on congruences on the tensor product of archimedean commutative semigroups,* Comment. Math. Univ. St. Paul. 20 (1971/72) 93-96.

[1973] *Certain fundamental congruences on the tensor product of commutative inverse semigroups,* Czechoslovak Math. J. 23 (98) (1973) 252-256.

Lal, H. [1972P] *A note on regular semigroups. II,* Math. Education 6 (1972) A128.

[1972R] *On Fuchs radical of a semigroup,* Aligarh Bull. Math. 2 (1972) 25-32.

Lal, H. (cont'd) [1975] *Commutative semi-primary semigroups,* Czechoslovak Math. J. 25 (100) (1975) 1-3.

Lal, R.N. [1967] *Pseudogroup,* Proc. Nat. Acad. Sci. India Sect. A 37 (1967) 32-36.

Lallement, G. [1971] *On a theorem of Malcev,* Proc. Amer. Math. Soc. 30 (1971) 49-54.

Laplaza, M.L. [1964] *Imbedding of cancellative semigroups in difference semigroups,* Rev. Mat. Hisp.-Amer. (4) 24 (1964) 103-112.

[1974] *A note on commutative semigroups of finite length* [Spanish], Rev. Mat. Hisp.-Amer. (4) 34 (1974) 109-111.

Lau, G. [1979] *Finite abelian semigroups represented into the power set of finite groups,* Czechoslovak Math. J. 29 (104) (1979) 159-162.

Lawson, J.D. [1996] *The earliest semigroup paper?,* Semigroup Forum 52 (1996) 55-60.

Leech, J. [1975] *$\mathcal{H}$-coextensions of monoids,* Mem. Amer. Math. Soc. no.157 (1975) 1-66.

Lenzi, D. [1980] *The paracancellation law in commutative semigroups,* Acta Sci. Math. Acad. Sci. Hungar. 36 (1980) 65-69.

Lepetit, C. [1973] *Décomposition en somme directe de monoïdes libres abéliens,* C. R. Acad. Sci. Paris Sér. A-B, 277 (1973) A877-A880.

Lesieur, L. [1955] *Sur les idéaux irréductibles d'un demi-groupe,* Rend. Sem. Mat. Univ. Padova 24 (1955) 29-36.

Lesohin, M.M. [also spelled **Lesokhin**] [1958] *Certain properties of generalized characters of semigroups,* [Russian], Leningrad. Gos. Ped. Inst. Učen. Zap. 183 (1958) 277-286.

[1961] *On the regularity of systems with external multiplication and the simplicity of their components,* [Russian], Leningrad. Gos. Ped. Inst. Učen. Zap. 218 (1961) 23-37.

[1963A] *Homomorphisms into regular semigroups* [Russian], Sibirsk Mat. Ž. 4 (1963) 1431-1432.

[1963M] *Regularity of systems with exterior multiplication with regular first component* [Russian], Izv. Vysš. Učebn. Zaved. Mat. 1963 no.4, 89-91.

[1963S] *On the completeness of systems with exterior multiplication* [Russian], Izv. Vysš. Učebn. Zaved. Mat. 1963 no.5, 59-62.

[1964G] *Galois connections in systems with exterior multiplication* [Russian], Izv. Vysš. Učebn. Zaved. Mat. 1964 no.1, 84-86.

[1964M] *On the semigroup of a multiplicative commutative regular semigroup* [Russian], Izv. Vysš. Učebn. Zaved. Mat. 1964 no.3, 84-87.

[1965] *On the duality of complex characters and semi-characters of commutative semigroups* [Russian], Mat. Sb. (N.S.) 66 (108) (1965) 378-383.

[1966A] *Approximability of semigroups and separability of subsemigroups* [Russian], Interuniv. Sci. Sympos. General Algebra [Russian] 52-62, Tartu Gos. Univ., Tartu 1966.

[1966D] *Duality of commutative semigroups and of sets with binary relations* [Russian], Herzen Lectures, XIX, Math. [Russian] 6-7, Leningrad. Gos. Ped. Inst., Leningrad 1966.

[1966H] *Additive semigroups of homomorphisms of regular semigroups* [Russian], Trudy Naučn. Ob'ed. Prepodav. Fiz.-Mat. Fak. Ped. Inst. Dal'n. Vostok. 7 (1966) 54-67.

[1967A] *The approximation of semigroups* [Russian], Leningrad. Gos. Ped. Inst. Učen. Zap. 328 (1967) 147-171.

[1967D] *Distributive operations on semigroups* [Russian], Leningrad. Gos. Ped. Inst. Učen. Zap. 302 (1967) 101-115.

[1967E] *Distributive operations on semigroups, II* [Russian], Leningrad. Gos. Ped. Inst. Učen. Zap. 328 (1967) 172-180.

[1967S] *Systems with exterior multiplication which are complete and quasi-complete with respect to the first and second component* [Russian], Leningrad. Gos. Ped. Inst. Učen. Zap. 302 (1967) 85-100.

[1968A] *Finitary approximability of semigroups and finite separability of subsemigroups* [Russian], Leningrad. Gos. Ped. Inst. Učen. Zap. 387 (1968) 134-158.

[1968C] *Characters of commutative semigroups* [Russian], Symposium on Semigroup Theory and Appl. [Russian], Smolensk 1968, 17-27.

[1968D] *Absolute duality of commutative semigroups* [Russian], Leningrad. Gos. Ped. Inst. Učen. Zap. 387 (1968) 159-170.

[1969] *Characters and bicharacters of semigroups* [Russian], First All-Union Symposium on the Theory of Semigroups [Russian], 43-52. Ural. Gos. Univ., Sverdlovsk 1969.

[1970] *Characters of commutative semigroups, I* [Russian], Izv. Vysš. Učebn. Zaved. Mat. 1970 no.8, 67-74.

[1971A] *Approximation of semigroups with respect to predicates* [Russian], Leningrad. Gos. Ped. Inst. Učen. Zap. 404 (1971) 191-219.

[1971B] *Bicharacters of commutative semigroups* [Russian], Leningrad. Gos. Ped. Inst. Učen. Zap. 464 (1971) no. 2, 4

[1971C] *Characters of commutative semigroups, II* [Russian], Izv. Vysš. Učebn. Zaved. Mat. 1971 no.2, 71-77.

[1971H] *Homomorphisms of systems with exterior multiplication* [Russian], Leningrad. Gos. Ped. Inst. Učen. Zap. 404 (1971) 220-232.

[1971R] *Reflexive semigroups* [Russian], Theory of semigroups and its applications, No.2 [Russian], 36-40, 152. Izdat. Saratov. Univ., Saratov 1971.

[1972] *Examination of semigroups by characters* [Russian], Mini-Conference on Semigroup Theory (Szeged, 1972) 38-40. Jószef Attila Univ., Szeged 1973.

[1974A] *The finite approximation of semigroups* [Russian], Theory of semigroups and its applications, No.3 [Russian], 39-46, 152. Izdat. Saratov. Univ., Saratov 1974.

[1974I] *The finitary I-approximability of commutative semigroups* [Russian], Modern Algebra, No.1 [Russian], 124-129. Leningrad. Gos. Ped. Inst., Leningrad 1974.

[1974S] *The separability of subsemigroups by characters* [Russian], Mat. Časopis Sloven. Akad. Vied 24 (1974) 129-138.

[1975] *The finitary approximability of semigroups with respect to predicates* [Russian], Modern Algebra, No.3 [Russian], 87-98. Leningrad. Gos. Ped. Inst., Leningrad 1975.

[1976] *Finite approximability of semigroups of bilinear mappings* [Russian], Modern Algebra, No.5 [Russian], 88-92. Leningrad. Gos. Ped. Inst. im Gercena, Leningrad 1976.

[1977] *The local finiteness of semigroups of bilinear complex functionals* [Russian], Modern Algebra, No.6 [Russian], 132-135. Leningrad. Gos. Ped. Inst., Leningrad 1977.

[1983] *Separability of subsemigroups by complex characters* [Russian], Algebraic actions and orderings [Russian], 79-81. Leningrad. Gos. Ped. Inst., Leningrad 1983.

[1992] *Duality of regular commutative semigroups* [Russian], Partitions and holomorphic mappings of semigroups [Russian], 64-75. "Obrazovanie", St. Petersburg 1992.

Lesohin, M.M. & Golubov, È.G. [1966] *Finitary approximability of commutative semigroups* [Russian], Ural. Gos. Univ. Mat. Zap. 5 (1966) tetrad' 3, 82-90.

Lesohin, M.M. & Kupcov, A.I. [1980] *Duality and quasiduality of semigroups* [Russian], Modern algebra [Russian] 99-102, Leningrad. Gos. Ped. Inst., Leningrad 1980.

Lesohin, M.M. & Popyrin, A.V. [1987] *Bicharacters of semigroups,* Semigroup Forum 35 (1987) 253-264.

Lettl, G. [1988] *Subsemigroups of finitely generated groups with divisor-theory,* Monatsh. Math. 106 (1988) 205-210.

Levin, R.G. [1968] *On commutative nonpotent archimedean semigroups,* Pacific J. Math. 27 (1968) 365-371.

Levin, R.G. & Tamura, T. [1970] *Notes on commutative power joined semigroups,* Pacific J. Math. 35 (1970) 673-679.

Li, S.Zh. [1986] *On the tensor products in some subcategories of the category of semigroups,* J. Math. Res. Exposition (English ed.) 1986, no.2, 15-18.

[1987] *The tensor product of cancellation semigroups* [Chinese], Yangzhou Shiyuan Xuebao Ziran Kexue Ban 1987, no.4, 33-37.

Li, S.Zh. & Liu, Zh.H. [1992] *A note on injections of abelian groups* [Chinese], Qufi Shifan Daxue Xuebao Ziran Kexue Ban 18 (1992) no.4, 28-30.

Li, S.Zh. & Zhang, Y.F. [1993] *Tensor products of separable semigroups* [Chinese], J. Math. (Wuhan) 13 (1993) 283-288.

Liberman, A.A. [1980] *Lattice isomorphisms of cyclic semigroups* [Russian], Ural. Gos. Univ. Mat. Zap. 12, no.2, 100-111, ii (1980).

Libih, A.L. [1974] *Local automorphisms of commutative monomorphic inverse semigroups* [Russian], Studies in algebra, No.4 [Russian], 56-69, 125. Izdat. Saratov. Univ., Saratov 1974.

Lidl, R. & Müller, W.B. [1986] *On commutative semigroups of polynomials with respect to composition,* Monatsh. Math. 102 (1986) 139-153.

Lipkovski, A. [1988] *Newton polyhedra and irreducibility,* Math. Z. 199 (1988) 119-127.

Ljubenova, L. & Cvetkova, G. [1967] *The subsemigroups of a certain free abelian semigroup of finite rank* [Bulgarian], Godišnik Vysš. Tehn. Učebn. Zaved. Mat. 4 (no.1, 13-26.

Lomadze, D.D. [1992A] *Approximation of semigroups with respect to Green equivalences* [Russian], Soobšč. Akad. Nauk. Gruzii 145 (1992) 265-268.

[1992B] *SH-approximation of semigroups by finite commutative regular semigroups* [Russian], Partitions and holomorphic mappings of semigroups [Russian], 105-112. "Obrazovanie", St. Petersburg 1992.

Lord, J.M. [1978] *The structure of commutative Archimedean semigroups without idempotent,* Doct. Diss., Univ. California Davis, 1978.

[1979] *Isomorphism criterion and structure group description for $\mathfrak{N}$-semigroups,* Proc. Japan Acad. Ser. A Math. Sci. 55 (1979) 252-254.

Lorenzen, P. [1939] *Abstrakte Begründung der multiplikativen Idealtheorie,* Math. Z. 45 (1939) 533-553.

[1949] *Über halbgeordnete Gruppen,* Math. Z. 52 (1949) 483-526.

[1952] *Teilbarkeitstheorie in Bereichen,* Math. Z. 55 (1952) 269-275.

Luedemann, J.K. [1976] *The generalized translational hull of a semigroup,* Pacific J. Math. 67 (1976) 489-507.

Lugowski, H. [1966] *Die Charackterisierung gewisser geordneter Halbmoduln mit Hilfe der Erweiterungstheorie,* Publ. Math. Debrecen 13 (1966) 237-248.

Lyapin, E.S. [1950S] *Simple commutative associative systems* [Russian], Izv. Akad. Nauk SSSR Ser. Mat. 14 (1950) 275-282.

[1950R] *Semisimple commutative associative systems* [Russian], Izv. Akad. Nauk SSSR Ser. Mat. 14 (1950) 367-380.

Lyapin, E.S. & Evseev, A.E. [1997] *The theory of partial algebraic operations,* Kluwer Acad. Publ., Dordrecht 1997.

MacKenzie, R.E. [1954] *Commutative semigroups,* Duke Math. J. 21 (1954) 471-477.

Maclachlan, C. [1971] *Weierstrass points on compact Riemann surfaces,* J. London Math. Soc. (2) 3 (1971) 722-724.

MacLane, S. [1963] *Homology,* Springer-Verlag, New York, 1963.

[1971] *Categories for the working mathematician,* Springer-Verlag, New York, 1971. 2nd. Ed. 1998.

Malcev, A.I. [also spelled **Mal'tsev**] [1958] *On homomorphisms onto finite groups* [Russian], Uch. Zap. Ivanov. Gos. Ped. Inst. 18 (1958) 49-60.

Mannepalli, V.L. [1976A] *Multiplication semigroups,* Semigroup Forum 11 (1975/76) 310-327.

[1976B] *Multiplication semigroups. II,* Semigroup Forum 12 (1976) 145-164.

Mannepalli, V.L. & Satyanarayana, M. [1974] *Monoids of the Dedekind type,* Semigroup Forum 9 (1974/75) 19-27.

Manukjanc, M.G. [1974] *Construction of the semigroup of endomorphisms of certain commutative regular semigroups,* [Russian], Izv. Vysš. Učebn. Zaved. Mat. 1974 no.8, 65-71.

[1976A] *The structure of a semigroup of multiplications of a class of commutative regular semigroups* [Russian], Theory of functions. Differential Equations and their applications, no.1 [Russian], 3-7, 220. Kalmyck. Gos. Univ., Elista, 1976.

[1976B] *Semigroups of multiplications of certain commutative semigroups* [Russian], Theory of functions. Differential Equations and their applications, no.1 [Russian], 56-72, 216. Kalmyck. Gos. Univ., Elista, 1976.

Marcus, D.A. [1974] *Direct decompositions in free commutative monoids,* J. Combinatorial Theory Ser. A, 16 (1974) 286-312.

Markov, S.M. [1995] *On the foundations of interval arithmetic,* Scientific computing and validated numerics (Wuppertal 1995), 307-313, Akademie Verlag, Berlin 1996.

Matsuda, R. [1997] *The Krull-Akizuki theorem for semigroups,* Math. J. Ibaraki Univ. 29 (1997) 55-56.

[1998A] *Some theorems for semigroups,* Math. J. Ibaraki Univ. 30 (1998) 1-7.

[1998B] *Note on overrings without a specified element,* Math. J. Ibaraki Univ. 30 (1998) 9-14.

[1999] *Note on Macaulay semigroups,* Tsukuba J. Math. 23 (1999) 189-199.

Maury, G. [1958] *Une caractérisation des demi-groupes noethériens intégralement clos,* C. R. Acad. Sci. Paris 247 (1958) 254-255.

Maury, G. (cont'd) [1959] *Une caractérisation des demi-groupes noethériens intégralement clos,* C. R. Acad. Sci. Paris 248 (1959) 3260-3261.

McAlister, D.B. [1968] *Characters on commutative semigroups,* Quart. J. Math. Oxford Ser.(2) 19 (1968) 141-157.

McAlister, D.B. & O'Carroll, L. [1970] *Finitely generated commutative semigroups,* Glasgow Math. J. 11 (1970) 134-151.

[1970] *Maximal homomorphic images of commutative semigroups,* Glasgow Math. J. 12 (1971) 12-17.

McMorris, F.R. [1971] *The quotient semigroup of a semigroup that is a semilattice of groups,* Glasgow Math. J. 12 (1971) 18-23.

[1972] *The maximal quotient semigroup,* Semigroup Forum 4 (1972) 360-364.

McMorris, F.R. & Satyanarayana, M. [1972] *Categorical semigroups,* Proc. Amer. Math. Soc. 33 (1972) 271-277.

McNeil, P.E. [1971] *Group extensions of null semigroups,* Duke Math. J. 38 (1971) 491-497.

[1972] *Finite commutative subdirectly irreducible semigroups,* Trans. Amer. Math. Soc. 172 (1972) 57-67.

Megyesi, L. & Pollák, G. [1968] *Über die Struktur der Hauptidealhalbgruppen, I,* Acta Sci. Math. (Szeged) 29 (1968) 261-270.

[1977] *Über die Struktur der Hauptidealhalbgruppen, II,* Acta Sci. Math. (Szeged) 39 (1977) 103-108.

Migliorini, F. & Szép, J. [1979] *Equivalences, congruences, and decompositions in semigroups,* Riv. Mat. Univ. Parma (4) 5 (1979) 745-752.

[1981] *On* Γ*-decompositions of semigroups,* Recent developments in [...] semigroups (Oberwolfach, 1981), 385-403, Lecture Notes in Math. no.998, Springer, Berlin 1983.

Mitrović, Ž.M. [1976] *Interval semigroups,* Univ. Beograd. Publ. Elektrotehn. Fak. Ser. Mat. Fiz. no. 544-576 (1976) 127-143.

[1977] *Contribution to the theory of interval arithmetic, and its applications* [Serbo-Croatian], Univ. Beograd. Publ. Elektrotehn. Fak. Ser. Mat. Fiz. no. 601 (1977).

Mitsch, H. [1983] *Semigroups and their lattice of congruences,* Semigroup Forum 26 (1983) 1-63.

[1986] *A natural order for semigroups,* Proc. Amer. Math. Soc. 97 (1986) 384-388.

[1994] *Semigroups and their natural order,* Math. Slovaca 44 (1994) 445-462.

Mogiljanskaja, E.M. [1971] *The semigroup of the subsemigroups of a commutative semigroup* [Russian], Leningrad. Gos. Ped. Inst. Učen. Zap. 404 (1971) 250-256.

[1972] *Definability of certain holoid semigroups by means of the semigroups of all their subsets and subsemigroups* [Russian], Leningrad. Gos. Ped. Inst. Učen. Zap. 496 (1972) 37-48.

Moh, T.T. & Heinzer, W. [1982] *On the Lüroth semigroup and Weierstrass canonical divisors,* J. Algebra 77 (1982) 62-73.

Monzo, R.A.R. [1973] *On categorical semigroups,* Bull. Austral. Math. Soc. 9 (1973) 61-68.

Moore, E.H. [1902] *A definition of abstract groups,* Trans. Amer. Math. Soc. 3 (1902) 485-492.

Morel, A.C. [1979] *Locally cyclic semigroups,* Algebra Universalis 9 (1979) 391-394.

Moszner, Z. [1979] *Sur l'inégalité de translation,* Demonstratio Math. 11 (1978) 1095-1100 (1979).

Motzkin, T.S. & Selfridge, J.L. [1956] *Semigroups of order five,* Bull. Amer. Math. Soc. 62 (1956) 14.

Murata, K. [1980] *A note on arithmetics in semigroups,* Proc. Japan Acad. Ser. A Math. Sci. 56 (1980) 133-135.

Murphy, G.J. [1996] *C*-algebras generated by commuting isometries,* Rocky Mountain J. Math. 26 (1996) 237-267.

Murty, A.S.R. [1983] *Noetherian S-spaces and primary decompositions,* Math. Sem. Notes Kobe Univ. 11 (1983) 231-238.

Mushtaq, Q. & Yusuf, S.M. [1987] *On LA-semigroup defined by a commutative regular semigroup,* J. Natur. Sci. and Math. 27 (1987) 11-15.

[1988] *On LA-semigroup defined by a commutative inverse semigroup,* Mat. Vesnik 40 (1988) 59-62.

Nagy, A. [2001] *Special classes of semigroups,* Kluwer Acad. Publ., Dordrecht 2001.

Nakajima, S. [1977] *The endomorphism semigroup of a semigroup and its application,* Sûrikaisekikenkyûsho Kôkyûroku No.292 (1977) 1-8.

Narendran, P., O'Dunlaing, C., & Rolletschek, H. [1985] *Complexity of certain decision problems about congruential languages,* J. Comput. System Sci. 30 (1985) 343-358.

Narendran, P. & O'Dunlaing, C. [1989] *Cancellativity in finitely presented semigroups,* J. Symbolic Comput. 7 (1989) 457-472.

Nasirov, S.N. [1977] *Representations of semigroups in Boolean algebras,* Tashkent. Gos. Univ. Sb. Naučn. Trudov No.623 Mat. Analiz i Geometrija (1980) 56-61, 94.

Nathanson, M.B. [2000] *Growth of sumsets in abelian semigroups,* Semigroup Forum 61 (2000) 149-153.

Neeb, K.-H. [1992] *Toric varieties and algebraic monoids,* Sem. Sophus Lie 2 (1992) 159-187.

Nguyen, H.K. [1981] *Schreiersche Erweiterungen von Halbgruppen,* Seminarberichte no. 34, Humboldt Univ., Sekt. Math., Berlin 1981.

Niculescu, C.P. [1995] *A combinatorial property of abelian semigroups,* Rev. Roumaine Math. Pures Appl. 40 (1995) 669-671.

[1996] *Birkhoff recurrence theorem and combinatorial properties of abelian semigroups,* Rev. Roumaine Math. Pures Appl. 41 (1996) 675-686.

Nikolaev, A.M. [1971] *A tensor product of semigroups* [Russian], Leningrad. Gos. Ped. Inst. Učen. Zap. 464 (1971) no.2, 122-127.

Niven, I. [1971] *A characterization of complementing sets of pairs of integers,* Duke Math. J. 38 (1971) 193-203.

Nöbauer, W. [1986] *Some remarks on permutable chains of polynomials,* Contributions to general algebra 5 (Salzburg 1986), 247-256. Hölder-Pichler-Tempsky, Vienna 1987.

[1987] *Vertauschbare Polynome: an den Grenzen der Koeffizientenvergleichsmethode,* Österreich. Akad. Wiss. Math.-Natur. Kl. Sitzungsber. II, 196 (1987) 403-417.

Nordahl, T. [1973] *Commutative semigroups whose proper subsemigroups are power joined,* Semigroup Forum 6 (1973) 35-41.

[1977] *Abelian groups and N-semigroups, II,* Proc. Japan Acad. 53 (1977) 8-10.

Novikov, B.V. [1990] *Commutative semigroups with cancellation of dimension 1* [Russian], Mat. Zametki 48 (1990) 148-149.

Oda, T. [1988] *Convex bodies and algebraic varieties,* Springer-Verlag, Berlin-New York, 1988.

Okniński, J. [1984] *Semigroup algebras,* Marcel Dekker, New York, 1991.

[1986] *Commutative monoid rings with Krull dimension,* Semigroups, theory and applications (Oberwolfach, 1986), 251-259, Lecture Notes in Math. no.1320, Springer, Berlin-New York 1988.

Ovsyannikov, A.Ya. [also spelled **Ovsjannikov**] [1977A] *Structural isomorphisms of commutative semigroups with one defining relation* [Russian], Izv. Vysš. Učebn. Zaved. Mat. 1977 no.4, 136-138.

[1977B] *Lattice isomorphisms of commutative semigroups with a single defining relation* [Russian], Ural. Gos. Univ. Mat. Zap. 10, no.3 Issled. po Sovremen. Algebre, 138-172, 218 (1977).

[1981] *Structural isomorphisms of commutative semigroups with one defining relation. II* [Russian], Ural. Gos. Univ. Mat. Zap. 12, no.3, 90-107, iii (1981).

Painter, P. [1966] *Isomorphisms between certain subsemigroups of the positive integers under multiplication,* Amer. Math. Monthly 73 (1966) 62-64.

[1967] *Isomorphisms between semigroups of integers,* Proc. Amer. Math. Soc. 18 (1967) 1100-1101.

Páles, Z. [1989] *A Stone-type theorem for abelian semigroups,* Arch. Math. (Basel) 52 (1989) 265-268.

Parízek, B. [1957] *Note on structure of multiplicative semigroup of residue classes,* Mat.-Fyz. Časopis Sloven. Akad. Vied 7 (1957) 183-185.

Parízek, B. & Schwarz, Š. [1961] *Semicharacters of the multiplicative semigroup of integers modulo m,* Mat.-Fyz. Časopis Sloven. Akad. Vied 11 (1961) 63-74.

Parker, E.T. [1954] *On multiplicative semigroups of residue classes,* Proc. Amer. Math. Soc. 5 (1954) 612-616.

Patil, D.P. [1993] *Minimal sets of generators for the relation ideals of certain monomial curves,* Manuscripta Math. 80 (1993) 239-248.

Patil, D.P. & Singh, B. [1990] *Generators for the derivation modules and the relation ideals of certain curves,* Manuscripta Math. 68 (1990) 327-335.

Paxia, G., Raciti, G., & Ragusa, A. [1992] *On the Lüroth semigroup of curves lying on a smooth quadric,* Manuscripta Math. 75 (1992) 225-246.

Pedersen, F.D. & Sizer, W.S. [1978] *Certain semigroups of complexes of an abelian group,* Proceedings of the Conference on Semigroups in Honor of Alfred H. Clifford (New Orleans, 1978), 215-221; Tulane Univ., New Orleans 1979.

Perel, W.M. [1957] *Principal representations in commutative semigroups,* Proc. Amer. Math. Soc. 8 (1957) 957-960.

Perić, V. [1966] *Zu den fastinvertierbaren Idealen in kommutativen Ringen und Halbgruppen,* Glasnik Mat. Ser. III 1 (21) (1966) 139-146.

Perrot, J.-F. [1965] *Sur quelques familles de parties des monoïdes abéliens libres,* C. R. Acad. Sci. Paris 261 (1965) 3008-3011.

Petrich, M. [1962] *Semicharacters of the Cartesian product of two semigroups,* Pacific J. Math. 12 (1962) 679-683.

[1963C] *Some theorems concerning the dual of a commutative semigroup,* Duke Math. J. 30 (1963) 25-32.

[1963F] *The maximal semilattice decomposition of a semigroup,* Bull. Amer. Math. Soc. 69 (1963) 342-344.

[1964] *On the structure of a class of commutative semigroups,* Czechoslovak Math. J. 14 (1964) 147-153.

[1973] *Introduction to semigroups,* Charles E. Merrill, Columbus OH, 1973.

[1973E] *L'enveloppe de translations d'un demi-treillis de groupes,* Canad. J. Math. 25 (1973) 164-177.

[1977] *Commutative semigroups,* Rend. Sem. Mat. Fis. Milano 47 (1977) 61-66.

Pfister, G. & Steenbrink, J.H.M. [1992] *Reduced Hilbert schemes for irreducible curve singularities,* J. Pure Appl. Algebra 77 (1992) 103-116.

Philip, J.M. [1974] *A proof of Isbell's zigzag theorem,* J. Algebra 32 (1974) 328-331.

Plappert, P. [1995] *A sandwich theorem for monotone additive functions,* Semigroup Forum 51 (1995) 347-355.

Plemmons, R.J. [1966] *Cayley tables for all semigroups of order* $N \leqq 6$, Auburn Univ. 1966.

Pondělíček, B. [1966] *On the characters of semigroups whose idempotents form a chain* [Czech], Časopis Pěst. Mat. 91 (1966) 4-7.

[1983] *Atomicity of tolerance lattices of commutative semigroups,* Czechoslovak Math. J. 33 (108) (1983) 485-498.

[1985A] *Modularity and distributivity of tolerance lattices of commutative separative semigroups,* Czechoslovak Math. J. 35 (110) (1985) 146-157.

[1985B] *Modularity and distributivity of tolerance lattices of commutative separative semigroups,* Czechoslovak Math. J. 35 (110) (1985) 333-337.

[1988A] *Note on the congruence lattice of a commutative separative semigroup,* Časopis Pěst. Mat. 113 (1988) 74-79.

[1988B] *Commutative semigroups whose lattice of tolerances is Boolean,* Czechoslovak Math. J. 38 (113) (1988) 226-230.

[1988C] *Principal tolerance trivial commutative semigroups,* Acta Sci. Math. (Szeged) 52 (1988) 29-33.

[1988D] *Direct decomposability of tolerances and congruences on semigroups,* Czechoslovak Math. J. 38 (113) (1988) 701-704.

Ponizovsky, I.S. [1962] *A remark on commutative semigroups* [Russian], Dokl. Akad. Nauk SSSR 142 (1962) 1258-1260.

[1970] *The matrix representations of finite commutative semigroups* [Russian], Sibirsk. Mat. Ž. 11 (1970) 1098-1106, 1197-1198.

Popyrin, A.V. [1984] *Properties of bilinear mappings of commutative semigroups* [Russian], Properties of semigroups [Russian], 122-130. Leningrad. Gos. Ped. Inst., Leningrad 1984.

[1985] *Weak duality of commutative semigroups* [Russian], Algebraic systems with one action and relation [Russian], 100-113. Leningrad. Gos. Ped. Inst., Leningrad 1985.

[1986] *Separability of commutative semigroups by bicharacters* [Russian], Semigroups with additional structures [Russian], 101-108. Leningrad. Univ., Leningrad 1986.

Popyrin, A.V. (cont'd) [1988] *Nondegenerate bilinear mappings of commutative semigroups* [Russian], Homomorphisms [Russian], 111-115. Leningrad. Gos. Ped. Inst., Leningrad 1988.

Porubský, Š. [1977] *Commutative semi-primary x-semigroups,* Czechoslovak Math. J. 27 (102) (1977) 467-472. Correction, ibid. 28 (103) (1978) 505.

Poyatos, F. [1966] *On irreducible compositions of abelian semigroups* [Spanish], Proc. Seventh Annual Meeting of Spanish Mathematicians (Valladolid 1966) [Spanish], 85-91. Univ. Valladolid, Valladolid 1967.

[1967] *On irreducible decompositions of commutative semigroups* [Spanish], Rev. Mat. Hisp.-Amer. (4) 27 (1967) 39-54.

[1968] *Formulas of length for commutative semigroups* [Spanish], Rev. Mat. Hisp.-Amer. (4) 28 (1968) 95-127.

[1985] *Archimedean decomposition of left S-semimodules and semirings,* Studia Sci. Math. Hungar. 20 (1985) 323-324.

[1988] *Archimedean decomposition of commutative semigroups with operators,* Studia Sci. Math. Hungar. 23 (1988) 339-342.

Preston, G.B. [1975] *Rédei's characterization of congruences on finitely generated free commutative semigroups,* Acta Sci. Math. Acad. Sci. Hungar. 26 (1975) 337-342.

Puš, V. [1991] *Combinatorial properties of products of graphs,* Czechoslovak Math. J. 41 (116) (1991) 269-277.

[1992] *On multiplicative bases in commutative semigroups,* European J. Combin. 13 (1992) 201-211.

Putcha, M. [1971] *Commutative semigroups whose homomorphic images in groups are groups,* Semigroup Forum 3 (1971) 51-57.

[1973] *Positive quasi-orders on semigroups,* Duke Math. J. 40 (1973) 857-869.

[1975] *Maximal cancellative subsemigroups and cancellative congruences,* Proc. Amer. Math. Soc. 47 (1975) 49-52.

[1976] *Positive functions from $\mathcal{S}$-indecomposable semigroups into partially ordered sets,* Czechoslovak Math. J. 26 (101) (1976) 161-170.

[1980] *Cancellative subsemigroups of Archimedean semigroups,* Semigroup Forum 21 (1980) 187-193.

Putcha, M. & Tamura, T. [1976] *Homomorphisms of commutative cancellative semigroups into nonnegative real numbers,* Trans. Amer. Math. Soc. 221 (1976) 147-157.

Putcha, M. & Weissglass, J. [1973] *Aplications of semigroup algebras to ideal extensions of semigroups,* Semigroup Forum 6 (1973) 283-294.

Rachunek, J. [1987] *Prime ideals in autometrized algebras,* Czechoslovak Math. J. 37 (112) (1987) 188-194.

Radelecki, S. [1986] *On the properties of decompositions of commutative semigroups,* Mathematica (Cluij) 28 (51) (1986) 163-165.

Rankin, S.A., Reis, C.M., & Wang, M.L. [1976] *Globally abelian transformation semigroups,* Semigroup Forum 13 (1976/77) 221-228.

Ratschek, H. [1972] *Teilbarkeitskriterien der Intervallarithmetik,* J. Reine Angew. Math. 252 (1972) 128-138.

Ratschek, H. & Schröder, G. [1977] *Representation of semigroups as systems of compact convex sets,* Proc. Amer. Math. Soc. 65 (1977) 24-28.

Rédei, L. [1952] *Die Verallgemeinerung der Schreierschen Erweiterungstheorie,* Acta Sci. Math. (Szeged) 14 (1952) 252-273.

[1956] *The theory of finitely generated commutative semigroups,* Akad. Kiadó, Budapest 1956. English translation, Pergamon Press, Oxford, 1965.

Rees, D. [1940] *On semi-groups,* Proc. Cambridge Phil. Soc. 36 (1940) 387-400.

Repnitsky, V.B. [1994] *On subsemigroup lattices without nontrivial identities,* Algebra Universalis 31 (1994) 256-265.

Repnitsky, V.B. & Katsman, S.I. [1988] *Commutative semigroups the lattice of whose subsemigroups satisfies a nontrivial identity* [Russian], Mat. Sb. (N.S.) 137 (179) (1988) 462-482, 575.

Ressel, P. & Ricker, W.J. [1998] *Semigroup representations, positive definite functions and abelian* C*-*algebras,* Proc. Amer. Math. Soc. 126 (1998) 2949-2955.

Rim, D.S. & Vitulli, M.A. [1977] *Weierstrass points and monomial curves,* J. Algebra 48 (1977) 454-476.

Rödseth, Ö.J. [1978] *On a linear Diophantine problem of Frobenius,* J. Reine Angew. Math. 301 (1978) 171-178.

Roĭz, E.N. & Schein, B.M. [1978] *Radicals of semigroups,* Semigroup Forum 16 (1978) 299-344.

Rompke, J. [1975] *Regular, commutative, maximal semigroups of quotients,* Canad. Math. Bull. 18 (1975) 99-104.

Rosales, J.C. [1995] *Function minimum associated to a congruence on integral n-tuple space,* Semigroup Forum 51 (1995) 87-95.

[1996A] *An algorithmic method to compute a minimal relation for any numerical semigroup,* Internat. J. Algebra Comput. 6 (1996) 441-455.

[1996N] *On numerical semigroups,* Semigroup Forum 52 (1996) 307-318.

[1996S] *On symmetric numerical semigroups,* J. Algebra 182 (1996) 422-434.

[1997] *On presentations of subsemigroups of* $\mathbb{N}^n$, Semigroup Forum 55 (1997) 152-159.

Rosales, J.C. & García-García, J.I. [1999] *Hereditarily finitely generated commutative monoids,* J. Algebra 221 (1999) 723-732.

Rosales, J.C. & García-Sánchez, P.A. [1995] *On complete intersection affine semigroups,* Comm. Algebra 23 (1995) no.14, 5395-5412.

[1998C] *On Cohen-Macaulay and Gorenstein simplicial affine semigroups,* Proc. Edinburgh Math. Soc. (2) 41 (1998) 517-537.

[1998N] *On numerical semigroups with high embedding dimension,* J. Algebra 203 (1998) 567-578.

[1999] *Finitely generated commutative semigroups,* Nova Science Publishers, Commack NY, 1991.

[1999F] *On free affine semigroups,* Semigroup Forum 58 (1999) 367-385.

[1999N] *On normal affine semigroups,* Linear Algebra Appl. 286 (1999) 175-186.

[1999P] *Presentations for subsemigroups of finitely generated commutative semigroups,* Israel J. Math. 113 (1999) 269-283.

Rosales, J.C., García-Sánchez, P.A. & Urbano-Blanco, J.M. [1998] *On Cohen-Macaulay subsemigroups of* $\mathbb{N}^2$, Comm. Algebra 26 (1998) no.8, 2543-2558.

[1999] *On presentations of commutative monoids,* Internat. J. Algebra Comput. 9 (1999) 539-553.

Ross, K.A. [1959] *A note on extending semicharacters on semigroups,* Proc. Amer. Math. Soc. 10 (1959) 579-583.

[1961] *Extending characters on semigroups,* Proc. Amer. Math. Soc. 12 (1961) 988-990.

Ruiz, J.M. [1985] *Pythagorean real curve germs,* J. Algebra 94 (1985) 126-144.

Sagastume Berra, A.E. [1957] *Theory of ova,* Math. Notae 16 (1957/58) 43-59, 65-77.

Sankaran, N. [1961] *On some permutable processes in semi-groups,* J. Madras Univ. Sect. B, 31 (1961) 97-107.

Sasaki, M. [1964] *On semigroups of positive integer vectors,* Math. Japon. 9 (1964) 31-47.

[1965] *On the isomorphism problem of certain semigroups constructed from indexed groups,* Proc. Japan Acad. 41 (1965) 763-766.

[1966] *Commutative nonpotent archimedean semigroups with cancellation law, II,* Math. Japon. 11 (1966) 153-165.

[1978] *Commutative semigroups obtained by an abelian group and a generalized $\mathcal{S}$-function,* Annual Rep. Fac. Ed. Iwate Univ. 38 (1978) no.3, 1-6.

[1982] *Supplement to the paper: "Commutative semigroups obtained by an abelian group and a generalized $\mathcal{S}$-function",* Annual Rep. Fac. Ed. Iwate Univ. 42 (1982) no.1, 51-54.

[1983] *$\tilde{\mathfrak{N}}$-semigroups with an identity element,* Annual Rep. Fac. Ed. Iwate Univ. 43 (1983) no.1, 57-68.

[1984] *Non-Archimedean $\tilde{\mathfrak{N}}$-semigroups without idempotents,* Annual Rep. Fac. Ed. Iwate Univ. 44 (1984) no.1, 1-12.

[1988] *Isomorphism problem of power semigroups of integer semigroups,* Annual Rep. Fac. Ed. Iwate Univ. 48 (1988) no.1, 123-126.

Sasaki, M. & Tamura, T. [1971] *Positive rational semigroups and commutative power joined cancellative semigroups without idempotent,* Czechoslovak Math. J. 21 (1971) 567-576.

Sasaki, M., Spake, R., & Tamura, T. [1986] *Archimedean components of the power semigroup of the group of rational numbers,* Proc. 10th Sympos. on Semigroups (Sakado 1986) 31-35, Jōsai Univ., Sakado 1987.

Sathaye, A. [1977] *On planar curves,* Amer. J. Math. 99 (1977) 1105-1135.

Sato, S. [Satoh, S.], Yama K., & Tokizawa, M. [1991] *Semigroups of order 8,* Proc. 15th Sympos. on Semigroups, Shimane Univ. (1991) 7-16. See also: Semigroup Forum 48 (1994) 1-23.

Satyanarayana, M. [1971A] *A class of commutative semigroups in which the idempotents are linearly ordered,* Czechoslovak Math. J. 21 (96) (1971) 633-637.

[1971B] *Commutative semigroups in which primary ideals are prime,* Math. Nachr. 48 (1971) 107-111.

[1972] *Commutative primary semigroups,* Czechoslovak Math. J. 22 (97) (1972) 509-516.

[1973] *On semigroups admitting ring structure. II,* Semigroup Forum 6 (1973) 189-197.

[1975] *On 0-radical of a semigroup,* Math. Nachr. 66 (1975) 231-234.

[1977] *On commutative semigroups which are unions of a finite number of principal ideals,* Czechoslovak Math. J. 27 (102) (1977) 61-68.

[1978] *Structure and ideal theory of commutative semigroups,* Czechoslovak Math. J. 28 (103) (1978) 171-180. Correction, ibid. 29 (104) (1979) 662-663.

Schein, B.M. [also spelled Šaĭn] [1962] *On subdirectly irreducible semigroups* [Russian], Dokl. Akad. Nauk SSSR 144 (1962) 999-1002.

[1965] *A class of commutative semigroups* [Russian], Publ. Math. Debrecen 12 (1965) 87-88.

[1966] *Homomorphisms and subdirect decompositions of semigroups,* Pacific J. Math. 17 (1966) 529-547.

[1969] *Commutative semigroups where congruences form a chain,* Bull. Acad. Polon. Sci. Sér. Sci. Math. Astron. Phys. 17 (1969) 523-527; Errata, ibid. 23 (1975) 1247-1248.

[1974] *Injectives in certain classes of semigroups,* Semigroup Forum 9 (1974/75) 159-171.

[1975] *Commutative semigroups all of whose subsemigroups are densely embedded,* Acta Math. Acad. Sci. Hungar. 26 (1975) 327-330.

[1976] *Injectives commutative semigroups,* Algebra Universalis 6 (1976) 395-397.

[1981C] *Pseudosimple commutative semigroups,* Monatsh. Math. 91 (1981) 77-78.

[1981I] *On two papers by B.M. Schein,* Semigroup Forum 23 (1981) 87-88.

Schmidt, J. [1975A] *Quasi-decompositions, exact sequences, and triple sums of semigroups, I. General theory,* Contributions to universal algebra (Szeged, 1975) 365-397. Colloq. Math. Soc. János Bolyai, North-Holland, Amsterdam, 1977. See also Proc. Lattice Theory Conf., 180-206. Univ. Ulm, Ulm 1975.

[1975B] *Quasi-decompositions, exact sequences, and triple sums of semigroups, II. Applications,* Contributions to universal algebra (Szeged, 1975) 399-428. Colloq. Math. Soc. János Bolyai, North-Holland, Amsterdam, 1977.

Schreier, O. [1926] *Über die Erweiterungen von Gruppen. I.,* Monatsch. Math. Phys. 34 (1926) 165-180.

Schützenberger, M.P. [1957] *$\overline{\mathcal{D}}$-représentation des demi-groupes,* C. R. Acad. Sci. Paris 244 (1957) 1994-1996.

Schwarz, Š. [1943] *On the theory of semigroups* [Slovakian], Sborník prác Prírodovedekej Fak. Slovensk. Univ. Bratislava No.6 (1943).

[1953] *Contribution to the theory of torsion semigroups* [Russian], Czechoslovak Math. J. 3 (1953) 7-21.

[1954A] *The theory of characters of finite commutative semigroups* [Russian], Czechoslovak Math. J. 4 (1954) 219-247.

[1954B] *Characters of finite commutative semigroups as class functions* [Russian], Czechoslovak Math. J. 4 (1954) 291-295.

[1954C] *On a Galois connection in the theory of characters of commutative semigroups* [Russian], Czechoslovak Math. J. 4 (1954) 291-295.

[1977] *A counting theorem in the semigroup of Boolean circulant matrices,* Czechoslovak Math. J. 27 (102) (1977) 504-510.

Seitz, K. & Blickle, T. [1974] *The structure of systems,* Report DM 1974-2, Dept. of Math., Karl Marx Univ. of Economics, Budapest 1974.

Seitz, K., Blickle, T., & Grega, B. [1975] *On semigroups of type α with special emphasis on certain applications in chemical engineering,* Report DM 1975-4, Dept. of Math., Karl Marx Univ. of Economics, Budapest 1975.

Selmer, E.S. [1977] *On the linear Diophantine problem of Frobenius,* J. Reine Angew. Math. 293 (1977) 1-17.

Shershin, A.C. & Moore, J.T. [1965] *Direct summands of abelian monoids,* Math. Notae 20 (1965) 109-116.

Shevrin, L.N. [also spelled **Ševrin**] [1961] *Semigroups with certain types of subsemigroup lattices* [Russian], Dokl. Akad. Nauk SSSR 138 (1961) 796-798.

[1963] *Commutative semigroups of finite rank* [Russian], Uspehi Mat. Nauk 18 (1963) 201-204.

[1966] *Elementary lattice properties of semigroups* [Russian], Sibirsk. Mat. Ž. 7 (1966) 664-684.

Shevrin, L.N. (ed.) [1989] *The Sverdlovsk Tetrad: unsolved problems in the theory of semigroups* [Russian], Sverdlovsk, 1989.

Shevrin, L.N. & Ovsyannikov, A.Ya. [1983] *Semigroups and their subsemigroup lattices* [Russian], Semigroup Forum 27 (1983) 1-154.

[1990] *Finitely assembled semigroups and ascending chain condition for subsemigroups,* Monash Conf. on Semigroup Theory (Melbourne, 1990) 269-284. World Scientific Publ., River Edge NJ, 1991.

Shirota, T. [1952] *A generalization of a theorem of I. Kaplansky,* Osaka Math. J. 4 (1952) 121-132.

Shiryaev, V.M. [1985] *Semigroups with semidistributive subsemigroup lattices* [Russian], Semigroup Forum 27 (1983) 1-154.

Shleifer, F.G. [1981] *Identities, quasi-identities, and algorithms* [Russian], Modern algebra [Russian] 128-143, Leningrad. Gos. Ped. Univ., Leningrad 1981.

Shoji, K. [1989] *Commutative semigroups which are amalgamation bases,* Proc. 13th Sympos. on semigroups (Kyoto 1989) 4-8. Kyoto Sangyo Univ., Kyoto 1990.

Shortt, R.M. & Rao, K.P.S.B. [1989] *The dual of a refinement algebra,* General topology and applications (Staten Island 1989) 335-367. Marcel Dekker, New York 1991.

Simmons, H. [1976] *The word and torsion problems for commutative Thue systems,* Word problems, II (Conf. on Decision problems in algebra, Oxford 1976) 395-400, North-Holland, Amsterdam-New York 1988.

Šimon, J. [1979] *The lattice of normal subsemigroups of an abelian semigroup* [Czech], Sb. Praci Ped. Fac. v Ostravě Ser. A 14 (1979) no.62, 3-8.

Simons, S. [1963] *On the product of semigroups* [Russian], J. London Math. Soc. 38 (1963) 66-70.

Sitnikov, V.M. [1983] *Semigroups with a Boolean lattice of tolerances* [Russian], Ural. Gos. Univ. Mat. Zap. 13, no.3, 146-158 (1983).

Skula, L. [1975] *Extension of a partial x-operator to an x-operator* [Czech], Knižnice Odborn. Věd. Spisu Vysoké. Učeni Tech. v Brně B-56 (1975) 125-130.

[1976] *On extensions of partial x-operators,* Czechoslovak Math. J. 26 (101) (1976) 477-505.

Spake, R. [1986A] *Idempotent-free Archimedean components of the power semigroup of the group of integers. I,* Math. Japon. 31 (1986) 791-810.

[1986F] *The semigroup of nonempty finite subsets of integers,* Internat. J. Math. Math. Sci. 9 (1986) 605-616.

[1988Q] *The semigroup of nonempty finite subsets of rationals,* Internat. J. Math. Math. Sci. 11 (1988) 81-86.

[1988Z] *Semilattice decomposition of the power semigroup of the group of integers.* Math. Japon. 33 (1988) 609-625.

Spake, R. & Hanlon, B. [1993] *Archimedean components of* $\mathcal{P}(\mathbb{Z})$, Math. Japon. 38 (1993) 55-57.

Speed, T.P. [1968] *A note on commutative semigroups,* J. Austral. Math. Soc. 8 (1968) 731-736.

Stambolieva, G.C. [1972] *Projectors in free commutative semigroups with a finite number of generators* [Bulgarian], Godišnik Vysš. Tehn. Učebn. Zaved. Mat. 8 (1972) no.4, 97-105.

Stenström, B. [1971] *Flatness and localization over monoids,* Math. Nachr. 48 (1971) 315-334.

Storrer, H.H. [1976] *An algebraic proof of Isbell's zigzag theorem,* Semigroup Forum 12 (1976) 83-88.

Strecker, R. [1969] *Verallgemeine Schreiersche Halbgruppenerweiterungen,* Monatsb. Deutsch. Akad. Wiss. Berlin 11 (1969) 325-328.

Sylvester, J.J. [1884] *Mathematical questions with their solutions,* Educational Times 41 (1884) 21.

Szász, G. [1976] *Unitary subsemigroups of commutative semigroups,* Acta Sci. Math. (Szeged) 38 (1976) 379-381.

Szép, J. [1969] *On the structure of finite semigroups,* Report DM 1969-4, Dept. of Math., Karl Marx Univ. of Economics, Budapest 1969.

[1970] *On the structure of finite semigroups. II,* Report DM 1970-9, Dept. of Math., Karl Marx Univ. of Economics, Budapest 1970.

[1973] *On the structure of finite semigroups. III,* Report DM 1973-3, Dept. of Math., Karl Marx Univ. of Economics, Budapest 1973.

Szimtenings, D. [1970] *Primitivmengen in additiven abelschen Gruppen,* J. Reine Angew. Math. 243 (1970) 17-31.

Taiclin, M.A. [1974] *The isomorphism problem for commutative semigroups solved affirmatively* [Russian], Theory of models and its applications [Russian], 75-81. Kazah. Gos. Univ., Alma-Ata, 1980.

[1980] *The isomorphism problem for commutative semigroups,* [Russian], Mat. Sb. (N.S.) 93 (135) (1974) 102-128, 152.

Takahashi, M. [1984A] *On semimodules, I,* Kobe J. Math. 1 (1984) 67-97.

[1984B] *On semimodules, II,* Kobe J. Math. 1 (1984) 177-190.

[1985] *On semimodules, III,* Kobe J. Math. 2 (1985) 131-141.

[1987] *Structures of semimodules,* Kobe J. Math. 4 (1987) 79-101.

Tamura, T. [1953] *Some remarks on semigroups and all types of orders 2,3,* J. Gakugei Tokushima Univ. 3 (1953) 1-11.

[1954] *Notes on finite semigroups and determination of semigroups of order 4,* J. Gakugei Tokushima Univ. 5 (1954) 17-27.

Tamura, T. (cont'd) [1957] *Commutative nonpotent archimedean semigroup with cancelation law,* J. Gakugei Tokushima Univ. 8 (1957) 5-11.

[1963] *Minimal commutative divisible semigroups,* Bull. Amer. Math. Soc. 69 (1963) 713-716.

[1966] *Notes on commutative Archimedean semigroups. I, II,* Proc. Japan Acad. 42 (1966) 35-40, 545-548.

[1968] *Construction of trees and commutative archimedean semigroups,* Math. Nachr. 36 (1968) 257-287.

[1969] *Commutative semigroups whose lattice of congruences is a chain,* Bull. Soc. Math. France 97 (1969) 369-380.

[1970C] *Commutative archimedean cancellative semigroups without idempotent,* Sém. Dubreil 1969/1970, Fasc. 2, Exp. DG5.

[1970F] *Abelian groups and N-semigroups,* Proc. Japan Acad. 46 (1970) 212-216.

[1970P] *Finite union of commutative power joined semigroups,* Semigroup Forum 1 (1970) 75-83.

[1971] *On commutative exclusive semigroups,* Semigroup Forum 2 (1971) 181-187.

[1973C] *N-congruences on N-semigroups,* J. Algebra 27 (1973) 11-30.

[1973S] *Finitely generated steady N-semigroups,* Proc. Amer. Math. Soc. 41 (1973) 425-430.

[1974B] *Basic study of N-semigroups and their homomorphisms,* Semigroup Forum 8 (1974) 21-50.

[1974B] *Irreducible N-semigroups,* Math. Nachr. 63 (1974) 71-88.

[1975] *Basic study of N-semigroups and their homomorphisms II,* Semigroup Forum 10 (1975) 250-261.

[1976] *Putcha's problem on maximal cancellative subsemigroups,* Proc. Japan Acad. 52 (1976) 91-94.

[1977N] *Commutative Archimedean cancellative idempotent-free semigroups with an infinite cyclic group as structure group,* J. Algebra 45 (1977) 343-390.

[1977S] *Commutative subarchimedean semigroups,* Sûrikaisekikenkyûsho Kôkyûroku No.292 (1977) 98-126.

[1977T] *N-semigroups and their translation semigroups,* J. Austral. Math. Soc. Ser. A 24 (1977) 184-202.

[1978] *Recent results and problems on commutative archimedean semigroups,* Proceedings of the 2nd Seminar on Semigroups, Tokyo Gakugei Univ., Tokyo, 1978, 69-79.

[1982N] *Nil-orders of commutative nil-bounded semigroups,* Semigroup Forum 24 (1982) 255-262.

[1982R] *Commutative cancellative semigroups with nontrivial homomorphisms into nonnegative real numbers,* J. Algebra 76 (1982) 25-41.

[1984C] *On chains whose power semigroups are lattices,* Semigroup Forum 30 (1984) 13-16.

[1984Z] *On the power semigroup of the group of integers,* Proc. Japan Acad. Ser. A Math. Sci. 60 (1984) 388-390.

[1985] *Archimedean components of the power semigroup of the infinite cyclic group,* Proc. 8th Sympos. on Semigroups, Shimane Univ. (1985) 22-25.

Tamura, T., Dehara, K., Iwata, T., Saito, H., & Tsukumo, K. [1960] *Semigroups of order 5, 6, 7, 8 whose greatest c-homomorphic image are unipotent semigroups with groups,* J. Gakugei Tokushima Univ. 11 (1960) 53-66.

Tamura, T. & Etterbeek, W. [1966] *The lattice of congruences of locally cyclic semigroups,* Proc. Japan Acad. 42 (1966) 682-684.

Tamura, T. & Hamilton, H.B. [1971] *Commutative semigroups with greatest group-homomorphic image,* Trans. Amer. Math. Soc. 173 (1972) 401-419; see also Proc. Japan Acad. 47 (1971) 671-675.

[1972] *Is minimal group congruence smallest?* Semigroup Forum 4 (1972) 173-176.

Tamura, T. & Kimura, N. [1954] *On decompositions of a commutative semigroup,* Kōdai Math. Sem. Rep. 6 (1954) 109-112.

Tamura, T. & Sasaki, M. [1963] *Semigroups of positive integer vectors,* Proc. Japan Acad. 39 (1963) 289-293.

Tamura, T. & Shafer, J. [1967] *Power semigroups,* Math. Japon. 12 (1967) 25-32; errata, ibid. 29 (1984) 679.

Tanabe, T. & Matsuda, R. [1999] *Note on Kaplansky's commutative rings,* Nihonkai Math. J. 10 (1999) 31-61.

Taylor, M. [1999] *The generalized equation of bisymmetry: solutions based on cancellative abelian monoids,* Aequationes Math. 57 (1999) 288-302.

Teissier, B. [1973] *Appendix to Zariski* [1973], *Le problème des modules pour les branches planes,* Ecole Polytechnique, Paris 1973.

Teissier, M. [1951] *Sur les équivalences régulières dans les demi-groupes,* C. R. Acad. Sci. Paris 232 (1951) 1987-1989.

Tetsuya, K., Hashimoto, T., Akazawa, T., Shibata, R., Inui, T., and Tamura, T. [1955] *All semigroups of order at most 5,* J. Gakugei Tokushima Univ. 6 (1955) 19-39 and Erratum.

Thành, D.T. [1992] *d-isomorphic semigroups of continuous functions in locally compact spaces,* Acta Math. Hungar. 60 (1992) 103-105.

Thierrin, G. [1953] *Quelques properiétés des sous-groupoïdes consistants d'un demi-groupe abélien,* C. R. Acad. Sci. Paris 236 (1953) 1837-1839.

[1954] *Sur quelques properiétés de certaines classes de demi-groupes,* C. R. Acad. Sci. Paris 239 (1954) 1335-1337.

[1956] *Sur la structure des demi-groupes,* Publ. Sci. Univ. Alger. Sér.A 3 (1956) 161-171.

Thoma, A. [1996] *Affine semigroups and monomial varieties,* Comm. Algebra 24 (1996) no.7, 2463-2471.

Thron, R. [1981] *Über das Zutreffen einer Eigenschaft einer Menge auf die Elemente der Potenzmenge,* Beiträge Algebra Geom. no.11 (1981) 41-50.

Tichy, R.F. [1979] *Polynomial functions over monoids,* Semigroup Forum 18 (1979) 371-380.

[1981] *A remark on polynomial functions over finite monoids,* Semigroup Forum 22 (1981) 391-392.

Toader, Gh. [1990] *A hierarchy of supermultiplicity of sequences in a semigroup,* Studia Univ. Babes-Bolyai Math. 35 (1990) no.4, 3-8.

Toca, A. [1977] *Generalized monoids of fractions,* Studia Univ. Babes-Bolyai Math. 22 (1977) no.1, 12-13.

Toffalori, C. [1980] *Existentially closed Archimedean semigroups* [Italian], Atti Acad. Lincei Rend. Cl. Sci. Fis. Mat. Natur. (8) 67 (1979) 162-167 (1980).

Tokunaga, H. & Yoshihara, H. [1995] *Degree of irrationality of abelian surfaces,* J. Algebra 97 (1994) 51-71.

Torres, F. [1994] *Weierstrass points and double coverings of curves. With application: symmetric numerical semigroups which cannot be realized as Weierstrass semigroups,* Manuscripta Math. 83 (1994) 39-58.

[1995] *On certain N-sheeted coverings of curves and numerical semigroups which cannot be realized as Weierstrass semigroups,* Comm. Algebra 23 (1995) no.11, 4211-4228.

[1997] *On γ-hyperelliptic numerical semigroups,* Semigroup Forum 55 (1997) 364-379.

Trepetin, M.S. [1971] *Semirings of endomorphisms of commutative nilpotent semigroups* [Russian], Tartu Riikl. Ül. Toimetised Vih. 277 (1971) 20-36; letter to the editors, ibid. 300 (1972) 300.

[1974] *The determinability of monogenic nilsemigroups by their semigroups of endomorphisms* [Russian], Tartu Riikl. Ül. Toimetised Vih. 336 (1974) 160-177.

Tretjakova, E.G. [1993] *Commutative semigroups which are embeddable into epigroups,* Semigroup Forum 46 (1993) 78-84.

Trnková, V. [1975A] *Representation of semigroups by products in a category,* J. Algebra 34 (1975) 191-204. see also Comment. Math. Univ. Carolinae 16 (1975) 393-394.

[1975B] *On a representation of commutative semigroups,* Semigroup Forum 10 (1975) 203-214.

[1976] *Isomorphisms of products and representations of commutative semigroups,* Algebraic theory of semigroups (Szeged, 1976), 657-683. Colloq. Math. Soc. János Bolyai, North-Holland, Amsterdam, 1979.

[1978] *Cardinal multiplication of relational structures,* Algebraic methods in graph theory, Vol. I,II (Szeged, 1978), 763-792. Colloq. Math. Soc. János Bolyai, North-Holland, Amsterdam, 1971.

Trueman, D.C. [1979] *Direct products of cyclic semigroups,* Semigroups, Academic Press, New York 1980, 103-110.

[1980] *Finite direct products of finite cyclic semigroups and their characterization,* Proc. Roy. Soc. Edinburgh Sect. A 85 (1980) 337-351.

[1983] *The lattice of congruences of direct products of cyclic semigroups and certain other semigroups,* Proc. Roy. Soc. Edinburgh Sect. A 95 (1983) 203-214.

[1985] *Lattices of congruences on free finitely generated commutative semigroups and direct products of cyclic monoids,* Proc. Roy. Soc. Edinburgh Sect. A 100 (1985) 175-179.

Trung, N.V. & Hoa, L.T. [1986] *Affine semigroups and Cohen-Macaulay rings generated by monomials,* Trans. Amer. Math. Soc. 298 (1986) 145-167.

Tully, E.J., Jr. [1966] *A class of naturally partly ordered commutative archimedean semigroups with maximal condition,* Proc. Amer. Math. Soc. 17 (1966) 1133-1139.

[1969] *Semigroups in which every ideal is a retract,* J. Austral. Math. Soc. 9 (1969) 239-245.

[1974] *A class of finite commutative archimedean semigroups,* Math. Systems Theory 7 (1974) 289-293.

Turnwald, G. [1986] *Permutable chains of polynomials of odd degree,* Contributions to general algebra 5 (Salzburg 1986), 407-413. Hölder-Pichler-Tempsky, Vienna 1987.

Tutalar, H.I. [1987] *The Weierstrass nongap sequence beginning with 4, and a construction of its algebraic function field,* Doğa Mat. 11 (1987) 135-140.

Utumi, Y. [1956] *On quotient rings,* Osaka Math. J. 8 (1956) 1-18.

Vandiver, H.S. [1940] *On the imbedding of one semi-group in another, with applications to semi-rings,* Amer. J. Math. 62 (1940) 72-78.

Vinárek, J. [1982] *Representations of commutative semigroups by products of metric 0-dimensional spaces,* Comment. Math. Univ. Carolinae 23 (1982) 715-726.

Waldi, R. [1972] *Wertehalbgruppe und Singularität einer ebenen algebroiden Kurve,* Doct. Diss., Regensburg Univ., Regensburg 1972.

[1980] *Zur Konstruktion von Weierstrasspunkten mit vorgegebener Halbgruppe,* Manuscripta Math. 30 (1980) 257-278.

Wallace, A.D. [1953] *A note on mobs, II,* Anais Acad. Brasil Ci. cc 25 (1953) 335-336.

Wang, H.X. [1988] *A note on endomorphism semirings of semimodules,* Kobe J. Math. 5 (1988) 155-160.

Ward, M. & Dilworth, R.P. [1939] *The lattice theory of ova,* Ann of Math. 40 (1939) 712-730.

Warne, R.J. & Williams, L.K. [1961] *Characters on inverse semigroups,* Czechoslovak Math. J. 11 (1961) 150-155.

Warner, S. [1960] *Mathematical induction in commutative semigroups,* Amer. Math. Monthly 67 (1960) 533-537.

Wegmann, H. [1966] *Beiträge zur Zahlentheorie auf freie Halbgruppen. II.* J. Reine Angew. Math. 221 (1966) 150-159.

Wehrung, F. [1996] *Tensor products of structures with interpolation,* Pacific J. Math. 176 (1996) 267-285.

[1998D] *The dimension monoid of a lattice,* Algebra Universalis 40 (1998) 247-411.

[1998E] *Embedding simple commutative monoids into simple refinement monoids,* Semigroup Forum 56 (1998) 104-129.

Wells, C. [1978] *Extension theories for monoids,* Semigroup Forum 16 (1978) 13-35.

Wiegandt, R. [1958] *On complete semi-modules,* Acta Sci. Math. (Szeged) 19 (1958) 219-123.

Woan, W.J. [1975] *Minimal group congruences,* Semigroup Forum 11 (1975/76) 178-180.

Yamada, M. [1955] *On the greatest semilattice decomposition of a semigroup,* Kōdai Math. Sem. Rep. 7 (1955) 59-62.

[1964] *Construction of commutative z-semigroups,* Proc. Japan Acad. 40 (1964) 94-98.

[1965] *Construction of finite commutative semigroups,* Bull. Shimane Univ. Natur. Sci. no.15 (1965) 1-11.

[1968] *Commutative ideal extensions of null semigroups,* Mem. Fac. Lit. Sci. Shimane Univ. Natur. Sci. 1 (1968) no.1, 8-22.

Yamada, M. & Tamura, T. [1969] *Note on finite commutative nil-semigroups,* Portugal. Math. 28 (1969) 189-203.

Yap, C.-K. [1991] *A new lower bound construction for commutative Thue systems with applications,* J. Symbolic Comput. 12 (1991) 1-27.

Youssfi, E.H. [1998] *Harmonic analysis on conelike bodies and holomorphic functions on tube domains,* J. Funct. Anal. 155 (1998) 381-435.

Zanardo, P. & Zannier, U. [1975] *The class semigroup of orders in number fields,* Math. Proc. Cambridge Phil. Soc. 115 (1994) 379-391.

Zariski, O. [1973] *Le problème des modules pour les branches planes,* Ecole Polytechnique, Paris 1973.

Zelinka, B. [1975] *Tolerance relations on semilattices,* Comment. Math. Univ. Carolinae 16 (1975) 333-338.

[1977] *Tolerance relations on periodic commutative semigroups,* Czechoslovak Math. J. 27 (102) (1977) 167-169.

Zhang, Y.F., Li, S.Zh., & Wang, D.Sh. [1993] *An abstract characterization of Grothendieck functor,* Northeast. Math. J. 9 (1993) 477-482.

Zheng, B.D. & Qi, Zh.K. [1997] *Primary decompositions of commutative monoids* [Chinese], J. Harbin. Inst. Tech. 29 (1997) no.6, 46-49.

AUTHOR INDEX

NOTATION

INDEX

Advances in Mathematics

1. A. Nagy: *Special Classes of Semigroups*. 2001 ISBN 0-7923-6890-8
2. P.A. Grillet: *Commutative Semigroups*. 2001 ISBN 0-7923-7067-8

SPRINGER-SCIENCE+BUSINESS MEDIA, B.V.

GPSR Compliance
The European Union's (EU) General Product Safety Regulation (GPSR) is a set of rules that requires consumer products to be safe and our obligations to ensure this.

If you have any concerns about our products, you can contact us on

ProductSafety@springernature.com

In case Publisher is established outside the EU, the EU authorized representative is:

Springer Nature Customer Service Center GmbH
Europaplatz 3
69115 Heidelberg, Germany

www.ingramcontent.com/pod-product-compliance
Ingram Content Group UK Ltd.
Pitfield, Milton Keynes, MK11 3LW, UK
UKHW021504200726
13877UKWH00027B/483
* 9 7 8 1 4 7 5 7 3 3 9 0 7 *